Series and Products in the Development of Mathematics

Volume 2

This is the second volume of a two-volume work that traces the development of series and products from 1380 to 2000 by presenting and explaining the interconnected concepts and results of hundreds of unsung as well as celebrated mathematicians. Some chapters deal with the work of primarily one mathematician on a pivotal topic, and other chapters chronicle the progress over time of a given topic. This updated second edition of *Sources in the Development of Mathematics* adds extensive context, detail, and primary source material, with many sections rewritten to more clearly reveal the significance of key developments and arguments. Volume 1, accessible even to advanced undergraduate students, discusses the development of the methods in series and products that do not employ complex analytic methods or sophisticated machinery. Volume 2 examines more recent results, including de Branges's resolution of Bieberbach's conjecture and Nevanlinna's theory of meromorphic functions.

RANJAN ROY (1947–2020) was the Ralph C. Huffer Professor of Mathematics and Astronomy at Beloit College, where he was a faculty member for 38 years. Roy published papers and reviews on Riemann surfaces, differential equations, fluid mechanics, Kleinian groups, and the development of mathematics. He was an award-winning educator, having received the Allendoerfer Prize, the Wisconsin MAA teaching award, and the MAA Haimo Award for Distinguished Mathematics Teaching, and was twice named Teacher of the Year at Beloit College. He coauthored *Special Functions* (2001) with George Andrews and Richard Askey and coauthored chapters in the *NIST Handbook of Mathematical Functions* (2010); he also authored *Elliptic and Modular Functions from Gauss to Dedekind to Hecke* (2017) and the first edition of this book, *Sources in the Development of Mathematics* (2011).

Ranjan Roy 1948–2020

Series and Products in the Development of Mathematics

Second Edition

Volume 2

RANJAN ROY

Beloit College

CAMBRIDGE
UNIVERSITY PRESS

CAMBRIDGE
UNIVERSITY PRESS

University Printing House, Cambridge CB2 8BS, United Kingdom

One Liberty Plaza, 20th Floor, New York, NY 10006, USA

477 Williamstown Road, Port Melbourne, VIC 3207, Australia

314-321, 3rd Floor, Plot 3, Splendor Forum, Jasola District Centre, New Delhi - 110025, India

103 Penang Road, #05-06/07, Visioncrest Commercial, Singapore 238467

Cambridge University Press is part of the University of Cambridge.

It furthers the University's mission by disseminating knowledge in the pursuit of education, learning and research at the highest international levels of excellence.

www.cambridge.org
Information on this title: www.cambridge.org/9781108709378
DOI: 10.1017/9781108671620

First edition © Ranjan Roy 2011
Second edition © Ranjan Roy 2021

First published as *Sources in the Development of Mathematics*, 2011
Second edition 2021

A catalogue record for this publication is available from the British Library

ISBN – 2-Volume Set 978-1-108-70943-9 Paperback
ISBN – Volume 1 978-1-108-70945-3 Paperback
ISBN – Volume 2 978-1-108-70937-8 Paperback

Contents

Contents of Volume 1

Preface

Sources in the Development of Mathematics: Series and Products from the Fifteenth to the Twenty-first Century, my book of 2011, was intended for an audience of graduate students or beyond. However, since much of its mathematics lies at the foundations of the undergraduate mathematics curriculum, I decided to use portions of my book as the text for an advanced undergraduate course. I was very pleased to find that my curious and diligent students, of varied levels of mathematical talent, could understand a good bit of the material and get insight into mathematics they had already studied as well as topics with which they were unfamiliar. Of course, the students could profitably study such topics from good textbooks. But I observed that when they read original proofs, perhaps with gaps or with slightly opaque arguments, students gained very valuable insight into the process of mathematical thinking and intuition. Moreover, the study of the steps, often over long periods of time, by which earlier mathematicians refined and clarified their arguments revealed to my students the essential points at the crux of those results, points that may be more difficult to discern in later streamlined presentations. As they worked to understand the material, my students witnessed the difficulty and beauty of original mathematical work, and this was a source of great enjoyment to many of them. I have now thrice taught this course, with extremely positive student response.

In order for my students to follow the foundational mathematical arguments in *Sources*, I was often required to provide additional material, material actually contained in the original works of the mathematicians being studied. I therefore decided to expand my book, as a second edition in two volumes, to make it more accessible to readers, from novices to accomplished mathematicians. This second edition contains about 250 pages of new material, including more details within the original proofs, elaborations and further developments of results, and additional results that may give the reader a better perspective. Furthermore, to give the material greater focus, I have limited this second edition to the topics of series and products, areas that today permeate both applied and pure mathematics; the second edition is thus entitled *Series and Products in the Development of Mathematics*.

The first volume of my work discusses the development of the fundamental though powerful and essential methods in series and products that do not employ complex analytic methods or sophisticated machinery such as Fourier transforms. Much of this material would be accessible, perhaps with guidance, to advanced undergraduate students. The second volume deals with more recent work and requires considerable mathematical background. For example, in volume 2, I discuss Weil's 1949 paper on solutions of equations in finite fields and de Branges's conquest of the Bieberbach conjecture. Each volume contains the same complete bibliography.

The exercises at the end of the chapters present many additional original results and may be studied simply for the supplementary theorems they contain. The exercises are accompanied by references to the original works, as an aid to further research. Readers may attempt to prove the results in the problems and, by use of the references, compare their own solutions with the originals. Moreover, many of the exercises can be tackled by methods similar to those given in the text, so that some exercises can be realistically assigned to a class as homework. I assigned many exercises to my classes, and found that the students enjoyed and benefited from their efforts to find solutions. Thus, the exercises may be useful as problems to be solved, and also for the results they present.

Detailed study of original mathematical works provides a point of entry into the minds of the creators of powerful theories, and thus into the theories themselves. But tracing the discovery and evolution of mathematical ideas and theorems entails the examination of many, many papers, letters, notes, and monographs. For example, in this work I have discussed the work of more than three hundred mathematicians, including arguments and theorems contained in approximately one hundred works and letters of Euler alone. Locating, studying, and grasping the interconnections among such original works and results is a ponderous, complex, and rewarding effort. In this second edition, I have added numerous footnotes and almost five hundred works to the bibliography. My hope is that the detailed footnotes and the expanded bibliography, containing both original works and works of distinguished expositors and historians of mathematics, may encourage and facilitate the efforts of those who wish to search out and study the original sources of our inherited mathematical wealth.

I first wish to thank my wife, who typeset and edited this work, made innumerable corrections and refinements to the text, and devotedly assisted me with translations and locating references. I am also very grateful to NFN Kalyan for his encouragement and for creating the eloquent artwork for the cover of these volumes. I greatly appreciate Maitreyi Lagunas's unflagging support and interest. I thank Bruce Atwood who cheerfully constructed the nice diagrams contained in this work, and Paul Campbell who generously provided expert technical support and advice. I am grateful to my student Shambhavi Upadhyaya, who has an unusual ability to proofread very accurately, for spending so much time giving useful suggestions for improvement. I am indebted to my students whose questions and enthusiasm helped me refine this second edition. I also thank the very capable librarians at Beloit College, especially Chris Nelson and Cindy Cooley. Finally, I wish to acknowledge the inspiration provided me by my friend, the late Dick Askey.

25

q-Series

25.1 Preliminary Remarks

The theory of q-series in modern mathematics plays a significant role in partition theory and modular functions as well as in some aspects of Lie algebras and statistical mechanics. This subject began quietly, however, with two combinatorial problems posed in a September 1740 letter from Phillipe Naudé (1684–1747) to Euler. Naudé was a mathematician of French origin working in Berlin. In general, his question was how to find the number of ways in which a given number could be expressed as the sum of a fixed number, first of distinct integers and then without the requirement that the integers in the sum be distinct. For example, in how many ways can 50 be expressed as a sum of 7 distinct/not necessarily distinct integers?[1]

As an example of both these problems, 7 can be expressed as a sum of three distinct integers in one way, $1 + 2 + 4$; whereas it can be expressed as a sum of three integers in four ways: $1 + 1 + 5$, $1 + 2 + 4$, $1 + 3 + 3$, $2 + 2 + 3$. Euler received Naudé's letter in St. Petersburg, just before he moved to Berlin. Within two weeks, in a reply to Naudé, Euler outlined a solution and soon after that he presented his complete solution to the Petersburg Academy.[2] In 1748, he devoted a whole chapter to this topic in his *Introductio in Analysin Infinitorum*.[3] The essential idea in Euler's solution was that the coefficient of $q^k x^m$ in the series expansion of the infinite product

$$f(q,x) = (1 + qx)(1 + q^2 x)(1 + q^3 x) \cdots \tag{25.1}$$

gave the number of ways of writing k as a sum of m distinct positive integers. Euler used the functional relation

$$f(q,x) = (1 + qx) f(q,qx) \tag{25.2}$$

---

[1] See Eu. 1-2 pp. 163–193, especially § 19, E 158 § 19 and Weil (1983) pp. 276–277.
[2] Eu. 1-2 pp. 163–193, E 158.
[3] Euler (1988) chapter 16, especially pp. 256–270.

to prove that

$$f(q,x) = \sum_{m=0}^{\infty} \frac{q^{\frac{m(m+1)}{2}} x^m}{(1-q)(1-q^2)\cdots(1-q^m)}.$$ (25.3)

He noted that

$$\frac{1}{(1-q)(1-q^2)\cdots(1-q^m)}$$
$$= (1+q+q^{1+1}+\cdots)(1+q^2+q^{2+2}+\cdots)\cdots(1+q^m+q^{m+m}+\cdots)$$
$$= \sum_{n=0}^{\infty} a_n q^n,$$ (25.4)

where the middle product showed that a_n, the coefficient of q^n, was the number of ways of writing n as a sum of integers chosen from the set $1, 2, \ldots, m$. This implied that the coefficient of $q^k x^m$ on the right-hand side of (25.3) was the number of ways of writing $k - \frac{m(m+1)}{2}$ as a sum of integers from the set $1, 2, \ldots, m$. Thus, Euler stated the theorem: The number of different ways in which the number n can be expressed as a sum of m different numbers is the same as the number of different ways in which $n - \frac{m(m+1)}{2}$ can be expressed as the sums of the numbers $1, 2, 3, \ldots, m$.

For the second problem, Euler used the product

$$g(q,x) = \prod_{n=1}^{\infty} (1-q^n x)^{-1}$$ (25.5)

and obtained the corresponding series and theorem in a similar way. Euler here used functional relations to evaluate the product as a series, just as he earlier employed functional relations to evaluate the beta integral as a product. Of course, this method goes back to Wallis.

Euler also considered the case $x = 1$. In that case (in modern notation), we have

$$\prod_{n=1}^{\infty} (1-q^n)^{-1} = (1+q+q^{1+1}+\cdots)(1+q^2+q^{2+2}+\cdots)(1+q^3+q^{3+3}+\cdots)\cdots$$

$$= \sum_{n=0}^{\infty} p(n) q^n,$$ (25.6)

where $p(n)$ is the number of partitions of n, or the number of ways in which n can be written as a sum of positive integers. For example, $p(4) = 5$ because 4 has the five partitions

$$1+1+1+1,\ 2+1+1,\ 2+2,\ 3+1,\ 4.$$

The product in (25.6) also led Euler to consider its reciprocal, $\prod_{n=1}^{\infty}(1-q^n)$. He attempted to expand this as a series but it took him nine years to completely resolve this difficult problem. In his first attempt presented in 1741 but published ten years later,[4] he multiplied a large number of terms of the product to find that

$$\prod_{n=1}^{\infty}(1-q^n) = 1 - q - q^2 + q^5 + q^7 - q^{12} - q^{15} + q^{22} + q^{26} - q^{35} - q^{40} + q^{51} + \cdots .$$

(25.7)

He quickly found a general expression for the exponents, $\frac{m(3m\pm1)}{2}$. He most probably did this by considering the differences in the sequence of exponents; note that the sequence of exponents is

$$0, 1, 2, 5, 7, 12, 15, 22, 26, 35, 40, 51, \ldots .$$

Observe that the sequence of differences is then

$$1, 1, 3, 2, 5, 3, 7, 4, 9, 5, 11, \ldots .$$

The pattern of this sequence suggests that one should group the sequence of exponents into two separate sequences, first taking the exponents of the odd-numbered terms and then the exponents of the even-numbered terms.[5] For example, the sequence of exponents of the odd-numbered terms is $0, 2, 7, 15, 26, 40, \ldots$, and their differences are $2, 5, 8, 11, 14, \ldots$. Since the differences of these differences are 3 in every case, we may apply the formula of Zhu Shijie and Montmort, given in Section 10.3, to perceive that the $(n + 1)$th term of the sequence of odd-numbered exponents will be given by

$$0 + 2n + \frac{3n(n-1)}{2} = \frac{n(3n+1)}{2}.$$

Similarly, the nth term in the sequence of even-term exponents is $\frac{n(3n-1)}{2}$. In the *Introductio*, Euler wrote, "If we consider this sequence with some attention we will note that the only exponents which appear are of the form $\frac{(3n^2\pm n)}{2}$ and that the sign of the corresponding term is negative when n is odd, and the sign is positive when n is even."[6] Thus, Euler made the conjecture

$$\prod_{n=1}^{\infty}(1-q^n) = \sum_{m=-\infty}^{\infty}(-1)^m q^{\frac{m(3m+1)}{2}} = 1 + \sum_{m=1}^{\infty}(-1)^m q^{\frac{m(3m\pm1)}{2}},$$

(25.8)

and finally found a proof of this in 1750. He immediately wrote Goldbach about the details of the proof,[7] explaining that it depended on the algebraic identity:

$$(1 - \alpha)(1 - \beta)(1 - \delta) \text{ etc.}$$
$$= 1 - \alpha - \beta(1 - \alpha) - \gamma(1 - \alpha)(1 - \beta) - \delta(1 - \alpha)(1 - \beta)(1 - \gamma) - \text{etc.}$$

---

[4] Eu. 1-2 pp. 163–193, especially p. 193. E 158 § 37.
[5] Eu. 1-2 pp. 241–253. E 175 § 8.
[6] Euler (1988) p. 274.
[7] Fuss (1968) vol. 1, pp. 522–524. See also Eu. 1-2 pp. 390–398. E 244.

This identity is easy to check, since the first three terms on the right-hand side add up to

$$1 - \alpha - \beta(1 - \alpha) = (1 - \alpha)(1 - \beta),$$

and when this is added to the fourth term, we get

$$(1 - \alpha)(1 - \beta) - \gamma(1 - \alpha)(1 - \beta) = (1 - \alpha)(1 - \beta)(1 - \gamma),$$

and so on. An interesting feature of the series (25.8) is that the exponent of q is a quadratic in m, the index of summation. Surprisingly, series of this kind had already appeared in 1690 within Jakob Bernoulli's works on probability theory,[8] but he was unable to do much with them. Over a century later, Gauss initiated a systematic study of these series. Entry 58 of Gauss's mathematical diary, dated February 1797, gives a continued fraction expansion of one of Bernoulli's series:[9]

$$1 - a + a^3 - a^6 + a^{10} - \cdots$$

$$= \cfrac{1}{1 + \cfrac{a}{1 + \cfrac{a^2 - a}{1 + \cfrac{a^3}{1 + \cfrac{a^4 - a^2}{1 + \cfrac{a^5}{1 + \text{etc.}}}}}} \tag{25.9}$$

In his diary, Gauss added the comment, "From this all series where the exponents form a series of the second order are easily transformed." About a year later, he raised the problem of expressing $1 + q + q^3 + q^6 + q^{10} + \cdots$ as an infinite product. Gauss came upon series of this type around 1794 in the context of his work on the arithmetic-geometric mean, that he had been studying since 1791.[10] This latter work was absorbed into his theory of elliptic functions. Series (25.8) and (25.9) are actually examples of the special kind of q-series called theta functions. Theta functions also arose naturally in Fourier's 1807 study of heat conduction.

Unfortunately, Gauss did not publish any of his work on theta or elliptic functions, and it remained for Abel and Jacobi to independently rediscover much of this work, going beyond Gauss in many respects. Around 1805–1808, Gauss began to view q-series in a different way. For example, his 1808 paper[11] on q-series dealt with a generalization of the binomial coefficient and the binomial series. In particular, he defined the Gaussian polynomial

---

[8] Bernoulli and Sylla (2006) pp. 176–180.
[9] See Dunnington (2004) p. 474.
[10] Peters (1860–1865) vol. 1, p. 125. See also Gauss (1863–1927) vol. 3, pp. 361–371, also vol. 10, part 2, p. 18 of Schlesinger's article on Gauss's work in function theory.
[11] Gauss (1981) pp. 463–495.

$$(m,\mu) = \frac{(1-q^m)(1-q^{m-1})(1-q^{m-2})\cdots(1-q^{m-\mu+1})}{(1-q)(1-q^2)(1-q^3)\cdots(1-q^\mu)}. \qquad (25.10)$$

Note that Gauss wrote x instead of q. Observe that as $q \to 1$

$$(m,\mu) \to \binom{m}{\mu}. \qquad (25.11)$$

This work led to an unexpected byproduct: an evaluation of the Gauss sum $\sum_{k=0}^{n-1} e^{\frac{2\pi i k^2}{n}}$ where n was an odd positive integer. This sum had already appeared naturally in Gauss's theory of the cyclotomic equation $x^n - 1 = 0$, to which he had devoted the final chapter of his 1801 *Disquisitiones Arithmeticae*.[12] There Gauss had computed the square of the Gauss sum, but he was unable to determine the correct sign of the square root. Already in 1801, he knew that it was important to find the exact value of the sum; he expended considerable effort over the next four years to compute the Gauss sum, and it was a complete surprise for him when the result dropped out of his work on q-series. In September 1805, he wrote his astronomer friend, Wilhelm Olbers,[13]

> What I wrote there [Disqu. Arith. section 365] ..., I proved rigorously, but I was always annoyed by what was missing, namely, the determination of the sign of the root. This gap spoiled whatever else I found, and hardly a week may have gone by in the last four years without one or more unsuccessful attempts to unravel this knot - just recently it again occupied me much. But all the brooding, the searching, was to no avail, and I had sadly to lay down my pen again. A few days ago, I finally succeeded - not by my efforts, but by the grace of God, I should say. The mystery was solved the way lightning strikes, I myself could not find the connection between what I knew previously, what I investigated last, and the way it was finally solved.

He recorded these events in his diary:[14]

> (May 1801) A method for proving the first fundamental theorem has been found by means of a most elegant theorem in the division of the circle, thus
>
> $$\sum \frac{\sin}{\cos} \frac{nn}{a} P = + \begin{array}{c|c|c|c} \sqrt{a} & 0 & 0 & +\sqrt{a} \\ \sqrt{a} & +\sqrt{a} & 0 & 0 \end{array} \qquad (25.12)$$
>
> according as $a \equiv 0, 1, 2, 3 \pmod 4$ substituting for n all numbers from 0 to $(a-1)$. (August 1805) The proof of the most charming theorem recorded above, May 1801, which we have sought to prove for 4 years and more with every effort, at last perfected.

Conceptually, this was a major achievement, since it served to connect cyclotomy with the reciprocity law. Gauss may have initially considered the polynomial $\sum_{k=0}^m (m,k)x^k$ as a possible analog of the finite binomial series. In any case, he expressed the sum as a finite product when $x = -1$ and when $x = \sqrt{q}$, and these formulas finally yielded the correct value of the Gauss sum. It is interesting to note that the polynomial

---

[12] For an interesting commentary on Gauss's work in cyclotomy, see Neumann (2007a)) and (2007b).
[13] See Bühler (1981) p. 31.
[14] Dunnington (2004) p. 481.

$\sum_{k=0}^{m}(m,k)x^k$ played a key role in Szegő's theory of orthogonal polynomials on the unit disc.

Gauss found the appropriate q-extension of the terminating binomial theorem, perhaps around 1808, but he did not publish it. In 1811, Heinrich A. Rothe (1773–1841) first published this result in the preface of his *Systematisches Lehrbuch der Arithmetik*[15] as the formula

$$\sum_{k=0}^{m} \frac{1-q^m}{1-q} \cdot \frac{1-q^{m-1}}{1-q^2} \cdot \cdots \cdot \frac{1-q^{m-k+1}}{1-q^k} \cdot q^{\frac{k(k+1)}{2}} x^{m-k} y^k$$

$$= (x+y)(x+qy)\cdots(x+q^{k-1}y). \tag{25.13}$$

Although this was the most important result in the book, Rothe excluded it from the body of text, apparently in order to keep the book within the size required by the publisher. Gauss's paper and Rothe's formula indicated a direction for further research on q-series relating to the extension of the binomial theorem. This path was not pursued until the 1840s, except in Schweins's *Analysis* of 1820.[16] This work presented a q-extension of Vandermonde's identity (25.64).

In the 1820s, Jacobi investigated q-series in connection with his work on theta functions, a byproduct of his researches on elliptic functions. His most remarkable discovery in this area was the triple product identity. Jacobi's famous *Fundamenta Nova* of 1829 stated the formula as[17]

$$(1+qz)(1+q^3z)(1+q^5z)\cdots\left(1+\frac{q}{z}\right)\left(1+\frac{q^3}{z}\right)\left(1+\frac{q^5}{z}\right)\cdots$$

$$= \frac{1+q\left(z+\frac{1}{z}\right)+q^4\left(z^2+\frac{1}{z^2}\right)+q^9\left(z^3+\frac{1}{z^3}\right)+\cdots}{(1-q^2)(1-q^4)(1-q^6)(1-q^8)\cdots}. \tag{25.14}$$

Jacobi regarded this identity as his most important formula in pure mathematics. He gave several very important applications. In one of these, he derived an identity, giving the number of representations of an integer as a sum of four squares. In another, he obtained an important series expression for the square root of the period of some elliptic functions, allowing him to find a new derivation of the following transformation of a theta function, originally due to Cauchy and Poisson:

$$1+2\sum_{n=1}^{\infty} e^{-n^2\pi x} = \frac{1}{\sqrt{x}}\left(1+2\sum_{n=1}^{\infty} e^{-\frac{n^2\pi}{x}}\right). \tag{25.15}$$

Jacobi also published a long paper on those series whose powers are quadratic forms; the triple product identity formed the basis for this. In the 1820s, when Gauss

[15] Rothe (1811).
[16] Schweins (1820) pp. 292–293.
[17] Jacobi (1969) vol. 1, p. 234.

learned of Jacobi's work, he informed Jacobi that he had already found (25.14) in 1808. Legendre, on very friendly terms with Jacobi, refused to believe that Gauss had anticipated his friend. In a letter to Jacobi,[18] Legendre wrote, "Such outrageous impudence is incredible in a man with enough ability of his own that he should not have to take credit for other people's discoveries."[19] Then again, Legendre had had his own priority disputes with Gauss with regard to quadratic reciprocity and the method of least squares.

In the early 1840s, papers on q-series appeared in quick succession by Cauchy in France and Eisenstein, Jacobi, and E. Heine in Germany. As a second-year student at Berlin in 1844, Eisenstein presented twenty-five papers for publication to *Crelle's Journal*. One of them, "Neuer Beweis und Verallgemeinerung des binomischen Lehrsatzes,"[20] It began with the statement and proof of the Rothe–Gauss theorem; it then applied Euler's approach to the proof of the binomial theorem to obtain a version of the q-binomial theorem. Some details omitted by Euler in his account were treated in Eisenstein's paper.

Jacobi and Cauchy stated and proved the q-binomial theorem in the form

$$1 + \frac{v-w}{1-q}z + \frac{(v-w)(v-qw)}{(1-q)(1-q^2)}z^2 + \frac{(v-w)(v-qw)(v-q^2w)}{(1-q)(1-q^2)(1-q^3)}z^3 + \cdots$$
$$= \frac{(1-wz)(1-qwz)(1-q^2wz)(1-q^3wz)\cdots}{(1-vz)(1-qvz)(1-q^2vz)(1-q^3vz)\cdots}. \tag{25.16}$$

The idea in this proof was the same as the one used by Euler to prove (25.3), clearly a particular case. Jacobi also went on to give a q-extension of Gauss's $_2F_1$ summation formula. At that time, it was natural for someone to consider the q-extension of a general $_2F_1$ hypergeometric series; E. Heine did just that, and we discuss his work in Chapter 27.

25.2 Jakob Bernoulli's Theta Series

It is interesting that the series with quadratic exponents, normally arising in the theory of elliptic functions, occurred in Bernoulli's work in probability. In 1685, he proposed the following two problems in the *Journal des Savans*:

Let there be two players A and B, playing against each other with two dice on the condition that whoever first throws a 7 will win. There are sought their expectations if they play in one of these orders:

(1) A once, B once, A twice, B twice, A three times, B three times, A four times, B four times, etc.

(2) A once, B twice, A three times, B four times, A five times, etc.

---

[18] Jacobi (1969) vol. 1, pp. 396–399, especially p. 398.

[19] For the English translation, see Remmert (1998) p. 29.

[20] Eisenstein (1975) vol. 1, pp. 117–121.

In his *Ars Conjectandi*, Bernoulli gave a solution, saying that in May 1690, when no solution to this problem had yet appeared, he communicated a solution to *Acta Eruditorum*.[21] In the first case, Bernoulli gave the probability for A to win as

$$1 - m + m^2 - m^4 + m^6 - m^9 + m^{12} - m^{16} + m^{20} - m^{25} + \text{etc.} \qquad (25.17)$$

In the second case, the probability for A to win was

$$1 - m + m^3 - m^6 + m^{10} - m^{15} + m^{21} - m^{28} + m^{36} - m^{45} + \text{etc.} \qquad (25.18)$$

In both cases, $m = \frac{5}{6}$. To make the quadratic exponents explicit, write the two series as

$$1 + \sum_{n=1}^{\infty} m^{n(n+1)} - \sum_{n=1}^{\infty} m^{n^2} \text{ and } \sum_{n=0}^{\infty} (-1)^n m^{\frac{n(n+1)}{2}}.$$

Bernoulli remarked that the summation of these series was difficult because of the unequal jumps in the powers of m. He noted that numerical approximation to any degree of accuracy was easy and for $m = \frac{5}{6}$, the value of the second series was 0.52393; we remark that this value is inaccurate by only one in the last decimal place. Jakob Bernoulli was very interested in polygonal and figurate numbers; in fact, he worked out the sum of the reciprocals of triangular numbers. Here he had series with triangular and square numbers as exponents. Gauss discovered a way to express these series as products. Euler found the product expansion of a series with pentagonal numbers as exponents.

25.3 Euler's *q*-Series Identities

In response to the problems of Naudé, Euler proved the two identities:

$$(1 + qx)(1 + q^2x)(1 + q^3x)(1 + q^4x) \cdots$$

$$= 1 + \frac{q}{1-q}x + \frac{q^3}{(1-q)(1-q^2)}x^2 + \cdots + \frac{q^{\frac{m(m+1)}{2}}}{(1-q)\cdots(1-q^m)}x^m + \cdots, \qquad (25.19)$$

$$\frac{1}{(1 - qx)(1 - q^2x)(1 - q^3x)\cdots}$$

$$= 1 + \frac{q}{1-q}x + \frac{q^2}{(1-q)(1-q^2)}x^2 + \cdots + \frac{q^m}{(1-q)\cdots(1-q^m)}x^m + \cdots. \qquad (25.20)$$

Euler's argument for the first identity was outlined in the opening remarks of this chapter. His proof of his second identity ran along similar lines. We here follow Euler's

---

presentation from his *Introductio*,[22] noting that Euler wrote x for our q and z for our x. Note also that the term q-series came into use only in the latter half of the nineteenth century, appearing in the works of Cayley, Rogers, and others. Jacobi may possibly have been the first to use the symbol q in the context of elliptic functions, though he did not use the term q-series. Euler let Z denote the infinite product on the left of (25.20) and he assumed that Z could be expanded as a series:

$$Z = 1 + Px + Qx^2 + Rx^3 + Sx^4 + \cdots . \qquad (25.21)$$

When x was replaced by qx in Z, he got

$$\frac{1}{(1 - q^2x)(1 - q^3x)(1 - q^4x)\cdots} = (1 - qx)Z.$$

Making the same substitution in (25.21), he obtained

$$(1 - qx)Z = 1 + Pqx + Qq^2x^2 + Rq^3x^3 + Sq^4x^4 + \cdots . \qquad (25.22)$$

When the series for Z was substituted in (25.22) and the coefficients of the various powers of x were equated, the result was

$$P = \frac{q}{1 - q}, \quad Q = \frac{Pq}{1 - q^2}, \quad R = \frac{Qq}{1 - q^3}, \quad S = \frac{Rq}{1 - q^4}, \text{ etc.}$$

and this proved (25.20).

25.4 Euler's Pentagonal Number Theorem

Pentagonal numbers can be generated by the exponents $\frac{m(3m\pm1)}{2}$ in Euler's formula (25.8)

$$\prod_{n=1}^{\infty}(1 - x^n) = 1 + \sum_{m=1}^{\infty}(-1)^m x^{\frac{m(3m\pm1)}{2}} . \qquad (25.23)$$

This identity is often referred to as the pentagonal number theorem.

Recall that Euler had conjectured this result in 1741; he was convinced that this formula was valid, but he could not prove it. He was so confident of his conjecture that in 1747, he used this formula to prove a remarkable theorem on the sum of the divisors of an integer.[23] Concerning this theorem, he remarked in section 25.9 of his paper, "Indeed, I have no other proof."

To understand the 1747 theorem in which he used his conjecture, let n be a nonzero integer and let $\sigma(n) = \sum_{d|n} d$. Observe that if n were a negative integer,

[22] Euler (1988) pp. 361–363.
[23] See Euler's letter to Goldbach: Fuss (1968) vol. I, pp. 407–408. Also see his paper E 175.

then $\sigma(n) = 0$. We remark that Euler's own notation for $\sigma(n)$ was $\int n$. Thus, for n a positive integer, Euler's formula was

$$\sigma(n) = \sigma(n-1) + \sigma(n-2) - \sigma(n-5) - \sigma(n-7) + \cdots, \qquad (25.24)$$

where if $n = \frac{m(3m \mp 1)}{2}$, then $\sigma(0) = n$. Here the numbers $1, 2, 5, 7 \ldots$ in (25.24) are the pentagonal numbers. In his proof, here given in modernized and more brief notation, Euler first took the logarithmic derivative of

$$\sum_{n=0}^{\infty} (-1)^n x^{\frac{n(3n \mp 1)}{2}} = \prod_{m=1}^{\infty} (1 - x^m) \qquad (25.25)$$

to obtain

$$\sum_{n=1}^{\infty} (-1)^n \frac{n(3n \mp 1)}{2} x^{\frac{n(3n \mp 1)}{2}} = -\left(\sum_{n=0}^{\infty} (-1)^n x^{\frac{n(3n \mp 1)}{2}} \right) \sum_{m=1}^{\infty} \frac{m x^m}{1 - x^m}. \qquad (25.26)$$

He noted that the last sum on the right-hand side of (25.26) could be written as

$$\sum_{m=1}^{\infty} m x^m (1 + x^m + x^{2m} + \cdots) = \sum_{m=1}^{\infty} m \sum_{k=1}^{\infty} x^{mk}. \qquad (25.27)$$

He next observed that the coefficient of x^n would contain an m for each $mk = n$. This meant that when the order of summation in (25.27) was changed, the coefficient of x^n would be given by $\sum_{m|n} m = \sigma(n)$. Therefore

$$\sum_{m=1}^{\infty} \frac{m x^n}{1 - x^m} = \sum_{n=1}^{\infty} \sigma(n) x^n,$$

and (25.26) could be rewritten as

$$\left(\sum_{n=0}^{\infty} (-1)^n x^{\frac{n(3n \mp 1)}{2}} \right) \sum_{n=1}^{\infty} \sigma(n) x^n - \sum_{m=1}^{\infty} (-1)^m \frac{n(3n \mp 1)}{2} x^{\frac{m(3n \mp 1)}{2}} = 0. \qquad (25.28)$$

Euler then multiplied the two sums in (25.28) and equated coefficients to obtain

$$\sigma(n) - \sigma(n-1) - \sigma(n-2) + \sigma(n-5) + \sigma(n-7) - \cdots = 0, \qquad (25.29)$$

where the last nonzero term on the right-hand side of (25.29) wold be $\pm n$ if n were a pentagonal number. This proved the formula, on the assumption that (25.25) was true.

Euler's proof of the pentagonal number theorem is elementary and employs simple algebra in an ingenious way; we present it almost exactly as it appeared in Euler's June 1750 letter to Goldbach.[24] He began with the algebraic identity mentioned earlier

$$(1 - \alpha)(1 - \beta)(1 - \delta)(1 - \gamma) \text{ etc.}$$

$$= 1 - \alpha - \beta(1 - \alpha) - \delta(1 - \alpha)(1 - \beta) - \gamma(1 - \alpha)(1 - \beta)(1 - \delta) - \cdots.$$

From this he had

$$(1 - x)(1 - x^2)(1 - x^3)(1 - x^4)(1 - x^5) \text{ etc.} = S$$

$$= 1 - x - x^2(1 - x) - x^3(1 - x)(1 - x^2) - x^4(1 - x)(1 - x^2)(1 - x^3) - \text{ etc.}$$

He set $S = 1 - x - Axx$, where

$$A = 1 - x + x(1 - x)(1 - x^2) + x^2(1 - x)(1 - x^2)(1 - x^3) + \text{ etc.}$$

Multiplying out by the factor $1 - x$ in each term, he obtained

$$A = 1 - x \qquad - x^2(1 - x^2) \qquad - x^3(1 - x^2)(1 - x^3) - \text{ etc.}$$
$$+ x(1 - x^2) + x^2(1 - x^2)(1 - x^3) + x^3(1 - x^2)(1 - x^3)(1 - x^4) + \text{ etc.}$$
$$= 1 - x^3 \qquad - x^5(1 - x^2) \qquad - x^7(1 - x^2)(1 - x^3) - \text{ etc.}$$

He set $A = 1 - x^3 - Bx^5$, where

$$B = 1 - x^2 + x^2(1 - x^2)(1 - x^3) + x^4(1 - x^2)(1 - x^3)(1 - x^4) + \text{ etc.}$$

After multiplying out by the factor $1 - x^2$, appearing in each term of B, he arrived at

$$B = 1 - x^2 \qquad - x^4(1 - x^3) \qquad - x^6(1 - x^3)(1 - x^4) - \text{ etc.}$$
$$+ x^2(1 - x^3) + x^4(1 - x^3)(1 - x^4) + x^6(1 - x^3)(1 - x^4)(1 - x^5) + \text{ etc.}$$
$$= 1 - x^5 \qquad - x^8(1 - x^3) \qquad - x^{11}(1 - x^3)(1 - x^4) - \text{ etc.}$$

Euler then set $B = 1 - x^5 - x^8 C$, where $C = 1 - x^3 + x^3(1 - x^3)(1 - x^4) + x^6(1 - x^3)(1 - x^4)(1 - x^5) + $ etc. Multiplying out by $1 - x^3$,

$$C = 1 - x^3 \qquad - x^6(1 - x^4) \qquad - x^9(1 - x^4)(1 - x^5) - \text{ etc.}$$
$$+ x^3(1 - x^4) + x^6(1 - x^4)(1 - x^5) + x^9(1 - x^4)(1 - x^5)(1 - x^6) + \text{ etc.}$$
$$= 1 - x^7 \qquad - x^{11}(1 - x^4) \qquad - x^{15}(1 - x^4)(1 - x^5) - \text{ etc.}$$

When this process was continued, he got

$$C = 1 - x^7 - x^{11}D, \quad D = 1 - x^9 - x^{14}E, \quad E = 1 - x^{11} - x^{17}F.$$

---

[24] Fuss (1968) vol. 1, pp. 522–524.

This completed Euler's proof. To describe it more succinctly, write $S = P_0, A = P_1$, $B = P_2$, $C = P_3$ and so on. If he had completed the inductive step, Euler would have shown that

$$P_{n-1} = 1 - x^{2n-1} - x^{3n-1} P_n, \qquad (25.30)$$

where

$$P_n = \sum_{k=0}^{\infty} x^{kn}(1 - x^n)(1 - x^{n+1}) \cdots (1 - x^{n+k}). \qquad (25.31)$$

Since Euler's method of proving (25.30) is useful in establishing other identities in q-series, we describe it in the general situation. The first step is to break up each term of P_n into two parts

$$x^{kn}(1 - x^{n+1}) \cdots (1 - x^{n+k}) - x^{(k+1)n}(1 - x^{n+1}) \cdots (1 - x^{n+k});$$

in the second step, take the second (negative) part and add it to the first part of the next term of P_n:

$$-x^{(k+1)n}(1 - x^{n+1}) \cdots (1 - x^{n+k}) + x^{(k+1)n}(1 - x^{n+1}) \cdots (1 - x^{n+k+1})$$

$$= -x^{(k+2)n+k+1}(1 - x^{n+1}) \cdots (1 - x^{n+k}).$$

It can now be seen that

$$P_n = 1 - x^{2n+1} - x^{3n+2} \sum_{k=0}^{\infty} x^{k(n+1)}(1 - x^{n+1}) \cdots (1 - x^{n+1+k})$$

$$= 1 - x^{2n+1} - x^{3n+2} P_{n+1},$$

proving (25.30) by induction. Euler's method was used by Gauss, and then in 1884 Cayley applied it to prove an interesting identity of Sylvester, as discussed in Chapter 26. Rogers and Ramanujan independently employed the idea to prove the Rogers–Ramanujan identities. Recently, Andrews has further developed this method.

A repeated application of (25.30) converts the infinite product in (25.23) to the required sum:

$$(1 - x)(1 - x^2)(1 - x^3)(1 - x^4) \cdots$$

$$= 1 - x - x^2(1 - x^3) + x^{2+5}(1 - x^5) - x^{2+5+8}(1 - x^7) + \cdots$$

$$+ (-1)^{n-1} x^{2+5+\cdots+3n-4}(1 - x^{2n-1}) + (-1)^n x^{2+5+\cdots+3n-1}(1 - x^{2n+1}) + \cdots$$

$$= 1 - (x + x^2) + (x^5 + x^7) - (x^{12} + x^{15}) + \cdots + (-1)^n \left(x^{\frac{n(3n-1)}{2}} + x^{\frac{n(3n+1)}{2}} \right) + \cdots.$$

As Euler noted, this series can also be written as

$$\cdots + x^{26} - x^{15} + x^7 + x^0 - x^1 + x^5 - x^{12} + \cdots = \sum_{n=-\infty}^{\infty} (-1)^n x^{\frac{n(3n-1)}{2}}.$$

This result of Euler was quite remarkable and its proof ingenious; it made quite an impression on the young Gauss who continued Euler's work in new directions.

25.5 Gauss: Triangular and Square Numbers Theorem

We saw Euler's algebraic virtuosity in his proof of the pentagonal number theorem. It is therefore interesting to see Gauss's extremely skillful performance in his similar evaluation of series with triangular and square numbers as exponents. Gauss too divided each term of an appropriate series into two parts and added the second part of each term to the first part of the next term. These formulas for triangular and square exponents are particular cases of the triple product identity. Gauss knew this, but he was sufficiently fond of ingenious calculations to make a brief note of his method in a paper published only in his collected works, "Zur Theorie der transscendenten Functionen Gehörig."[25] He gave a proof of the formula

$$\frac{1-x}{1+x} \cdot \frac{1-xx}{1+xx} \cdot \frac{1-x^3}{1+x^3} \cdot \frac{1-x^4}{1+x^4} \cdot \text{etc.}$$

$$= 1 - 2x + 2x^4 - 2x^9 + 2x^{16} - \text{etc. for } |x| < 1.$$

Gauss started with the series:

$$P = 1 + \sum_{k=1}^{\infty} \frac{x^{kn}}{1+x^n} \cdot \frac{(1-x^{2n+k})(1-x^{n+1})(1-x^{n+2})\cdots(1-x^{n+k-1})}{(1+x^{n+1})(1+x^{n+2})(1+x^{n+3})\cdots(1+x^{n+k})},$$

$$Q = \frac{x^n}{1+x^n} + \frac{x^{2n}}{1+x^n} \cdot \frac{1-x^{n+1}}{1+x^{n+1}} + \frac{x^{2n}}{1+x^n} \cdot \frac{1-x^{n+1}}{1+x^{n+1}} \cdot \frac{1-x^{n+2}}{1+x^{n+2}} + \text{etc.},$$

$$R = P - Q.$$

He evaluated R in two different ways. First, he subtracted the kth term in Q from the kth term in P for each k to get

$$R = \frac{1}{1+x^n} + \sum_{k=1}^{n} \frac{x^{kn}}{1+x^n} \cdot \frac{(1-x^n)(1-x^{n+1})\cdots(1-x^{n+k-1})}{(1+x^{n+1})(1+x^{n+2})\cdots(1+x^{n+k})}. \tag{25.32}$$

He denoted this series for R by $\phi(x,n)$. To find another series for R, Gauss subtracted the kth term in Q from the $(k+1)$th term in P for each k to get

---

[25] Gauss (1863–1927) vol. 3, pp. 437–439.

$$R = 1 - \frac{x^{2n+1}}{1+x^{n+1}} - \frac{x^{2n+2}}{1+x^{n+1}} \cdot \frac{1-x^{n+1}}{1+x^{n+2}} - \frac{x^{2n+3}}{1+x^{n+1}} \cdot \frac{1-x^{n+1}}{1+x^{n+2}} \cdot \frac{1-x^{n+2}}{1+x^{n+3}} - \text{etc.}$$

He concluded from this relation that

$$R = 1 - x^{2n+1} \cdot \phi(x, n+1)$$

or

$$\phi(x,n) = 1 - x^{2n+1} \cdot \phi(x, n+1).$$

Gauss noted that the relation was true for $n \geq 1$, and for such n

$$\phi(x,n) = 1 - x^{2n+1} + x^{4n+4} - x^{6n+9} + x^{8n+16} - \text{etc.}$$

Note that the series P, Q, and R with $n \geq 1$ are absolutely convergent and that the terms can be rearranged. When $n = 0$, the series for P and Q are divergent and care must be exercised. It is clear from the definition of $\phi(x,n)$ given by (25.32) that $\phi(x,0) = \frac{1}{2}$. For clarity, we now employ notation not used by Gauss. Let $p_1, p_2, p_3, \ldots$ and $q_1, q_2, q_3, \ldots$ denote the consecutive terms of P and Q when $n = 0$. Then

$$\phi(x,0) = \lim_{m \to \infty} ((p_1 - q_1) + (p_2 - q_2) + \cdots + (p_m - q_m))$$

$$= \lim_{m \to \infty} (p_1 + (p_2 - q_1) + (p_3 - q_2) + \cdots + (p_m - q_{m-1})) - \lim_{m \to \infty} q_m.$$

Gauss denoted the second limit, $\lim_{m \to \infty} q_m$, by T and called it the last term of the series Q, with $n = 0$. He observed that the first limit could be expressed as $1 - x\phi(x,1)$. Thus, he had $T = 1 - x\phi(x,1) - \phi(x,0)$ or

$$\phi(x,0) = 1 - x\phi(x,1) - T = 1 - x + x^4 - x^9 + x^{16} - \cdots - T.$$

From the definition of T, Gauss could see that

$$T = \frac{1}{2} \frac{1-x}{1+x} \cdot \frac{1-xx}{1+xx} \cdot \frac{1-x^3}{1+x^3} \cdots,$$

and since $\phi(x,0) = \frac{1}{2}$, Gauss finally had

$$2T = \frac{1-x}{1+x} \cdot \frac{1-xx}{1+xx} \cdot \frac{1-x^3}{1+x^3} \cdots = 1 - 2x + 2x^4 - 2x^9 + 2x^{16} - \cdots.$$

He gave an abbreviated form of the argument for the series with triangular numbers. We reproduce Gauss's calculation exactly:

$$P_1 = \frac{1-x^{2n+2}}{1-x^{n+1}} + \frac{x^n \cdot 1 - x^{2n+4} \cdot 1 - x^{n+2}}{1-x^{n+1} \cdot 1 - x^{n+3}}$$

$$+ \frac{x^{2n} \cdot 1 - x^{2n+6} \cdot 1 - x^{n+2} \cdot 1 - x^{n+4}}{1-x^{n+1} \cdot 1 - x^{n+3} \cdot 1 - x^{n+5}} + \text{etc.,}$$

$$Q_1 = \frac{x^n \cdot 1 - x^{n+2}}{1 - x^{n+1}} + \frac{x^{2n} \cdot 1 - x^{n+2} \cdot 1 - x^{n+4}}{1 - x^{n+1} \cdot 1 - x^{n+3}}$$

$$+ \frac{x^{3n} \cdot 1 - x^{n+2} \cdot 1 - x^{n+4} \cdot 1 - x^{n+6}}{1 - x^{n+1} \cdot 1 - x^{n+3} \cdot 1 - x^{n+5}} + \text{etc.},$$

$$R_1 = \frac{1 - x^n}{1 - x^{n+1}} + \frac{x^n \cdot 1 - x^n \cdot 1 - x^{n+2}}{1 - x^{n+1} \cdot 1 - x^{n+3}} + \frac{x^{2n} \cdot 1 - x^n \cdot 1 - x^{n+2} \cdot 1 - x^{n+4}}{1 - x^{n+1} \cdot 1 - x^{n+3} \cdot 1 - x^{n+5}} + \text{etc.},$$

where R_1 was obtained by subtracting Q_1 termwise from P_1. Gauss denoted this series for R_1 as $\psi(x, n)$. Then, by subtracting the kth term of Q_1 from the $(k + 1)$th term of P_1, Gauss had

$$R_1 = 1 + x^{n+1} + \frac{x^{2n+3} \cdot 1 - x^{n+2}}{1 - x^{n+3}} + \frac{x^{2n+3} \cdot x^{n+2} \cdot 1 - x^{n+2} \cdot 1 - x^{n+4}}{1 - x^{n+3} \cdot 1 - x^{n+5}} + \text{etc.}$$

$$= 1 + x^{n+1} + x^{2n+3}\psi(x, n + 2) = \psi(x, n),$$

when $n \geq 1$. Therefore,

$$\psi(x, n) = 1 + x^{n+1} + x^{2n+3} + x^{3n+6} + x^{4n+10} + \text{etc.}$$

In the case $n = 0$, $\psi(x, 0) = 0$. Moreover,

$$\psi(x, 0) = 1 + x + x^3 \psi(x, 2) - \frac{1 - x^2}{1 - x} \cdot \frac{1 - x^4}{1 - x^3} \cdot \frac{1 - x^6}{1 - x^5} \text{ etc.}$$

Hence, the required result followed:

$$\frac{1 - x^2}{1 - x} \cdot \frac{1 - x^4}{1 - x^3} \cdot \frac{1 - x^6}{1 - x^5} \cdots = 1 + x + x^3 + x^6 + x^{10} + \text{etc.}$$

25.6 Gauss Polynomials and Gauss Sums

In his paper of 1808, republished in 1811,[26] Gauss defined the q-extension of a binomial coefficient by

$$(m, \mu) = \frac{(1 - q^m)(1 - q^{m-1})(1 - q^{m-2}) \cdots (1 - q^{m-\mu+1})}{(1 - q)(1 - q^2)(1 - q^3) \cdots (1 - q^\mu)}. \tag{25.33}$$

He noted the easily verified formula

$$(m, \mu + 1) = (m - 1, \mu + 1) + q^{m-\mu-1}(m - 1, \mu). \tag{25.34}$$

Note that it follows from this that (m, μ) is a polynomial in q when m is a positive integer. These polynomials are now called Gaussian polynomials and are extensions

---

[26] Gauss (1981) pp. 463–495.

of the binomial coefficients $\binom{m}{\mu}$. We remark that we are using the familiar symbol q, although Gauss used x.

Let us now see how Gauss evaluated the polynomial $\sum_{\mu=0}^{m}(m,\mu)x^{\mu}$ for $x = -1$ and $x = \sqrt{q}$. For $x = -1$, Gauss used (25.34) to show that

$$f(q,m) = 1 - (m,1) + (m,2) - (m,3) + (m,4) - \cdots$$

satisfied the functional relation

$$f(q,m) = (1 - q^{m-1})f(q,m-2). \tag{25.35}$$

Since $f(q,0) = 1$ and $f(q,1) = 0$, he deduced that

$$\begin{aligned} f(q,m) &= (1-q)(1-q^3)\cdots(1-q^{m-1}), &&\text{for } m \text{ even,} \\ &= 0, &&\text{for } m \text{ odd.} \end{aligned} \tag{25.36}$$

For $x = \sqrt{q}$, Gauss wrote

$$F(q,m) = 1 + q^{\frac{1}{2}}(m,1) + q(m,2) + q^{\frac{3}{2}}(m,3) + \cdots$$

$$= q^{\frac{m}{2}} + q^{\frac{m-1}{2}}(m,1) + q^{\frac{m-2}{2}}(m,2) + q^{\frac{m-3}{2}}(m,3) + \cdots . \tag{25.37}$$

Note that the second (finite) series is identical to the first one, but is in reverse order. Gauss then multiplied the second series by $q^{\frac{m+1}{2}}$, and added the result to the first series, yielding

$$\begin{aligned} (1 + q^{\frac{m+1}{2}})F(q,m) &= 1 + q^{\frac{1}{2}}(m,1) + q(m,2) + q^{\frac{3}{2}}(m,3) + \cdots \\ &\quad + q^{\frac{1}{2}} \cdot q^m + q \cdot q^{m-1}(m,1) + q^{\frac{3}{2}} \cdot q^{m-2}(m,2) + \cdots \\ &= 1 + q^{\frac{1}{2}}(q^m + (m,1)) + q((m,2) + q^{m-1}(m,1)) \\ &\quad + q^{\frac{3}{2}}((m,3) + q^{m-2}(m,2)) + \cdots . \end{aligned}$$

By (25.34), he concluded that $(1+q^{\frac{m+1}{2}})F(q,m) = F(q,m+1)$; since $F(q,0) = 1$, he had the required result

$$F(q,m) = (1 + q^{\frac{1}{2}})(1 + q)(1 + q^{\frac{3}{2}})\cdots(1 + q^{\frac{m}{2}}). \tag{25.38}$$

Gauss used formulas (25.36) and (25.38) to show that

$$\sum_{k=0}^{n-1} e^{\frac{2\pi i k^2}{n}} = \frac{1 + i^{-n}}{1 + i^{-1}}\sqrt{n}. \tag{25.39}$$

We note that the expression on the left side is called a quadratic Gauss sum. He proved this formula in four separate exhaustive cases: for $n = 0, 1, 2, 3 \pmod 4$. Gauss explained how to convert the expression $F(q, m)$ into a Gauss sum. He set $q^{\frac{1}{2}} = -y^{-1}$ so that

$$F(y^{-2}, m) = 1 - y^{-1}\frac{1 - y^{-2m}}{1 - y^{-2}} + y^{-2}\frac{(1 - y^{-2m})(1 - y^{-2m+2})}{(1 - y^{-2})(1 - y^{-4})} - \cdots. \quad (25.40)$$

He then set $m = n - 1$ and took y to be a primitive root of $y^n - 1 = 0$, to get

$$\frac{1 - y^{-2m}}{1 - y^{-2}} = \frac{1 - y^2}{1 - y^{-2}} = -y^2 \; ; \; \frac{1 - y^{-2m+2}}{1 - y^{-4}} = -y^4 \; ; \; \frac{1 - y^{-2m+4}}{1 - y^{-6}} = -y^6 \cdots.$$

Thus, he found

$$F(y^{-2}, m) = 1 + y^{-1} \cdot y^2 + y^{-2} \cdot y^2 \cdot y^4 + y^{-3} \cdot y^2 \cdot y^4 \cdot y^6 + \cdots$$

$$= 1 + y + y^4 + y^9 + \cdots + y^{(n-1)^2}. \quad (25.41)$$

Observe that for $y = e^{\frac{2\pi i}{n}}$, the expression (25.41) was the Gauss sum. From (25.38) and (25.41), it followed that

$$1 + y + y^4 + y^9 + \cdots + y^{(n-1)^2} = (1 - y^{-1})(1 + y^{-2})(1 - y^{-3}) \cdots (1 \pm y^{-n+1}), \quad (25.42)$$

when y was a primitive root of $y^n - 1 = 0$. Gauss showed that when $y = e^{\frac{2\pi i}{n}}$, the product in (25.42) reduced to the expression on the right-hand side of (25.39). The case $n = 4s + 2$ is elementary. In fact, for this case he observed that for any primitive root y, $y^{2s+1} = -1$, so that $y^{(2s+1)^2} = -1$. Moreover, for any integer t,

$$y^{(2s+1+t)^2} = y^{(2s+1)^2 + (4s+2)t + t^2} = -y^{t^2}.$$

Therefore, by cancellation of terms, Gauss found the sum (25.41) to be zero. Turning to the case $n = 4s$, he applied (25.42) to evaluate (25.39). Now $y^{(2s+t)^2} = y^{t^2}$, and hence

$$1 + y + y^4 + \cdots + y^{(n-1)^2} = 2(1 + y + y^4 + \cdots + y^{(2s-1)^2}). \quad (25.43)$$

By taking $m = \frac{1}{2}n - 1 = 2s - 1$ in (25.40) and using the calculations leading up to (25.42), Gauss had

$$1 + y + y^4 + \cdots + y^{(2s-1)^2} = (1 - y^{-1})(1 + y^{-2})(1 - y^{-3}) \cdots (1 - y^{-2s+1}). \quad (25.44)$$

Then $y^{2s} = -1$, and hence $1 + y^{-2k} = -y^{2s-2k}(1 - y^{-2s+2k})$. He applied this to the product in (25.44), so that by (25.43)

$$F \equiv 1 + y + y^4 + \cdots + y^{(n-1)^2}$$

$$= 2(-1)^{s-1}y^{s^2-s}(1 - y^{-1})(1 - y^{-2}) \cdots (1 - y^{-2s+1}). \quad (25.45)$$

Again, from the fact that $1 - y^{-k} = -y^{-k}(1 - y^{-4s+k})$, Gauss got

$$(1 - y^{-1})(1 - y^{-2})(1 - y^{-3}) \cdots (1 - y^{-2s+1})$$

$$= (-1)^{2s-1} y^{-2s^2+s} (1 - y^{-2s-1})(1 - y^{-2s-2})(1 - y^{-2s-3}) \cdots (1 - y^{-4s+1}).$$
(25.46)

Therefore, by (25.45) and (25.46)

$$F = 2(-1)^{3s-2} y^{-s^2} (1 - y^{-2s-1})(1 - y^{-2s-2}) \cdots (1 - y^{-4s+1}).$$
(25.47)

Next, Gauss took the product of (25.45) and (25.47) and multiplied by $1 - y^{-2s}$ to obtain

$$(1 - y^{-2s}) F^2 = 4(-1)^{4s-3} y^{-s} (1 - y^{-1})(1 - y^{-2}) \cdots (1 - y^{-4s+1}). \quad (25.48)$$

So Gauss could conclude that $F^2 = 2y^s n = \pm 2in$, since $y^{2s} = -1$ or $y^s = \pm i$. Note that he also made use of the fact that the product

$$(1 - y^{-1})(1 - y^{-2}) \cdots (1 - y^{-4s+1})$$

was equal to n, because $y^{-1}, \ldots, y^{-4s+1}$ were all the nontrivial nth roots of unity. By taking square roots, he obtained

$$F = 1 + y + y^4 + \cdots + y^{(n-1)^2} = \pm(1 + i)\sqrt{n}. \quad (25.49)$$

To determine the sign when $y = e^{\frac{2\pi i}{n}}$, Gauss set $y = p^2$ in (25.44) and used $p^n = -1$ to get

$$F = 2(1 + p^{n-2})(1 + p^{-4})(1 + p^{n-6})(1 + p^{-8}) \cdots (1 + p^{-n+4})(1 + p^2).$$

He rewrote this equation as

$$F = 2(1 + p^2)(1 + p^{-4})(1 + p^6) \cdots (1 + p^{-n+4})(1 + p^{n-2})$$

and observed that $1 + p^{\pm 2k} = 2p^{\pm k} \cos(\frac{k\pi}{n})$, finally concluding that

$$F = 2^{2s} p^s \cos \frac{\pi}{n} \cos \frac{2\pi}{n} \cos \frac{3\pi}{n} \cdots \cos \frac{(2s - 1)\pi}{n}.$$

Now $p^s = \cos \frac{\pi}{4} + i \sin \frac{\pi}{4} = \frac{1+i}{\sqrt{2}}$ and since all the cosine values were positive, Gauss determined that the sign in (25.49) was positive. This concluded his proof of the case $n = 4s$.

The other two cases of (25.39), where n is odd, or $n = 4s + 1$ or $4s + 3$, are the most important because they lead to the proof of the quadratic reciprocity theorem. For these cases, Gauss first gave a detailed derivation using (25.36), although he indicated that (25.38) could also serve the purpose. So it remained for Gauss to prove that the Gauss sum in (25.49) was equal to $\sqrt{n}$ when n was of the form $4m + 1$, and equal to

$i\sqrt{n}$ when n took the form $4m + 3$. He took $n = s + 1$ where s was even, x was a primitive root of $x^n - 1 = 0$, and $q = x^{-2}$. Then

$$\frac{1 - q^{s-j}}{1 - q^{j+1}} = \frac{1 - x^{-2(s-j)}}{1 - x^{-2j-2}} = \frac{1 - x^{-2(n-1-j)}}{1 - x^{-2j-2}} = \frac{1 - x^{2j+2}}{1 - x^{-2j-2}} = -x^{2(j+1)}$$

and the Gaussian polynomial was given by

$$(s, k) = (-1)^k x^{2(1+2+\cdots+k-1)} = (-1)^k x^{k(k-1)}.$$

Using this in (25.36), he had

$$\sum_{k=1}^{n} x^{k(k-1)} = (1 - x^{-2})(1 - x^{-6}) \cdots (1 - x^{-2(n-2)})$$

$$= x^{\frac{1}{4}(n-1)^2}(x - x^{-1})(x^3 - x^{-3}) \cdots (x^{n-2} - x^{-(n-2)}). \qquad (25.50)$$

Since x was an nth root of unity and since

$$\frac{1}{4}(n - 1)^2 + k(k - 1) = \frac{1}{4}(n^2 - 2n + (2k - 1)^2),$$

$$x^{\frac{1}{4}(n-1)^2+k(k-1)} = x^{\frac{1}{4}(n-(2k-1))^2} = x^{(e-k)^2}, \quad \text{where } n + 1 = 2e.$$

Thus Gauss could rewrite (25.50) as

$$W \equiv \sum_{k=0}^{n-1} x^{k^2} = (x - x^{-1})(x^3 - x^{-3}) \cdots (x^{n-2} - x^{-(n-2)}). \qquad (25.51)$$

We note that Gauss also worked out a derivation of this formula using (25.38). Now $x^{n-2} - x^{-(n-2)} = -(x^2 - x^{-2})$ etc. implied that

$$W = (-1)^{\frac{n-1}{2}}(x^2 - x^{-2})(x^4 - x^{-4}) \cdots (x^{n-1} - x^{-n+1}). \qquad (25.52)$$

By multiplying (25.51) and (25.52), he obtained

$$W^2 = (-1)^{\frac{n-1}{2}}(x - x^{-1})(x^2 - x^{-2})(x^3 - x^{-3}) \cdots (x^{n-1} - x^{-n+1}).$$

When n was of the form $4s + 1$, the factor $(-1)^{\frac{n-1}{2}}$ became $+1$ and when n was of the form $4s + 3$, it became -1. Thus, he arrived at

$$W^2 = \pm x^{\frac{1}{2}(n-1)n}(1 - x^{-2})(1 - x^{-4}) \cdots (1 - x^{-2(n-1)}).$$

Using an argument similar to the one for (25.49), Gauss concluded that $W = \pm n$, where the $+$ sign applied to $n = 4s + 1$ and the $-$ sign to $4s + 3$. Note that Gauss had already arrived at this point in 1801, but by a different route. The problem remaining in 1805 was to choose the correct sign for the square root to obtain W. To find that, Gauss set $x = e^{\frac{2\pi i}{n}}$ in (25.51) to get

$$W = (2i)^{\frac{n-1}{2}} \sin \frac{2\pi}{n} \sin \frac{6\pi}{n} \cdots \sin \frac{2(n-2)\pi}{n}.$$

Whether $n = 4s + 1$ or $n = 4s + 3$, Gauss saw that there were clearly s negative factors in the sine product. Thus, he could conclude that $W = \sqrt{n}$ for $n = 4s + 1$ and $W = i\sqrt{n}$ for $n = 4s + 3$.

25.7 Gauss's q-Binomial Theorem and the Triple Product Identity

Gauss wrote a paper "Hundert Theoreme über die neuen Transcendenten," but he did not publish it. In this paper, he derived a form of the terminating q-binomial theorem. He then wrote the result in a symmetric form and by an ingenious argument derived the triple product identity.[27] We follow Gauss's notation and proof: He stated the terminating q-binomial theorem in the form

$$1 + \frac{a^n - 1}{a - 1} t + \frac{a^n - 1 \cdot a^n - a}{a - 1 \cdot aa - 1} tt + \frac{a^n - 1 \cdot a^n - a \cdot a^n - aa}{a - 1 \cdot aa - 1 \cdot a^3 - 1} t^3 + \text{etc.}$$

$$= (1 + t)(1 + at)(1 + aat) \cdots (1 + a^{n-1}t). \tag{25.53}$$

Recall that we would write q instead of a. To prove the formula inductively, he denoted the sum as T and multiplied it by $(1 + a^n t)$ to obtain a series of the same form with n changed to $n + 1$. The reader may work out this calculation. Gauss next observed that by taking $T = \theta(n)$, one could see that $T(1 + a^n t) = \theta(n + 1)$. Thus, the terminating q-binomial theorem was proved inductively.

To prove the triple-product identity, he wrote his result in a symmetric form. He took n even, set $y = a^{\frac{n-1}{2}} t$ and $x^2 = a$ to transform (25.53) into

$$1 + \frac{1 - x^n}{1 - x^{n+2}} x \left(y + \frac{1}{y} \right) + \frac{1 - x^n}{1 - x^{n+2}} \cdot \frac{1 - x^{n-2}}{1 - x^{n+4}} \cdot x^4 \left(yy + \frac{1}{yy} \right)$$

$$+ \frac{1 - x^n}{1 - x^{n+2}} \cdot \frac{1 - x^{n-2}}{1 - x^{n+4}} \cdot \frac{1 - x^{n-4}}{1 - x^{n+6}} \cdot x^9 \left(y^3 + \frac{1}{y^3} \right) + \cdots$$

$$= \frac{1 - xx}{1 - x^{n+2}} \cdot \frac{1 - x^4}{1 - x^{n+4}} \cdot \frac{1 - x^6}{1 - x^{n+6}} \cdots \cdots \frac{1 - x^n}{1 - x^{2n}}$$

$$\cdot (1 + xy)(1 + x^3 y) \cdots \cdots (1 + x^{n-1}y) \left(1 + \frac{x}{y} \right)$$

$$\cdot \left(1 + \frac{x^3}{y} \right) \cdots \cdots \left(1 + \frac{x^{n-1}}{y} \right). \tag{25.54}$$

---

[27] Gauss (1863–1927) vol. 3, pp. 461–464.

Next, he took $|x| < 1$, so that $x^n \to 0$ as $n \to \infty$. The result was the triple product identity:

$$1 + x(y + y^{-1}) + x^4(yy + y^{-2}) + x^9(y^3 + y^{-3}) + \cdots$$
$$= (1 - xx)(1 - x^4)(1 - x^6) \cdots (1 + xy)(1 + x^3 y)(1 + x^5 y)$$
$$\cdots (1 + xy^{-1})(1 + x^3 y^{-1})(1 + x^5 y^{-1}) \cdots . \tag{25.55}$$

Now let us examine the algebraic steps Gauss used to get the required symmetric form (25.54). Note that the last term in the left-hand side of (25.53) was $a^{\frac{n(n-1)}{2}} t^n = y^n$. Combining the first and last terms of the sum, then the second to the last but one, and so on, he arrived at

$$1 + y^n + \frac{1 - a^n}{1 - a} t \left(1 + y^{n-2}\right) + \frac{1 - a^n}{1 - a} \cdot \frac{1 - a^{n-1}}{1 - aa} \cdot att(1 + y^{n-4}) + \cdots$$
$$+ \frac{1 - a^n}{1 - a} \cdot \frac{1 - a^{n-1}}{1 - aa} \cdots \frac{1 - a^{\frac{1}{2}n+2}}{1 - a^{\frac{1}{2}n-1}} \cdot a^{\frac{1}{2}\left(\frac{1}{2}n-1\right)\left(\frac{1}{2}n-2\right)} t^{\frac{1}{2}n-1}(1 + yy)$$
$$+ \frac{1 - a^n}{1 - a} \cdot \frac{1 - a^{n-1}}{1 - aa} \cdots \frac{1 - a^{\frac{1}{2}n+1}}{1 - a^{\frac{1}{2}n}} \cdot a^{\frac{1}{2} \cdot \frac{1}{2}n \cdot \frac{1}{2}n - 1} t^{\frac{1}{2}n}.$$

He set $a = x^2$, denoted the last term by A and took it out as a common factor to get

$$A\left(1 + \frac{1 - x^n}{1 - x^{n+2}} \cdot x(y + y^{-1}) + \frac{1 - x^n}{1 - x^{n+2}} \cdot \frac{1 - x^{n-2}}{1 - x^{n+4}} x^4(yy + y^{-2}) \right.$$
$$\left. + \frac{1 - x^n}{1 - x^{n+2}} \cdot \frac{1 - x^{n-2}}{1 - x^{n+4}} \cdot \frac{1 - x^{n-4}}{1 - x^{n+6}} \cdot x^9(y^3 + y^{-3}) + \cdots \right),$$

where A could be written as

$$\frac{1 - x^{n+2}}{1 - xx} \cdot \frac{1 - x^{n+4}}{1 - x^4} \cdot \frac{1 - x^{n+6}}{1 - x^6} \cdots \frac{1 - x^{2n}}{1 - x^n} \cdot \frac{y^{\frac{1}{2}n}}{x^{\frac{1}{4}nn}}.$$

He rewrote the product $(1 + t)(1 + at)(1 + aat) \cdots (1 + a^{n-1}t)$ as

$$\left(1 + \frac{y}{x^{n-1}}\right)\left(1 + \frac{y}{x^{n-3}}\right) \cdots \left(1 + \frac{y}{x}\right)(1 + yx)(1 + yx^3) \cdots (1 + yx^{n-1}).$$

To complete the calculations necessary for the symmetric form (25.54), it was sufficient for Gauss to observe that the first half of the product could be rewritten as

$$\frac{y^{\frac{1}{2}n}}{x^{\frac{1}{4}nn}} \left(1 + \frac{x^{n-1}}{y}\right)\left(1 + \frac{x^{n-3}}{y}\right) \cdots \left(1 + \frac{x}{y}\right).$$

It is interesting to note that the triple product formula (25.55) contains a plethora of important special cases. Euler's pentagonal numbers identity follows on taking $x = q^{\frac{3}{2}}$ and $y = -q^{\frac{1}{2}}$. Gauss's formula for triangular numbers, derived earlier, follows by taking $x = q^{\frac{1}{2}}$ and $y = q^{\frac{1}{2}}$.

It is surprising that Gauss did not publish his work related to the triple product. Gauss's 1808 paper correctly noted the significance of the Gaussian polynomial (m, μ) and later work of O. Rodrigues, P. MacMahon, and others revealed the combinatorial import of the Gaussian polynomial and its generalization. In addition, Gaussian polynomials played an important role in Cayley and Sylvester's development of invariant theory. It remained for Jacobi to rediscover the triple product formula and use it in his theory of elliptic functions.

25.8 Jacobi: Triple Product Identity

In his work in the theory of elliptic functions, Jacobi encountered numerous infinite products, a large number of which were particular cases of the product side of the triple-product identity. And the product side of this identity was composed of two infinite products, first elucidated by Euler, of the form (25.1). Jacobi gave two proofs oof the triple product identity in his *Fundamenta Nova*; we present the second.[28] Because Jacobi wished to convert his products in elliptic function theory into series, it was only natural for him to start with Euler's formula (25.3). Change q to q^2 and x to $\frac{z}{q}$ to get

$$(1 + qz)(1 + q^3 z)(1 + z^5 z)(1 + q^7 z) \cdots$$

$$= 1 + \frac{qz}{1 - q^2} + \frac{q^4 z^2}{(1 - q^2)(1 - q^4)} + \frac{q^9 z^3}{(1 - q^2)(1 - q^4)(1 - q^6)} + \cdots .$$

Jacobi then multiplied this equation by one in which z was replaced by $\frac{1}{z}$, to obtain

$$(1 + qz)(1 + q^3 z)(1 + q^5 z) \cdots \left(1 + \frac{q}{z}\right)\left(1 + \frac{q^3}{z}\right)\left(1 + \frac{q^5}{z}\right) \cdots$$

$$= 1 + \frac{qz}{1 - q^2} + \frac{q^4 z^2}{(1 - q^2)(1 - q^4)} + \frac{q^9 z^3}{(1 - q^2)(1 - q^4)(1 - q^6)} + \cdots$$

$$\times \left(1 + \frac{q}{1 - q^2}\frac{1}{z} + \frac{q^4}{(1 - q^2)(1 - q^4)}\frac{1}{z^2} + \frac{q^9}{(1 - q^2)(1 - q^4)(1 - q^6)}\frac{1}{z^3} + \cdots \right) ..$$

[28] Jacobi (1969) vol. 1, pp. 232–234.

Jacobi observed that the coefficient of $z^n + \frac{1}{z^n}$ in the product on the right-hand side was

$$\frac{q^{nn}}{(1-q^2)(1-q^4)\cdots(1-q^{2n})}$$

$$\times \left(1 + \frac{q^2}{1-q^2}\cdot\frac{q^{2n}}{1-q^{2n+2}} + \frac{q^8}{(1-q^2)(1-q^4)}\cdot\frac{q^{4n}}{(1-q^{2n+2})(1-q^{2n+4})}\right.$$

$$\left. + \frac{q^{18}}{(1-q^2)(1-q^4)(1-q^6)}\cdot\frac{q^{6n}}{(1-q^{2n+2})(1-q^{2n+4})(1-q^{2n+6})} + \cdots\right).$$

$$(25.56)$$

It seems that Jacobi had some trouble simplifying this expression and this delayed him for quite a while. But he succeeded in resolving the problem by proving that

$$\prod_{n=1}^{\infty}(1-q^n z)^{-1} = \sum_{n=1}^{\infty}\frac{q^{n^2}z^n}{(1-q)(1-q^2)\cdots(1-q^n)(1-qz)\cdots(1-q^n z)}. \quad (25.57)$$

He replaced q by q^2 and then set $z = q^{2n}$ to sum the series in (25.56). He thus found the coefficient of $z^n + \frac{1}{z^n}$ to be

$$\frac{q^{nn}}{(1-q^2)\cdots(1-q^{2n})}.$$

This proved the triple product identity. To prove (25.57), Jacobi assumed that the product on the left-hand side could be expressed as a sum of terms of the form

$$\frac{A_n z^n}{((1-qz)\cdots(1-q^n z))}.$$

For A^n, he applied the standard procedure of changing z to qz to get a functional relation. Obviously, the difficult point here was to conceive of that form of the series in which the variable z would also appear in the denominator. Neither Euler nor Gauss came up with such a series.

Jacobi's formula (25.57) is very interesting. Note that the product on the left-hand is the same as the product in Euler's second formula (25.20) but the series on the right, though similar in appearance, has an additional factor in the denominator of each term of the sum. Jacobi may have asked whether it was possible to directly transform one series into the other. This suggests a transformation theory of q-series similar to that for hypergeometric series. Heinrich Eduard Heine (1821-1881) paved the way for the study of transformations of q-series in his 1846 theory of the q-hypergeometric series. We also note that in 1843, Cauchy gave a generalization of (25.57).

25.9 Eisenstein: *q*-Binomial Theorem

Eisenstein's "Neuer Beweis und Verallgemeinerung des binomischen Lehrsatzes"[29] was one of three papers he submitted to *Crelle's Journal* in May 1844. In this paper, he proved the general *q*-binomial theorem, although Eisenstein wrote *p* instead of *q*, and deduced from it the ordinary binomial theorem. His proof was based on an idea of Euler and employed the multiplication of series. Eisenstein did not refer to Euler, but mentioned Dirichlet and Martin Ohm, who may have discussed Euler's idea in their lectures. Eisenstein first proved the finite case of the *q*-binomial theorem. For this he defined for a positive integer α,

$$\phi(x,\alpha) = (1+x)(1+qx)(1+q^2x)\cdots(1+q^{\alpha-1}x). \tag{25.58}$$

He proved Rothe's formula without reference to Rothe:

$$\phi(x,\alpha) = \sum_{t=0}^{\alpha} A_t x^t, \tag{25.59}$$

where

$$A_t = \frac{q^\alpha - 1}{q - 1} \cdot \frac{q^{\alpha-1} - 1}{q^2 - 1} \cdots \frac{q^{\alpha-t+1} - 1}{q^t - 1} q^{\frac{1}{2}t(t-1)}. \tag{25.60}$$

Note that this was done in the standard way by using the relation

$$(1+q^\alpha x)\phi(x,\alpha) = (1+x)\phi(qx,\alpha).$$

Eisenstein stated the general *q*-binomial theorem in the form

$$\phi(x,\alpha) \equiv \sum_{t=0}^{\infty} A_t x^t = \frac{(1+x)(1+qx)(1+q^2x)\cdots}{(1+q^\alpha x)(1+q^{\alpha+1}x)(1+q^{\alpha+2}x)\cdots}, \tag{25.61}$$

where $|q| < 1$, and α was any number. To prove this, he first wished to show that

$$\phi(x,\alpha+\beta) = \phi(x,\alpha)\phi(q^\alpha x,\beta). \tag{25.62}$$

For this purpose, he demonstrated that

$$C_t = A_t + A_{t-1}B_1q^\alpha + A_{t-2}B_2q^{2\alpha} + \cdots + B_tq^{t\alpha}, \tag{25.63}$$

where B_t and C_t were obtained from (25.60) by replacing α by β and α by $\alpha + \beta$, respectively. He noted that (25.62) and (25.63) were clearly true when α and β were positive integers. Eisenstein then set $u = q^\alpha$ and $v = q^\beta$ and observed that both sides of (25.63) were equal for infinitely many values of u and v, and thus (25.63) was identically true. At this point, Eisenstein noted that the proof could be completed in the usual manner and referred to Dirichlet and Ohm. From Chapter 4, one may see

[29] Eisenstein (1975) pp. 117–121.

that Eisenstein intended to use (25.62) to prove (25.61) for all integers α, and then for all rational numbers, and finally (by continuity) for all real α.

25.10   Jacobi's q-Series Identity

In 1846, Jacobi proved the q-binomial theorem[30] and obtained an extension of the Vandermonde identity, as well as an extension of Gauss's $_2F_1$ summation formula. Recall Gauss's summation formula:

$$\sum_{n=0}^{\infty} \frac{(a)_n (b)_n}{n!\,(c)_n} = \frac{\Gamma(c)\Gamma(c-a-b)}{\Gamma(c-a)\Gamma(c-b)}$$

when $\mathrm{Re}(c-a-b) > 0$. Note that when $a = -m$, a negative integer, we have Vandermonde's identity

$$\sum_{n=0}^{m} \frac{(-m)_n (b)_n}{n!\,(c)_n} = \frac{(c-b)_m}{(c)_m}. \tag{25.64}$$

This identity is not difficult and follows immediately from the Gregory–Newton interpolation formula; it can also be obtained by multiplying two binomial series and equating coefficients. In about 1975,[31] while studying Needham's excellent work,[32] Richard Askey saw that around 1301, the Chinese mathematician Chu Shih-Chieh (also Zhu Shijie) discovered two equations; when they are combined, they yield the Vandermonde identity, found by Vandermonde in 1772.[33] Later, when Chu Shih-Chieh's work was fully translated,[34] Askey and I studied it very carefully and found only one of the required equations. Thus, denoting Vandermonde's identity as the Chu–Vandermonde identity might possibly be an exaggeration.

In Jacobi's notation, the q-binomial theorem was stated as

$$[w, v] \equiv 1 + \frac{v - w}{1 - x} z + \frac{(v - w)(v - xw)}{(1 - x)(1 - x^2)} z^2 + \frac{(v - w)(v - xw)(v - x^2 w)}{(1 - x)(1 - x^2)(1 - x^3)} z^3 + \cdots$$

$$= \frac{(1 - wz)(1 - xwz)(1 - x^2 wz)(1 - x^3 wz) \cdots}{(1 - vz)(1 - xvz)(1 - x^2 vz)(1 - x^3 vz) \cdots}. \tag{25.65}$$

Let $\phi(z)$ denote the product. In his proof, Jacobi assumed that

$$\phi(z) = 1 + A_1 z + A_2 z^2 + A_3 z^3 + A_4 z^4 + \cdots$$

[30] Jacobi (1969) vol. 6, pp. 163–173.
[31] Askey (1975) pp. 59–60.
[32] Needham (1959) p. 138.
[33] Vandermonde (1772).
[34] Hoe (2007).

and observed that $\phi(z)$ satisfied the functional relation

$$\phi(z) - \phi(xz) = v\phi(z) - w\phi(xz).$$

Thus, the coefficients A_1, A_2, A_3, ... satisfied the equations

$$(1-x)A_1 = v - w, \; (1-x^2)A_2 = (v - xw)A_1, \; (1-x^3)A_3 = (v - x^2w)A_2, \cdots.$$

By induction, this gives the desired result.

Now note that from the product expression for $[w, v]$ it is easy to see that $[w, v][v, 1] = [w, 1]$. If the corresponding series are substituted in this equation and the coefficient of z^p equated on the two sides, then the result is a q-extension of the Vandermonde identity. Jacobi wrote that he saw this result in Schweins's *Analysis*:[35]

$$\frac{(1-w)(1-xw)(1-x^2w)\cdots(1-x^{p-1}w)}{(1-x)(1-x^2)(1-x^3)\cdots(1-x^p)}$$

$$= \sum_{k=0}^{p} \frac{(v-w)(v-xw)\cdots(v-x^{k-1}w)}{(1-x)(1-x^2)\cdots(1-x^k)} \cdot \frac{(1-v)(1-xv)\cdots(1-x^{p-k-1}v)}{(1-x)(1-x^2)\cdots(1-x^{p-k})}.$$

$$(25.66)$$

Note that the empty products occurring in the sum have the value 1. Now, it is possible to prove Gauss's formula from the Vandermonde or Chu–Vandermonde identity, but it is not easy, and such a proof was not known in Jacobi's time. But in a beautiful argument, Jacobi used (25.66), the q-extension of Vandermonde, to prove a q-extension of Gauss's formula. He divided both sides of the equation by the first term on the right-hand side, to get (after a change of variables)

$$\frac{(1-u)(1-xu)(1-x^2u)\cdots(1-x^{p-1}u)}{(1-r)(1-xr)(1-x^2r)\cdots(1-x^{p-1}r)}$$

$$= 1 + \sum_{k=1}^{p}(-1)^k \frac{(u-r)(u-xr)\cdots(u-x^{k-1}r)}{(1-x)(1-x^2)\cdots(1-x^k)} \tag{25.67}$$

$$\times \frac{(1-x^p)(1-x^{p-1})\cdots(1-x^{p-k+1})}{(1-r)(1-xr)\cdots(1-x^{k-1}r)} x^{\frac{k(k-1)}{2}}. \tag{25.68}$$

Jacobi stated the extension of Gauss's formula in the form

$$1 + \frac{(1-s)(1-t)}{(1-x)(1-r)}r + \frac{(1-s)(x-s)(1-t)(x-t)}{(1-x)(1-x^2)(1-r)(1-xr)}r^2$$

$$+ \frac{(1-s)(x-s)(x^2-s)(1-t)(x-t)(x^2-t)}{(1-x)(1-x^2)(1-x^3)(1-r)(1-xr)(1-x^2r)}r^3 + \cdots$$

$$= \frac{(1-sr)(1-tr)}{(1-r)(1-str)} \cdot \frac{(1-xsr)(1-xtr)}{(1-xr)(1-xstr)} \cdot \frac{(1-x^2sr)(1-x^2tr)}{(1-x^2r)(1-x^2str)} \cdots. \tag{25.69}$$

[35] Schweins (1820).

He showed that when $t = x^p$, and $p = 0, 1, 2, \ldots$, this was reduced to the identity (25.68). Thus, (25.69) was true for an infinite number of values of t and, by symmetry, for an infinite number of values of s. He then observed that

$$\frac{(1 - sr)(1 - tr)}{(1 - r)(1 - str)} = 1 + c_1 r + c_2 r^2 + c_3 r^3 + \cdots$$

where $c_1, c_2, c_3, \ldots$ were polynomials in s and t. This implied that the product on the right-hand side of (25.69) was of the form

$$(1 + c_1 r + c_2 r^2 + c_3 r^3 + \cdots) \cdot (1 + c_1 x r + c_2 x^2 r^2 + \cdots) \cdot (1 + c_1 x^2 r + c_2 x^4 r^2 + \cdots) \cdots.$$

This product would then be of the form $1 + b_1 r + b_2 r^2 + b_3 r^3 + \cdots$, where $b_1, b_2, b_3, \ldots$ were polynomials in s and t. To complete the proof, Jacobi wrote the left-hand side of (25.69) in powers of r as $1 + k_1 r + k_2 r^2 + \cdots$, so that $k_1, k_2, \ldots$ were polynomials in s and t. Jacobi concluded that $b_i = k_i$ because it held for an infinite number of values of s and t. This completed the proof of (25.69). Jacobi also observed, without giving a precise definition, that the products on the right-hand side of (25.69) could be considered q-analogs of the gamma functions in Gauss's formula. Very soon after this, Heine obtained a nearly correct definition.

25.11 Cauchy and Ramanujan: The Extension of the Triple Product

In 1843, Augustin-Louis Cauchy published an important paper[36] containing the first statement and proof of the general q-binomial theorem and an extension of the triple product identity. To be clear and succinct in stating the results of Cauchy and Ramanujan, we introduce the following modern notation: Let

$$(a; q)_n = (1 - a)(1 - aq) \cdots (1 - aq^{n-1}), \qquad \text{for } n \geq 1,$$
$$= 1, \qquad \text{for } n = 0,$$
$$= \frac{1}{(1 - q^{-1}a)(1 - q^{-2}a) \cdots (1 - q^{-n}a)}, \qquad \text{for } n \leq 0.$$

And $(a; q)_\infty = (1 - a)(1 - qa)(1 - q^2 a) \cdots$. Using this notation, the q-binomial theorem can be stated as

$$\sum_{n=0}^{\infty} \frac{(a; q)_n}{(q; q)_n} x^n = \frac{(ax; q)_\infty}{(x; q)_\infty}.$$

For convergence we require $|x| < 1$, $|q| < 1$. Cauchy's extension of the triple product identity can now be stated for $0 < |bx| < 1$:

$$\sum_{n=-\infty}^{\infty} \left(\frac{a}{b}; q\right)_n b^n x^n = \frac{(ax; q)_\infty \left(\frac{q}{ax}; q\right)_\infty (q; q)_\infty}{(bx; q)_\infty \left(\frac{bq}{a}; q\right)_\infty}.$$

[36] Cauchy (1843a).

Here, Cauchy failed to find the better result, called the Ramanujan $_1\psi_1$ sum, generalizing the q-binomial theorem as well as the triple product identity. G. H. Hardy found this theorem without proof in Srinivasa Ramanujan's (1887–1920) notebooks and published it in his 1940 lectures on Ramanujan's work.[37] This formula provides the basis for the study of bilateral q-series.

Ramanujan's theorem can be stated as

$$\sum_{n=-\infty}^{\infty} \frac{(a;q)_n}{(b;q)_n} x^n = \frac{(ax;q)_\infty \left(\frac{q}{ax};q\right)_\infty (q;q)_\infty \left(\frac{b}{a};q\right)_\infty}{(x;q)_\infty \left(\frac{b}{ax};q\right)_\infty (b;q)_\infty \left(\frac{q}{a};q\right)_\infty},$$

where $|q| < 1$ and $\left|\frac{b}{a}\right| < |x| < 1$.

25.12 Rodrigues and MacMahon: Combinatorics

Olinde Rodrigues and Percy Alexander MacMahon made important contributions to combinatorial problems connected with Gaussian polynomials and their generalizations. Olinde Rodrigues (1794–1851) was a French mathematician whose ancestors most probably left Spain, fleeing the persecution of the Jews. He studied at the Lycée Impérial in Paris and then at the new Université de Paris. He published six mathematical papers during 1813–16, one of which contains his well-known formula for Legendre polynomials. He did not pursue an academic career, perhaps because of religious discrimination. In fact, he apparently gave up mathematical research for over two decades, returning to it in 1838; he then produced papers on combinatorics and an important work on rotations.

Rodrigues's theorem from 1839[38] gave the generating function for the number of permutations $Z(n,k)$ of n distinct objects with k inversions; this was the number of permutations $a_1, a_2, \ldots, a_n$, of $1, 2, 3, \ldots, n$ with k pairs (a_i, a_j), such that $i < j$ and $a_i > a_j$. The values of k range from 0 to $\frac{n(n-1)}{2}$. To find the generating function of $Z(n,k)$, Rodrigues argued that $Z(n,k)$ was the number of integer solutions of the equation

$$x_0 + x_1 + x_2 + \cdots + x_{n-1} = k,$$

where $0 \leq x_i \leq i$ for $i = 0, 1, \ldots, n - 1$. This implied that the $Z(n,k)$ was the coefficient of t^k in the product

$$(1 + t)(1 + t + t^2)(1 + t + t^2 + t^3) \cdots (1 + t + t^2 + \cdots + t^{n-1}).$$

As immediate corollaries, Rodrigues had

$$Z(n,0) + Z(n,1) + \cdots + Z(n,n-1) = n!,$$

$$Z(n,0) - Z(n,1) + Z(n,2) - \cdots + (-1)^{n-1} Z(n,n-1) = 0.$$

---

[37] Hardy (1940) p. 222.
[38] Rodrigues (1839).

The first relation also answered a question posed by Stern on the sum of all the inversions in the permutations of n letters. Note that we can write Rodrigues's result as

$$\sum_{k=0}^{\frac{n(n-1)}{2}} Z(n,k)q^k = \frac{(1-q)(1-q^2)\cdots(1-q^n)}{(1-q)^n}. \tag{25.70}$$

Then, we see that the expression on the right-hand side is the q-extension of $n!$.

In 1913, MacMahon found another important way of classifying permutations by defining the greater index of a permutation.[39] For a permutation $a_1, a_2, \ldots a_n$ of $1, 2, 3, \ldots n$, MacMahon defined the greater index to be the sum $\sum_{i=1}^{n-1} \lambda(a_i)$, where $\lambda(a_i) = i$ if $a_i > a_{i+1}$, and $\lambda(a_i) = 0$ otherwise. Let $G(n,k)$ denote the number of permutations for which the greater index is equal to k. MacMahon proved that

$$\sum_{k=0}^{\frac{n(n-1)}{2}} G(n,k)q^k = \frac{(1-q)(1-q^2)\cdots(1-q^n)}{(1-q)^n}. \tag{25.71}$$

This immediately gave him the result

$$G(n,k) = Z(n,k). \tag{25.72}$$

In fact, MacMahon proved his theorems even more generally, for permutations of multisets. In a multiset, the elements need not be distinct. For example, $1^{m_1} 2^{m_2} \ldots r^{m_r}$ denotes a multiset with m_1 ones, m_2 twos, and so on. The concepts of inversion and greater index can be extended in an obvious way to multisets. So if $Z(m_1, m_2, \ldots, m_r; k)$ and $G(m_1, m_2, \ldots m_r; k)$ denote the number of permutations with k inversions and the number of permutations with greater index k, then MacMahon had

$$\sum Z(m_1, m_2, \ldots, m_r; k)q^k = \sum G(m_1, m_2, \ldots, m_r; k)q^k$$

$$= \frac{(1-q)(1-q^2)\cdots(1-q^{m_1+m_2+\cdots+m_r})}{(1-q)\cdots(1-q^{m_1})(1-q)\cdots(1-q^{m_2})\cdots(1-q)\cdots(1-q^{m_r})}. \tag{25.73}$$

Note that when $r = 2$, the expression on the right is the Gaussian polynomial $(m_1 + m_2, m_1)$, in Gauss's notation. Just as the Gaussian polynomial is the q-binomial coefficient, we can see that (25.73) is the q-multinomial coefficient.

MacMahon (1854–1929) studied at the military academy at Woolwich. He became a lieutenant in 1872, captain in 1881, and major in 1889. He returned to Woolwich as an instructor in 1882. This teaching post, along with his friendship with the mathematician George Greenhill, set the scene for MacMahon to exercise his mathematical talents. Starting in the early 1880s, he contributed numerous important papers to the subject of combinatorics and related topics, including symmetric functions and invariants. He was also a fast arithmetical calculator and constructed a table of partitions of integers up through 200. By studying this table, Ramanujan was able to

---

[39] MacMahon (1978) vol. 1, pp. 508–563.

discover the arithmetical properties of the partition function. MacMahon's calculations played a crucial role in Ramanujan's research, influential even today.

25.13 Exercises

(1) Prove Bernoulli's formulas (25.17) and (25.18) for the probabilities to win.

(2) Prove Euler's first identity (25.19).

(3) For (m, μ) defined by (25.33), prove Gauss's formulas (25.34), (25.35), and (25.36).

(4) Following F. H. Jackson, set $[\alpha] = \frac{1-q^{\alpha}}{1-q}, [n]! = [1][2] \cdots [n]$, and

$$(1 - x)^{(-\alpha)} = \sum_{k=1}^{\infty} \left(\frac{1 - q^k x}{1 - q^{k+\alpha} x} \right).$$

Show that

$$(1 - x)^{(-\alpha)} = 1 + \frac{[\alpha]}{[1]!} x + \frac{[\alpha][\alpha + 1]}{[2]!} q x^2 + \frac{[\alpha][\alpha + 1][\alpha + 2]}{[3]!} q^3 x^3 + \cdots.$$

See Jackson (1910).

(5) Let $u'(x) \equiv \Delta u(x) = \frac{u(x)-u(qx)}{x-qx}$ and $\Delta^{-1}u(x) \equiv \int u(x) d_q x$. Show that

(a) $\int f(x)u'(x)d_q x = u(x)v(x) - \int u(qx)v'(x)d_q x$.

(b) (i) $\Delta(1 - x)^{(n+1)} = [n + 1](1 - qx)^{(n)}$.

(ii)

$$\int x^m (1 - qx)^{(n)} d_q x = -\frac{x^m}{[n + 1]}(1 - x)^{(n+1)}$$

$$+ \frac{[m]}{[n + 1]} \int x^{m-1}(1 - qx)^{(n+1)} d_q x.$$

(iii) $\int_0^1 x^m (1 - qx)^{(n)} d_q x = \frac{[m]}{[n+1]} \int_0^1 x^{m-1}(1 - qx)^{(n+1)} d_q x$

$$= \frac{\Gamma_q(m + 1)(\Gamma_q(n + 1)}{\Gamma_q(m + n + 2)} \equiv B_q(m + 1, n + 1).$$

(iv) $\int_0^1 t^{\beta-1}(1 - qt)^{(\gamma-\beta-1)}(1 - q^{\alpha}tx)^{\alpha} d_q t$

$$= B_q(\beta, \gamma - \beta) \left(1 + \frac{(1 - q^{\alpha})(1 - q^{\beta})}{(1 - q)(1 - q^{\gamma})} \right) x$$

$$+ \frac{(1 - q^{\alpha})(1 - q^{\alpha+1})(1 - q^{\beta})(1 - q^{\beta+1})}{(1 - q)(1 - q^2)(1 - q^{\gamma})(1 - q^{\gamma+1})} x^2 + \cdots.$$

(c)
$$1 + \frac{(1-q^\alpha)(1-q^\beta)}{(1-q)(1-q^\gamma)} q^{\gamma-\alpha-\beta}$$
$$+ \frac{(1-q^\alpha)(1-q^{\alpha+1})(1-q^\beta)(1-q^{\beta+1})}{(1-q)(1-q^2)(1-q^\gamma)(1-q^{\gamma+1})} q^{2(\gamma-\alpha-\beta)} + \cdots$$
$$= \frac{B_q(\beta, \gamma - \alpha - \beta)}{B_q(\beta, \gamma - \beta)}.$$

(d) $\int_0^\infty \frac{t^{m-1}}{(1+qy)^{(l+m)}} d_q t = \frac{\Gamma_q(m)\Gamma_q(l)}{\Gamma_q(m+l)}$, provided $l + m$ is an integer. See Jackson (1910).

(6) Prove Cauchy's formula

$$\frac{(ax;q)_\infty}{(bx;q)_\infty} = \sum_{n=0}^\infty \frac{(b-a)(bq-a)\cdots(bq^{n-1}-a)q^{\binom{n}{2}} x^n}{(q;q)_n (bx;q)_n}.$$

See Cauchy (1882–1974) vol. 8, series 1, pp. 42–50.

(7) Prove Ramanujan's quintuple product identity

$$H(x) \equiv \prod_{n=1}^\infty (1-q^n)(1-xq^n)\left(1 - \frac{q^{n-1}}{x}\right)(1 - x^2 q^{2n-1})\left(1 - \frac{q^{2n-1}}{x^2}\right)$$
$$= \sum_{n=-\infty}^\infty (x^{3n} - x^{-3n-1}) q^{\frac{n(3n+1)}{2}}.$$

One method of proof is to assume $H(x) = \sum_{n=-\infty}^\infty c(n)x^n$. Then compute $\frac{H(qx)}{H(x)}$ and $\frac{H(\frac{1}{x})}{H(x)}$ to determine $c(n)$ in terms of $c(0)$. To find $c(0)$, specialize x. This formula was discovered several times. It is possible that Weierstrass was aware of it, since it follows from a three-term relation for sigma functions, a part of elliptic functions theory, presented by Weierstrass in his lectures. This formula appears explicitly in a 1916 book on elliptic functions by R. Fricke. Again, Ramanujan found it around that same time and made extensive use of it. In this exercise, we name the formula after Ramanujan. For a detailed history of the formula and several proofs, see Cooper (2006). Also see the remarks in Berndt (1985–1998) Part III, p. 83.

(8) Prove the septuple product identity of Farkas and Kra:

$$(1+x)(1-x)^2 \prod_{n=1}^\infty (1-q^n)^2(1-q^n x)\left(1 - \frac{q^n}{x}\right)(1 - q^n x^2)\left(1 - \frac{q^n}{x^2}\right)$$
$$= \sum_{-\infty}^\infty (-1)^n q^{\frac{5n^2+n}{2}} \left(\sum_{-\infty}^\infty (-1)^n q^{\frac{5n^2+3n}{2}} x^{5n+3} + \sum_{-\infty}^\infty (-1)^n q^{\frac{5n^2-3n}{2}} x^{5n} \right)$$
$$- \sum_{-\infty}^\infty (-1)^n q^{\frac{5n^2+n}{2}} \left(\sum_{-\infty}^\infty (-1)^n q^{\frac{5n^2+n}{2}} x^{5n+2} + \sum_{-\infty}^\infty (-1)^n q^{\frac{5n^2-n}{2}} x^{5n+1} \right).$$

This result generalizes the quintuple product identity. For a proof, see Farkas and Kra (2001) p. 271.

25.14 Notes on the Literature

A history of the quadratic Gauss sum is presented in Patterson (2007). For more papers on related topics, see Goldstein, Schappacher, and Schwermer (2007). Altmann and Ortiz (2005) gives interesting information about Rodrigues. For recent developments and proofs connected with the triple product identity, see Andrews (1986b) pp. 63–64, Foata and Han (2001) and Wilf (2001).

26

Partitions

26.1 Preliminary Remarks

Naudé's problems posed to Euler marked the beginning of research in the theory of partitions, as discussed in Chapter 25. Euler then solved these problems using generating functions, employing the same idea to prove the following remarkable theorem:[1]

> The number of different ways a given number can be expressed as the sum of different whole numbers is the same as the number of ways in which that same number can be expressed as the sum of odd numbers, whether the same or different.

For example, the number of ways 6 can be expressed as a sum of different whole numbers is four:

$$6, \quad 5+1, \quad 4+2, \quad 3+2+1.$$

And 6 can be expressed as a sum of odd numbers in the following four ways:

$$5+1, \quad 3+3, \quad 3+1+1+1, \quad 1+1+1+1+1+1.$$

To prove this in general, Euler gave the generating function for the number of partitions with distinct parts:

$$(1+q)(1+q^2)(1+q^3)(1+q^4)(1+q^5)(1+q^6)\cdots.$$

Observe that 4 is the coefficient of q^6 in the power series expansion of this product, for q^6 can be obtained as $q^6, q^5 q, q^4 q^2$, and $q^3 q^2 q$. On the other hand, Euler noted that the generating function for odd parts was

$$\frac{1}{(1-q)(1-q^3)(1-q^5)\cdots} = (1+q+q^{1+1}+\cdots)(1+q^3+q^{3+3}+\cdots)(1+q^5+\cdots).$$

[1] Euler (1988) pp. 275–276.

To prove the theorem, Euler showed that the generating functions and therefore the coefficients of their series expansions were identical:

$$(1+q)(1+q^2)(1+q^3)(1+q^4)\cdots = \frac{1-q^2}{1-q}\cdot\frac{1-q^4}{1-q^2}\cdot\frac{1-q^6}{1-q^3}\cdot\frac{1-q^8}{1-q^4}\cdots$$

$$= \frac{1}{1-q}\cdot\frac{1}{1-q^3}\cdot\frac{1}{1-q^5}\cdots . \tag{26.1}$$

We also noted in Chapter 25 that in 1741 Euler conjectured and eight years later proved the pentagonal number theorem

$$\prod_{n=1}^{\infty}(1-q^n) = \sum_{n=-\infty}^{\infty}(-1)^n q^{\frac{n(3n-1)}{2}}. \tag{26.2}$$

Though Euler did not give a combinatorial interpretation of this identity, A. M. Legendre found one and included it in the 1830 edition of his number theory book.[2] To understand Legendre's interpretation, consider how q^6 would arise in the power series expansion of the infinite product

$$(-q)(-q^2)(-q^3) = -q^6, \ (-q)(-q^5) = +q^6, \ (-q^2)(-q^4) = +q^6, \ (-q^6) = -q^6.$$

When the partition of 6 contains an odd number of parts (e.g., $6 = 1 + 2 + 3$) then a corresponding -1 is contributed to the coefficient of q^6 in the series. When the number of parts is even, then $+1$ is contributed. Hence the coefficient of q^6 in the series is 0. Thus, if we denote by $p_e(n), p_0(n)$ the number of partitions of n with an even/odd number of distinct parts, then Legendre's theorem states that

$$p_e(n) - p_0(n) = (-1)^m, \quad \text{when} \quad n = \frac{m(3m\pm 1)}{2},$$

$$= 0, \qquad \text{when} \quad n \neq \frac{m(3m\pm 1)}{2}.$$

Before Euler, Leibniz conceived of the problem of partitioning a positive integer. In a brief letter of July 1699 to Johann Bernoulli,[3] Leibniz enquired whether he had considered the difficult problem of partitioning a given number into two, three, or more parts. According to Mahnke, in a manuscript dated September 1674, Leibniz gave examples of partitions: 3 into two parts as $2 + 1$ and into three parts as $1 + 1 + 1$; 4 into two parts in two ways, $2 + 2$ and $3 + 1$, and into three parts in only one way, $2 + 1 + 1$. And in another manuscript, Leibniz pointed out the connection of partitions with symmetric functions: For example, the three partitions of 3, namely $3, 2 + 1, 1 + 1 + 1$, correspond to the symmetric functions $\sum a^3, \sum a^2 b, \sum abc$.[4]

After Euler, J. J. Sylvester (1814–1897) was the next mathematician to make major contributions to the theory of partitions. Sylvester entered St. John's College,

[2] Legendre (1830) pp. 131–133.

[3] Leibniz and Bernoulli (1745) pp. 461–462 or Leibniz (1971) vol. 3, part 2, p. 601.

[4] Mahnke (1912–13) p. 37. Also see Knuth (2011) pp. 505–506.

Cambridge, in 1833 and came out as Second Wrangler in 1837. The great applied mathematician George Green was fourth. Sylvester, of Jewish heritage, was unwilling to sign the thirty-nine articles; consequently, he was unable to take a degree, to obtain a fellowship, or to compete for one of the Smith's prizes. It was only in 1855 that he received a professorship of mathematics at the Royal Military Academy at Woolwich. Unfortunately, in 1870 he was retired early from this position, when his mathematical creativity was at its peak. In 1875, when Johns Hopkins University was founded in Baltimore, Sylvester was elected the first professor of mathematics (1876–83). Sylvester enjoyed a happy and productive late career in Baltimore; he there founded the *American Journal of Mathematics* whose first volume appeared in 1878. Moreover, Sylvester very successfully trained a number of excellent mathematicians, inaugurating serious mathematical research in America. It is not surprising that many of these American mathematicians contributed to the theory of partitions, since research in that topic required abundant ingenuity but more limited background.

Sylvester's interest in partitions arose fairly early. In 1853, he published a paper on his friend Cayley's quick method for determining the degree of a symmetric function expressed as a polynomial in elementary symmetric functions. For that purpose, Cayley had employed a result Sylvester attributed to Euler:[5] "To wit, that the number of ways of breaking up a number n into parts is the same, whether we impose the condition that the number of parts in any partitionment shall not exceed m, or that the magnitude of any one of the parts should not exceed m." To understand this last result, consider that the generating function for the number of partitions of an integer into at most m parts, with each part $\leq n$, can be inductively demonstrated to be equal to the Gaussian polynomial

$$\frac{(1 - q^{m+n})(1 - q^{m+n-1}) \cdots (1 - q^{m+1})}{(1 - q^n)(1 - q^{n-1}) \cdots (1 - q)}.$$

This polynomial remains unchanged when m and n are interchanged; hence follows the result used by Sylvester. The Gaussian polynomial also cropped up in the work of Cayley and Sylvester in invariant theory. As we see in Chapter 30, they related the coefficients of the polynomial to the number of independent seminvariants. Cayley and Sylvester took an interest in partitions as a result of their researches on invariants. Though they both contributed to partition theory, Sylvester made the subject his own domain by establishing fundamental ideas and producing new researchers, in the form of his students.

A graphical proof of Euler's theorem would start out by representing a given partition as a graph. For example, write the partition $5 + 2 + 1$ of eight as

---

[5] Sylvester (1853b) p. 200.

and then enumerate by columns. Thus, one obtains the conjugate partition $3 + 2 + 1 + 1 + 1$ of eight. It is immediately clear that if we have a partition of an integer N into n parts of which the largest is m, then its conjugate is a partition of N into m parts of which the largest is n. This at once gives us the theorem: The number of partitions of any integer N into exactly n parts with the largest part m and the number of partitions of N into at most n parts with the largest part at most m both remain the same when m and n are interchanged. The proof of this theorem using the generating function method is less illuminating, illustrating the power of the graphical method. Sylvester remarked that he learned the technique from its originator, N. M. Ferrers. In a footnote to his paper, Sylvester wrote, "I learn from Mr Ferrers that this theorem was brought under his cognizance through a Cambridge examination paper set by Mr Adams of Neptune notability."[6] Here Sylvester was referring to the astronomer John Couch Adams, discoverer of Neptune.

It was within this very concrete graphical method that Sylvester and his American students, including Fabian Franklin, William Durfee, and Arthur Hathaway, made their original and important contributions to the theory of partitions. It is interesting to note that the other significant results obtained by American mathematicians at around the same time were in abstract algebra. At that time, this too was a topic requiring a minimal amount of background knowledge, unlike subjects such as the theory of abelian functions. Early American results in abstract algebra included Benjamin Peirce's (1809–1880) paper on linear associative algebras dating from 1869, published posthumously in 1881 by his son Charles Saunders Peirce (1839–1914) in Sylvester's new journal.[7] B. Peirce introduced the important concepts of nilpotent and idempotent elements and the paper starts with his famous dictum "Mathematics is the science which draws necessary conclusions." C. S. Peirce added an appendix to the paper, proving a significant theorem of his own on finite dimensional algebras over the real numbers. In modern language, the theorem states: The only division algebras algebraic over the real numbers are the fields of real and complex numbers and the division ring of quaternions.

The German mathematician G. Frobenius (1849–1917) also discovered this theorem at about the same time as Peirce, though he published it in 1877.[8] The Frobenius-Peirce theorem and Franklin's beautiful proof of Euler's pentagonal number theorem are the earliest major contributions by Americans to mathematics. We shall see details of Franklin's work later in this chapter; concerning C. S. Peirce, we simply note that he made outstanding contributions to mathematical logic and to some aspects of philosophy. The systematic philosopher Justus Buchler, who edited Peirce's philosophical writings, stated in the introduction, "Even to the most unsympathetic, Peirce's thought cannot fail to convey something of lasting value. It has a peculiar property, like that of the Lernean hydra: discover a weak point, and two strong ones spring up beside it. Despite the elaborate architectonic planning of its creator, it is everywhere uncompleted, often distressingly so. There are many who have small

[6] ibid. p. 201.
[7] Peirce (1881).
[8] Frobenius (1878).

regard for things uncompleted, and no doubt what they value is much to be valued. In his quest for magnificent array, in his design for a mighty temple that should house his ideas, Peirce failed. He succeeded only in advancing philosophy."[9]

After the researches of Sylvester and his young American students, P. A. MacMahon (1854–1929) dominated the topic of partitions. One of MacMahon's results was connected with Ramanujan's 1910 rediscovery of two identities, first found by Rogers in the 1890s[10] during his work on q-series:

$$\sum_{m=0}^{\infty} \frac{q^{m^2}}{(q;q)_m} = \prod_{m=0}^{\infty} \left(1 - q^{5m+1}\right)^{-1}\left(1 - q^{5m+4}\right)^{-1}, \qquad (26.3)$$

$$\sum_{m=0}^{\infty} \frac{q^{m(m+1)}}{(q;q)_m} = \prod_{m=0}^{\infty} \left(1 - q^{5m+2}\right)^{-1}\left(1 - q^{5m+3}\right)^{-1}. \qquad (26.4)$$

In 1913, Srinivasa Ramanujan communicated these identities to G. H. Hardy, although by this time Rogers's work was forgotten. Ramanujan had no proof; Hardy unsuccessfully sought a proof, showing the identities to his colleagues. MacMahon was among those who saw the formulas. An expert in symmetric functions, invariant theory, partitions, and combinatorics, he had known Sylvester and his work. Thus, it was natural that MacMahon conceived of an interpretation of the identities in terms of partitions. By expanding $(1 - q^{5m+1})^{-1}$ and $(1 - q^{5m+4})^{-1}$ as geometric series, the coefficient of q^n in the expression on the right-hand side of the first identity is clearly equivalent to the number of partitions of n into parts $\equiv 1$ or $4 \pmod 5$. For the left-hand side, observe that

$$m^2 = (2m - 1) + (2m - 3) + \cdots + 5 + 3 + 1,$$

or the sum of the first m odd parts. We can find a partition of n if $n - m^2$ is partitioned into at most m parts with the largest part added to $2m - 1$, the next to $2m - 3$ and so on. The parts in this partition of n differ by at least 2. Moreover, the partitions of n associated with a specific m are enumerated by

$$\frac{q^{m^2}}{(1 - q)(1 - q^2) \cdots (1 - q^m)},$$

and the sum of these terms yields all the partitions of this form. We therefore have MacMahon's theorem, presented in his 1915 *Combinatory Analysis*:[11] The number of partitions of n in which the difference between any two parts is at least 2, equals the number of partitions of n into parts $\equiv 1$ or $4 \pmod 5$. We note that in MacMahon's own statement of the theorem, instead of specifying that the parts differ by at least 2, he wrote that there were neither repetitions nor sequences. In a similar way, the second identity states: The number of partitions of n in which the least part is ≥ 2 and the

[9] Buchler (1955) p. xvi.
[10] Rogers (1894) § 5.
[11] Macmahon (1915–16) vol. 2, pp. 32–36.

difference between any two parts is at least 2, is equal to the number of partitions of n into parts $\equiv 2$ or 3 (mod 5). This arises out of the relation $m(m + 1) = 2 + 4 + 6 + \cdots + 2m$.

Several proofs of the Rogers–Ramanujan identities have been given and they have been generalized both combinatorially and analytically. Issai Schur independently discovered the Rogers–Ramanujan identities and their partition theoretic interpretation; in 1917 he gave two proofs, one of which was combinatorial.[12] However, as Hardy wrote in 1940, it is only natural to seek an argument that sets up a one-to-one correspondence between the two sets of partitions. No such bijective proof was known in Hardy's time, and it was not until 1981 that Adriano Garsia and Stephen Milne, working on the foundation established by Schur, published a proof of the MacMahon-Schur theorem, equivalent to the Rogers–Ramanujan identities.[13] We note that Schur's combinatorial proof also motivated Basil Gordon's 1961 partition-theoretic generalization.[14] See the exercises.

Issai Schur (1875–1941) was born in Russia but studied at the University of Berlin under Georg Frobenius who had a great influence on him. Schur made fundamental contributions to representation theory, to the related theory of symmetric functions, and also to topics in analysis such as the theory of commutative differential operators. A great teacher, he founded an outstanding school of algebra in Berlin. Dismissed from his chair by the Nazi government, he took a position in 1938 at the Hebrew University in Jerusalem.

Garsia and Milne's bijective proof of the Rogers–Ramanujan identities is based on their involution principle: Let $C = C^+ \cup C^-$, where $C^+ \cap C^- = \phi$, be the disjoint union of two finite components C^+ and C^-. Let α and β be two involutions on C, each of whose fixed points lie in C^+. Let F_α (resp F_β) denote the fixed-point set of α (resp β). Suppose $\alpha(C^+ - F_\alpha) \subset C^-$ and $\alpha(C^-) \subset C^+$ and similarly $\beta(C^+ - F_\beta) \subset C^-$ and $\beta(C^-) \subset C^+$. Then a cycle of the permutation $\Delta = \alpha\beta$ contains either fixed points of neither α nor β, or exactly one element of F_α and one of F_β. This powerful involution principle has been successfully applied to several q-series identities. Garsia and Milne's proof of Rogers-Ramanujan was very long but soon afterward David Bressoud and Doron Zeilberger found a shorter proof.[15]

Now observe that in Euler's theorem the parts are distinct and hence differ by at least one, whereas in MacMahon's theorem the parts differ by at least two. If we denote by $q_{d,m}(n)$ the number of partitions of n into parts differing by at least d, each part being greater than or equal to m, the Euler and MacMahon theorems take the form

$$q_{d,m}(n) = p_{d,m}(n),$$

where $p_{d,m}(n)$ is the number of partitions of n into parts taken from a fixed set $S_{d,m}$. H. L. Alder observed that for $d = 1$, m could be taken to be any positive integer. In fact, the number of partitions of n into distinct parts, with each part $\geq m$, was equal to

[12] Schur (1917).
[13] Garsia and Milne (1981).
[14] Gordon (1961).
[15] Bressoud and Zeilberger (1982).

the number of partitions of n into parts taken from the set $\{m, m+1, \ldots, 2m-1, 2m+1, 2m+3, \ldots\}$.

In 1946 D. H. Lehmer proved for $m = 1$, and in 1948 Alder proved for the general case: The number $q_{d,m}(n)$ is not equal to the number of partitions of n into parts taken from any set of integers whatsoever unless $d = 1$ or $d = 2$, $m = 1, 2$. Now the generating function for $q_{d,m}(n)$ is easily seen to be

$$\sum_{k=0}^{\infty} \frac{q^{\frac{mk+dk(k-1)}{2}}}{(1-q)(1-q^2)\cdots(1-q^k)}, \tag{26.5}$$

while the generating for partitions with parts from a fixed set $\{a_1, a_2, a_3, \ldots\}$ is

$$\frac{1}{\prod_{k=1}^{\infty}(1-q^{a_k})}$$

Alder's proof consisted in showing that no matter how the a_k were chosen, the two generating functions could not be equal for the values of m and d excluded by the theorem.[16]

When MacMahon interpreted the Rogers–Ramanujan identity in terms of partitions, Hardy and Ramanujan may have been spurred to examine the asymptotic behavior of $p(n)$, the number of partitions of n. MacMahon assisted them in this work by constructing a table of $p(n)$ for $n = 1, 2, \ldots, 200$. We later consider the impact of this on the work of Hardy and Ramanujan. For now, we note that this table was created by means of Euler's formula

$$p(n) = p(n-1) + p(n-2) - p(n-5) - p(n-7) + \cdots$$
$$+ (-1)^{m-1} p\left(n - \frac{1}{2}m(3m-1)\right) - (-1)^{m-1} p\left(n - \frac{1}{2}m(3m+1)\right) \cdots. \tag{26.6}$$

Note that $p(k) = 0$ for k negative. This formula is quite efficient for numerical work. Ramanujan enjoyed numerical computation and could do it with unusual rapidity and accuracy. It is therefore interesting that in his obituary notice of Ramanujan, Hardy wrote, "There is a table of partitions at the end of our paper .... This was, for the most part, calculated independently by Ramanujan and Major MacMahon; and Major MacMahon was, in general, slightly the quicker and more accurate of the two."[17]

J. E. Littlewood once remarked that every positive integer was one of Ramanujan's personal friends.[18] Thus, Ramanujan noticed in the tables something missed by others, the arithmetical properties of partitions. In his 1919 paper on partitions he wrote,[19]

---

[16] For this history and for good references, see Alder (1969).
[17] Ramanujan (2000) p. xxxv.
[18] Littlewood (1986) p. 61.
[19] Ramanujan (1919b).

On studying the numbers in this table I observed a number of curious congruence properties, apparently satisfied by $p(n)$. Thus

(1) $p(4)$, $p(9)$, $p(14)$, $p(19)$,   ... $\equiv 0$ $(\mathrm{mod}\,5)$,

(2) $p(5)$, $p(12)$, $p(19)$, $p(26)$,   ... $\equiv 0$ $(\mathrm{mod}\,7)$,

(3) $p(6)$, $p(17)$, $p(28)$, $p(39)$,   ... $\equiv 0$ $(\mathrm{mod}\,11)$,

(4) $p(24)$, $p(49)$, $p(74)$, $p(99)$,   ... $\equiv 0$ $(\mathrm{mod}\,25)$,

(5) $p(19)$, $p(54)$, $p(89)$, $p(124)$,   ... $\equiv 0$ $(\mathrm{mod}\,35)$,

(6) $p(47)$, $p(96)$, $p(145)$, $p(194)$,   ... $\equiv 0$ $(\mathrm{mod}\,49)$,

(7) $p(39)$, $p(94)$, $p(149)$,   ... $\equiv 0$ $(\mathrm{mod}\,55)$,

(8) $p(61)$, $p(138)$,   ... $\equiv 0$ $(\mathrm{mod}\,77)$,

(9) $p(116)$,   ... $\equiv 0$ $(\mathrm{mod}\,121)$,

(10) $p(99)$,   ... $\equiv 0$ $(\mathrm{mod}\,125)$.

From these data I conjectured the truth of the following theorem: If $\delta = 5^a 7^b 11^c$ and $24\lambda \equiv 1$ $(\mathrm{mod}\,\delta)$ then

$$p(\lambda), p(\lambda + \delta), p(\lambda + 2\delta), \ldots \quad \equiv 0 \quad (\mathrm{mod}\,\delta).$$

Ramanujan gave very simple proofs of $p(5m + 4) \equiv 0$ (mod 5) and $p(7m + 5) \equiv 0$ (mod 7), using only Euler's pentagonal number theorem and Jacobi's formula for $\prod_{n=1}^{\infty}(1 - q^n)^3$. Ramanujan's further efforts, to prove $p(25m + 24) \equiv 0$ (mod 25) and $p(49m + 47) \equiv 0$ (mod 49), led him deeper into the theory of modular functions. In particular, he found the following two remarkable identities:

$$p(4) + p(9)q + p(14)q^2 + \cdots = 5\,\frac{\{(1 - q^5)(1 - q^{10})(1 - q^{15})\cdots\}^5}{\{(1 - q)(1 - q^2)(1 - q^3)\cdots\}^6}, \quad (26.7)$$

$$p(5) + p(12)q + p(19)q^2 + \cdots = 7\frac{\{(1 - q^7)(1 - q^{14})(1 - q^{21})\cdots\}^3}{\{(1 - q)(1 - q^2)(1 - q^3)\cdots\}^4}$$

$$+ 49q\,\frac{\{(1 - q^7)(1 - q^{14})(1 - q^{21})\cdots\}^7}{\{(1 - q)(1 - q^2)(1 - q^3)\cdots\}^8}. \quad (26.8)$$

The rest of Ramanujan's conjecture concerning the divisibility of the partition function by $5^a 7^b 11^c$ is not completely correct. In 1934, on the basis of the extended tables for $p(n)$ constructed by Hansraj Gupta, Sarvadaman Chowla observed[20] that $p(243)$ was not divisible by 7^3, though $24 \cdot 243 \equiv 1$ (mod 7^3). However, $p(243)$ is divisible by 7^2. The correct reformulation of Ramanujan's conjecture would state: Let $\delta = 5^a 7^b 11^c, \delta' = 5^a 7^{b'} 11^c$, where $b' = b$, if $b = 0, 1, 2$, and $b' = \lfloor (b + 2)/2 \rfloor$, if $b > 2$. If $24\lambda \equiv 1$ (mod δ), then

$$p(\lambda + n\delta) \equiv 0 \quad (\mathrm{mod}\,\delta'), \quad n = 0, 1, 2, \ldots. \quad (26.9)$$

In an unpublished manuscript, Ramanujan outlined a proof of his conjecture for arbitrary powers of 5. He may have had a proof for the powers of 7 as well, since

[20] Chowla (1934).

he apparently began writing it down.[21] George N. Watson's proof[22] of Ramanujan's conjecture for powers of 5 is identical with the one contained in the unpublished manuscript. Watson also gave a proof of the corrected version for the powers of 7. In 1967, A. O. L. Atkin provided a proof for powers of 11,[23] based on work of Joseph Lehner from the 1940s. Atkin and Lehner's proofs require the use of modular equations, a topic in which Ramanujan was a great expert. It is remarkable that he was able to conjecture an essentially correct result on so little numerical evidence, especially in higher powers.

The fact that $p(5n + 4) \equiv 0 \pmod 5$ suggests that partitions of $5n + 4$ should be divisible into five classes with the same number of partitions in each class. Freeman Dyson got this idea around 1940 when he was in high school; as a second year student at Cambridge University, he found a way of making this division.[24] For this purpose, he defined the concept of the rank of a partition: the largest part minus the number of parts. He checked this concept, applying it to the three cases $p(4), p(9)$, and $p(14)$ and found it accurate; he also found that it worked for $p(5)$ and $p(12)$. He conjectured its truth for all $p(5n + 4)$ and for $p(7n + 5)$, but was unable to prove it. A decade later, Atkin and Peter Swinnerton-Dyer found a proof[25] involving combinatorial arguments combined with ideas from modular function theory. Mock theta functions also made an appearance; Atkin and Swinnerton-Dyer rediscovered and used a number of identities for mock theta functions. Unbeknownst to them and the rest of the world, these identities were contained in Ramanujan's lost notebook later discovered by Andrews, then buried under a mountain of paper on the floor of Watson's study.

The rank of a partition can be defined graphically as the signed difference between the number of nodes in the first row and number of nodes in the first column. Consider the ranks of the partitions of 5:

Partition	Rank	
5	$5\text{-}1 \equiv 4$	$\pmod 7$
4+1	$4\text{-}2 \equiv 2$	$\pmod 7$
3+2	$3\text{-}2 \equiv 1$	$\pmod 7$
3+1+1	$3\text{-}3 \equiv 0$	$\pmod 7$
2+2+1	$2\text{-}3 \equiv 6$	$\pmod 7$
2+1+1+1	$2\text{-}4 \equiv 5$	$\pmod 7$
1+1+1+1+1	$1\text{-}5 \equiv 3$	$\pmod 7$.

Dyson found that the concept of rank failed to classify the partitions of $11n + 6$; he conjectured the existence of a crank for this purpose. Almost half a century later, a day after the 1987 Centenary Conference at the University of Illinois, celebrating the work of Ramanujan, Andrews and Frank G. Garvan discovered the crank:[26] The crank of a partition is the largest part in the partition if it has no ones; otherwise,

[21] Ramanujan (1988) pp. 238–243. Also see Berndt and Ono (2001).
[22] Watson (1938).
[23] Atkin (1967).
[24] Dyson (1944).
[25] Atkin and Swinnerton-Dyer (1954).
[26] Andrews and Garvan (1988a).

it is the number of parts greater than the number of ones, minus the number of ones. A nice property of the crank is that it works for 5, 7, and 11. Amazingly, Ramanujan discovered the generating functions for both the rank and the crank, and his results can again be found in his lost notebook,[27] though he did not use these names.

Concerning the congruence properties of partitions, Ramanujan wrote, "It appears that there are no equally simple properties for any moduli involving primes other than these three."[28] As we shall see, Ramanujan's intuition has been shown to be correct. However, in the late 1960s, Atkin found some more complicated congruences involving other primes. For example, he showed that[29]

$$p(11^3.13n + 237) \equiv 0 \pmod{13},$$

$$p(23^3.17n + 2623) \equiv 0 \pmod{17}.$$

Atkin used computers to do the numerical work necessary for constructing these examples. In fact, Atkin was among the pioneers in the use of computers for number theory research. Concerning this aspect of his work, he wrote in his 1968 paper, "it is often more difficult to discover results in this subject than to prove them, and an informed search on the machine may enable one to find out precisely what happens." Atkin's aim was to understand partition identities, including Ramanujan's, from the more general viewpoint of modular function theory. His student Margaret Ashworth (1944–73) shared this perspective, although her researches were halted much too soon. Thus, Atkin and Ashworth did not succeed in fully developing their approach. Atkin himself made important contributions to the theory of modular forms and in 1970, Atkin and Lehner conceived the fundamental idea of new forms.[30] These are eigenforms for Hecke operators, on the space of cusp forms for Hecke subgroups of the modular group.

In fact, it was only recently that Ken Ono developed a theory of the kind Atkin may have been seeking. In 2000, Ono was able to prove that for any prime $l \geq 5$, there exist infinitely many congruences of the form $p(An + B) \equiv 0 \pmod{l}$. Soon after this, Scott Ahlgren extended the congruence to the case in which l is replaced by l^k. Subsequently, Ono and Ahlgren jointly extended these results and wrote a historical essay explaining that their work "provides a theoretical framework which explains every known partition function congruence."[31] Ono and Ahlgren based their work on results in modular forms from the 1960s and 1970s due to Goro Shimura, Jean-Pierre Serre, and Pierre Deligne.

Confirming another conjecture of Ramanujan, in 2003[32] Ahlgren and Matthew Boylan proved that if l is prime and $0 \leq \beta \leq l$ is any integer for which

$$p(ln + \beta) \equiv 0 \pmod{l} \quad \text{for all } n \geq 0,$$

[27] Ramanujan (1988) pp. 179–182.
[28] Ramanujan (1919b).
[29] Atkin (1968).
[30] Atkin and Lehner (1970).
[31] Ahlgren and Ono (2001).
[32] Ahlgren and Boylan (2003).

then

$$(l, \beta) \epsilon \{(5,4), (7,5), (11,6)\}.$$

We note that all these cases of simple congruence were found by Ramanujan; his intuition that no other cases exist has been verified. In 2005, Karl Mahlburg succeeded in extending the partition congruences to the crank function.[33] Let $M(m, N, n)$ be the number of partitions of n whose rank equals m (mod N). Mahlburg's theorem states that for every prime $l \geq 5$ and integer $i \geq 1$, there are infinitely many nonnested arithmetical progressions $An + B$ such that simultaneously for every $0 \leq m \leq l^j - 1$

$$M(m, l^j, An + B) \equiv 0 \ (\text{mod } l^i).$$

It is clear from the definition of M that

$$p(n) = M(0, N, n) + M(1, N, n) + \cdots + M(N - 1, N, n).$$

Therefore, Mahlburg's theorem implies the corresponding result for $p(n)$.

MacMahon and Hardy greatly admired Ramanujan's generating function for $p(5n + 4)$. A number of proofs of this and the generating function for $p(7n + 5)$ have subsequently been found. A recent proof by Hershel Farkas and Irwin Kra is based on the theory of Riemann surfaces and theta functions.[34]

In the final year of his life, Ramanujan introduced a new type of series, mock theta functions. These q-series, convergent in $|q| < 1$, also have connections with the theory of partitions, although Ramanujan's motivation was to study their asymptotic properties as q approached a root of unity. Ramanujan noted that the asymptotic behavior of theta series such as

$$\sum_{n=0}^{\infty} \frac{q^{n^2}}{((1-q)(1-q^2)\cdots(1-q^n))^2}$$

and

$$\sum_{n=0}^{\infty} \frac{q^{n^2}}{(1-q)(1-q^2)\cdots(1-q^n)}$$

could be expressed in a neat and closed exponential form as q approached roots of unity. He conceived mock theta functions as those series with similar asymptotic properties, without being theta functions. He gave seventeen examples of mock theta functions, dividing them into four groups, named mock theta functions of orders 3, 5, 5, and 7. One of the third-order functions he mentioned was defined by

$$f(q) = 1 + \frac{q}{(1+q)^2} + \frac{q^4}{(1+q)^2(1+q^2)^2} + \cdots.$$

---

[33] Mahlburg (2005).
[34] Farkas and Kra (2001) chapter 5.

He noted that when $q = -e^{-t}$ and $t \to 0$

$$f(q) + \sqrt{\frac{\pi}{t}} \exp\left(\frac{\pi^2}{24t} - \frac{t}{24}\right) \to 4.$$

Ramanujan also stated a few identities connecting some of these functions with each other. For example, he mentioned the third-order function

$$\chi(q) = 1 + \frac{q}{1 - q + q^2} + \frac{q^4}{(1 - q + q^2)(1 - q^2 + q^4)} + \cdots$$

and the relation

$$4\chi(q) - f(q) = \frac{(1 - 2q^3 + 2q^{12} - \cdots)^2}{(1 - q)(1 - q^2)(1 - q^3) \cdots}.$$

After Ramanujan, G. N. Watson (1886–1965) was the first to study these functions. The title of his 1936 paper on this topic,[35] "The Final Problem: An Account of the Mock Theta Functions," was borrowed from an Arthur Conan Doyle story. In this paper, Watson introduced three new third-order functions, and proved identities such as

$$f(q) \prod_{n=1}^{\infty}(1 - q^n) = 1 + 4 \sum_{n=1}^{\infty} \frac{(-1)^n q^{\frac{n(3n+1)}{2}}}{1 + q^n}.$$

Watson employed the identities to show that the third-order mock theta functions had the asymptotic properties asserted by Ramanujan and that they were not theta functions. A year later, Watson proved that the fifth-order functions listed by Ramanujan had the asymptotic properties; he did not succeed in showing that they were not theta functions. Watson's proofs of some of the identities were long, and he wrote that he counted the number of steps in the longest to be twenty-four instead of the thirty-nine he had hoped for as a student of John Buchan.

Watson's papers motivated Atle Selberg (1917–2007) to prove asymptotic formulas for seventh-order functions. Selberg had been drawn to a study of Ramanujan's work by a 1934 article by Carl Störmer in a periodical of the Norwegian Mathematical Society. The next year, Selberg started reading Ramanujan's *Collected Papers*. In 1987, he described his impressions:[36] "So I got a chance to browse through it for several weeks. It seemed quite like a revelation – a completely new world to me, quite different from any mathematics book I had ever seen – with much more appeal to the imagination, I must say. And frankly, it still seems very exciting to me and also retains that air of mystery which I felt at the time. It was really what gave the impetus which started my own mathematical work. I began on my own, experimenting with what is often referred to as q-series and identities and playing around with them."

[35] Watson (1936).
[36] Selberg (1989) vol. 1, pp. 695–706, especially p. 696.

In the 1960s, Andrews began his extensive work on mock theta functions. His work was further facilitated by his dramatic 1976 discovery of Ramanujan's Lost Notebook in the Trinity College library of Cambridge University. For example, among myriad formulas in this notebook, Ramanujan gave ten identities for the fifth-order functions. In 1987, Andrews and Garvan showed that these ten identities could be reduced to two conjectures on partitions.[37] To state these conjectures, let $R_a(n)$ denote the number of partitions of n with rank congruent to a (mod 5). The first conjecture stated that for every positive integer n, $R_1(5n) - R_0(5n)$ was equal to the number of partitions of n with unique smallest part and all other parts less than or equal to the double of the smallest part. The second stated that $2R_2(5n+3) - R_1(5n+3) - R_0(5n+3) - 1$ was equal to the number of partitions of n with unique smallest part and all other parts less than or equal to one plus the double of the smallest part. A year later Dean Hickerson proved these conjectures.[38]

We mention in passing that as that as a biproduct of his work on mock theta functions, Andrews discovered the identity[39]

$$\left(\sum_{n=0}^{\infty} q^{\binom{n+1}{2}}\right)^3 = \sum_{n=0}^{\infty} \sum_{j=0}^{2n} \frac{q^{2n^2+2n-\binom{j+1}{2}}(1+q^{2n})}{(1-q^{2n+1})}.$$

An immediate consequence of this formula is that every positive integer can be expressed as a sum of at most three triangular numbers. This theorem was first stated by Fermat, who said he had a proof. The first published proof appeared in Gauss's *Disquisitiones*.

Though mock theta functions were shown to have connections with several areas of mathematics, it was not clear how they fit into any known general framework. The work of Sander Zwegers, Don Zagier, Ken Ono, and Kathrin Bringmann, 2002–2007, has shown that Ramanujan's twenty-two mock theta functions are examples of infinite families of weak Maass forms of weight $\frac{1}{2}$. This understanding has led to further new results.

26.2 Sylvester on Partitions

In 1882, Sylvester collected together the investigations he and his students had done on partitions dating from 1877–1882 and published them in his newly founded journal as a long paper,[40] "A Constructive Theory of Partitions, Arranged in Three Acts, an Interact and an Exodion." He presented Franklin's proof of Euler's pentagonal number theorem (26.2). Sylvester placed the smallest part at the top of his graphical

[37] Andrews and Garvan (1988a).
[38] Hickerson (1988).
[39] Andrews (1986a).
[40] Sylvester (1882) § 12.

representation. We present the proof in his own words, illustrating his habit of using periods very sparingly.

If a regular graph represent a partition with unequal elements, the lines of magnitude must continually increase or decrease. Let the annexed figures be such graphs written in ascending order from above downward:

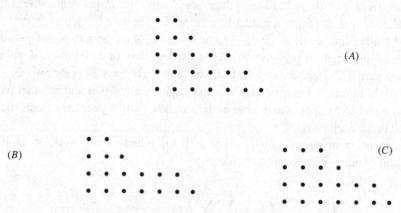

In (A) and (B) the graphs may be transformed without altering their content or regularity by removing the nodes at the summit and substituting for them a new slope line at the base. In C the new slope line at the base may be removed and made to form a new summit; the graphs so transformed will be as follows:

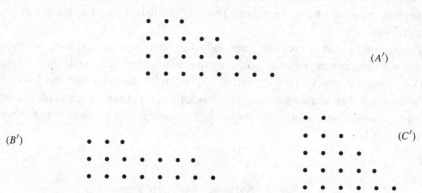

A' and B' may be said to be derived from A, B by a process of contraction, and C' from C by one of protraction.

Contraction could not now be applied to A' and B', nor protraction to C' without destroying the regularity of the graph; but the inverse processes may of course be applied, namely, of protraction to A' and B' and contraction to C', so as to bring back the original graph A, B, C.

In general (but as will be seen not universally), it is obvious that when the number of nodes in the summit is inferior or equal to the number in the base-slope, contraction may be applied, and when superior to that number, protraction: each process alike will alter the number of parts from even to odd or from odd to even, so that barring the exceptional cases which remain to be considered where neither protraction nor contraction is feasible, there will be a one-to-one correspondence between the partitions of n into an odd number and the partitions of n into an even number of unrepeated parts; the exceptional cases are those shown below where the summit meets the base-slope line, and contains either the same number or one more than the number of nodes in that line; in which case neither protraction nor contraction will be possible, as seen in the annexed figures which are written in regular order of succession, but may be indefinitely continued:

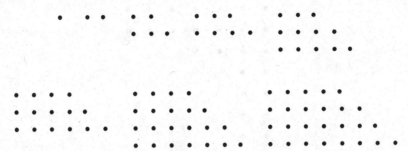

for the protraction process which *ought*, for example, according to the general rule, to be applicable to the last of the above graphs, cannot be applied to it, because on removing the nodes in the slope line and laying them on the summit, in the very act of so doing the summit undergoes the loss of a node and is thereby incapacitated to be surmounted by the nodes in the slope, which will have not now a less, but the same number of nodes as itself; and in like manner, in the last graph but one, the nodes in the summit cannot be removed and a slope line be added on containing the same number of nodes without the transformed graph ceasing to be regular, in fact it would take the form

and so the last graph transformed according to rule [by protraction] would become:

which, although regular, would cease to represent a partition into unlike numbers. The excepted cases then or unconjugate partitions are those where the number of parts being j, the successive parts form one or the other of the two arithmetical series

$$j, j + 1, j + 2, \ldots, 2j - 1 \quad \text{or} \quad j + 1, j + 2, \ldots, 2j,$$

in which cases the contents are $\frac{3j^2-j}{2}$ and $\frac{3j^2+j}{2}$ respectively, and consequently since in the product of $1 - x \cdot 1 - x^2 \cdot 1 - x^3 \cdots$ the coefficient of x^n is the number of ways of composing n with an even less the number of ways of composing it with an odd number of parts, the product will be completely represented by $\sum_{j=-\infty}^{\infty} (-1)^j x^{\frac{3j^2+j}{2}}$.

Sylvester's student Durfee introduced the important concept of the Durfee square for the purpose of studying self-conjugate graphs. These graphs remain unchanged when rows of nodes are changed to columns. Sylvester gave the partition of $27 = 7 + 7 + 4 + 3 + 2 + 2 + 2$ as an example;[41] it has the self-conjugate graph:

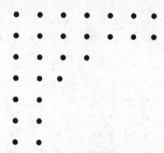

Note that the largest square in this graph is of size 3×3 in the upper left corner and the remaining nodes form two graphs with nine nodes each, partitioned into identical partitions, $3 + 2 + 2 + 2$, provided the nodes on the right-hand side of the square are read column-wise. The number of partitions of 9 in which the largest part is at most 3 is the coefficient of

$$x^9 \quad \text{in} \quad \frac{1}{(1-x)(1-x^2)(1-x^3)},$$

and this is the same as the coefficient of

$$x^{18} \quad \text{in} \quad \frac{1}{(1-x^2)(1-x^4)(1-x^6)}.$$

Sylvester applied this analysis to find the number of self-conjugate partitions of n. He considered all the partitions that could be dissected into a square of size m^2. The number of such partitions would be the coefficient of

$$x^{n-m^2} \quad \text{in} \quad \frac{1}{(1-x^2)(1-x^4)\cdots(1-x^{2m})}$$

or the coefficient of

$$x^n \quad \text{in} \quad \frac{x^{m^2}}{(1-x^2)(1-x^4)\cdots(1-x^{2m})}.$$

Thus, the number of self-conjugate partitions of n was the coefficient of x^n in the series

$$1 + \frac{x}{1-x^2} + \frac{x^4}{(1-x^2)(1-x^4)} + \frac{x^9}{(1-x^2)(1-x^4)(1-x^6)} + \cdots.$$

There is yet another manner in which a self-conjugate partition can be dissected: by counting the number of nodes in the m angles or bends, as Sylvester called them.

41 ibid. § 27.

Thus, for the self-conjugate partitions of 27, there are three bends. The outermost right angle has thirteen nodes; the second has eleven; and the third three. It is easy to see that the number of nodes in each right angle of a self-conjugate partition will always be an odd number. Moreover, different right angles in the same partition will have different numbers of nodes. Thus, the number of self-conjugate partitions of n will be the coefficient of x^n in $(1+x)(1+x^3)(1+x^5)\cdots$. Therefore

$$\prod_{n=0}^{\infty}(1+x^{2n+1}) = \sum_{n=0}^{\infty} \frac{x^{n^2}}{(1-x^2)(1-x^4)\cdots(1-x^{2n})}.$$

Sylvester generalized this analysis of self-conjugate partitions[42] by introducing an additional parameter a, whose exponent registered the number of parts in a partition, concluding that the coefficient of $x^n a^j$ in $(1+ax)(1+ax^3)\cdots(1+ax^{2j-1})\cdots$ was the same as in

$$\frac{x^{j^2}a^j}{(1-x^2)(1-x^4)\cdots(1-x^{2j})}.$$

Thus, he had Euler's formula

$$\prod_{n=0}^{\infty}(1+ax^{2n+1}) = \sum_{n=0}^{\infty} \frac{x^{n^2}a^n}{(1-x^2)(1-x^4)\cdots(1-x^{2n})},$$

but by a combinatorial argument.

By means of a Durfee square analysis, Sylvester also obtained the identity needed by Jacobi to complete his proof of the triple product identity.[43] Thus, in Jacobi's formula

$$\prod_{n=1}^{\infty} \frac{1}{(1-aq^n)} = \sum_{m=0}^{\infty} \frac{q^{m^2}a^m}{(1-q)(1-q^2)\cdots(1-q^m)} \cdot \frac{1}{(1-aq)(1-aq^2)\cdots(1-aq^m)},$$

the factor

$$\frac{q^{m^2}a^m}{(1-q)(1-q^2)\cdots(1-q^m)}$$

accounted for the square and the nodes to the right of it, while

$$\frac{1}{(1-aq)(1-aq^2)\cdots(1-aq^m)}$$

did the same for the nodes and the number of rows below the square. It is an interesting and instructive exercise to work out the details.

[42] ibid. § 28.
[43] ibid. § 33.

Sylvester demonstrated that graphical analysis could also be used as a tool for the discovery of new identities. As an example, he presented[44]

$$\sum_{n=0}^{\infty} \frac{(1+ax)(1+ax^2)\cdots(1+ax^{n-1})(1+ax^{2n})a^n x^{\frac{n(3n-1)}{2}}}{(1-x)(1-x^2)\cdots(1-x^n)} = \prod_{m=1}^{\infty}(1+ax^n).$$

(26.10)

Briefly, Sylvester considered all partitions with distinct parts to account for the product on the right-hand side. To obtain the series on the left-hand side, he considered a graph of an arbitrary partition of n with distinct parts. He supposed that the Durfee square (the largest square of nodes in the upper left corner) had θ^2 nodes. Again, there were two subgraphs, called by Sylvester appendages: one to the right of the square with either θ or $\theta - 1$ rows and with unrepeated parts; and one below the square with $j-\theta$ rows and with unrepeated parts. Moreover, since the parts were distinct, Sylvester observed that the subgraph below the square had the largest part, at most θ or $\theta - 1$, depending on whether the subgraph to the right had θ or $\theta - 1$ rows. In the first case, because $1 + 2 + \cdots + \theta = \frac{\theta(\theta+1)}{2}$, the number of distributions was the coefficient of $x^{n-\theta^2}a^{j-\theta}$ in

$$\frac{x^{\frac{(\theta^2+\theta)}{2}}}{(1-x)(1-x^2)\cdots(1-x^\theta)} \cdots (1+ax)(1+ax^2)\cdots(1+ax^\theta);$$

in the second case, it was the coefficient of $x^{n-\theta^2}a^{j-\theta}$ in

$$\frac{x^{\frac{(\theta^2-\theta)}{2}}}{(1-x)(1-x^2)\cdots(1-x^{\theta-1})} \cdot (1+ax)(1+ax^2)\cdots(1+ax^{\theta-1}).$$

By adding these two expressions, Sylvester obtained the θth term of his series and this proved the formula. Note that Euler's pentagonal number theorem follows from Sylvester's formula when one takes $a = -1$. Sylvester commented, "Such is one of the fruits among a multitude arising out of Mr. Durfee's ever-memorable example of the dissection of a graph (in the case of a symmetrical one) into a square, and two regular graph appendages."

26.3 Cayley: Sylvester's Formula

In 1882, Sylvester's mathematical correspondent and comrade, Cayley, responded to his friend's great paper on partitions by showing how the "very beautiful formula" (26.10) could be proved by an interesting analytic method. He expressed the series side of the formula as[45]

---

[44] ibid. § 35.
[45] See Cayley (1889–1898) vol. 12, pp. 217–219.

$$\Omega = 1 + P + Q(1 + ax) + R(1 + ax)(1 + ax^2)$$
$$+ S(1 + ax)(1 + ax^2)(1 + ax^3) + \cdots \qquad (26.11)$$

where

$$P = \frac{(1 + ax^2)xa}{1}, \quad Q = \frac{(1 + ax^4)x^5 a^2}{1 \cdot 2},$$

$$R = \frac{(1 + ax^6)x^{12} a^3}{1 \cdot 2 \cdot 3}, \quad S = \frac{(1 + ax^8)x^{22} a^4}{1 \cdot 2 \cdot 3 \cdot 4}, \text{ etc.} \qquad (26.12)$$

where the numbers in bold $1, 2, 3, 4, \ldots$ denoted $1 - x, 1 - x^2, 1 - x^3, 1 - x^4, \ldots$. Cayley observed that the x exponents $1, 5, 12, 22, \ldots$ were the pentagonal numbers $\frac{(3n^2 - n)}{2}$. Cayley then set

$$P' = \frac{ax^2}{1}, \quad Q' = \frac{ax^3}{1} + \frac{a^2 x^7}{1 \cdot 2}, \quad R' = \frac{ax^4}{1} + \frac{a^2 x^9}{1 \cdot 2} + \frac{a^3 x^{15}}{1 \cdot 2 \cdot 3}, \text{ etc.,}$$

where the x exponents were

$$2; \ 3, 3 + 4; \ 4, 4 + 5, 4 + 5 + 6; \quad \text{etc.}$$

He then noted that it was easily verified that

$$1 + P = (1 + ax)(1 + P'),$$
$$1 + P' + Q = (1 + ax^2)(1 + Q'),$$
$$1 + Q' + R = (1 + ax^3)(1 + R'),$$
$$1 + R' + S = (1 + ax^4)(1 + S'), \quad \text{etc.}$$

Cayley concluded from these relations that

$$\Omega \div (1 + ax) = 1 + P' + Q + R(1 + ax^2) + S(1 + ax^2)(1 + ax^3) + \cdots,$$
$$\Omega \div (1 + ax)(1 + ax^2) = 1 + Q' + R + S(1 + ax^3) + T(1 + ax^3)(1 + ax^4) + \cdots,$$
$$\Omega \div (1 + ax)(1 + ax^2)(1 + ax^3) = 1 + R' + S + T(1 + ax^4) + \cdots,$$

and so on. When this was done infinitely often, the right-hand side would become 1, giving Cayley the complete proof of Sylvester's theorem.

Andrews pointed out that Cayley's method is more easily understood in terms of Euler's method of proving the pentagonal number theorem. In Sylvester's series, take the factor $1 + aq^{2n}$ in the $(n + 1)$th term, counting 1 as the first term, and split it into two parts:

$$1 + ax^{2n} = 1 - x^n + x^n(1 + ax^n). \qquad (26.13)$$

Interestingly, this breaks the $(n + 1)$th term into two parts, so that one can associate the second part of the term with the first part of the succeeding term. The result of this association is

$$\frac{(1+ax)(1+ax^2)\cdots(1+ax^{n-1})(1+ax^n)x^na^nx^{\frac{n(3n-1)}{2}}}{(1-x)(1-x^2)\cdots(1-x^n)}$$

$$+\frac{(1+ax)(1+ax^2)\cdots(1+ax^n)a^{n+1}x^{\frac{(n+1)(3n+2)}{2}}}{(1-x)(1-x^2)\cdots(1-x^n)}$$

$$= (1+ax)\cdot\frac{(1+ax^2)\cdots(1+ax^n)}{(1-x)\cdots(1-x^n)}\cdot a^n x^{\frac{n(3n+1)}{2}}(1+ax^{2n+1}). \tag{26.14}$$

Now observe that the factor multiplying $1 + ax$ is the $(n + 1)$th term of Sylvester's series except that a has been replaced by ax. So, if we denote Sylvester's series by $f(a)$, then by (26.14) we have

$$f(a) = (1+ax)f(ax)$$
$$= (1+ax)(1+ax^2)f(ax^2)$$
$$= (1+ax)(1+ax^2)(1+ax^3)\cdots.$$

Note that for convergence we would require $|x| < 1$, and therefore $x^n \to 0$ as $n \to \infty$. This implies

$$\lim_{n\to\infty} f(ax^n) = f(0) = 1.$$

We note that Gauss also used this method on various occasions and that it is possible that Cayley rediscovered Euler's method. We shall see in the next section how Ramanujan made brilliant use of this technique to prove the Rogers–Ramanujan identities.

26.4 Ramanujan: Rogers–Ramanujan Identities

Ramanujan discovered a new proof of the Rogers–Ramanujan identity after he saw Rogers's original proof, presented in Chapter 27. Ramanujan communicated the proof to Hardy in a letter of April 1919 and Hardy had it published in a 1919 paper.[46] In this paper, Hardy also included another proof, sent by Rogers to MacMahon in October 1917. Ramanujan's proof started with a series very similar to Sylvester's series (26.10):

$$G(x) = 1 + \sum_{n=1}^{\infty}(-1)^n x^{2n}q^{n(5n-1)}(1-xq^{2n})\frac{(1-xq)(1-xq^2)\cdots(1-xq^{n-1})}{(1-q)(1-q^2)\cdots(1-q^n)}.$$

$$\tag{26.15}$$

---

[46] Ramanujan (1919a).

He split this series into two parts, exactly as Cayley had done with Sylvester's series, by applying

$$1 - xq^{2n} = 1 - q^n + q^n(1 - xq^n)$$

to transform (26.15) into

$$G(x) = \sum_{n=1}^{\infty} x^{2n-2} q^{\frac{(5n^2-9n+4)}{2}} (1 - x^2 q^{2(2n-1)}) \frac{(1-xq)\cdots(1-xq^{n-1})}{(1-q)\cdots(1-q^{n-1})}, \qquad (26.16)$$

where the empty product when $n = 1$ was set equal to 1. Ramanujan then set

$$H(x) = \frac{G(x)}{1 - xq} - G(xq)$$

and used the value of $G(x)$ from (26.16) and of $G(xq)$ from (26.15) to obtain

$$H(x) = xq - \frac{x^2 q^3}{1-q}((1-q) + xq^4(1-xq^2))$$
$$+ \frac{x^4 q^{11}(1-xq^2)}{(1-q)(1-q^2)}((1-q^2) + xq^7(1-xq^3))$$
$$- \frac{x^6 q^{24}(1-xq^2)(1-xq^3)}{(1-q)(1-q^2)(1-q^3)}((1-q^3) + xq^{10}(1-xq^4)) + \cdots.$$

Again, as in Cayley's argument, Ramanujan associated the second part of each term with the first part of the succeeding term to arrive at the relation

$$H(x) = xq(1 - xq^2)G(xq^2)$$

or

$$G(x) = (1 - xq)G(xq) + xq(1 - xq)(1 - xq^2)G(xq^2).$$

Setting

$$F(x) = \frac{G(x)}{(1-xq)(1-xq^2)(1-xq^3)\cdots},$$

Ramanujan obtained the relation

$$F(x) = F(xq) + xqF(xq^2). \qquad (26.17)$$

He observed that it readily followed that

$$F(x) = 1 + \frac{xq}{1-q} + \frac{x^2 q^4}{(1-q)(1-q^2)} + \frac{x^3 q^9}{(1-q)(1-q^2)(1-q^3)} + \cdots, \qquad (26.18)$$

an equation that follows from

$$F(x) = 1 + A_1(q)x + A_2(q)x^2 + A_3(q)x^3 + \cdots.$$

Applying (26.17), he obtained

$$A_n(q) = \frac{q^{2n-1}}{1-q^n} A_{n-1}(q), \quad n = 1, 2, 3, \ldots,$$

where $A_0(q) = 1$. This implied (26.18). Ramanujan obtained the required identities by taking $x = 1$ and $x = q$ in (26.18). For $x = 1$, he got

$$1 + \frac{q}{1-q} + \frac{q^4}{(1-q)(1-q^2)} + \cdots = \frac{G(1)}{(1-q)(1-q^2)(1-q^3)\cdots}$$

$$= \frac{1 - q^2 - q^3 + q^9 + q^{11} - \cdots}{(1-q)(1-q^2)(1-q^3)\cdots}.$$

The series can be converted to a product by the triple product identity and the result follows.

26.5   Ramanujan's Congruence Properties of Partitions

Ramanujan was the first mathematician to study the divisibility properties of the partition function. In a paper published in 1919,[47] he gave fairly simple proofs of the congruence relations $p(5m + 4) \equiv 0 \pmod 5$ and $p(7m + 5) \equiv 0 \pmod 7$. He started with Euler's generating function for $p(n)$,

$$\frac{q}{(1-q)(1-q^2)(1-q^3)\cdots} = \sum_{n=1}^{\infty} p(n-1)q^n, \qquad (26.19)$$

and observed that the first congruence would follow if the coefficient of q^{5m} on the right-hand side were divisible by 5. Thus, it was sufficient to show that the same was true for the coefficient of q^{5m} in

$$\frac{q(1-q^5)(1-q^{10})(1-q^{15})\cdots}{(1-q)(1-q^2)(1-q^3)\cdots} = \frac{q(1-q^5)(1-q^{10})\cdots}{\{(1-q)(1-q^2)\cdots\}^5}\{(1-q)(1-q^2)\cdots\}^4.$$

Ramanujan then noted that $1 - q^{5m} \equiv (1 - q^m)^5 \pmod 5$ and hence

$$\frac{(1-q^5)(1-q^{10})(1-q^{15})\cdots}{\{(1-q)(1-q^2)(1-q^3)\cdots\}^5} \equiv 1 \pmod 5.$$

[47] Ramanujan (1919b).

Thus, to prove that $p(5m + 4) \equiv 0 \pmod 5$, it was enough to show that the coefficient of q^{5m} in

$$q\{(1-q)(1-q^2)(1-q^3)\cdots\}^4$$
$$= q\{(1-q)(1-q^2)(1-q^3)\cdots\}^3(1-q)(1-q^2)(1-q^3)\cdots$$
$$= q\sum_{m=0}^{\infty}(2m+1)q^{\frac{m(m+1)}{2}}\sum_{n=-\infty}^{\infty}(-1)^n q^{\frac{n(3n+1)}{2}}$$

was divisible by 5. Observe that in the last step, Ramanujan used Jacobi's identity and Euler's pentagonal number theorem. Jacobi's identity can be derived from the triple product identity. He then noted that the exponent of q in the double sum was divisible by 5 when

$$1 + \frac{m(m+1)}{2} + \frac{n(3n+1)}{2} \equiv 0 \pmod 5, \quad \text{or}$$
$$8 + 4m(m+1) + 4n(3n+1) \equiv 0 \pmod 5, \quad \text{or}$$
$$(2m+1)^2 + 2(n+1)^2 \equiv 0 \pmod 5.$$

Ramanujan noted that $(2m+1)^2 \equiv 0, 1, 4 \pmod 5$ and $2(n+1)^2 \equiv 0, 2, 3 \pmod 5$. So when the exponent of q was a multiple of 5, then $2m+1 \equiv 0 \pmod 5$ and $n+1 \equiv 0 \pmod 5$. Since the coefficient of this power of q was $2m + 1$, a multiple of 5, Ramanujan's proof was complete. He gave a similar proof for the congruence modulo 7; see the Exercises.

In the same paper, Ramanujan outlined a proof of (26.7). He intended to publish more details at a later date, but his premature death made this impossible. However, he wrote notes giving these details, later found and published as part of his lost notebook.[48] In fact, this proof used only the pentagonal number theorem, Jacobi's identity

$$\prod_{n=1}^{\infty}(1-q^n)^3 = \sum_{m=0}^{\infty}(-1)^m(2m+1)q^{\frac{m(m+1)}{2}} \tag{26.20}$$

and fifth roots of unity. By the pentagonal number theorem:

$$\prod_{n=1}^{\infty}(1-q^{\frac{n}{5}}) = \sum_{n=-\infty}^{\infty}(-1)^n q^{\frac{n(3n+1)}{10}}. \tag{26.21}$$

Partition the series into five parts according to whether $n \equiv 0, \pm1, \pm2 \pmod 5$. For example, for $n \equiv 0 \pmod 5$, we have $n = 5m$ and the part of the series corresponding to these values of n would be given by

$$\sum_{m=-\infty}^{\infty}(-1)^m q^{\frac{m(15m+1)}{2}}.$$

---

[48] Ramanujan (1988) pp. 238–239.

Note that the subseries corresponding to $n = 5m - 1$ can once again be expressed a product:

$$- \sum_{m=-\infty}^{\infty} (-1)^m q^{\frac{(5m-1)(15m-2)}{10}} = -q^{\frac{1}{5}} \sum_{m=-\infty}^{\infty} (-1)^m q^{\frac{5m(3m-1)}{2}}$$

$$= -q^{\frac{1}{5}} \prod_{n=1}^{\infty} (1 - q^{5n}).$$

Thus, (26.21) can be written as

$$\prod_{n=1}^{\infty} (1 - q^{\frac{n}{5}}) = \sum_{m=-\infty}^{\infty} (-1)^m q^{\frac{m(15m+1)}{2}} + \sum_{m=-\infty}^{\infty} (-1)^m q^{\frac{(3m-1)(5m-2)}{2}}$$

$$+ q^{\frac{2}{5}} \left(\sum_{m=-\infty}^{\infty} (-1)^m q^{\frac{(3m+2)(5m+1)}{2}} - \sum_{m=-\infty}^{\infty} (-1)^m q^{\frac{m(15m+7)}{2}} \right)$$

$$- q^{\frac{1}{5}} \prod_{n=1}^{\infty} (1 - q^{5n}).$$

Dividing by $\prod_{n=1}^{\infty} (1 - q^{5n})$ gives

$$\prod_{n=1}^{\infty} \left(\frac{1 - q^{\frac{n}{5}}}{1 - q^{5n}} \right) = \xi_1 - q^{\frac{1}{5}} - \xi q^{\frac{2}{5}}, \tag{26.22}$$

where ξ and ξ_1 are power series in q. Ramanujan applied Jacobi's identity to show that $\xi \xi_1 = 1$. So cube both sides of (26.22) and use Jacobi's identity (26.20) to get

$$\frac{\sum_{n=0}^{\infty} (-1)^n (2n + 1) q^{\frac{n(n+1)}{10}}}{\prod_{n=1}^{\infty} (1 - q^{5n})^3} = \left(\xi_1 - q^{\frac{1}{5}} - \xi q^{\frac{2}{5}} \right)^3. \tag{26.23}$$

Since the exponent of q, given by $n(n + 1)$, is either 0, 2, or 6 (mod 10), it follows that no power of q is of the form $\frac{2}{5} +$ an integer. This implies that the term

$$3q^{\frac{2}{5}} \xi_1 - 3\xi_1^2 \xi q^{\frac{2}{5}} = 3q^{\frac{2}{5}} \xi_1 (1 - \xi \xi_1)$$

on the right-hand side of (26.23) must be zero. This in turn implies that $\xi_1 = \xi^{-1}$, and we can write

$$\prod_{n=1}^{\infty} \left(\frac{1 - q^{5n}}{1 - q^{\frac{n}{5}}} \right) = \frac{1}{\xi^{-1} - q^{\frac{1}{5}} - \xi q^{\frac{2}{5}}}. \tag{26.24}$$

Consider the expression $\lambda^{-1} - \lambda - 1$, where $\lambda = \xi q^{\frac{1}{5}} w$, and w is a fifth root of unity. Observe that if $\lambda^{-1} - \lambda = 1$, then by an elementary calculation $\lambda^{-5} - \lambda^5 = 11$. Thus,

$$\xi^{-5} - 11q - \xi^5 q^2 = \prod_{k=0}^{4} \left(\xi^{-1} - q^{\frac{1}{5}} w^k - \xi q^{\frac{2}{5}} w^{2k} \right).$$

It is now easy to check by long division that

$$\prod_{n=1}^{\infty} \left(\frac{1 - q^{5n}}{1 - q^{\frac{n}{5}}} \right)$$

$$= \frac{\xi^{-4} - 3q\xi + q^{\frac{1}{5}}(\xi^{-3} + 2q\xi^2) + q^{\frac{2}{5}}(2\xi^{-2} - q\xi) + q^{\frac{3}{5}}(3\xi^{-1} + q\xi^4) + 5q^{\frac{4}{5}}}{\xi^{-5} - 11q - q^2\xi^5}.$$

Now multiply across by $q^{\frac{1}{5}}$ and replace $q^{\frac{1}{5}}$ by $q^{\frac{1}{5}} e^{\frac{2\pi i k}{5}}$, $k = 1, 2, 3, 4$, to obtain five identities. Next, apply

$$q^{\frac{1}{5}} \prod_{n=1}^{\infty} (1 - q^{\frac{n}{5}})^{-1} = \sum_{n=1}^{\infty} p(n-1) q^{\frac{n}{5}}$$

and add the five identities to get

$$\prod_{n=1}^{\infty} (1 - q^{5n}) \sum_{n=0}^{\infty} p(5n + 4) q^n = \frac{5}{\xi^{-5} - 11q - q^2\xi^5}. \tag{26.25}$$

By replacing $q^{\frac{1}{5}}$ by $e^{\frac{2\pi i k}{5}} q^{\frac{1}{5}}$, $k = 0, 1, 2, 3, 4$, and multiplying the five equations together, Ramanujan arrived at

$$\prod_{n=1}^{\infty} \left(\frac{1 - q^{5n}}{1 - q^n} \right)^6 = \frac{1}{\xi^{-5} - 11q - q^2\xi^5}. \tag{26.26}$$

Combining (26.25) and (26.26) gives the necessary result.

In his paper, Ramanujan also noted that

$$\xi^{-1} = \prod_{n=0}^{\infty} \frac{(1 - q^{5n+2})(1 - q^{5n+3})}{(1 - q^{5n+1})(1 - q^{5n+4})}. \tag{26.27}$$

This can be proved by using (26.24), the pentagonal number theorem and the quintuple product identity. Ramanujan observed in his paper that (26.7) implied that $p(25m+24)$ was divisible by 25. He argued that by (26.7),

$$\frac{p(4)x + p(9)x^2 + p(14)x^3 + \cdots}{5((1 - x^5)(1 - x^{10})(1 - x^{15}) \cdots)^4}$$

$$= \frac{x}{(1 - x)(1 - x^2)(1 - x^3) \cdots} \frac{(1 - x^5)(1 - x^{10})(1 - x^{15}) \cdots}{((1 - x)(1 - x^2)(1 - x^3) \cdots)^5},$$

and since the coefficient of x^{5n} on the right-hand side was a multiple of 5, it followed that $p(25m + 24)$ was divisible by 25. It is interesting to see how Ramanujan used (26.27) to compute the Rogers–Ramanujan continued fraction as well as the generating function for $p(5m + 4)$.

26.6 Exercises

(1) Prove that the number of partitions of n into parts not divisible by d is equal to the number of partitions of n of the form $n = n_1 + n_2 + \cdots + n_s$, where $n_i \geq n_{i+1}$ and $n_i \geq n_{i+d-1} + 1$. See Glaisher (1883). James Whitehead Lee Glaisher (1848–1928) single-handedly edited two journals for over forty years: *Messenger of Mathematics* and *Quarterly Journal*. The *Messenger* carried the first published papers of many English mathematicians and physicists of the late nineteenth century, including H. F. Baker, E. W. Barnes, W. Burnside, G. H. Hardy, J. J. Thompson, and J. Jeans. Glaisher published almost four hundred papers, many of them in his own journals. G. H. Hardy wrote in his obituary notice, "He wrote a great deal of very uneven quality, and he was 'old fashioned' in a sense which is most unusual now; but the best of his work is really good." This best work included results in number theory and, in particular, the representation of numbers as sums of squares. See vol. 7 of Hardy (1966–1979).

(2) Complete the following number theoretic proof, due to Glaisher, of Euler's theorem that the number of partitions of n into odd parts equals the number of partitions of n into distinct parts. Let

$$n = f_1 \cdot 1 + f_3 \cdot 3 + \cdots + f_{2m-1} \cdot (2m - 1).$$

Here $f_1, f_3, \ldots$ represent the number of times $1, 3, \ldots$, respectively, occur in the partition of n into odd parts. Now write $f_1, f_3, \ldots$ in powers of two:

$$f_1 = 2^{a_1} + 2^{a_2} + \cdots + 2^{a_l},$$
$$f_3 = 2^{b_1} + 2^{b_2} + \cdots + 2^{b_l}, \cdots.$$

Then

$$n = 2^{a_1} + 2^{a_2} + \cdots + 2^{a_l} + 2^{b_1} \cdot 3 + 2^{b_2} \cdot 3 + \cdots$$
$$+ 2^{b_l} \cdot 3 + \cdots + 2^{r_1} \cdot (2m - 1) + \cdots + 2^{r_s} \cdot (2m - 1)$$

gives a partition of n into distinct parts. See Glaisher (1883).

(3) Show that the number of partitions of n into odd parts, where exactly k distinct parts appear, is equal to the number of partitions of n into distinct parts, where exactly k sequences of consecutive integers appear. Show that this correspondence is one-to-one. This result was published by Sylvester in 1882. See Sylvester (1973) vol. 4, p. 45.

(4) Let $p_{k,r}(n)$ denote the number of partitions of n into parts not congruent to 0, $\pm r \pmod{2k+1}$, where $1 \le r \le k$. Let $q_{k,r}(n)$ denote the number of partitions of n of the form $n = n_1 + n_2 + \cdots + n_s$ where $n_1 \ge n_{i+1}$, $n_i \ge n_{i+k-1} + 2$ and with 1 appearing as a part at most $r - 1$ times. Prove that then

$$p_{k,r}(n) = q_{k,r}(n).$$

See Gordon (1961).

(5) Prove that if $p_m(n)$ denotes the number of partitions of n with rank m, then

$$\sum_{n=1}^{\infty} p_m(n)q^n = \frac{1}{(q,q)_\infty} \sum_{n=1}^{\infty} (-1)^{n-1} \left(q^{\frac{1}{2}n(3n-1)} - q^{\frac{1}{2}n(3n+1)} \right) q^{|m|n}.$$

See Atkin and Swinnerton-Dyer (1954).

(6) Derive Szekeres's combinatorial interpretation of the Rogers–Ramanujan continued fraction. Let $B(n.k)$, $n \ge 1$, $k \ge 0$ represent the number of sequences of integers $b_1 \le b_2 \le \cdots \le b_n = n$ with $b_i > i$ for $1 \le i < n$ and $b_1 + b_2 + \cdots + b_{n-1} = \binom{n}{2} + k$. Observe that $B(n,k) = 0$ for $0 \le k < n - 1$ and for $k > \binom{n}{2}$. Also note that $B(n, n-1) = B\left(n, \binom{n}{2}\right) = 1$, $B(1,0) = 1$. Now show that

$$\frac{x}{1+} \frac{qx}{1+} \frac{q^2 x}{1+} \cdots = \sum_{0 \le n-1 \le k} (-1)^{n+1} B(n,k) x^n q^k.$$

See Szekeres (1968). George Szekeres (1911–2005) was trained as a chemical engineer in Hungary but his association with Paul Erdős, Esther Klein, and Paul Turán turned his interest to mathematics. In 1935, Erdős and Szekeres wrote a paper laying the foundation of Ramsey theory. This arose out of Szekeres's efforts to solve a problem (proposed by Klein who later became his wife): For all n there exists N such that for any N points in a plane there are n which form a convex n-gon. Szekeres and his family escaped to China from the Nazi government in Germany and moved to Australia after the war. His presence gave a boost to the development and teaching of mathematics in Australia where he was greatly admired.

(7) Andrews gave an interpretation of the Rogers–Ramanujan continued fraction different from that of Szekeres. Let

$$C(q) = 1 + \frac{q}{1+} \frac{q^2}{1+} \cdots = \sum_{m \ge 0} c_m q^m.$$

Also let $B_{k,a}(n)$ denote the number of partitions of n of the form $n = b_1 + b_2 + \cdots + b_s$, where $b_i \ge b_{i+1}$, $b_i - b_{i+k-1} \ge 2$ and at most $a - 1$ of the b_i are equal to one. Then prove that

$$c_{5m} = B_{37,37}(m) + B_{37,13}(m-4),$$

$$c_{5m+1} = B_{37,32}(m) + B_{37,7}(m-6),$$

$$c_{5m+2} = -(B_{37,23}(m-1) + B_{37,2}(m-8)),$$

$$c_{5m+3} = -(B_{37,28}(m) + B_{37,22}(m-1)),$$
$$c_{5m+4} = -(B_{37,17}(m-2) + B_{37,8}(m-5)).$$

Show that it follows that, in particular, $c_2 = c_4 = c_9 = 0$ and that the remaining c_n satisfy

$$c_{5m} > 0,\ c_{5m+1} > 0,\ c_{5m+2} < 0,\ c_{5m+3} < 0,\ c_{5m+4} < 0.$$

See Andrews (1981).

(8) Prove Ramanujan's result that $p(7m + 5) \equiv 0 \pmod 7$ by the following method. Square Jacobi's identity to obtain

$$x^2 \prod_{n=1}^{\infty} (1-x^n)^6 = \sum_{\mu=-\infty}^{\infty} \sum_{\nu=-\infty}^{\infty} (-1)^{\mu+\nu}(2\mu+1)(2\nu+1)x^{2+\frac{1}{2}\mu(\mu+1)+\frac{1}{2}\nu(\nu+1)}.$$

Now show that the coefficient of x^{7n} in the sum is divisible by 49. Next observe that

$$\frac{1-x^7}{(1-x)^7} \equiv 1 \pmod 7$$

and deduce that the coefficient of x^{7n} in $\frac{x^2}{\prod_{n=1}^{\infty}(1-x^n)}$ is a multiple of 7. See Ramanujan (2000) p. 212.

(9) For a partition π of n, let $\lambda(\pi)$ denote the largest part of π; let $\mu(\pi)$ denote the number of ones in π; and let $\nu(\pi)$ denote the number of parts of π larger than $\mu(\pi)$. The crank $c(\pi)$ is defined as

$$c(\pi) = \lambda(\pi)\ \text{if}\ \mu(\pi) = 0;\quad c(\pi) = \nu(\pi) - \mu(\pi)\ \text{if}\ \mu(\pi) > 0.$$

Let $M(m,n)$ denote the number of partitions of n with crank m. Then prove that

$$\sum_{m=-\infty}^{\infty} \sum_{n=0}^{\infty} M(m,n)a^m q^n = \frac{(q;q)_\infty}{(aq;q)_\infty(\frac{q}{a};q)_\infty}.$$

See Andrews and Garvan (1988a).

26.7　Notes on the Literature

For more on partitions, see Andrews (1998). Ramanujan (2000) contains his published papers on partitions; Berndt has also added seventy pages of helpful commentary at the end of the book, where one will find references to other works on Ramanujan's papers.

The theory of modular forms has been extensively used in recent years to study the partition function; K. Ono and his students and collaborators have been leaders in this area. Modular forms are also important in the study of many arithmetical problems. For example, Fermat's last theorem is a consequence of the Shimura-Taniyama conjecture. See Shimura (2008) for an account of how he arrived at this conjecture.

27

q-Series and q-Orthogonal Polynomials

27.1 Preliminary Remarks

In the early nineteenth century, q-series proved their worth with their broad applicability to number theory, elliptic and modular functions, and combinatorics. Nevertheless, as late as 1840, no general framework for the study of q-series had been established; although q-series had been used to solve problems in other areas, it had not become a subject of its own. Finally, q-series came into its own when it was viewed as an extension of the hypergeometric series. As discussed in Chapter 26, in the 1840s Cauchy, Eisenstein, Jacobi, and Heine each presented the q-binomial theorem for general exponents. In 1846, Jacobi wrote a paper stating a q-extension of Gauss's $_2F_1$ summation formula.[1] His interesting proof was based on Schweins's q-extension of the Vandermonde identity, a terminating form of Gauss's summation. The latter result gave the value of the series in terms of gamma functions. So Jacobi suggested a q-analog of $\Gamma(a)$:

$$\Omega(q,a) = \frac{(1-q)(1-q^2)(1-q^3)\cdots}{(1-q^{a+1})(1-q^{a+2})(1-q^{a+3})\cdots}. \tag{27.1}$$

It is interesting that at the end of his 1846 paper, Jacobi wrote a lengthy historical note mentioning the 1729 letter from Euler to Goldbach, containing Euler's description of his discovery of the gamma function by the use of infinite products. Jacobi had a keen interest in the history of mathematics, and some of his papers contain very helpful historical information. With Jacobi's work, the stage was set to obtain the q-extension of the hypergeometric series and this was soon accomplished by Heinrich Eduard Heine (1821–1881) who studied in Göttingen under Gauss and Stern and in Berlin under Dirichlet. Heine received his doctoral degree under the supervision of Dirksen and Ohm in 1842. He then spent a year in Königsberg with Jacobi and Franz Neumann. It is most likely that Jacobi encouraged Heine to work on hypergeometric series and its q-extension; Heine later edited a posthumous paper of Jacobi on this subject.

---

[1] Jacobi (1846).

In his 1847 paper defining the q-hypergeometric series,[2] Heine developed properties of the series $\phi(\alpha, \beta, \gamma, q, x)$ defined by

$$1 + \frac{(1 - q^\alpha)(1 - q^\beta)}{(1 - q)(1 - q^\gamma)}x + \frac{(1 - q^\alpha)(1 - q^{\alpha+1})(1 - q^\beta)(1 - q^{\beta+1})}{(1 - q)(1 - q^2)(1 - q^\gamma)(1 - q^{\gamma+1})}x^2 + \cdots.$$

(27.2)

Observe that as $q \to 1^-$, this series converges term by term to

$$1 + \frac{\alpha \cdot \beta}{1 \cdot \gamma}x + \frac{\alpha(\alpha + 1) \cdot \beta(\beta + 1)}{1 \cdot 2 \cdot \gamma(\gamma + 1)}x^2 + \cdots,$$

the hypergeometric series $F(\alpha, \beta, \gamma, x)$. Heine took many results from Gauss's 1813 paper on hypergeometric series and extended them to the q-series ϕ. He listed the contiguous relations for ϕ, from which he derived continued fractions expansions for ratios of q-hypergeometric series, and he gave a very simple proof of the q-binomial theorem. The notation $\Omega(q, a)$ for the gamma function analog is also due to Heine, and as an analog of Gauss's $F(a, b, c, 1)$ sum, he presented

$$\phi(\alpha, \beta, \gamma, q, q^{\gamma-\alpha-\beta}) = \frac{\Omega(q, \gamma - 1)\,\Omega(q, \gamma - \alpha - \beta - 1)}{\Omega(q, \gamma - \alpha - 1)\,\Omega(q, \gamma - \beta - 1)}.$$

(27.3)

Heine applied the q-binomial theorem to obtain an important transformation now known as Heine's transformation:

$$
\begin{aligned}
&\frac{(1 - q^\gamma x)(1 - q^{\gamma+1}x)\cdots}{(1 - q^\beta x)(1 - q^{\beta+1}x)\cdots}\left(1 + \frac{(1 - q^\alpha)(1 - q^\beta x)}{(1 - q)(1 - q^\gamma x)} \cdot z \right.\\
&\quad + \left. \frac{(1 - q^\alpha)(1 - q^{\alpha+1})(1 - q^\beta x)(1 - q^{\beta+1}x)}{(1 - q)(1 - q^2)(1 - q^\gamma x)(1 - q^{\gamma+1}x)} \cdot z^2 + \cdots \right)\\
&= \frac{(1 - q^\alpha z)(1 - q^{\alpha+1}z)\cdots}{(1 - z)(1 - qz)\cdots}\left(1 + \frac{(1 - q^{\gamma-\beta})(1 - z)}{(1 - q)(1 - q^\alpha z)} \cdot q^\beta x \right.\\
&\quad + \left. \frac{(1 - q^{\gamma-\beta})(1 - q^{\gamma-\beta+1})(1 - z)(1 - qz)}{(1 - q)(1 - q^2)(1 - q^\alpha z)(1 - q^{\alpha+1}z)} \cdot q^{2\beta}x^2 + \cdots \right).
\end{aligned}
$$

(27.4)

He also defined a q-difference operator and found the second-order difference equation of which the hypergeometric equation was a limiting case. However, he did not define a q-integral.

The origin of the transformation (27.4) remained a puzzle; to which result in hypergeometric series did this correspond? C. Johannes Thomae (1840–1921) answered this question. Thomae studied at the University of Halle and was inspired by Heine to devote himself to function theory. Thomae moved to Göttingen in 1862 with the intention of working under Riemann who soon fell seriously ill. Thomae stayed on and in 1864, he earned his doctoral degree under Schering, one of the editors of Gauss's collected works. Thomae then returned to teach at Halle for a few years before

---

[2] Heine (1847).

moving to Freiburg and then to Jena. Thomae wrote his paper on Heine's series[3] in 1869 while at Halle. In this paper he defined the q-integral and showed that Heine's transformation was actually the q-extension of Euler's integral representation of the hypergeometric series. Thomae defined the q-integral by

$$\int_x^{xq^n} f(x)\frac{\Delta x}{x(q-1)} = \sum_{s=0}^{n-1} f(xq^s). \tag{27.5}$$

He explained that the integral was the inverse of the difference operator

$$\Delta f(x) = f(xq) - f(x), \tag{27.6}$$

also noting that it was better to define the q-gamma, or more correctly, the q-Π function by

$$\Pi(\alpha, q) = (1-q)^{-\alpha} \Omega(q, \alpha). \tag{27.7}$$

Indeed, with this definition,

$$\Pi(\alpha, q) = \frac{1-q^\alpha}{1-q} \cdot \Pi(\alpha - 1, q), \tag{27.8}$$

and $\Pi(\alpha, q) \to \Pi(\alpha) = \Gamma(\alpha + 1)$ as $q \to 1^-$. He also observed that the q-binomial theorem was equivalent to

$$\int_1^{q^\infty} s^\alpha \, p(\beta, sq) \, \Delta s = -\frac{\Pi(\alpha, q) \, \Pi(\beta, q)}{\Pi(\alpha + \beta + 1, q)}, \tag{27.9}$$

where

$$p(\beta, x) = \frac{(1-x)(1-xq)(1-xq^2)\cdots}{(1-xq^\beta)(1-xq^{\beta+1})(1-xq^{\beta+2})\cdots}. \tag{27.10}$$

Moreover, he noted that as $q \to 1^-$, formula (27.9) reduced to Euler's beta integral formula

$$\int_0^1 s^\alpha (1-s)^\beta \, ds = \frac{\Pi(\alpha) \, \Pi(\beta)}{\Pi(\alpha + \beta + 1)}.$$

About forty years later, the able amateur mathematician Frank Hilton Jackson (1870–1960) redefined the q-integral and the q-gamma function.[4] He took the q-integral to be the inverse of the q-derivative

$$\Delta_q \phi(x) = \frac{\phi(qx) - \phi(x)}{qx - x}, \tag{27.11}$$

[3] Thomae (1869).
[4] Jackson (1910).

so that the q-integral amounted to

$$\int_0^a f(x)\, d_q x = \sum_{n=0}^{\infty} f(aq^n)(aq^n - aq^{n+1}). \qquad (27.12)$$

The reader may observe that the expression on the right is the Riemann sum for the division points $a, aq, aq^2, \ldots$ on $[0,a]$. Note that Fermat integrated $x^{\frac{m}{n}}$, where m and n were integers, by first evaluating its q-integral and then letting $q \to 1^-$. Jackson's notation for the q-gamma, $\Gamma_q(x)$, is still in use; he set

$$\Gamma_q(x) = \Pi(x - 1, q). \qquad (27.13)$$

Jackson was a British naval chaplain; he apparently had a little difficulty in publishing some of his earlier works because their significance was not quite clear to referees. Jackson's lifelong program was to systematically develop the theory of q-hypergeometric (or basic hypergeometric) series by proving analogs of summation and transformation formulas for generalized hypergeometric series. Jackson conceived of q as the base of the series, analogous to the base of the logarithm. His terminology is now widely used. In recent years, Vyacheslav Spiridonov has worked out another important and productive generalization of hypergeometric functions, elliptic hypergeometric functions.[5] A formal series, $\sum_{-\infty}^{\infty} c_n$, is called an elliptic hypergeometric series if $\frac{c_{n+1}}{c_n} = h(n)$, where $h(n)$ is some elliptic function of $n \in \mathbb{C}$.

Leonard James Rogers (1862–1933) gave a new direction to q-series theory through his researches in the early 1890s. Rogers studied at Oxford where his father was a professor of political economy. As a boy, Rogers was tutored in mathematics by A. Griffith, an Oxford mathematician with a strong interest in elliptic functions. Rogers's earliest work was in reciprocants, a topic in invariant theory. The second half of Sylvester's 1886 Oxford lectures on this subject were devoted to the work of Rogers. Around this same time, Rogers's interest turned to analysis and to the topic in which he did his most famous work, theta series and products and, more generally, q-series. In his Royal Society obituary notice of Rogers, A. L. Dixon recalled attending an 1887 course of lectures at Oxford in which Rogers manipulated q-series and products with great skill. In the period 1893–95, after a study of Heine's 1878 *Kugelfunctionen*,[6] Rogers published four important papers on q-extensions of Hermite and ultraspherical polynomials. In his book, Heine included extra material, printed in smaller type, on basic hypergeometric series, clearly expecting that the q-extensions of some results in the book would be important and fruitful. Rogers showed that Heine was right. Rogers was initially struck by a lack of symmetry in Heine's transformation formula. In order to present Rogers's work succinctly and with transparency, we introduce some modern notation different from that of Rogers, though he also employed abbreviations. Let

$$(x)_n \equiv (x; q)_n = (1 - x)(1 - qx)(1 - q^2 x) \cdots (1 - q^{n-1} x),$$
$$n = 1, 2, 3, \ldots, \infty. \qquad (27.14)$$

---

[5] Spiridonov (2013).
[6] Heine (1878).

We observe that Rogers wrote x_n for $1 - q^{n-1}x$; $x_n!$ for $(x)_n$; and (x) for $(x)_\infty$. We now replace $q^\alpha, q^\beta, q^\gamma$ in Heine's formula (27.4) by a, b, c and write it as

$$\phi(a, bx, cx, q, z) = \frac{(bx)_\infty (az)_\infty}{(cx)_\infty (z)_\infty} \phi\left(\frac{c}{b}, z, az, q, bx\right).$$

Here

$$\phi(a, b, c, q, x) = \sum_{n=0}^{\infty} \frac{(a)_n (b)_n}{(q)_n (c)_n} x^n. \qquad (27.15)$$

After reparametrization, we can write Heine's transformation as

$$\phi(a, b, c, q, x) = \frac{(ax)_\infty (b)_\infty}{(x)_\infty (c)_\infty} \phi\left(\frac{c}{b}, x, ax, q, b\right). \qquad (27.16)$$

Perhaps Rogers's earlier work on invariant theory had made him sensitive to symmetry, so as a first step he wrote the transformation in symmetric form.[7] He observed that, by the symmetry in a and b and also by a reapplication of Heine's transformation, he obtained two results from (27.16):

$$\phi(a, b, c, q, x) = \frac{(bx)_\infty \left(\frac{c}{b}\right)_\infty}{(x)_\infty (c)_\infty} \phi\left(b, \frac{abx}{c}, bx, q, \frac{c}{b}\right), \qquad (27.17)$$

$$\phi(a, b, c, q, x) = \frac{\left(\frac{abx}{c}\right)_\infty}{(x)_\infty} \phi\left(\frac{c}{a}, \frac{c}{b}, c, q, \frac{abx}{c}\right). \qquad (27.18)$$

Rogers set $a = \mu e^{-i\theta}$, $b = \gamma e^{-i\theta}$, $c = \mu\gamma$, and $x = \lambda e^{i\theta}$, observing that the last three formulas implied that

$$\psi(\lambda, \mu, \gamma, q, \theta) = (\lambda e^{i\theta})_\infty (\mu\gamma)_\infty \phi(\mu e^{-i\theta}, \gamma e^{-i\theta}, \mu\gamma, q, \lambda e^{i\theta})$$

was symmetric in λ, μ, γ and in θ and $-\theta$. He then went on to define a q-extension of the Hermite polynomials $A_n(\theta)$ by the relation

$$P(t) \equiv \frac{1}{\prod_{n=0}^{\infty}(1 - 2tq^n \cos\theta + t^2 q^{2n})} = \sum_{n=0}^{\infty} \frac{A_n(\theta)}{(q)_n} t^n. \qquad (27.19)$$

In a later paper,[8] Rogers defined a q-extension of the ultraspherical or Gegenbauer polynomials and denoted it by $L_n(\theta)$, using the equation

$$\frac{P(t)}{P(\lambda t)} \equiv \prod_{n=0}^{\infty} \left(\frac{1 - 2\lambda t q^n \cos\theta + \lambda^2 t^2 q^{2n}}{1 - 2t q^n \cos\theta + t^2 q^{2n}}\right) = \sum_{n=0}^{\infty} \frac{L_n(\theta)}{(q)_n} t^n. \qquad (27.20)$$

[7] Rogers (1893b).
[8] Rogers (1894).

In connection with the q-Hermite polynomials, Rogers raised the question: Suppose

$$a_0 + a_1 A_1(\theta) + a_2 A_2(\theta) + \cdots = b_0 + 2b_1 \cos\theta + 2b_2 \cos 2\theta + \cdots.$$

How are the coefficients $a_0, a_1, a_2, \ldots$ and $b_0, b_1, b_2, \ldots$ related to each other? He solved the problem and then applied the solution to the function

$$\prod_{n=1}^{\infty} \left(1 + 2q^{n-\frac{1}{2}}\cos\theta + q^{2n-1}\right).$$

From the triple product identity, the Fourier cosine expansion of this function was already known; Rogers found the expansion in terms of $A_r(\theta)$, and in particular, he got

$$a_0 = \sum_{n=0}^{\infty} \frac{q^{n^2}}{(q)_n}, \qquad a_1 = \frac{q^{\frac{1}{2}}}{1-q} \sum_{n=0}^{\infty} \frac{q^{n(n+1)}}{(q)_n}.$$

When he expressed a_0, a_1 in terms of the bs, he obtained a series convertible to products by the triple product identity. The final results emerged as the Rogers–Ramanujan identities. We give Ramanujan's derivation of these formulas in Chapter 26. Although Rogers discovered these remarkable identities in 1894, they remained unnoticed until Ramanujan rediscovered them without proof. Then, in 1917, quite by chance, Ramanujan came across Rogers's paper while browsing through old journals. The ensuing correspondence between Ramanujan and Rogers led them both to new proofs of the identity now known by their names. Somewhat surprisingly, around 1980, the physicist R. J. Baxter rediscovered the Rogers–Ramanujan identities in the course of his work on the hard hexagon model.[9]

Rogers left unanswered the question of the orthogonality of his q-extensions of the Hermite and ultraspherical polynomials. He had found three-term recurrence relations for these polynomials, but it was not until the 1930s that these relations were widely understood to imply that the polynomials were orthogonal with respect to some positive weights. Consider the statement of the spectral theorem for orthogonal polynomials: Suppose that a sequence of monic polynomials $\{P_n(x)\}$ with real coefficients satisfies a three-term recurrence relation

$$x P_n(x) = P_{n+1}(x) + \alpha_n P_n(x) + \beta_n P_{n-1}(x), \quad n \geq 1, \tag{27.21}$$

with $P_0(x) = 1$, $P_1(x) = x - \alpha_0$, α_{n-1} real and $\beta_n > 0$. Then there exists a distribution function μ, corresponding to a positive and finite Borel measure on the real line, such that

$$\int_{-\infty}^{\infty} P_m(x) P_n(x) \, d\mu(x) = \zeta_n \delta_{m,n}, \tag{27.22}$$

---

[9] Baxter (1980).

where

$$\zeta_n = \beta_1 \beta_2 \cdots \beta_n. \tag{27.23}$$

The converse is also true and straightforward to prove. This theorem is sometimes known as Favard's theorem since Jean Favard (1902–1965) published a proof in the *Comptes Rendus* (Paris) in 1935. However, the theorem appeared earlier in the works of O. Perron, M. Stone,[10] and A. Wintner.[11] In fact, in 1934 J. Meixner applied this theorem to his work on orthogonal polynomials,[12] with a reference to a result of Perron.[13] Stieltjes's famous 1895 paper on continued fractions also contains a result yielding the spectral theorem.[14] It is unlikely that Rogers was aware of Stieltjes's work. We note that A. L. Dixon wrote of Rogers that he had only a vague notion of the work of other mathematicians. Probably an abstract result such as the spectral theorem for orthogonal polynomials would not have interested Rogers who loved special functions and formulas. He would have wanted to use the actual weight function, needed for computational purposes.

The Hungarian mathematician Gabor Szegő (1895–1985) took the first step toward finding an explicit weight function. Szegő studied in Hungary and Germany under Fejér, Frobenius, and Hilbert. His most famous book, *Problems and Theorems in Analysis*, was originally written in German in 1924, in collaboration with George Pólya. Szegő made significant contributions to orthogonal polynomials, on which he also wrote a very influential book first published in 1939. He founded the study of orthogonality on the unit circle; for a probability measure $\alpha(\theta)$, he defined this orthogonality by

$$\int_0^{2\pi} \phi_n(e^{i\theta}) \overline{\phi_m(e^{i\theta})} \, d\alpha(\theta) = 0, \quad m \neq n,$$

where

$$\phi_n(z) = \sum_{k=0}^n a_{k,n} z^k.$$

Szegő was the first to appreciate the general program Rogers had in mind, as opposed to its specific though very important corollaries, such as the Rogers–Ramanujan identities. Rogers's papers inspired Szegő to discover the first nontrivial example of orthogonal polynomials on the circle, where

$$a_{k,n} = (-1)^k q^{-\frac{k}{2}} \frac{(1-q^n)(1-q^{n-1})\cdots(1-q^{n-k+1})}{(1-q)(1-q^2)\cdots(1-q^k)}.$$

---

[10] Stone (1932).
[11] Wintner (1929).
[12] Meixner (1934).
[13] Perron (1929).
[14] Stieltjes (1993) vol. 2, pp. 628–630.

Here note the connection with the Gaussian polynomial. Recall also that Gauss had expressed $\phi(-q)$ and $\phi(\sqrt{q})$ as finite products and evaluated the general quadratic Gauss sum from these expressions. Szegő found that the weight function $f(\theta)\,d\theta = d\alpha(\theta)$ in this case was

$$f(\theta) = \sum_{n=-\infty}^{\infty} q^{\frac{n^2}{2}} e^{in\theta},$$

and he applied the triple product identity to prove it. The weight function for Rogers's q-Hermite polynomial is also a theta function; the proof of orthogonality in that case also uses the triple product identity.

The q-ultraspherical polynomials were independently rediscovered by Feldheim and I. L. Lanzewizky in 1941. The Hungarian mathematician Ervin Feldheim (1912–1944) studied in Paris, since he was not admitted to the university in Budapest. His thesis was in probability theory but on his return to Hungary he contributed important results to the classical theory of orthogonal polynomials. One of these results was contained in a letter to Fejér written by Feldheim shortly before his tragic death at the hands of the Nazis. The letter was later found by Paul Turán, who described the incident,[15] "Thus the letter had been resting among Fejér's letters for some 15 years. ... *On the next day* I received a letter from Szegő, in which he raised just a problem solved in Feldheim's letter! I sent this letter to Szegő and he published it with applications."

The origin of Feldheim and Lanzewizky's papers was the work of Fejér on a generalization of Legendre polynomials; this generalization also included the ultraspherical polynomials. Feldheim and Lanzewizky wished to determine those generalized Legendre polynomials that were also orthogonal. They used the spectral theorem and found conditions under which the generalized Legendre polynomials satisfied the appropriate three-term recurrence relation. At the end of his paper, Feldheim raised the problem of determining the weight or distribution function for orthogonality but was unable to resolve the question. Earlier works of Stieltjes and Markov could have helped him here, and from a remark in his paper, it seems that he may have been aware of their work. We should remember, however, that Feldheim was working under extremely difficult circumstances during the war. So, like Rogers himself, Feldheim and Lanzewizky did not give the relevant orthogonality relations and, in fact, they did not write the polynomials as q-extensions of the ultraspherical polynomials. Around 1980, Richard Askey and Mourad Ismail finally established the explicit orthogonality relation.

In his paper containing the Rogers–Ramanujan identities, Rogers also observed that the series

$$\chi(\lambda^2) = 1 + \frac{\lambda^2 q^2}{1-q} + \frac{\lambda^4 q^6}{(1-q)(1-q^2)} + \cdots \qquad (27.24)$$

---

[15] Turán (1990) vol. 3, p. 2626.

satisfied the relation

$$\chi(\lambda^2) - \chi(\lambda^2 q) = \lambda^2 q^2 \chi(\lambda^2 q^2). \qquad (27.25)$$

This can be easily verified and implies the continued fraction expansion

$$\frac{\chi(\lambda^2)}{\chi(\lambda^2 q)} = \frac{1}{1+} \frac{\lambda^2 q^2}{1+} \frac{\lambda^2 q^3}{1+} \frac{\lambda^2 q^4}{1+} \cdots. \qquad (27.26)$$

This is now known as the Rogers–Ramanujan continued fraction because Ramanujan rediscovered and went much farther with it. In his first letter to Hardy, of January 16, 1913, Ramanujan stated without proof that

$$\frac{1}{1+} \frac{e^{-2\pi}}{1+} \frac{e^{-4\pi}}{1+} \frac{e^{-6\pi}}{1+} \cdots = \left(\sqrt{\frac{5 + \sqrt{5}}{2}} - \frac{\sqrt{5} + 1}{2} \right) e^{\frac{2\pi}{5}}, \qquad (27.27)$$

$$\frac{1}{1-} \frac{e^{-\pi}}{1+} \frac{e^{-2\pi}}{1-} \frac{e^{-3\pi}}{1+} \cdots = \left(\sqrt{\frac{5 - \sqrt{5}}{2}} - \frac{\sqrt{5} - 1}{2} \right) e^{\frac{\pi}{5}}, \qquad (27.28)$$

and that

$$\frac{1}{1+} \frac{e^{-\pi \sqrt{n}}}{1+} \frac{e^{-2\pi \sqrt{n}}}{1+} \frac{e^{-3\pi \sqrt{n}}}{1+} \cdots$$

could be exactly determined if n were any positive rational quantity. In his next letter of February 27, 1913, Ramanujan wrote

$$\text{If } F(x) = \frac{1}{1+} \frac{x}{1+} \frac{x^2}{1+} \frac{x^3}{1+} \cdots, \text{ then}$$

$$\left\{ \frac{\sqrt{5}+1}{2} + e^{-\frac{2\alpha}{5}} F(e^{-2\alpha}) \right\} \left\{ \frac{\sqrt{5}+1}{2} + e^{-\frac{2\beta}{5}} F(e^{-2\beta}) \right\} = \frac{5 + \sqrt{5}}{2}$$

with the condition $\alpha\beta = \pi^2, \ldots$. This theorem is a particular case of a theorem on the continued fraction

$$\frac{1}{1+} \frac{ax}{1+} \frac{ax^2}{1+} \frac{ax^3}{1+} \cdots,$$

which is a particular case of the continued fraction

$$\frac{1}{1+} \frac{ax}{1 + bx+} \frac{ax^2}{1 + bx^2+} \frac{ax^3}{1 + bx^3+} \cdots,$$

which is a particular case of a general theorem on continued fractions.

Hardy was very impressed with Ramanujan's results on continued fractions. Concerning formulas (27.27) and (27.28) he wrote,[16] "I had never seen anything in the least

[16] Hardy (1978) p. 9.

like them before. A single look at them is enough to show that they could only be written down by a mathematician of the highest class. They must be true because, if they were not true, no one would have had the imagination to invent them."

27.2   Heine's Transformation

Heine proved his transformation for the q-hypergeometric series by a judicious application of the q-binomial theorem.[17] He proved the q-binomial theorem by the use of contiguous relations; as we mentioned in Exercise 4.13, Gauss may have employed this method to prove the binomial theorem. Heine required the contiguous relation

$$\phi(\alpha+1,\beta,\gamma,q,x) - \phi(\alpha,\beta,\gamma,q,x) = q^\alpha x \frac{1-q^\beta}{1-q^\gamma} \phi(\alpha+1,\beta,\gamma,q,x) \quad (27.29)$$

and the q-difference relation

$$\phi(\alpha,\beta,\gamma,q,x) - \phi(\alpha,\beta,\gamma,q,qx) = \frac{(1-q^\alpha)(1-q^\beta)}{1-q^\gamma} x\phi(\alpha+1,\beta+1,\gamma+1,q,x).$$

$$(27.30)$$

Note that the second relation is the analog of the derivative equation

$$\frac{d}{dx} F(a,b,c,x) = \frac{a \cdot b}{c} F(a+1,b+1,c+1,x).$$

Heine then supposed that $\beta = \gamma = 1$ so that he had the q-binomial series

$$\phi(\alpha,x) = \phi(\alpha,1,1,q,x) = 1 + \frac{1-q^\alpha}{1-q} x + \frac{(1-q^\alpha)(1-q^{\alpha+1})}{(1-q)(1-q^2)} x^2 + \cdots . \quad (27.31)$$

By (27.29),

$$\phi(\alpha+1,x) = \frac{1}{1-q^\alpha x} \phi(\alpha,x). \quad (27.32)$$

Combined with (27.30), this produced

$$\phi(\alpha,x) = \frac{1-q^\alpha x}{1-x} \phi(\alpha,qx) = \frac{(1-q^\alpha x)(1-q^{\alpha+1}x)\cdots(1-q^{\alpha+n}x)}{(1-x)(1-qx)\cdots(1-q^n x)} \phi(\alpha,q^{n+1}x).$$

$$(27.33)$$

The q-binomial theorem followed, since it was assumed that $|q| < 1$ and

$$\phi(\alpha,q^{n+1}x) \to \phi(\alpha,0) = 1 \quad \text{as} \quad n \to 0.$$

[17] Heine (1847).

Then, to obtain the transformation formula (27.32), Heine started with the series

$$S = 1 + \frac{(1-q^\alpha)(1-q^\beta x)}{(1-q)(1-q^\gamma x)}z + \frac{(1-q^\alpha)(1-q^{\alpha+1})(1-q^\beta x)(1-q^{\beta+1}x)}{(1-q)(1-q^2)(1-q^\gamma x)(1-q^{\gamma+1}x)}z^2 + \cdots ,$$

$$(27.34)$$

where he assumed $|q| < 1$, $|x| < 1$, and $|z| < 1$ for convergence. He multiplied both sides by $\phi(\gamma - \beta, q^\beta x)$ and used the product expression for this function, from the q-binomial theorem, to get

$$\phi(\gamma - \beta, q^\beta x)S = \phi(\gamma - \beta, q^\beta x) + \frac{1-q^\alpha}{1-q} z \cdot \phi(\gamma - \beta, q^{\beta+1}x)$$

$$+ \frac{(1-q^\alpha)(1-q^{\alpha+1})}{(1-q)(1-q^2)}z^2 \cdot \phi(\gamma - \beta, q^{\beta+2}x) + \cdots .$$

$$(27.35)$$

In the next step, Heine expanded each of the ϕ on the right as series. We here employ the abbreviated notation given in (27.14).

$$1 + \frac{(q^{\gamma-\beta};q)_1}{(q;q)_1} \cdot q^\beta x + \frac{(q^{\gamma-\beta};q)_2}{(q;q)_2} \cdot q^{2\beta}x^2 + \cdots$$

$$+ z\left(\frac{(q^\alpha;q)_1}{(q;q)_1} + \frac{(q^{\gamma-\beta};q)_1}{(q;q)_1} \cdot \frac{(q^\alpha;q)_1}{(q;q)_1}q^{\beta+1}x + \frac{(q^{\gamma-\beta};q)_2}{(q;q)_2} \cdot \frac{(q^\alpha;q)_1}{(q;q)_1}q^{2\beta+2}x^2 + \cdots\right)$$

$$+ z^2\left(\frac{(q^\alpha;q)_2}{(q^2;q)_2} + \frac{(q^{\gamma-\beta};q)_1}{(q;q)_1} \cdot \frac{(q^\alpha;q)_2}{(q;q)_2}q^{\beta+2}x + \cdots\right) + \cdots .$$

He changed the order of summation and used the q-binomial theorem to obtain

$$\phi(\alpha, z) + \frac{(q^{\gamma-\beta};q)_1}{(q;q)_1} \cdot q^\beta x \cdot \phi(\alpha, qz) + \frac{(q^{\gamma-\beta};q)_2}{(q;q)_2} \cdot q^{2\beta}x^2 \phi(\alpha, q^2z) + \cdots$$

$$= \phi(\alpha, z)\left(1 + \frac{(q^{\gamma-\beta};q)_1(z;q)_1}{(q;q)_1(q^\alpha z;q)_1} \cdot q^\beta x + \frac{(q^{\gamma-\beta};q)_2(z;q)_2}{(q;q)_2(q^\alpha z;q)_2} \cdot q^{2\beta}x^2 + \cdots\right).$$

Finally, Heine substituted this expression into the right-hand side of (27.35) and replaced $\phi(\alpha, z)$ and $\phi(\gamma - \beta, q^\beta x)$ by their product expressions, to arrive at the transformation (27.4). Heine found the q-extension of Gauss's summation of $F(\alpha, \beta, \gamma, 1)$ by taking $x = 1$, $z = q^{\gamma-\alpha-\beta}$ in his transformation:

$$\phi(\alpha, \beta, \gamma, q^{\gamma-\alpha-\beta})$$

$$= \frac{(q^\beta;q)_\infty(q^{\gamma-\beta};q)_\infty}{(q^\gamma;q)_\infty(q^{\gamma-\alpha-\beta};q)_\infty} \cdot \phi(\gamma - \alpha - \beta, 1, 1, q^\beta) \qquad (27.36)$$

$$= \frac{(q^{\gamma-\alpha};q)_\infty(q^{\gamma-\beta};q)_\infty}{(q^\gamma;q)_\infty(q^{\gamma-\alpha-\beta};q)_\infty}.$$

Note that this derivation is analogous to the method used to derive Gauss's formula from Euler's integral for the hypergeometric function. This suggests that Heine's transformation is the q-analog of Euler's integral formula; indeed, recall that Thomae proved this after defining the q-integral.

27.3   Rogers: Threefold Symmetry

Rogers applied his knowledge and experience of elliptic functions and invariant theory to develop the theory of q-Hermite and q-ultraspherical polynomials. From invariant theory, he brought a sense of symmetry and expertise in applying infinite series/products of operators. We may recall that in the first half of the nineteenth century, Arbogast, Franais, Murphy, D. Gregory, and Boole made extensive use of operational calculus. Cayley and Sylvester appropriated these methods for invariant theory. Then, in the 1880s, J. Hammond and P. MacMahon applied these techniques to combinatorial and invariant theoretic problems. It is interesting to observe Rogers's use of algebra, combinatorics and analysis as he conceived of and solved new problems in analysis. In his second paper of 1893, Rogers converted Heine's trans-formation into an equation with threefold symmetry.[18] He showed that the function

$$\psi(\lambda,\mu,\nu,q,\theta) = \phi(\mu e^{-i\theta}, \nu e^{-i\theta}, \mu\nu, q, \lambda e^{i\theta})(\lambda e^{i\theta})_\infty (\mu\nu)_\infty \qquad (27.37)$$

was symmetric in λ, μ, and ν, and also symmetric in θ and $-\theta$. He then set

$$\chi(\lambda,\mu,\nu,q,\theta) = \frac{\psi(\lambda,\mu,\nu,q,\theta)}{P(\lambda)\,P(\mu)\,P(\nu)}, \qquad (27.38)$$

where

$$P(\lambda) = \prod_{n=0}^{\infty}(1 - 2\lambda q^n \cos\theta + \lambda^2 q^{2n})^{-1} = \frac{1}{(\lambda e^{i\theta})_\infty (\lambda e^{-i\theta})_\infty}. \qquad (27.39)$$

Also in his 1847 paper, Heine discussed such products; for particular values of λ they are ubiquitous in elliptic function theory. Rogers defined a q-extension of the Hermite polynomials $A_r(\theta)$ as the coefficient of $\frac{\lambda^r}{(q)_r}$ in the series expansion of $P(\lambda)$:

$$P(\lambda) = \sum_{r=0}^{\infty} \frac{A_r(\theta)}{(q)_r} \lambda^r. \qquad (27.40)$$

From Euler's expansion, (25.20), Rogers had

$$\frac{1}{(\lambda e^{i\theta})_\infty (\lambda e^{-i\theta})_\infty} = \left(1 + \frac{\lambda e^{i\theta}}{1-q} + \frac{\lambda^2 e^{2i\theta}}{(q)_2} + \cdots\right)\left(1 + \frac{\lambda e^{-i\theta}}{1-q} + \frac{\lambda^2 e^{-2i\theta}}{(q)_2} + \cdots\right).$$

$$(27.41)$$

---

[18] Rogers (1893b).

Hence Rogers obtained his expression for the q-Hermite polynomial:

$$A_r(\theta) = \sum_{n=0}^{r} \frac{(q)_r e^{i\theta(r-2n)}}{(q)_n (q)_{r-n}} = \sum_{n=0}^{r} \frac{(q)_r \cos(r-2n)\theta}{(q)_n (q)_{r-n}}. \tag{27.42}$$

Note that the last step followed because $A_r(\theta)$ was an even function of θ. In his 1894 paper, Rogers noted the three-term recurrence relation for $A_r(\theta)$:

$$2\cos\theta\, A_{r-1}(\theta) = A_r(\theta) + (1 - q^{r-1})A_{r-2}(\theta), \tag{27.43}$$

obtained as a consequence of the relation

$$P(\lambda q) = (1 - 2\lambda\cos\theta + \lambda^2)\, P(\lambda), \tag{27.44}$$

or

$$\sum_{r=0}^{\infty} \frac{A_r(\theta)\lambda^r q^r}{(q)_r} = (1 - 2\lambda\cos\theta + \lambda^2) \sum_{r=0}^{\infty} \frac{A_r(\theta)\lambda^r}{(q)_r}. \tag{27.45}$$

In his 1893 paper, Rogers raised the problem of expanding $\chi(\lambda, \mu, \nu, q, \theta)$ in (27.38) as a series in q-Hermite polynomials. The series would take the form $A_0 + A_1 H_1 + A_2 H_2 + \cdots$ and the problem was to determine H_r. He showed that H_r, a homogeneous symmetric function of degree r in λ, μ, and ν, was the coefficient of k^r in the series expansion of

$$\frac{1}{(k\lambda)_\infty (k\mu)_\infty (k\nu)_\infty}. \tag{27.46}$$

Rogers gave an interesting proof of this expansion and we sketch it very briefly. He observed that for the function χ defined by (27.38),

$$\delta_\lambda \chi = \delta_\mu \chi = \delta_\nu \chi, \tag{27.47}$$

where δ was the difference operator defined by

$$\delta_q f(x) = \frac{f(x) - f(qx)}{x}. \tag{27.48}$$

From (27.47), he was able to deduce that

$$\delta_\lambda H_r(\lambda, \mu, \nu) = \delta_\mu H_r(\lambda, \mu, \nu) = \delta_\mu H_r(\lambda, \mu, \nu). \tag{27.49}$$

He denoted the coefficient of $\lambda^\alpha \mu^\beta \nu^\gamma$ in H_r by $a_{\alpha,\beta,\gamma}$. Note that $\alpha + \beta + \gamma = r$. Rogers next showed that (27.49) implied the recurrence relations

$$(1 - q^{\alpha+1})a_{\alpha+1,\beta,\gamma} = (1 - q^{\beta+1})a_{\alpha,\beta+1,\gamma} = (1 - q^{\gamma+1})a_{\alpha,\beta,\gamma+1}. \tag{27.50}$$

Combined with the initial condition $a_{r,0,0} = \frac{1}{(q)_r}$, the relations (27.50) uniquely defined the coefficients. At this point, Rogers remarked that because of uniqueness it was sufficient to produce a set of coefficients satisfying these conditions. He then quickly demonstrated that the coefficient of k^r in the series expansion of (27.46) was a homogeneous symmetric function of degree r in λ, μ, and ν whose coefficients satisfied the same initial condition and recurrence relations (27.50). This proved his result, though he gave no indication of how arrived at (27.46).

Rogers noted some interesting and important particular cases of his theorem. When $\lambda = 0$, he had

$$\frac{(\mu\nu)_\infty}{P(\mu)P(\nu)} = 1 + A_1(\theta)H_1(\mu,\nu) + A_2(\theta)H_2(\mu,\nu) + \cdots, \tag{27.51}$$

where $H_r(\mu,\nu)$ was the coefficient of k^r in

$$\frac{1}{(k\mu)_\infty (k\nu)_\infty} = \left(1 + \frac{k\mu}{1-q} + \frac{k^2\mu^2}{(q)_2} + \cdots\right)\left(1 + \frac{k\nu}{1-q} + \frac{k^2\nu^2}{(q)_2} + \cdots\right). \tag{27.52}$$

So

$$H_r(\mu,\nu) = \mu^r + \frac{(q)_r}{(q)_1(q)_{r-1}}\mu^{r-1}\nu + \frac{(q)_r}{(q)_2(q)_{r-2}}\mu^{r-2}\nu^2 + \cdots + \nu^r. \tag{27.53}$$

We note that when $\mu = x$ and $\nu = 1$,

$$H_r(x) = x^r + \frac{(q)_r}{(q)_1(q)_{r-1}}x^{r-1} + \frac{(q)_r}{(q)_2(q)_{r-2}}x^{r-2} + \cdots + 1. \tag{27.54}$$

The polynomials $H_r(x)$ (27.54) are now called Rogers–Szegő polynomials, because Szegő proved the orthogonality of $H_r(-\frac{x}{\sqrt{q}})$ with respect to a suitable measure on the unit circle. When $\mu = xe^{i\theta}$ and $\nu = xe^{-i\theta}$ in (27.51), Rogers got

$$\prod_{n=0}^{\infty}\frac{(x^2)_\infty}{(1 - 2xq^n\cos(\theta+\phi) + x^2q^{2n})(1 - 2xq^n\cos(\theta-\phi) + x^2q^{2n})}$$
$$= \sum_{n=0}^{\infty}\frac{A_n(\theta)A_n(\phi)}{(q)_n}x^n. \tag{27.55}$$

The result (27.55) is now known as the q-Mehler formula. Rogers also applied (27.51) to prove the useful linearization formula for q-Hermite polynomials:

$$\frac{A_m(\theta)A_n(\theta)}{(q)_m(q)_n} = \sum_{k=0}^{min(m,n)}\frac{A_{m+n-2k}(\theta)}{(q)_k(q)_{m-k}(q)_{n-k}}. \tag{27.56}$$

Rogers also noted that, by definition, $A_r(\frac{\pi}{2})$ was the coefficient of $\frac{k^r}{(q)_r}$ in $\prod(1 + k^2 q^{2n})^{-1}$, and hence

$$A_r\left(\frac{\pi}{2}\right) = (-1)^r(1-q)(1-q^3)\cdots(1-q^{r-1}) \qquad r \text{ even},$$

$$= 0 \qquad\qquad\qquad\qquad\qquad\qquad r \text{ odd}.$$

Recall that Gauss evaluated Gauss sums from this result; see the formula (25.36). Thus, though Rogers may not have known it, this result is due to Gauss.

27.4 Rogers: Rogers–Ramanujan Identities

In order to derive the Rogers–Ramanujan identities, Rogers expanded $P(\lambda)$ as a Fourier series and as a series in the q-Hermite polynomials. He then found a relation between the coefficients in these two series, yielding the famous identities. In his 1894 paper,[19] Rogers raised and solved the problem: Suppose a function $f(\theta)$ is expanded as a Fourier cosine series and as a series in q-Hermite polynomials $A_n(\theta)$:

$$f(\theta) = a_0 + a_1 A_1(\theta) + a_2 A_2(\theta) + \cdots = b_0 + 2b_1 \cos\theta + 2b_2 \cos 2\theta + \cdots .$$
$$(27.57)$$

Express the coefficient a_n in terms of a series of b_j and, conversely, b_n in terms of a series of a_j. Rogers found that

$$b_n = a_n + \frac{1-q^{n+2}}{1-q}a_{n+2} + \frac{(1-q^{n+4})(1-q^{n+3})}{(1-q)(1-q^2)}a_{n+4}$$
$$+ \frac{(1-q^{n+6})(1-q^{n+5})(1-q^{n+4})}{(1-q)(1-q^2)(1-q^3)}a_{n+6} + \cdots .$$
$$(27.58)$$

For the converse, he gave a_0, a_1 in terms of series of b, but for the general case he merely described the method by which the a_n could be obtained for higher values of n. He had

$$a_0 = b_0 - (1+q)b_2 + q(1+q^2)b_4 - \cdots + (-1)^r q^{\frac{r(r-1)}{2}}(1+q^r)b_{2r} + \cdots ,$$
$$(27.59)$$

$$(1-q)a_1 = (1-q)b_1 - (1-q^3)b_3$$
$$+ q(1-q^5)b_5 - \cdots + (-1)^r q^{\frac{r(r-1)}{2}}(1-q^{2r+1})b_{2r+1} + \cdots . \quad (27.60)$$

The derivation of (27.58) was simple. Rogers substituted the expression (27.42) for $A_n(\theta)$ in terms of $\cos k\theta$ on the left side of (27.57) and then equated the coefficients

[19] Rogers (1894).

of $\cos n\theta$ on both sides. He noted the formula only for even n, but the method also yields the case for odd n. Rogers's method for finding (27.59) and (27.60) was quite elaborate and he did not give the general formula. In his third paper of 1893, on the expansion of infinite products, Rogers supposed

$$f(\theta) = C_0 + C_1 A_1(\theta) + C_2 A_2(\theta) + \cdots$$

to be given. He then asked how to find $K_0, K_1, K_2, \ldots$ in the expansion

$$\frac{f(\theta)}{\prod_{n=0}^{\infty}(1 - 2\lambda q^n \cos\theta + \lambda^2 q^{2n})} = K_0 + K_1 A_1(\theta) + K_2 A_2(\theta) + \cdots. \qquad (27.61)$$

Rogers expressed the result symbolically, in terms of the difference operator

$$\delta_\lambda \phi(\lambda) = \frac{\phi(\lambda) - \phi(\lambda q)}{\lambda} :$$

$$K_0 + K_1 x + K_2 x^2 + \cdots = \frac{1}{(x\lambda)_\infty (x\delta_\lambda)_\infty}(C_0 + C_1\lambda + C_2\lambda^2 + \cdots). \qquad (27.62)$$

Note that he used an infinite product in the operator δ_λ.

Also in his 1894 paper, Rogers applied (27.62) to find the q-Hermite expansion of

$$P(\lambda) = \prod_{n=0}^{\infty}(1 - 2\lambda q^n \cos\theta + \lambda^2 q^{2n}). \qquad (27.63)$$

He noted that for an analytic function ϕ

$$\frac{1}{(x\delta_\lambda)_\infty}\phi(\lambda) = \frac{1}{(\lambda\delta_x)_\infty}\phi(x). \qquad (27.64)$$

He verified this by taking the special value $\phi(\lambda) = \lambda^m$. Since

$$\delta_\lambda \lambda^m = (1 - q^m)\lambda^{m-1}, \qquad (27.65)$$

$$\frac{1}{(x\delta_\lambda)_\infty}\lambda^m = \sum_{n=0}^{\infty} \frac{x^n \delta_\lambda^n}{(q)_n}\lambda^m = \sum_{n=0}^{m} \frac{x^n \lambda^{m-n}(1 - q^m) \cdots (1 - q^{m-n+1})}{(q)_n}$$

$$= \sum_{n=0}^{m} \frac{\lambda^n x^{m-n}(1 - q^m) \cdots (1 - q^{n+1})}{(q)_{m-n}}$$

$$= \frac{1}{(\lambda\delta_x)_\infty}x^m.$$

Therefore, (27.62) could be rewritten as

$$K_0 + K_1 x + K_2 x^2 + \cdots = \frac{1}{(\lambda x)_\infty} \cdot \frac{1}{(\lambda \delta x)_\infty}(C_0 + C_1 x + C_2 x^2 + \cdots),$$

or $C_0 + C_1 x + C_2 x^2 + \cdots = (\lambda \delta x)_\infty (\lambda x)_\infty (K_0 + K_1 x + \cdots).$ (27.66)

Rogers then argued that if $f(\theta) = P(\lambda) = C_0 + C_1 A_1(\theta) + \cdots$, then

$$K_0 + K_1 A_1(\theta) + \cdots = 1;$$

substituting in (27.66) gave him

$$C_0 + C_1 x + C_2 x^2 + \cdots = (\lambda \delta x)_\infty (\lambda x)_\infty$$

$$= \left(1 - \frac{\lambda \delta x}{1-q} + \frac{q\lambda^2 \delta_x^2}{(1-q)(1-q^2)} - \cdots\right)\left(1 - \frac{\lambda x}{1-q} + \frac{q\lambda^2 x^2}{(1-q)(1-q^2)} - \cdots\right).$$
(27.67)

The last step made use of Euler's special case of the q-binomial theorem, (25.19). Rogers replaced λ by x (27.65); he then applied the first expression in parentheses to the second expression in parentheses, in the right-hand side of equation (27.67), to conclude that the coefficient C_r of x^r was

$$\frac{(-1)^r q^{\frac{r(r-1)}{2}} \lambda^r}{(q)_r} \sum_{s=0}^\infty \frac{\lambda^{2s} q^{rs+s(s-1)}}{(q)_s}.$$

This gave the q-Hermite expansion of $P(\lambda)$. Changing λ to $-\lambda q$ yielded

$$(1 + 2\lambda q \cos\theta + \lambda^2 q^2)(1 + 2\lambda q^2 \cos\theta + \lambda^2 q^4)\cdots$$

$$= \chi(\lambda^2) + \frac{q\lambda}{1-q}\chi(\lambda^2 q)A_1(\theta) + \frac{q^3\lambda^2}{(q)_2}\chi(\lambda^2 q^2)A_2(\theta) + \frac{q^6\lambda^3}{(q)_3}\chi(\lambda^2 q^3)A_3(\theta) + \cdots,$$
(27.68)

where

$$\chi(\lambda^2) = 1 + \frac{\lambda^2 q^2}{1-q} + \frac{\lambda^4 q^6}{(1-q)(1-q^2)} + \frac{\lambda^6 q^{12}}{(1-q)(1-q^2)(1-q^3)} + \cdots.$$
(27.69)

From the triple product identity, Rogers knew the Fourier cosine expansion of the product in (27.68) when $\lambda = \frac{1}{\sqrt{q}}$:

$$(1 + 2q^{\frac{1}{2}}\cos\theta + q)(1 + 2q^{\frac{3}{2}}\cos\theta + q^3)\cdots$$

$$= \frac{1}{(q)_\infty}(1 + 2q^{\frac{1}{2}}\cos\theta + 2q^2 \cos 2\theta + 2q^{\frac{9}{2}}\cos 3\theta + \cdots).$$

He therefore set $\lambda = \frac{1}{\sqrt{q}}$ in (27.68) and applied (27.59) and (27.60). Thus, the Rogers–Ramanujan identities were discovered. From (27.59) he obtained

$$
\begin{aligned}
a_0 = \chi\left(\frac{1}{q}\right) &= 1 + \frac{q}{1-q} + \frac{q^4}{(1-q)(1-q^2)} + \frac{q^9}{(1-q)(1-q^2)(1-q^3)} + \cdots \\
&= \frac{1}{(q)_\infty}\left(1 - (1+q)q^2 + (1+q^2)q^9 - (1+q^3)q^{21} + \cdots\right) \\
&= \frac{1}{(q)_\infty}\left(1 + \sum_{m=1}^{\infty}(-1)^m q^{\frac{m(5m\pm1)}{2}}\right).
\end{aligned}
\tag{27.70}
$$

The sum could be evaluated by the triple product identity

$$
(x)_\infty \left(\frac{q}{x}\right)_\infty (q)_\infty = \sum_{n=-\infty}^{\infty}(-1)^n q^{\frac{n(n-1)}{2}} x^n.
$$

Replacing q by q^5 and setting $x = q^2$ yielded

$$
(q^2;q^5)_\infty (q^3;q^5)_\infty (q^5;q^5)_\infty = 1 + \sum_{n=1}^{\infty}(-1)^n q^{\frac{n(5n\pm1)}{2}}.
$$

So the right-hand side of the equation (27.70) was reduced to

$$
\frac{(1-q^2)(1-q^7)\cdots(1-q^3)(1-q^8)\cdots(1-q^5)(1-q^{10})\cdots}{(1-q)(1-q^2)(1-q^3)\cdots}
$$

$$
= \frac{1}{(1-q)(1-q^4)(1-q^6)(1-q^9)\cdots} = \prod_{n=0}^{\infty}\frac{1}{(1-q^{5n+1})(1-q^{5n+4})}.
$$

This completed the proof of the first identity. Similarly,

$$
\begin{aligned}
(1-q)a_1 &= q^{\frac{1}{2}}\left(1 + \frac{q^2}{1-q} + \frac{q^6}{(1-q)(1-q^2)} + \frac{q^{12}}{(1-q)(1-q^2)(1-q^3)} + \cdots\right) \\
&= \frac{1}{(q)_\infty}\left((1-q)q^{\frac{1}{2}} - (1-q^3)q^{\frac{9}{2}} + (q-q^6)q^{\frac{25}{2}} - \cdots\right).
\end{aligned}
$$

In this way, he obtained the second identity

$$
1 + \frac{q^2}{1-q} + \frac{q^6}{(1-q)(1-q^2)} + \cdots
$$

$$
= \frac{1}{(q)_\infty}(1 - q - q^4 + q^7 + q^{13} - q^{18} - \cdots)
\tag{27.71}
$$

$$
= \prod_{n=0}^{\infty}\frac{1}{(1-q^{5n+2})(1-q^{5n+3})}.
$$

To prove (27.62), Rogers observed that since

$$\frac{1}{\prod_{n=0}^{\infty}(1 - 2\lambda q^n \cos\theta + \lambda^2 q^{2n})} = 1 + \frac{\lambda}{1-q}A_1(\theta) + \frac{\lambda^2}{(1-q)(1-q^2)}A_2(\theta) + \cdots,$$

the linearization formula (27.56) for $A_n(\theta)$ implied that the left-hand side of (27.62) was a product of series in $A_n(\theta)$ and therefore was itself such a series. He also observed that the linearization of $A_m(\theta) A_n(\theta)$ contained a term independent of θ only when $m = n$, and in that case this term would be $(q)_n$. Hence, he had

$$K_0 = C_0 + C_1\lambda + C_2\lambda^2 + \cdots. \tag{27.72}$$

From the difference relation

$$\delta_\lambda\left(\frac{1}{P(\lambda)}\right) = \frac{1}{\lambda}\left(\frac{1}{P(\lambda)} - \frac{1}{P(\lambda q)}\right)$$

$$= \frac{1 - (1 - 2\lambda\cos\theta + \lambda^2)}{\lambda P(\lambda)} = \frac{2\cos\theta - \lambda}{P(\lambda)},$$

Rogers obtained

$$\delta_\lambda(K_0 + K_1 A_1(\theta) + K_2 A_2(\theta) + \cdots) = (C_0 + C_1 A_1 + \cdots)\frac{2\cos\theta - \lambda}{P(\lambda)}$$

$$= (2\cos\theta - \lambda)(K_0 + K_1 A_1(\theta) + \cdots).$$

He equated the coefficients of A_r on both sides and applied the recurrence relation (27.43)

$$A_{r+1} + (1 - q^r)A_{r-1} = 2\cos\theta\, A_r$$

to conclude

$$(1 - q^{r+1})K_{r+1} = \lambda K_r - K_{r-1} + \delta_\lambda K_r. \tag{27.73}$$

He showed inductively that (27.73) implied

$$K_r = H_r(\lambda, \delta_\lambda)\, K_0, \tag{27.74}$$

where H_r was defined as in (27.53). For $r = 0$, (27.73) gave him

$$(1 - q)K_1 = (\lambda + \delta_\lambda)K_0 = H_1(\lambda, \delta_\lambda)K_0.$$

He assumed the result true up to r so that (27.73) could be written as

$$(1 - q^{r+1})K_{r+1} = (\lambda + \delta_\lambda) H_r(\lambda, \delta) K_0 - H_{r-1}(\lambda, \delta) K_0. \tag{27.75}$$

He noted that when $m + n = r$, the coefficient of $\lambda^m \delta_\lambda^n$ in $H_r(\lambda, \delta_\lambda)$ was $\frac{1}{(q)_m (q)_{r-m}}$. He defined the operator η by $\eta f(\lambda) = f(q\lambda)$, so that

$$
\begin{aligned}
\delta_\lambda \lambda^m \delta_\lambda^n f(\lambda) &= \lambda^{m-1} \delta_\lambda^n f(\lambda) - \lambda^{m-1} q^m \eta \delta_\lambda^n f(\lambda) \\
&= \lambda^{m-1} \delta_\lambda^n f(\lambda) - q^m (\lambda^{m-1} \delta_\lambda^n f(\lambda) - \lambda^m \delta_\lambda^{n+1} f(\lambda)) \\
&= (1 - q^m) \lambda^{m-1} \delta^n f(\lambda) + q^m \lambda^m \delta_\lambda^{n+1} f(\lambda). \qquad (27.76)
\end{aligned}
$$

The first term in the expression in the third line of (27.76) cancelled with the term containing $\lambda^{m-1} \delta_\lambda^n$ in

$$
H_{r-1}(\lambda, \delta_\lambda).
$$

So (27.74) implied that

$$
\begin{aligned}
(1 - q^{r+1}) K_{r+1} &= \sum \left(\frac{q^m}{(q)_m (q)_{r-m}} + \frac{1}{(q)_{m-1} (q)_{r-m+1}} \right) \lambda^m \delta_\lambda^{n+1} K_0 \\
&= (1 - q^{r+1}) \sum \frac{\lambda^m \delta_\lambda^{r+1-m}}{(q)_m (q)_{r+1-m}} K_0 \\
&= (1 - q^{r+1}) H_{r+1}(\lambda, \delta_\lambda) K_0.
\end{aligned}
$$

Thus (27.74) was proved. Moreover, by (27.72), Rogers had

$$
\frac{C_0 + C_1 A_1 + C_2 A_2 + \cdots}{P(\lambda)} = (1 + H_1(\lambda, \delta_\lambda) A_1 + H_2(\lambda, \delta_\lambda) A_2 + \cdots)
$$
$$
\times (C_0 + C_1 \lambda + C_2 \lambda^2 + \cdots).
$$

This meant, remarked Rogers, that $K_0 + K_1 x + K_2 x^2 + \cdots$ was equal to

$$
\begin{aligned}
(1 + x H_1(\lambda, \delta_\lambda) &+ x^2 H_2(\lambda, \delta_\lambda) + \cdots)(C_0 + C_1 \lambda + C_2 \lambda^2 + \cdots) \\
&= \frac{1}{(x\lambda)_\infty (x\delta_\lambda)_\infty} (C_0 + C_1 \lambda + C_2 \lambda^2 + \cdots).
\end{aligned} \qquad (27.77)
$$

This completed the proof of (27.62).

27.5   Rogers: "Third Memoir"

Rogers defined q-ultraspherical polynomials and derived some of their properties in his 1895 "Third memoir on the expansion of certain infinite products."[20] His definition used their generating function

$$
\frac{P(\lambda x)}{P(x)} = \sum_{n=0}^{\infty} \frac{L_n(\theta)}{(q)_n} x^n. \qquad (27.78)
$$

[20] Rogers (1895).

He obtained the recurrence relation

$$L_r - 2\cos\theta \cdot L_{r-1}(1 - \lambda q^{r-1}) + L_{r-2}(1 - q^{r-1})(1 - \lambda^2 q^{r-2}) = 0$$

from the equation

$$\frac{P(\lambda x)}{P(x)}(1 - 2x\cos\theta + x^2) = \frac{P(\lambda q x)}{P(q x)}(1 - 2\lambda x\cos\theta + \lambda^2 x^2). \tag{27.79}$$

He observed that by the q-binomial theorem,

$$\frac{(\lambda x e^{i\theta})_\infty}{(x^{i\theta})_\infty} = \sum_{n=0}^{\infty} \frac{(\lambda)_n}{(q)_n} x^n e^{in\theta};$$

an expression for $L_n(\theta)$ could be obtained by multiplying this series with the series for

$$\frac{(\lambda x e^{-i\theta})_\infty}{(x e^{-i\theta})_\infty}.$$

He also noted the following particular cases: when

$$\lambda = 0, \; L_r = A_r;$$

$$\lambda = q, \; L_r = (q)_r \frac{\sin(r+1)\theta}{\sin\theta};$$

$$\lambda \to 1, \; \frac{1 - \lambda q^r}{1 - \lambda} L_r \to (q)_r \cdot 2\cos r\theta.$$

For yet another noteworthy result, Rogers supposed M_r to be the same function of μ as L_r of λ:

$$\frac{P(\mu x)}{P(x)} = 1 + \sum \frac{M_r x^r}{(q)_r}.$$

Then

$$\frac{L_r}{(q)_r} = \sum_{0 \le s \le \frac{r}{2}} \frac{M_{r-2s}(1 - q^{r-2s}\mu)}{(q)_{r-2s}} \frac{(\mu - \lambda)(\mu - q\lambda)\cdots(\mu - q^{s-1}\lambda)(\lambda)_{r-s}}{(q)_s(\mu)_{r-s+1}}. \tag{27.80}$$

Secondly, Rogers gave a formula now known as the linearization formula: For $s \le r$,

$$\frac{L_r L_s}{(q)_r (q)_s} = \sum_{t=0}^{s} \frac{(1 - \lambda q^{r+s-2t}) L_{r+s-2t}(\lambda)_{s-t}(\lambda)_t(\lambda)_{r-t}(\lambda^2)_{r+s-t}}{(\lambda^2)_{r+s-2t}(q)_{s-t}(q)_t(q)_{r-t}(\lambda)_{r+s-t+1}}. \tag{27.81}$$

Note that the modern definition of the q-ultraspherical polynomial $C_n(\cos\theta;\lambda|q)$ is slightly different:

$$\frac{P(x)}{P(\lambda x)} = \sum_{n=0}^{\infty} C_n(\cos\theta;\lambda|q)x^n.$$
(27.82)

27.6 Rogers–Szegő Polynomials

Rogers found q-extensions for two systems of orthogonal polynomials, but he did not prove their orthogonality; Szegő was the first to take a significant step in that direction. He considered polynomials orthogonal on the unit circle. In a 1921 paper he showed how to associate polynomials orthogonal on $-1 \leq x \leq 1$ of weight function $w(x)$ with polynomials orthogonal on the unit circle of weight function $f(\theta) = w(\cos\theta)|\sin\theta|$. However, Szegő had been able to find only a few simple examples; in 1926, Rogers's work motivated him to discover that the polynomials[21]

$$\phi_n(z) = a \sum_{k=0}^{n} \frac{(q)_n}{(q)_k(q)_{n-k}} (-q^{-\frac{1}{2}}z)^k,$$

where

$$a = \frac{(-1)^n q^{\frac{n}{2}}}{\sqrt{(q)_n}},$$

were orthogonal on the unit circle with respect to the weight function

$$f(\theta) = |D(e^{i\theta})|^2 \text{ and } D(z) = \sqrt{(q)_\infty} (-q^{\frac{1}{2}}z)_\infty.$$

Szegő proved this in his paper "Ein Beitrag zur Theorie der Thetafunktionen," by first observing that by the triple product identity

$$f(\theta) = \sum_{n=-\infty}^{\infty} q^{\frac{n^2}{2}} e^{in\theta};$$

hence

$$\frac{1}{2\pi} \int_0^{2\pi} f(\theta) e^{-in\theta} d\theta = q^{\frac{n^2}{2}}, \quad n = 0, \pm 1, \pm 2, \dots.$$
(27.83)

He then took $\phi_n(z) = \sum_{k=0}^{n} a_k z^k$ to determine a_k, by requiring the relations

$$\frac{1}{2\pi} \int_0^{2\pi} f(\theta) \phi_n(z) \overline{z^k} d\theta = 0, \quad z = e^{i\theta}, \quad k = 0, 1, \dots, n-1.$$
(27.84)

---

[21] Szegő (1926) or pp. 795–805 of Szegő (1982).

By (27.83), equation (27.84) gave $\sum_{s=0}^{n} a_s q^{\frac{(s-k)^2}{2}} = 0$ or

$$\sum_{s=0}^{n} a_s q^{-sk+\frac{s^2}{2}} = 0, \quad k = 0, 1, \ldots, n-1. \tag{27.85}$$

To solve this system of equations, Szegő recalled the Rothe–Gauss formula (25.13):

$$(1+qx)(1+q^2x)\cdots(1+q^nx) = \sum_{s=0}^{n} \frac{(q)_n}{(q)_s(q)_{n-s}} q^{\frac{s(s+1)}{2}} x^s.$$

He took $x = -q^{-k-1}$ to get

$$\sum_{s=0}^{n} \frac{(q)_n}{(q)_s(q)_{n-s}} (-1)^s q^{-s(k+\frac{1}{2})+\frac{s^2}{2}} = 0, \quad k = 0, 1, \ldots, n-1. \tag{27.86}$$

By comparing (27.85) and (27.86), he concluded that

$$a_s = a(-1)^s \frac{(q)_n}{(q)_s(q)_{n-s}} q^{-\frac{s}{2}}, \quad s = 0, 1, \ldots, n.$$

The factor a was then chosen so that

$$\frac{1}{2\pi} \int_0^{2\pi} f(\theta) |\phi_n(z)|^2 d\theta = 1.$$

We note that the triple product identity can be applied to also obtain the orthogonality of the q-Hermite polynomials $A_n(\theta)$. The orthogonality relation here would be given by

$$\frac{1}{2\pi} \int_0^\pi A_m(\theta) A_n(\theta) |(e^{2i\theta})_\infty|^2 d\theta = \frac{\delta_{mn}}{(q^{n+1})_\infty}. \tag{27.87}$$

27.7 Feldheim and Lanzewizky: Orthogonality of q-Ultraspherical Polynomials

The work of Feldheim[22] and Lanzewizky[23] arose out of the papers of Fejér and Szegő on some questions relating to generalized Legendre polynomials. To define Fejér's generalized Legendre polynomial, let

$$f(z) = a_0 + a_1 z + a_2 z^2 + \cdots \tag{27.88}$$

[22] Feldheim (1941).
[23] Lanzewizky (1941).

be analytic in a neighborhood of zero with real coefficients. Then

$$|f(re^{i\theta})|^2 = \sum_{n=0}^{\infty} a_n r^n e^{in\theta} \sum_{n=0}^{\infty} a_n r^n e^{-in\theta}$$

$$= \sum_{n=0}^{\infty} r^n \sum_{k=0}^{n} a_{n-k} a_k e^{i(n-2k)\theta}$$

$$= \sum_{n=0}^{\infty} r^n \sum_{k=0}^{n} a_{n-k} a_k \cos(n-2k)\theta.$$

The last step is valid because the left-hand side is real and the coefficients are real. The polynomials $p_n(x)$ defined by

$$p_n(\cos\theta) = \sum_{k=0}^{n} a_k a_{n-k} \cos(n-2k)\theta, \tag{27.89}$$

where $x = \cos\theta$, are the Fejér-Legendre polynomials; they have properties similar to those of Legendre polynomials. For example, Fejér and Szegő proved that under certain conditions on the coefficients, $p_n(x)$ had n zeros in the interval $(-1,1)$ and that the zeros of $p_n(x)$ and $p_{n+1}(x)$ separated each other. Feldheim and Lanzewizky showed that these polynomials were orthogonal when $f(z)$ in (27.88) was the q-binomial series. Their result was stated in a different form, and they did not give the orthogonality relation. Feldheim used the theorem: For a sequence of polynomials $P_n(x)$ ($n = 0, 1, 2, \dots$) to be orthogonal, it is necessary and sufficient that the recurrence relation

$$2b_n x P_n(x) = P_{n+1}(x) + \lambda_n P_{n-1}(x) \quad (n \geq 1, \lambda_n > 0) \tag{27.90}$$

hold true. Feldheim substituted (27.89) in (27.90) and applied the trigonometric identity $2x T_n(x) = T_{n+1}(x) + T_{n-1}(x)$ where $T_k(x) = \cos k\theta$, $x = \cos\theta$. He then wrote (27.90) as

$$b_n \sum_{k=0}^{n} a_k a_{n-k} \left(T_{n-2k+1}(x) + T_{n-2k-1}(x) \right)$$

$$= \sum_{k=0}^{n+1} a_k a_{n-k+1} T_{n-2k+1}(x) + \lambda_n \sum_{k=0}^{n-1} a_k a_{n-k-1} T_{n-2k-1}(x),$$

or, with $a_{-1} = 0$,

$$b_n \sum_{k=0}^{n} (a_k a_{n-k} + a_{k-1} a_{n-k+1}) T_{n-2k+1}(x)$$

$$= \sum_{k=0}^{n+1} (a_k a_{n-k+1} + \lambda_n a_{k-1} a_{n-k}) T_{n-2k+1}(x).$$

By equating coefficients,

$$b_n(a_k a_{n-k} + a_{k-1} a_{n-k+1}) = a_k a_{n-k+1} + \lambda_n a_{k-1} a_{n-k}.$$

Dividing by $a_{k-1} a_{n-k}$ and setting $b_n = \frac{a_{n+1}}{a_n}$, Feldheim obtained

$$\lambda_n = b_n(b_{k-1} + b_{n-k}) - b_{k-1} b_{n-k}, \quad k = 1, 2, \ldots, n. \tag{27.91}$$

He considered the three equations, obtained when $k = n, n-1$, and $n-2$:

$$\lambda_n = b_n(b_{n-1} + b_0) - b_{n-1} b_0 = b_n(b_{n-2} + b_1) - b_{n-2} b_1 = b_n(b_{n-3} + b_2) - b_{n-3} b_2.$$

He initially set $b_0 = 0$, since b_0 could be arbitrarily chosen, and solved for b_n to obtain

$$b_n = \frac{b_1 b_{n-2}}{b_1 + b_{n-2} - b_{n-1}} = \frac{b_2 b_{n-3}}{b_2 + b_{n-3} - b_{n-1}}. \tag{27.92}$$

With n replaced by $n-1$, (27.92) gave

$$b_{n-1} = \frac{b_1 b_{n-3}}{b_1 + b_{n-3} - b_{n-2}}.$$

Solving for b_{n-3} produced

$$b_{n-3} = \frac{b_{n-1}(b_{n-2} - b_1)}{b_{n-1} - b_1}$$

and therefore

$$b_{n-3} - b_{n-1} = \frac{b_{n-1}(b_{n-2} - b_{n-1})}{b_{n-1} - b_1}$$

and then

$$b_n = \frac{b_2 b_{n-1}(b_{n-2} - b_1)}{b_2(b_{n-1} - b_1) + b_{n-1}(b_{n-2} - b_{n-1})} = \frac{b_2 b_{n-1}(b_{n-2} - b_1)}{b_{n-1}(b_2 + b_{n-2} - b_{n-1}) - b_1 b_2}$$
$$= \frac{b_1 b_{n-2}}{b_1 + b_{n-2} - b_{n-1}}.$$

By simplifying,

$$(b_2 - b_1)(b_{n-2} - b_{n-1}) b_{n-1} b_{n-2} = b_1 b_2(b_{n-2} - b_{n-1})(b_{n-1} - b_1), \quad n = 3, 4, 5, \ldots.$$

From these relations, Feldheim expressed the value of b_n in the simpler form

$$b_n = \frac{b_1^2 b_2}{b_1 b_2 - (b_2 - b_1) b_{n-1}}, \quad n = 1, 2, 3, \ldots. \tag{27.93}$$

Considering the general case where b_0 was not necessarily zero, he noted that since b_0 and b_1 could be arbitrarily chosen, all the b_k in (27.93) could then be replaced by $b_k - b_0$. Feldheim then rewrote this relation as

$$\beta_n = \frac{c}{c - \beta_{n-1}} = \frac{1}{1 - \frac{\beta_{b-1}}{c}}, \quad \beta_0 = 1, \ \beta_1 = 1, \ n = 1, 2, 3, \ldots \qquad (27.94)$$

where

$$\beta_n = \frac{b_n - b_0}{b_1 - b_0} \quad \text{and} \quad c = \frac{b_2 - b_0}{b_2 - b_1}.$$

To solve the Riccati difference equation (27.94), Feldheim expanded $\delta_n = 1 - \frac{\beta_n}{c}$ as the continued fraction

$$\delta_n = \frac{R_n}{S_n} = 1 - \frac{\frac{1}{c}}{1-} \frac{\frac{1}{c}}{1-} \cdots \frac{\frac{1}{c}}{1-},$$

so that R_n and S_n satisfied the recurrence relation

$$t_n = t_{n-1} - \frac{1}{c} t_{n-2} \qquad (27.95)$$

with initial condition $R_0 = 1$, $S_0 = 1$; $R_1 = 1 - \frac{1}{c}$, $S_1 = 1$. The linear equation, (27.95) was solved by the quadratic

$$x^2 - x + \frac{1}{c} = 0 \quad \text{to obtain} \quad x = \frac{\sqrt{c} \pm \sqrt{c-4}}{2\sqrt{c}}.$$

For real solutions, it was required that $c \geq 4$ so that Feldheim could set $c = 4 \cosh^2 \xi$. Then

$$R_n = \frac{(1 + \tanh \xi)^{n+2} - (1 - \tanh \xi)^{n+2}}{2^{n+2} \tanh \xi},$$

$$S_n = \frac{(1 + \tanh \xi)^{n+1} - (1 - \tanh \xi)^{n+1}}{2^{n+1} \tanh \xi}.$$

Substituting for δ_n and β_n, he arrived at

$$b_n = b_1 + (b_1 - b_0) \frac{\sinh(n-1)\xi}{\sinh(n+1)\xi}, \quad n = 0, 1, 2, \ldots, \qquad (27.96)$$

where $\xi \geq 0$, and b_0 and b_1 were arbitrary. Feldheim applied this to (27.91) to show that $\lambda_n > 0$ if $b_1 > b_0$. Thus, he found the orthogonal generalized Legendre polynomials. He also observed that he could obtain the ultraspherical polynomials as special cases.

Lanzewizky's paper was very brief and gave only statements of his results. In his first theorem, he noted that if $C_n = \frac{a_n}{a_{n-1}} c$, then C_n satisfied the difference equation

$$C_{n+1} = \frac{C_1 C_n - C_2 C_{n-1}}{C_1 + C_n - C_2 - C_{n-1}}, \quad n = 3, 4, 5, \ldots. \tag{27.97}$$

He presented the solution to this difference equation as

$$C_{n+1} = C_2 + (C_2 - C_1)\frac{U_{n-2}(\xi)}{U_n(\xi)}, \tag{27.98}$$

where

$$\xi = \frac{1}{2}\sqrt{\frac{C_3 - C_1}{C_3 - C_2}} \quad \text{and} \quad U_n(\xi) = \frac{\sin(n+1)\arccos \xi}{\sin \arccos \xi}. \tag{27.99}$$

For orthogonality, he required either that $\xi \geq 1$ and $-\xi < \frac{C_1}{2C_2} < 1$; or that $\xi = i\eta$ with $\eta > 0$ and $-\eta^2 < \frac{C_1}{2C_2} < 1$; he observed that for $0 < \xi < 1$, orthogonality was not possible.

Askey has pointed out[24] that it is more convenient to write the solution of the difference equation (27.97) as

$$C_n = \frac{\alpha(1 - \beta q^{n-1})}{1 - q^n}, \tag{27.100}$$

where α, β are real constants and $|q| \leq 1$. For $|q| < 1$, we get

$$\frac{a_n}{a_0} = C_1 \cdot C_2 \cdots C_n = \alpha^n \frac{(1-\beta)(1-\beta q)\cdots(1-\beta q^{n-1})}{(1-q)(1-q^2)\cdots(1-q^n)}, \tag{27.101}$$

and hence

$$f(re^{i\theta}) = a_0 \sum \frac{(\beta)_n}{(q)_n} \alpha^n r^n e^{in\theta},$$

where $|\alpha r| < 1$ for convergence. Orthogonality is obtained if

$$\frac{(1 - q^{n+1})(1 - \beta^2 q^n)}{(1 - \beta q^n)(1 - \beta q^{n+1})} > 0.$$

So one may take $\alpha = 1$, yielding

$$p_n(\cos\theta) = \sum_{k=0}^{n} \frac{(\beta)_k(\beta)_{n-k}}{(q)_k(q)_{n-k}} \cos(n - 2k)\theta, \tag{27.102}$$

the q-ultraspherical polynomials denoted by $C_n(x; \beta|q)$ with $x = \cos\theta$.

In 1977, Richard Askey and James Wilson derived explicit orthogonality relations for some basic hypergeometric orthogonal polynomials.[25] These relations included the orthogonality relation for the q-ultraspherical polynomials of Rogers. But it was

---

[24] Andrews et al. (1999) pp. 336–337.
[25] Askey and J. (1985).

later, upon reading Rogers's papers, that Askey recognized the full significance of the
q-ultraspherical polynomials. Jointly with Mourad Ismail, he worked out the properties of these polynomials and discovered various methods for deriving their orthogonality relation:[26]

$$
\int_{-1}^{1} C_n(x;\beta|q)\, C_m(x;\beta|q)\, w_\beta(x)\, \frac{dx}{\sqrt{1-x^2}}
$$

$$
= \frac{2\pi(1-\beta)}{1-\beta q^n} \cdot \frac{(\beta^2)_n}{(q)_n} \cdot \frac{(\beta)_\infty(\beta q)_\infty}{(\beta^2)_\infty(q)_\infty}\, \delta_{mn}, \quad 0 < q < 1, \tag{27.103}
$$

where

$$
w_\beta(\cos\theta) = \frac{(e^{2i\theta})_\infty(e^{-2i\theta})_\infty}{(\beta e^{2i\theta})_\infty(\beta e^{-2i\theta})_\infty}, \quad -1 < \beta < 1. \tag{27.104}
$$

Interestingly, one of these methods employed Ramanujan's summation formula,
paralleling the use of the triple product identity in the derivation of the orthogonality
relation for the q-Hermite polynomials.

27.8 Exercises

(1) Show that

$$
\frac{1}{1-x} + \frac{z}{1-qx} + \frac{z^2}{1-q^2x} + \cdots = \frac{1}{1-z} + \frac{x}{1-qz} + \frac{x^2}{1-q^2z} + \cdots .
$$

See Heine (1847).

(2) Show that

$$
\prod_{m=0}^{n-1} \Omega\left(q^n, a - \frac{m}{n}\right) = c\,\Omega(q, na),
$$

where

$$
c = \frac{\left((1-q^n)(1-q^{2n})(1-q^{3n})\cdots\right)^n}{(1-q)(1-q^2)(1-q^3)\cdots}.
$$

See Heine (1847).

(3) Let

$$
\phi(q) = \prod_{n=0}^{\infty} (1-q^{5n+1})^{-1}(1-q^{5n+4})^{-1},
$$

$$
\psi(q) = \prod_{n=0}^{\infty} (1-q^{5n+2})^{-1}(1-q^{5n+3})^{-1}
$$

---

[26] Askey and Ismail (1980); Askey and Ismail (1983).

and then prove that

(a) $\phi(q) = \prod_{n=1}^{\infty}(1+q^{2n}) \cdot \sum_{n=0}^{\infty} \frac{q^{n^2}}{(1-q^4)(1-q^8)\cdots(1-q^{4n})}$,

(b) $\phi(q^4) = \prod_{n=1}^{\infty}(1-q^{2n-1}) \cdot \sum_{n=0}^{\infty} \frac{q^{n^2}}{(q)_{2n}}$,

(c) $\psi(q) = \prod_{n=1}^{\infty}(1+q^{2n}) \cdot \sum_{n=0}^{\infty} \frac{q^{n(n+2)}}{(1-q^4)\cdots(1-q^{4n})}$.

See Rogers (1894) pp. 330–331.

(4) Following Rogers, define $B_r(\theta)$ by

$$\prod_{n=1}^{\infty}(1+2xq^n\cos\theta + x^2q^{2n}) = \sum_{n=0}^{\infty}\frac{B_n(\theta)}{(q)_n}x^n.$$

Demonstrate that

(a)

$$B_{2n}(\theta) = q^{n(n+1)}\frac{(q)_{2n}}{(q)_n(q)_n}\left(1 + \frac{1-q^n}{1-q^{n+1}}\cdot 2q\cos 2\theta\right.$$
$$\left. + \frac{(1-q^n)(1-q^{n-1})}{(1-q^{n+1})(1-q^{n+2})}\cdot 2q^4\cos 4\theta + \cdots\right).$$

(b)

$$B_{2n+1}(\theta) = q^{(n+1)^2}\frac{(q)_{2n+1}}{(q)_n(q)_{n+1}}\left(2\cos\theta + \frac{1-q^n}{1-q^{n+1}}\cdot 2q^2\cos 3\theta\right.$$
$$\left. + \frac{(1-q^n)(1-q^{n-1})}{(1-q^{n+2})(1-q^{n+3})}\cdot 2q^6\cos 5\theta + \cdots\right).$$

(c)

$$(q)_\infty\left(\sum_{n=0}^{\infty}\frac{q^{-\frac{n}{2}}B_n(\theta)}{(q)_n}\right) = 1 + 2\sum_{n=1}^{\infty}q^{\frac{n}{2}}\cos n\theta.$$

See Rogers (1917) pp. 315–316.

(5) Replace $2\cos 2k\theta$ by

$$(-1)^k(1+q^k)q^{\frac{k(k-1)}{2}}$$

in the expression for $B_{2n}(\theta)$ in Exercise 4(a) and denote the result β_{2n}. Likewise, let β_{2n+1} denote the result of replacing $2\cos(2k+1)\theta$ by

$$(-1)^k(1-q^{2k+1})q^{\frac{k(k-1)}{2}}$$

in 4(b). Show that

(a) $\beta_{2n+1} = q^{n+1}(1 - q^{2n+1})\beta_{2n}$,

(b) $\beta_{2n+2} = q^{n+1}\dfrac{1 - q^{2n+2}}{1 - q^{n+1}}\beta_{2n+1}$.

In Exercise 4(c), equate terms containing even multiples of θ, and replace these cosines as indicated earlier in this exercise. Show that this process leads to

$$\frac{1}{(q)_\infty}(1 - q^2 - q^3 + q^9 + q^{11} - \cdots) = 1 + \frac{\beta_2 q^{-1}}{(q)_2} + \frac{\beta_4 q^{-2}}{(q)_4} + \cdots$$

$$= 1 + \frac{q}{1 - q} + \frac{q^4}{(q)_2} + \frac{q^9}{(q)_3} + \cdots.$$

Prove that the cosines of odd multiples of θ lead to the second Rogers–Ramanujan identity. See Rogers (1917) pp. 316–317. In the 1940s, W. N. Bailey elucidated the underlying structure of Rogers's method. In the 1980s, G. E. Andrews developed Bailey's idea into a powerful tool to handle q-series and mock theta functions. Andrews named this method Bailey chains and around the same time, P. Paule independently realized the significance of Bailey's method. See Andrews (1986b).

(6) Show that

$$\prod_{n=0}^{\infty}\left(\frac{1 - q^{n+1}}{1 - q^n x}\right) = \frac{1 - q}{1 - x} + q \cdot \frac{x - q}{(1 - x)(1 - qx)}(1 - q^3)$$

$$+ q^4 \cdot \frac{(x - q)(x - q^2)}{(1 - x)(1 - qx)(1 - q^2 x)}(1 - q^5) + \cdots.$$

Consider the cases $x = 0$, $x = q^{\frac{1}{2}}$ and $x = -1$. Prove that

$$\sum_{n=1}^{\infty}\frac{c^n q^n}{1 - q^n} = \sum_{n=1}^{\infty}\frac{c^n q^{n^2}}{1 - q^n} + \sum_{n=1}^{\infty}\frac{c^{(n+1)}q^{n(n+1)}}{1 - cq^n}.$$

See Rogers (1893a) p. 30.

27.9 Notes on the Literature

See Andrews (1986b) for an interesting and detailed discussion, with good references, of the work of Heine, Thomae, Rogers, and Ramanujan.

28

Dirichlet L-Series

28.1 Preliminary Remarks

A Dirichlet L-series can be seen, in general, as a series taking the form

$$\sum_{m=1}^{\infty} \frac{a_m}{m^s},$$

where $|a_m| = 1$ or 0 and $a_{m+n} = a_m$ for some positive integer n; also, it must be expressible as the product over primes (known as an Euler product):

$$\prod_p \left(1 - a_p p^{-s}\right)^{-1}.$$

Dirichlet studied L-series in the context of proving the existence of an infinite number of primes of the form $a + nd$, with a and d relatively prime integers. Euler had earlier verified this result for primes of the form $4n + 1$ and $4n + 3$.

In 1739, as we discuss in Chapter 16, Euler summed the infinite series

$$1 + \frac{1}{2^{2k}} + \frac{1}{3^{2k}} + \cdots + \frac{1}{n^{2k}} + \cdots = \frac{(-1)^{k-1} 2^{2k-1} \pi^{2k} B_{2k}}{(2k)!}; \qquad (28.1)$$

as mentioned earlier, this series is written in modern notation as $\zeta(2k)$, with $\zeta(s)$ defined by

$$\zeta(s) = 1 + \frac{1}{2^s} + \frac{1}{3^s} + \frac{1}{4^s} + \cdots \qquad (28.2)$$

for $s > 1$ or, with s a complex number, Re $s > 1$.

In 1737, Euler had shown[1] that (28.2) could be expressed as the infinite product

$$\zeta(s) = \prod_p (1-p^{-s})^{-1} = (1-2^{-s})^{-1}(1-3^{-s})^{-1}(1-5^{-s})^{-1}(1-7^{-s})^{-1} \cdots, \qquad (28.3)$$

---

[1] Eu. I-14 pp. 217–244. E 72, Theorem 8.

where $s > 1$. Next, apply the geometric series expansion

$$(1 - x)^{-1} = 1 + x + x^2 + x^3 + \cdots$$

to each term in the product (28.3) to arrive at

$$\left(1 + \frac{1}{2^s} + \frac{1}{2^{2s}} + \frac{1}{2^{3s}} + \cdots\right)\left(1 + \frac{1}{3^s} + \frac{1}{3^{2s}} + \cdots\right)\left(1 + \frac{1}{5^s} + \frac{1}{5^{2s}} + \cdots\right)\cdots$$

$$= 1 + \frac{1}{2^s} + \frac{1}{3^s} + \frac{1}{2^{2s}} + \frac{1}{5^s} + \frac{1}{2^s \cdot 3^s} + \cdots$$

and, by the unique factorization of integers,

$$= \zeta(s) \quad \text{for} \quad s > 1.$$

Euler later presented several other similar series converted into products over primes,[2] now called Euler products. For example Euler gave:

$$1 - \frac{1}{2^s} + \frac{1}{4^s} - \frac{1}{5^s} + \frac{1}{7^s} - \frac{1}{8^s} + \frac{1}{10^s} - \frac{1}{11^s} + \cdots$$

$$= \left(1 + \frac{1}{2^s}\right)^{-1}\left(1 + \frac{1}{5^s}\right)^{-1}\left(1 - \frac{1}{7^s}\right)^{-1}\left(1 + \frac{1}{11^s}\right)^{-1}\cdots \qquad (28.4)$$

and

$$1 + \frac{1}{3^s} - \frac{1}{5^s} - \frac{1}{7^s} + \frac{1}{9^s} + \frac{1}{11^s} - \frac{1}{13^s} - \frac{1}{15^s} + \cdots$$

$$= \left(1 - \frac{1}{3^s}\right)^{-1}\left(1 + \frac{1}{5^s}\right)^{-1}\left(1 + \frac{1}{7^s}\right)^{-1}\left(1 - \frac{1}{11^s}\right)^{-1}\left(1 + \frac{1}{13^s}\right)^{-1}\cdots.$$

$$(28.5)$$

We remark that Euler wrote the formula (28.5) for $s = 1$, since he knew that for $s = 1$ its value was $\frac{\pi}{2\sqrt{2}}$. Recall that Newton had given this series and its value in his second letter to Leibniz.

Observe that in the series in (28.4), all multiples of 3 are missing and the coefficients of 4^{-s}, 7^{-s}, 10^{-s}, $\ldots$ are all $+1$, while the coefficients of 2^{-s}, 5^{-s}, 8^{-s}, $\ldots$ are all -1. Thus, every term of the form $(3m + 1)^{-s}$ has $+1$ as coefficient, whereas every term of the form $(3m - 1)^{-s}$ has coefficient -1. Note also that the product of two numbers of the same form is of the form $3m + 1$, while the product of two numbers of different forms is of the form $3m - 1$. Numbers of the form $3m - 1$ can also take the form $3(m - 1) + 2$. If we let [1] and [2] denote the numbers of the form $3m + 1$ and $3m + 2$ respectively and if we associate with [1] and [2] the integers 1 and -1 respectively, then we can write

$$\psi([1]) = 1 \quad \text{and} \quad \psi([2]) = -1$$

2 Euler (1988) chapter 15.

and

$$\psi([mn]) = \psi([m])\psi([n]). \tag{28.6}$$

Thus, denoting the integers modulo n by $\mathbb{Z}_n$ and the integers prime to n by $\mathbb{Z}_n^\times$, we see that (28.6) defines a mapping

$$\dot\psi : \mathbb{Z}_3^\times \to \{1, -1\} \tag{28.7}$$

where

$$\psi([ab]) = \psi([a])\psi([b]).$$

The function ψ can then be extended to $\mathbb{Z}_n$ by setting

$$\psi([a]) = 0 \quad \text{if} \quad (a, n) = d > 1,$$

where (a, n) is the gcd of a and n.

With this notation, we can rewrite (28.4) as

$$\sum_{n=1}^{\infty} \frac{\chi(n)}{n^s} = \prod_p \left(1 - \frac{\chi(p)}{p^s}\right)^{-1}, \tag{28.8}$$

where the product is over all primes and

$$\chi : \mathbb{Z}_3 \to \{0, 1, -1\}.$$

Clearly, for (28.5) the function χ is defined on $\mathbb{Z}_8$. Thus

$$\chi(2) = \chi(4) = \chi(6) = \cdots = 0,$$

while

$$\chi(8m + 1) = \chi(8m + 3) = 1$$

and

$$\chi(8m - 1) = \chi(8m - 3) = -1.$$

We now see that (28.5) may be written in exactly the same way as (28.8). There are other series for which χ can be defined modulo 8. For example, Euler also gave

$$1 - \frac{1}{3^s} + \frac{1}{5^s} - \frac{1}{7^s} + \frac{1}{9^s} - \frac{1}{11^s} + \frac{1}{13^s} - \cdots$$
$$= \left(1 + \frac{1}{3^s}\right)^{-1} \left(1 - \frac{1}{5^s}\right)^{-1} \left(1 + \frac{1}{7^s}\right)^{-1} \left(1 + \frac{1}{11^s}\right)^{-1} \left(1 - \frac{1}{13^s}\right)^{-1} \cdots. \tag{28.9}$$

Two other series of this kind, expressible as products, are:

$$1 - \frac{1}{3^s} - \frac{1}{5^s} + \frac{1}{7^s} + \frac{1}{9^s} - \frac{1}{11^s} - \frac{1}{13^s} + \cdots$$

$$1 + \frac{1}{3^s} - \frac{1}{5^s} - \frac{1}{7^s} + \frac{1}{9^s} + \frac{1}{11^s} - \frac{1}{13^s} - \cdots.$$

It can be shown that there are exactly four series that can be expressed as products in this manner; this is because $\mathbb{Z}_8^\times$ has four elements.

For his work on primes in arithmetic progressions, Dirichlet had to consider complex-valued multiplicative functions χ, while Euler discussed only those series in which the value of χ was ± 1. Since $|\pm 1| = 1$, Dirichlet considered those complex $\chi(m)$ for which $|\chi(m)| = 1$ and $m \in \mathbb{Z}_n^\times$. As a simple example, consider the case in which $n = 5$, that is where χ is defined on $\mathbb{Z}_5^\times$. Observe that 3 generates $\mathbb{Z}_5^\times$ since, modulo 5, $3^1 = 3, 3^3 = 4, 3^3 = 2$, and $3^4 = 1$. Since χ is multiplicative, it is sufficient to define $\chi(3)$. Clearly, $\chi(3)$ is a root of the equation $x^4 = 1$; the four possible values of $\chi(3)$ are thus $\pm i, \pm 1$. The four series corresponding to these values are then

$$1 - \frac{i}{2^s} + \frac{i}{3^s} - \frac{1}{4^s} + \frac{1}{6^s} - \frac{i}{7^s} + \frac{i}{8^s} - \frac{1}{9^s} + \cdots, \tag{28.10}$$

$$1 + \frac{i}{2^s} - \frac{i}{3^s} - \frac{1}{4^s} + \frac{1}{6^s} + \frac{i}{7^s} - \frac{i}{8^s} - \frac{1}{9^s} + \cdots, \tag{28.11}$$

$$1 + \frac{1}{2^s} + \frac{1}{3^s} + \frac{1}{4^s} + \frac{1}{6^s} + \frac{1}{7^s} + \frac{1}{8^s} + \frac{1}{9^s} + \cdots,$$

$$1 - \frac{1}{2^s} - \frac{1}{3^s} + \frac{1}{4^s} + \frac{1}{6^s} - \frac{1}{7^s} - \frac{1}{8^s} + \frac{1}{9^s} + \cdots.$$

Observe that (28.10) can be written as a product

$$\left(1 + \frac{i}{2^s}\right)^{-1} \left(1 - \frac{i}{3^s}\right)^{-1} \left(1 + \frac{i}{7^s}\right)^{-1} \cdots,$$

where the prime 5 is missing from the product, just as it is from the series.

Dirichlet denoted series such as (28.8) through (28.11), when written as products, by the letter L, so we now call them L-series.[3] One notation for these series is $L(\chi, s)$, where χ is a multiplicative function

$$\chi : \mathbb{Z}_n^\times \to \Gamma,$$

where Γ denotes the complex numbers z, with $|z| = 1$.

28.2   Dirichlet's Summation of $L(1, \chi)$

The series known as the Madhava–Leibniz formula, $1 - \frac{1}{3} + \frac{1}{5} - \cdots = \frac{\pi}{4}$, discussed in Section 16.1, gives the value of $L(1, \chi)$ for the nontrivial character modulo 4, defined

---

[3] Dirichlet (1969) vol. 1, pp. 317–318.

by $\chi(4n \pm 1) = \pm 1$. Euler employed this series to prove that primes of the form $4n+1$ and of the form $4n - 1$ were both infinite in number. In the 1830s, J. P. G. Lejeune Dirichlet went further, giving a general evaluation of $L(1, \chi)$ to prove his results on quadratic forms and on primes in arithmetic progressions. Dirichlet first examined the case in which χ was a character modulo p, where p was a prime. We limit our discussion to this simple case. Dirichlet defined this character by taking any generator g of the cyclic group consisting of the integers modulo p without the zero element. Next, he let w be any $(p - 1)$th root of unity. For n not divisible by p, he set

$$\chi(n) = w^{\gamma_n} \quad \text{where } g^{\gamma_n} \equiv n \pmod{p}.$$

By convention, $\chi(mp) = 0$. More details on Dirichlet's theory of characters are given later in this chapter. To evaluate the L-series $\sum_{n=1}^{\infty} \frac{w^{\gamma_n}}{n^s}$ at $s = 1$ when $w \neq 1$, Dirichlet first expressed the series as an integral. Note that the terms in which n was a multiple of p were taken to be 0. In a paper of 1768 on series related to the zeta function,[4] Euler had used the idea of expressing this type of series as an integral. Like Euler, Dirichlet started with[5]

$$\int_0^1 x^{n-1} \left(\log \frac{1}{x} \right)^{s-1} dx = \frac{\Gamma(s)}{n^s}. \tag{28.12}$$

From the periodicity of the character w^{γ_n}, he had

$$L_w(s) = \sum_{n=1}^{\infty} \frac{w^{\gamma_n}}{n^s} = \sum_{k=1}^{p-1} w^{\gamma_k} \left(\frac{1}{k^s} + \frac{1}{(k+p)^s} + \frac{1}{(k+2p)^s} + \cdots \right)$$

$$= \frac{1}{\Gamma(s)} \sum_{k=1}^{p-1} w^{\gamma_k} \sum_{d=0}^{\infty} \int_0^1 x^{k+dp-1} \left(\log \frac{1}{x} \right)^{s-1} dx$$

$$= \frac{1}{\Gamma(s)} \sum_{k=1}^{p-1} w^{\gamma_k} \int_0^1 \frac{x^{k-1}}{1 - x^p} \left(\log \frac{1}{x} \right)^{s-1} dx$$

$$= \frac{1}{\Gamma(s)} \int_0^1 \frac{x^{-1} f(x)}{1 - x^p} \left(\log \frac{1}{x} \right)^{s-1} dx \tag{28.13}$$

where

$$f(x) = \sum_{k=1}^{p-1} w^{\gamma_k} x^k. \tag{28.14}$$

Unlike most eighteenth-century mathematicians, Dirichlet dealt carefully with convergence and term-by-term integration, so he summed only the first $(p-1)h$ terms of the series and then showed that this sum differed from the integral (28.13) by an

[4] Eu. I-15 pp. 91–130. E 393.
[5] Dirichlet (1969) vol. 1, pp. 313–342.

integral with the limit zero as $h \to \infty$. When $s = 1$, the logarithmic term in the integral (28.13) vanished. Thus, Dirichlet observed that $L_w(1)$, an integral of a rational function, could be computed in terms of logarithms and circular functions.

Dirichlet pointed out that the factors of $x^p - 1$ were of the form $x - e^{\frac{2m\pi i}{p}}$; hence,

$$\frac{x^{-1}f(x)}{x^p - 1} = \sum_{m=1}^{p-1} \frac{A_m}{x - e^{\frac{2m\pi i}{p}}},$$

where

$$A_m = \lim_{x \to e^{\frac{2m\pi i}{p}}} \frac{\left(x - e^{\frac{2m\pi i}{p}}\right) x^{-1} f(x)}{x^p - 1}.$$

This limit was the value of $\frac{x^{-1}f(x)}{px^{p-1}}$ at $x = e^{\frac{2m\pi i}{p}}$; thus, he found

$$A_m = \frac{1}{p} f\left(e^{\frac{2m\pi i}{p}}\right) = \frac{1}{p} \sum_{k=1}^{p-1} w^{\gamma k} e^{\frac{2km\pi i}{p}}. \tag{28.15}$$

Next, Dirichlet set $km \equiv h \pmod{p}$ so that $w^{\gamma k} = w^{-\gamma m} w^{\gamma h}$ and

$$f\left(e^{\frac{2m\pi i}{p}}\right) = w^{-\gamma m} \sum_{h=1}^{p-1} w^{\gamma h} e^{\frac{2h\pi i}{p}} = w^{-\gamma m} f(e^{\frac{2\pi i}{p}}).$$

In this manner, Dirichlet arrived at

$$L_w(1) = \sum_{n=1}^{\infty} \frac{w^{\gamma n}}{n} = -\frac{1}{p} f(e^{\frac{2\pi i}{p}}) \sum_{m=1}^{p-1} w^{-\gamma m} \int_0^1 \frac{dx}{x - e^{\frac{2m\pi i}{p}}}. \tag{28.16}$$

He next noted that the last integral could be expressed as

$$\log(1 - e^{-\frac{2m\pi i}{p}}) = \log\left(2\sin\frac{m\pi}{p}\right) + i\frac{\pi}{2}\left(1 - \frac{2m}{p}\right),$$

so that

$$\sum_{n=1}^{\infty} \frac{w^{\gamma n}}{n} = -\frac{1}{p} f\left(e^{\frac{2\pi i}{p}}\right) \sum_{m=1}^{p-1} w^{-\gamma m}\left(\log\left(2\sin\frac{m\pi}{p}\right) + i\frac{\pi}{2}\left(1 - \frac{2m}{p}\right)\right). \tag{28.17}$$

Dirichlet further observed that this formula took a much simpler form when $w = -1$. This corresponded to the quadratic character $(-1)^{\gamma n} = \left(\frac{n}{p}\right)$; he then had

$$\sum_{n=1}^{\infty} \left(\frac{n}{p}\right)\frac{1}{n} = -\frac{1}{p} f(e^{\frac{2\pi i}{p}}) \sum_{m=1}^{p-1} \left(\frac{m}{p}\right)\left(\log\left(2\sin\frac{m\pi}{p}\right) + i\frac{\pi}{2}\left(1 - \frac{2m}{p}\right)\right).$$

Since $\sum_{m=1}^{p-1}\left(\frac{m}{p}\right) = \sum_{m=1}^{p-1}(-1)^{\gamma_m} = 0$, he could simplify to obtain

$$\sum_{n=1}^{\infty}\left(\frac{n}{p}\right)\frac{1}{n} = -\frac{1}{p}f(e^{\frac{2\pi i}{p}})\sum_{m=1}^{p-1}\left(\frac{m}{p}\right)\left(\log\left(2\sin\frac{m\pi}{p}\right) - i\frac{m\pi}{p}\right). \qquad (28.18)$$

He then noted that

$$\left(\frac{p-m}{p}\right) = \left(\frac{-m}{p}\right) = \left(-\frac{1}{p}\right)\left(\frac{m}{p}\right) = (-1)^{\frac{p-1}{2}}\left(\frac{m}{p}\right) = \pm\left(\frac{m}{p}\right),$$

using plus if p took the form $4n+1$ and minus if p took the form $4n+3$. Note that when p is of the form $4n+1$, the imaginary part of the sum vanishes because $\sum m\left(\frac{m}{p}\right) = 0$ when $\left(\frac{p}{m}\right) = \left(\frac{p-m}{m}\right)$. Dirichlet could then conclude that

$$\sum_{n=1}^{\infty}\left(\frac{n}{p}\right)\frac{1}{p} = \frac{1}{p}f(e^{\frac{2\pi i}{p}})\log\frac{\prod\sin\frac{b\pi}{p}}{\prod\sin\frac{a\pi}{p}}, \qquad (28.19)$$

where a represented quadratic residues (mod p) and b nonresidues. Observe that for the case $p = 4n+3$,

$$\left(\frac{m}{p}\right)\log\left(2\sin\frac{m\pi}{p}\right) = -\left(\frac{p-m}{p}\right)\log\left(2\sin\frac{(m-p)\pi}{p}\right),$$

and hence the sum of these terms is zero and

$$\sum_{n=1}^{\infty}\left(\frac{n}{p}\right)\frac{1}{n} = \frac{\pi}{p}f(e^{\frac{2\pi i}{p}})\left(\sum a - \sum b\right)\sqrt{-1}. \qquad (28.20)$$

Moreover, the term $f(e^{\frac{2\pi i}{p}})$ is the quadratic Gauss sum

$$\sum_{k=1}^{p-1}(-1)^{\gamma_k}e^{\frac{2k\pi i}{p}} = \sum_{k=1}^{p-1}\left(\frac{k}{p}\right)e^{\frac{2k\pi i}{p}} = \begin{cases} \sqrt{p}, & p = 4n+1, \\ i\sqrt{p}, & p = 4n+3; \end{cases}$$

in this connection, see Section 19.7 or Section 25.6. We thus obtain Dirichlet's final formulas[6]

$$\sum_{n=1}^{\infty}\left(\frac{n}{p}\right)\frac{1}{n} = \frac{1}{\sqrt{p}}\log\frac{\prod\sin\frac{b\pi}{p}}{\prod\sin\frac{a\pi}{p}}, \qquad p \equiv 1 \ (\mathrm{mod}\,4), \qquad (28.21)$$

$$\sum_{n=1}^{\infty}\left(\frac{n}{p}\right)\frac{1}{n} = \frac{\pi}{p\sqrt{p}}\left(\sum b - \sum a\right), \qquad p \equiv 3 \ (\mathrm{mod}\,4). \qquad (28.22)$$

Dirichlet wrote that the last formula implied that for primes of the form $4n+3$, $\sum b > \sum a$, that is, the sum of the quadratic nonresidues was greater than the sum of the quadratic residues, and that it would be difficult to prove this in a different way.

[6] ibid. p. 327.

28.3   Eisenstein's Proof of the Functional Equation

The discovery of Eisenstein's proof of the functional equation began in 1964 when B. Artmann came across Eisenstein's old copy of Gauss's *Disquisitiones* in the Giessen University Mathematical Institute Library. This book had belonged to Ferdinand Eisenstein (1823–1852) and then to Eugen Netto (1848–1919), student of Weierstrass and Kummer, before arriving at the Library. The proof, in Eisenstein's hand and dated 1849, appeared on the last blank page of the book; with the help of Artmann and the librarian, André Weil was able to examine it and to publish it in a paper of 1989.[7] Eisenstein's proof started with the formula

$$\int_0^\infty e^{\sigma \psi i} \psi^{q-1} \, d\psi = \frac{\Gamma(q)}{(\pm \sigma)^q} e^{\pm \frac{q\pi i}{2}}. \tag{28.23}$$

This is in fact the Fourier transform of the function

$$f(\psi) = \begin{cases} \psi^{q-1} & \text{for } \psi > 0, \quad 0 < q < 1, \\ 0 & \text{for } \psi < 0. \end{cases}$$

For this formula, Eisenstein referred to a 1836 paper by Dirichlet on definite integrals. In that paper, Dirichlet noted that the formula was first found by Euler but that Poisson gave the proof, with the convergence condition $0 < q < 1$. Eisenstein then applied the Poisson summation formula

$$\sum_{n=-\infty}^{\infty} \phi(n) = \sum_{m=-\infty}^{\infty} \hat{\phi}(m), \tag{28.24}$$

where $\hat{\phi}$ was the Fourier transformation of ϕ, to the function

$$\phi(x) = \begin{cases} e^{2\pi \alpha (x-\beta) i} (x-\beta)^{q-1} & \text{for } x > \beta, 0 < \alpha < 1, 0 < \beta < 1, \\ 0 & \text{for } x < \beta. \end{cases}$$

He then had

$$\frac{e^{2\pi \alpha (1-\beta) i}}{(1-\beta)^{1-q}} + \frac{e^{2\pi \alpha (2-\beta) i}}{(2-\beta)^{1-q}} + \frac{e^{2\pi \alpha (3-\beta) i}}{(3-\beta)^{1-q}} + \cdots$$

$$= \sum_{\sigma=-\infty}^{\infty} \int_\beta^\infty e^{2\pi \alpha (\lambda - \beta) i} (\lambda - \beta)^{q-1} e^{2\pi i \sigma \lambda} \, d\lambda$$

$$= \sum_{\sigma=-\infty}^{\infty} \int_0^\infty e^{2\pi (\alpha + \sigma) \lambda i} e^{2\pi i \lambda \beta} \lambda^{q-1} \, d\lambda \tag{28.25}$$

$$= \frac{\Gamma(q)}{(2\pi)^q} e^{\frac{q\pi i}{2}} \sum_{\sigma=0}^{\infty} \frac{e^{2\pi i \sigma \beta}}{(\sigma + \alpha)^q} + \frac{\Gamma(q)}{(2\pi)^q} e^{-\frac{q\pi i}{2}} \sum_{\sigma=1}^{\infty} \frac{e^{2\pi i \sigma \beta}}{(\sigma - \alpha)^q},$$

---

[7] Weil (1989a).

where the last step followed from (28.23). By taking $\alpha = \beta = \frac{1}{2}$, he obtained the functional equation for the L-function

$$1 - \frac{1}{3^{1-q}} + \frac{1}{5^{1-q}} - \frac{1}{7^{1-q}} + \cdots = \frac{2^q \Gamma(q)}{\pi^q} \sin \frac{q\pi}{2} \left(1 - \frac{1}{3^q} + \frac{1}{5^q} + \frac{1}{7^q} + \cdots\right).$$

$$(28.26)$$

Eisenstein also observed at this point that when q was replaced by $1-q$ and the two formulas were multiplied, he got another proof of Euler's reflection formula, discussed in Section 17.1:

$$\Gamma(q)\Gamma(1 - q) = \frac{\pi}{\sin \pi q}.$$

28.4 Riemann's Derivations of the Functional Equation

It is thought that Eisenstein may have discussed his proof of the functional equation with Riemann, perhaps inspiring Riemann's 1859 paper on the number of primes less than a given number.[8] In this paper, Riemann used complex analysis to give two new proofs of the functional equation. One proof made use of contour integration and the second, deeper proof employed the transformation of a theta function. The latter method presaged a connection between modular forms and the corresponding Dirichlet series obtained by applying the Mellin transform.

The proof by contour integration started with two formulas due to Euler, though Riemann did not attribute them to anyone, perhaps regarding them as well known: For Re $s > 0$,

$$\int_0^\infty e^{-nx} x^{s-1}\, dx = \frac{\Gamma(s)}{n^s},$$

$$(28.27)$$

$$\Gamma(s)\zeta(s) = \int_0^\infty \frac{x^{s-1}}{e^x - 1}\, dx.$$

$$(28.28)$$

We here mention that Riemann used Gauss's notation for the gamma function: $\Pi(s - 1)$. Observe that the second formula follows from the first, using the geometric series expansion

$$\frac{1}{e^x - 1} = e^{-x} + e^{-2x} + e^{-3x} + \cdots.$$

Riemann next considered the integral

$$\int \frac{(-x)^{s-1}}{e^x - 1}\, dx$$

$$(28.29)$$

[8] Riemann (2004) pp. 135–143.

over a contour from $+\infty$ to $+\infty$ in the positive sense around the boundary of a region containing in its interior 0 but no other singularities of the integrand. He noted that this integral simplified to

$$\left(e^{-\pi s i} - e^{\pi s i}\right) \int_0^\infty \frac{x^{s-1}}{e^x - 1}\, dx, \tag{28.30}$$

provided that one used the branch of the many-valued function $(-x)^{s-1} = e^{(s-1)\log(-x)}$ for which $\log(-x)$ was real for negative values of x. From (28.28), (28.29), and (28.30), he concluded that

$$2 \sin s\pi \, \Gamma(s)\, \zeta(s) = i \int_\infty^\infty \frac{(-x)^{s-1}}{e^x - 1}\, dx. \tag{28.31}$$

Riemann pointed out that this integral defined $\zeta(s)$ as an analytic function of s with a singularity at $s = 1$. In addition, we note that since

$$\frac{x}{e^x - 1} = 1 - \frac{1}{2}x + B_2 \frac{x^2}{2!} - B_4 \frac{x^4}{4!} + B_6 \frac{x^6}{6!} - \cdots,$$

two of Euler's famous formulas are immediate corollaries, though Riemann noted only the first one:

$$\zeta(-2n) = 0, \quad \text{and} \quad \zeta(1 - 2n) = \frac{(-1)^n B_{2n}}{2n}, \quad n = 1, 2, \ldots. \tag{28.32}$$

To obtain the functional equation, Riemann remarked at this point that for Re $s < 0$, the contour for the integral in (28.31) could be viewed as if defined (with a negative orientation) as the boundary of the complementary region containing the singularities $\pm 2n\pi i$, $n > 0$ of the integrand. Since the residue at $2n\pi i$ was $(-n2\pi i)^{s-1}(-2\pi i)$, he obtained the equation

$$2 \sin s\pi \, \Gamma(s)\, \zeta(s) = (2\pi)^s \sum n^{s-1}\left((-i)^{s-1} + i^{s-1}\right). \tag{28.33}$$

Riemann noted that by using the known properties of the gamma function, (28.33) could be seen as equivalent to the statement that:

$$\Gamma\left(\frac{s}{2}\right) \pi^{-\frac{s}{2}} \zeta(s) \tag{28.34}$$

was invariant under the transformation $s \to 1-s$. And this was the functional equation for $\zeta(s)$. In his 1859 paper, Riemann noted only the first equation in (28.32), though he clearly knew the second one as well; when combined with the functional equation, this yields a new proof of Euler's formula

$$\zeta(2n) = \frac{(-1)^{n-1} 2^{2n-1} \pi^{2n} B_{2n}}{(2n)!}.$$

Riemann wrote that the expression in (28.34) and its invariance led him to consider the integral for $\Gamma(\frac{s}{2})$ and thus directed him to another important derivation of the functional equation. Since

$$\Gamma\left(\frac{s}{2}\right)\pi^{-\frac{s}{2}}\frac{1}{n^s} = \int_0^\infty x^{\frac{s}{2}-1}e^{-n^2\pi x}\,dx,$$

Riemann used term-by-term integration to find that

$$\Gamma\left(\frac{s}{2}\right)\pi^{-\frac{s}{2}}\zeta(s) = \int_0^\infty x^{\frac{s}{2}-1}\sum_{n=1}^\infty e^{-n^2\pi x}\,dx. \tag{28.35}$$

It was proved by Cauchy and Poisson, and a little later by Jacobi, that

$$\sum_{n=-\infty}^\infty e^{-n^2\pi x} = \frac{1}{\sqrt{x}}\sum_{n=-\infty}^\infty e^{-\frac{n^2\pi}{x}}. \tag{28.36}$$

In Riemann's notation, this was equivalent to

$$2\psi(x) + 1 = x^{-\frac{1}{2}}\left(2\psi\left(\frac{1}{x}\right) + 1\right),$$

where

$$\psi(x) = \sum_{n=1}^\infty e^{-n^2\pi x}.$$

Riemann referred to Jacobi's *Fundamenta Nova* for (28.36). We note that Jacobi's proof of (28.36) used elliptic functions, while Cauchy and Poisson employed Fourier analysis. See Section 34.10 for Jacobi's proof and Section 34.11 for Cauchy's proof. Next Riemann rewrote (28.35) as

$$\Gamma\left(\frac{s}{2}\right)\pi^{\frac{s}{2}}\zeta(s) = \int_1^\infty \psi(x)x^{\frac{s}{2}-1}dx + \int_0^1 \psi\left(\frac{1}{x}\right)x^{\frac{s-3}{2}}dx$$

$$+ \frac{1}{2}\int_0^1\left(x^{\frac{s-3}{2}} - x^{\frac{s}{2}-1}\right)dx \tag{28.37}$$

$$= \frac{1}{s(s-1)} + \int_1^\infty \psi(x)\left(x^{\frac{s}{2}-1} + x^{-\frac{1+s}{2}}\right)dx.$$

This reproved the functional equation because the right-hand side was invariant under $s \to 1-s$. Moreover, $\zeta(s)$ was once again defined for all complex $s \neq 1$. To emphasize the significance of the line $\mathrm{Re}\,s = \frac{1}{2}$, Riemann set $s = \frac{1}{2} + it$ and denoted the left-hand side of (28.37) as $\xi(t)$, so that he had

$$\xi(t) = \frac{1}{2} - \left(t^2 + \frac{1}{4}\right) \int_1^\infty \psi(x) x^{-\frac{3}{4}} \cos\left(\frac{1}{2}t \log x\right) dx, \qquad (28.38)$$

$$\xi(t) = 4 \int_1^\infty \frac{d}{dx}(x^{\frac{3}{2}}\psi'(x)) x^{-\frac{1}{4}} \cos\left(\frac{1}{2}t \log x\right) dx. \qquad (28.39)$$

28.5 Euler's Product for $\sum \frac{1}{n^s}$

In his 1859 paper giving the formula for the number of primes less than a given number, Riemann remarked that he had taken Euler's infinite product for the zeta function as the starting point for his investigations. Indeed, it was Euler's product representation for the zeta function that made it possible to perceive the connection between the zeta function and prime numbers.

In a 1737 paper, Euler showed how to convert the series for the zeta function, $\sum_{k=1}^\infty \frac{1}{k^n}$, into a product.[9] Euler's insightful argument, amounting to an application of the fundamental theorem of arithmetic, is here presented in its original form. Euler let

$$x = 1 + \frac{1}{2^n} + \frac{1}{3^n} + \frac{1}{4^n} + \frac{1}{5^n} + \frac{1}{6^n} + \text{ etc.}$$

Then

$$\frac{1}{2^n}x = \frac{1}{2^n} + \frac{1}{4^n} + \frac{1}{6^n} + \frac{1}{8^n} + \text{ etc.}$$

Removing all even numbers by subtraction, he got

$$\frac{2^n - 1}{2^n}x = 1 + \frac{1}{3^n} + \frac{1}{5^n} + \frac{1}{7^n} + \frac{1}{9^n} + \text{ etc.}$$

Multiplying by $\frac{1}{3^n}$, he obtained

$$\frac{2^n - 1}{2^n} \cdot \frac{1}{3^n}x = \frac{1}{3^n} + \frac{1}{9^n} + \frac{1}{15^n} + \text{ etc.}$$

Again, Euler removed by subtraction all multiples of 3 so that

$$\frac{2^n - 1}{2^n} \cdot \frac{3^n - 1}{3^n}x = 1 + \frac{1}{5^n} + \frac{1}{7^n} + \text{ etc.}$$

By continuing this process with each of the prime numbers, all numbers on the right-hand side except one were eliminated, yielding

$$x \cdot \frac{2^n - 1}{2^n} \cdot \frac{3^n - 1}{3^n} \cdot \frac{5^n - 1}{5^n} \cdot \text{ etc.} = 1,$$

---

[9] Eu. I-14 pp. 217–244. E 72 Theorem 8.

or

$$1 + \frac{1}{2^n} + \frac{1}{3^n} + \frac{1}{4^n} + \text{ etc.} = \frac{2^n}{2^n - 1} \cdot \frac{3^n}{3^n - 1} \cdot \frac{5^n}{5^n - 1} \cdot \text{ etc.}$$

Note that this was in essence the fundamental theorem of arithmetic in analytic form. Euler's 1748 book *Introductio in Analysin Infinitorum* made this connection more clear, as he expressed this infinite product in almost modern form:[10]

$$\frac{1}{\left(1 - \frac{1}{2^n}\right)\left(1 - \frac{1}{3^n}\right)\left(1 - \frac{1}{5^n}\right) \text{ etc.}}$$

To see the unique factorization theorem here, simply expand these fractions using the geometric series.

In 1837, Dirichlet defined L-functions for which he found an analogous infinite product. For example, in the case of characters modulo a prime p, he stated the result as[11]

$$\prod \frac{1}{1 - w^{\gamma_q}\frac{1}{q^s}} = \sum w^{\gamma_n} \cdot \frac{1}{n^s}.$$

The product was defined over all primes other than p, while w was a $(p-1)$th root of unity. Dirichlet used this formula in his proof of his famous theorem on primes in arithmetic progressions. Note also that the product formula shows that the series on the left-hand side of (28.22) has to be positive, justifying Dirichlet's remark on it.

28.6 Dirichlet Characters

Dirichlet's construction of characters was based on a theorem first observed by Euler and later completely proved by Gauss in his 1801 *Disquisitiones Arithmeticae*. Gauss showed that for any prime p, the multiplicative group modulo p, whose elements could be represented by the integers $1, 2, \ldots, p-1$, was a cyclic group.[12] This means that there is at least one g among these $p-1$ integers such that for any n, not a multiple of p, there exists an integer γ_n such that

$$g^{\gamma_n} \equiv n \pmod{p}. \tag{28.40}$$

This equation implies that for positive integers m and n not multiples of p,

$$g^{\gamma_{mn}} \equiv mn \equiv g^{\gamma_m} g^{\gamma_n} \equiv g^{\gamma_m + \gamma_n} \pmod{p},$$

and hence

$$\gamma_{mn} \equiv \gamma_m + \gamma_n \pmod{(p-1)}. \tag{28.41}$$

---

[10] Euler (1988) p. 244.
[11] Dirichlet (1969) vol. 1, p. 317.
[12] Gauss (1965) pp. 35–36.

So if w is a $(p-1)$th root of unity, we have

$$w^{\gamma_{mn}} = w^{\gamma_m + \gamma_n} = w^{\gamma_m} w^{\gamma_n}. \tag{28.42}$$

The complex number w can be written as $e^{2\pi i k/(p-1)}$ for $k = 1, 2, \ldots, p-1$. For any one of these $p-1$ complex numbers w, Dirichlet defined a character with values $w^{\gamma_1}, w^{\gamma_2}, \ldots, w^{\gamma_{p-1}}$ and with the property

$$\frac{w^{\gamma_n}}{n^s} \cdot \frac{w^{\gamma_m}}{m^s} = \frac{w^{\gamma_{mn}}}{(mn)^s}. \tag{28.43}$$

He observed that by (28.42)

$$\frac{1}{1 - w^{\gamma_q} \frac{1}{q^s}} = 1 + w^{\gamma_q} \cdot \frac{1}{q^s} + w^{\gamma_{q^2}} \cdot \frac{1}{q^{2s}} + \cdots$$

for $s > 1$. Then, by the unique factorization theorem,[13]

$$\prod \frac{1}{1 - w^{\gamma_q} \frac{1}{q^s}} = \sum w^{\gamma_n} \cdot \frac{1}{n^s}. \tag{28.44}$$

The infinite product was defined over all primes not equal to p, and the sum was taken over all positive integers not divisible by p. Note that this sum can be taken over all positive integers with the convention that $w^{\gamma_n} = 0$ when n is a multiple of p. When $w = -1$, we have $w^{\gamma_n} = \pm 1$, depending on whether γ_n is even or odd. If it is even, we can write the left-hand side of (28.40) as a square and hence n is a square modulo p, or rather, n is a quadratic residue. We can therefore write

$$(-1)^{\gamma_n} = \left(\frac{n}{p}\right), \tag{28.45}$$

where $\left(\frac{n}{p}\right)$ is the Legendre symbol; it is $+1$ when n is a quadratic residue (mod p) and -1 when n is a quadratic nonresidue. For this character, we can write (28.44) as

$$\prod_q \frac{1}{1 - \left(\frac{q}{p}\right) q^{-s}} = \sum_{n=1}^{\infty} \frac{\left(\frac{n}{p}\right)}{n^s},$$

where $\left(\frac{n}{p}\right) = 0$ when n is a multiple of p and the product is taken over all primes not equal to p.

In his 1837 paper on primes within any arithmetic progression, Dirichlet also defined characters modulo any positive integer m. For this purpose, he employed a result from Gauss's *Disquisitiones*: For any odd prime p and positive integer k, the multiplicative group modulo p^k, that is, the integers relatively prime to p and

---

[13] Dirichlet (1969) vol. 1, pp. 316–317.

represented by integers less than p^k, is a cyclic group.[14] This theorem enabled Dirichlet to define $\phi(p^k) = p^k - p^{k-1}$ different characters corresponding to the $\phi(p^k)$ values $e^{2\pi i m / \phi(p^k)}$, $m = 1, 2, \ldots, \phi(p^k)$. Letting w denote any one of these values, the value of the corresponding character at n where n was not divisible by p, would be w^{γ_n}. As before, γ_n was defined as in (28.40), with respect to a generator g of the multiplicative group modulo p^k.[15]

For powers of 2, the situation was slightly more complex. Clearly, the multiplicative groups mod 2 and mod 4 are cyclic. Another result from the *Disquisitiones* stated[16] that every relatively prime residue class mod 2^k, where $k \geq 3$, could be represented uniquely as $(-1)^\gamma 5^{\gamma'}$, where γ was defined to the modulus 2 and γ' to the modulus $\frac{1}{2}\phi(2^k) = 2^{k-2}$. Again, Dirichlet used Gauss's result to define the characters modulo powers of 2 by

$$w^\gamma (w')^{\gamma'}, \quad \text{where} \quad w^2 = 1 \quad \text{and} \quad (w')^{2^{k-2}} = 1.$$

Dirichlet noted that the number of such characters was

$$2^{k-1} = \phi(2^k).$$

Next, Dirichlet defined characters modulo

$$m = 2^k p_1^{k_1} p_2^{k_2} \cdots p_l^{k_l}.$$

He considered an integer n relatively prime to m and assumed

$$n = (-1)^\gamma 5^{\gamma'} \pmod{2^k} \quad \text{and} \quad n = g_j^{\gamma_{n,j}} \pmod{p_j^{k_j}}$$

where g_j was the generator of the relatively prime residue classes modulo $p_j^{k_j}$. Then he gave the value of an arbitrary character at n modulo m as

$$w^\gamma (w')^{\gamma'} w_1^{\gamma_{n,1}} w_2^{\gamma_{n,2}} \cdots w_l^{\gamma_{n,l}}. \tag{28.46}$$

Here w_j was a root of $w_j^{(p-1)p^{j-1}} - 1 = 0$ and there were

$$\phi(m) = m \prod_{p|m} \left(1 - \frac{1}{p}\right)$$

such characters. Dirichlet showed that with this general definition of a character, the product formula (28.44) would continue to hold. The operative idea behind the product formula was the multiplicative property of characters.

[14] Gauss (1965) pp. 55–59.
[15] Dirichlet (1969) pp. 333–337.
[16] Gauss (1965) pp. 59–61.

29

Primes in Arithmetic Progressions

29.1 Preliminary Remarks

One of the great theorems of number theory states that any arithmetic progression $l, l + k, l + 2k, \ldots$, where l and k are relatively prime, contains an infinite number of primes. Euler conjectured this result for the particular case $l = 1$, probably in the 1750s, though it appeared in print much later.[1] Apparently, the general form of this conjecture first appeared in Legendre's 1798 book on number theory and then again in later editions. Legendre thought he had proved the theorem,[2] but in 1801, Gauss remarked that the proof "does not yet seem to satisfy geometric rigor."[3] Again, in a paper of 1837,[4] Dirichlet pointed out that Legendre's proof was based on a lemma whose proof was inadequate, but Dirichlet verified the theorem for k prime. He published a demonstration of the general result two years later.[5] Interestingly, the germ of the central idea in Dirichlet's proof came from Euler. Note that in a paper of 1737,[6] Euler used the formula

$$1 + \frac{1}{2} + \frac{1}{3} + \frac{1}{4} + \cdots = \frac{1}{\left(1 - \frac{1}{2}\right)\left(1 - \frac{1}{3}\right)\left(1 - \frac{1}{5}\right) \cdots} \tag{29.1}$$

to prove that the series of the reciprocals of primes $\sum \frac{1}{p}$ was divergent. Of course, this implied that the number of primes was infinite. It is obvious that the series and product in Euler's formula are divergent but, as discussed in Chapter 16, this defect is easy to remedy. In the same paper, Euler studied numerous Dirichlet series and their infinite products, including

---

[1] Eu. I-4 pp. 146–162. E 596.
[2] Legendre (1808) pp. 399–406.
[3] Gauss (1965) pp. 461–462.
[4] Dirichlet (1969) vol. 1, p. 316.
[5] Dirichlet (1969) vol. 1, pp. 411–493.
[6] Eu. I-14 pp. 217–244, especially pp. 242–244. E 72.

$$\frac{\pi}{4} = 1 - \frac{1}{3} + \frac{1}{5} - \frac{1}{7} + \cdots = \frac{1}{\left(1 - \frac{1}{3}\right)\left(1 + \frac{1}{5}\right)\left(1 - \frac{1}{7}\right)\left(1 - \frac{1}{11}\right)\left(1 + \frac{1}{13}\right)\cdots},$$

(29.2)

where the primes of the form $4n + 3$ appeared with a negative sign and those of the form $4n + 1$ had a positive sign. This led him to the series

$$\frac{1}{3} - \frac{1}{5} + \frac{1}{7} + \frac{1}{11} - \frac{1}{13} - \frac{1}{17} + \frac{1}{19} + \cdots.$$

(29.3)

In a letter to Goldbach dated October 28, 1752,[7] Euler wrote that he had found the sum of this series to be approximately 0.334980, implying that the series $\sum \frac{1}{p}$ and $\sum \frac{1}{q}$, where p and q were primes of the form $4n + 1$ and $4n + 3$, respectively, were both divergent. Euler's results were published in a posthumous paper of 1785,[8] also containing his conjecture for the case $l = 1$, and in which he gave the sum of (29.3) as 0.3349812.

In spite of Gauss's and Dirichlet's remarks on the error in Legendre's reasoning, Legendre, not known for taking heed of the criticism of others, included his flawed proof in the 1808 and 1830 editions of his book. In a paper of 1838,[9] Dirichlet stated that he unsuccessfully tried to prove the troublesome lemma, finding it at least as difficult to prove as the theorem deduced from it. In 1859, Athanase Dupré (1808–1869) published his proof that Legendre's lemma was false,[10] for which he was awarded half the 1858 *Gran Prix* from the French Academy of Sciences.

In his 1837 paper presented to the Berlin Academy,[11] Dirichlet wrote that he based his ideas on chapter 15 of Euler's *Introductio in Analysin Infinitorum*. He expressed the sum of the reciprocals of the primes in the given arithmetic progression as an appropriate linear combination of the logarithms of the $p - 1$ L-series arising from the $p - 1$ characters modulo p. He then had to prove the divergence of this expression, based on the divergence of the series corresponding to the trivial character, $\ln L_0(1)$. Then, in order to maintain the singularity of $L_0(1)$, he had to show that the values $L_k(1)$ did not vanish. For L-series arising from complex characters, Dirichlet was easily able to do this. However, it was much more difficult to prove that the L-series produced by the real character, defined by the Legendre symbol, did not vanish. To tackle this problem, Dirichlet first reduced the infinite series to a finite sum and considered two cases of primes: those of the form $4m + 3$, and then $4m + 1$. The first case was relatively easy; for the second case, he used a result on Pell's equation, from the *Disquisitiones*. The appearance of Pell's equation may have alerted Dirichlet to the connection between L-functions for real characters and quadratic forms. In fact, this allowed him to prove in 1839 that the class number of the binary quadratic forms

[7] Fuss (1968) pp. 586–591, especially p. 587.
[8] E 596.
[9] Dirichlet (1969) vol. 1, pp. 357–374, especially p. 357.
[10] Dupré (1859).
[11] Dirichlet (1969) vol. 1, pp. 309–312, especially p. 310.

of a given determinant could be evaluated in terms of the value of the L-function at 1, implying that the function did not vanish.[12]

With these papers, based on Euler's work on series, Dirichlet established analytic number theory as a distinct new branch of mathematics; he applied infinite series to the derivation of the class number formula, to the problem of primes in arithmetic progression, and to the evaluation of Gauss sums, leading to a proof of the quadratic reciprocity law. Interestingly, Gauss wrote Dirichlet in 1838 that he had worked with similar ideas around 1801, but he regretted not finding the time to develop and publish them.[13] Indeed, an incomplete manuscript among Gauss's unpublished papers, now included in the second volume of Gauss's collected works,[14] gave a partial outline for the theory of the class number formula. According to Mathews:[15]

> From this [Gauss's manuscript] it appears that Gauss succeeded in determining the number of classes belonging to a determinant both for definite and indefinite forms; and with regard to definite forms it is possible to make out the method that was actually adopted.

We parenthetically note that the ancient Babylonians considered particular cases of Pell's equation $x^2 - ny^2 = 1$, where n is a nonsquare positive integer; in India, Brahmagupta in the 600s and Bhaskara in the 1100s gave procedures for solving it.[16] William Brouncker can be credited with giving a general method for its solution in a 1657 letter[17] to Wallis, in response to a challenge from Fermat; Lagrange finally gave a rigorous derivation in 1768.[18]

Dirichlet's proof of the nonvanishing of the L-series was somewhat roundabout, but a more direct proof was published by the Belgian mathematician Charles de la Vallée-Poussin (1866–1962) in his 1896 paper "Démonstration simplifée du théorème de Dirichlet sur la progression arithmétique."[19] Vallée-Poussin took the L-functions to be functions of a complex variable and then employed analytic function theory to give his elegant proof. Interestingly, he made use of a construction also given by Dirichlet. Vallée-Poussin, who made many contributions to various areas of analysis, studied at the university at Louvain under L. P. Gilbert, whom he succeeded as professor of mathematics at the age of 26.

As early as 1861–62, Hermann Kinkelin of Basel studied L-functions of complex variables,[20] proving their functional relation for characters modulo a prime power. And in 1889 Rudolf Lipschitz, using the Hurwitz zeta function, proved the latter result for general Dirichlet characters.[21] Between 1895 and 1899,[22] Franz Mertens gave proofs of the nonvanishing of the L-series by elementary methods, that is, without the

---

[12] Dirichlet (1969) vol. 1, pp. 499–502.

[13] For an English translation of this letter, see Scharlau and Opolka (1984) pp. 178–779.

[14] Gauss (1863–1927) vol. 2, p. 269.

[15] Mathews (1961) p. 230.

[16] Datta and Singh (1962) pp. 146–172.

[17] Wallis (1693–1699) vol. 2, p. 797.

[18] Lagrange (1867–1892) vol. 2, pp. 494–496.

[19] Valleé-Poussin (1890).

[20] Kinkelin (1861–1862).

[21] Lipschitz (1889).

[22] In particular, Mertens (1895), (1897).

use of quadratic forms or functions of a complex variable. One of these proofs used a technique from an 1849 paper of Dirichlet on the average behavior of the divisor function. This result now has many elementary proofs, but the simplest may be due to Paul Monsky in 1994,[23] based on the earlier elementary proof of A. Gelfond and Yuri Linnik, published in Russian in 1962.[24]

29.2 Euler: Sum of Prime Reciprocals

In his 1737 paper "Variae Observationes circa Series Infinitas,"[25] Euler showed that the sum of the reciprocals of primes

$$\frac{1}{2} + \frac{1}{3} + \frac{1}{5} + \frac{1}{7} + \frac{1}{11} + \frac{1}{13} + \text{etc.}$$

was of infinite magnitude and was, moreover, the logarithm of the harmonic series

$$1 + \frac{1}{2} + \frac{1}{3} + \frac{1}{4} + \frac{1}{5} + \text{etc.}$$

Taking the logarithm of (29.1), Euler got

$$\ln\left(1 + \frac{1}{2} + \frac{1}{3} + \frac{1}{4} + \cdots\right) = -\ln\left(1 - \frac{1}{2}\right) - \ln\left(1 - \frac{1}{3}\right) - \ln\left(1 - \frac{1}{5}\right) - \cdots,$$

$$= \frac{1}{2} + \frac{1}{3} + \frac{1}{5} + \cdots + \frac{1}{2}\left(\frac{1}{2^2} + \frac{1}{3^2} + \frac{1}{5^2} + \cdots\right)$$

$$+ \frac{1}{3}\left(\frac{1}{2^3} + \frac{1}{3^3} + \frac{1}{5^3} + \cdots\right) + \cdots$$

$$\equiv A + \frac{1}{2}B + \frac{1}{3}C + \cdots.$$

He could express this relation as

$$e^{A + \frac{1}{2}B + \frac{1}{3}C + \frac{1}{4}D + \text{etc.}} = 1 + \frac{1}{2} + \frac{1}{3} + \frac{1}{4} + \frac{1}{5} + \frac{1}{6} + \frac{1}{7} + \text{etc.}$$

He then observed that since the harmonic series diverged to ∞ and the series B, C, D, etc. were finite, the series

$$\frac{1}{2}B + \frac{1}{3}C + \frac{1}{4}D + \text{etc.}$$

was negligible and hence

$$e^A = 1 + \frac{1}{2} + \frac{1}{3} + \frac{1}{4} + \frac{1}{5} + \text{etc.}$$

[23] Monsky (1994).
[24] English translation: Gelfond and Linnik (1966).
[25] Eu. I-14 pp. 217–244, especially pp. 242–244. E 72.

By taking the logarithm of both sides, he obtained his result:

$$\frac{1}{2} + \frac{1}{3} + \frac{1}{5} + \frac{1}{7} + \frac{1}{11} + \frac{1}{13} + \frac{1}{17} + \text{ etc.}$$

$$= \ln\left(1 + \frac{1}{2} + \frac{1}{3} + \frac{1}{4} + \frac{1}{5} + \text{ etc.}\right).$$

Euler then noted that the harmonic series summed to $\ln \infty$ and hence the sum of the reciprocals of the primes was $\ln \ln \infty$. To understand this, recall that $\sum_{k=1}^{n} \frac{1}{k} \sim \ln n$; see equation (20.4).

29.3   Dirichlet: Infinitude of Primes in an Arithmetic Progression

As discussed in our Section 13.3, in 1758–1759, Waring and Simpson, starting with $f(x) = \sum_{n=0}^{\infty} a_n x^n$, used roots of unity to obtain an expression for $\sum_{n=0}^{\infty} a_{mn+l} x^n$. In other words, they used characters of the additive group $\mathbb{Z}_m$ to extract from the power series the subsequence of terms in an arithmetic progression. Since the L-functions were multiplicative, Dirichlet had to define and use characters of the multiplicative group. An additional complication for Dirichlet was that he had to work with the logarithm of the L-functions and therefore had to prove their nonvanishing. In the case where $m = p$ was a prime, using some results of Gauss, he found an intricate proof of this fact, published in his 1837 paper "Beweiss des Satzes, dass jede unbegrenzte arithmetische Progression."[26] Dirichlet supposed p to be a prime and set

$$\Omega^k = e^{\frac{2\pi i k}{p-1}}, k = 0, 1, \ldots, p - 1.$$

He let L_k denote the L-function defined by the product

$$L_k(s) = \Pi \left(1 - \frac{\omega^{\gamma_q}}{q^s}\right)^{-1}, \quad \text{where} \quad \omega = \Omega^k = e^{\frac{2\pi i k}{p-1}},$$

and the product was taken over all primes $q \neq p$. Note that γ_q is defined by means of a generator of the multiplicative cyclic group of the integers modulo p. Then

$$\log L_k = \sum \frac{\omega^{\gamma_q}}{q^s} + \frac{1}{2} \sum \frac{\omega^{2\gamma_q}}{q^{2s}} + \frac{1}{3} \sum \frac{\omega^{3\gamma_q}}{q^{3s}} + \cdots.$$

To extract the primes in the arithmetic progression identical to 1 modulo p, Dirichlet first observed that for any integer h

$$1 + \Omega^{h\gamma} + \Omega^{2h\gamma} + \cdots + \Omega^{(p-2)h\gamma} = \begin{cases} p - 1, & h\gamma \equiv 0 \pmod{p - 1}, \\ 0 & h\gamma \not\equiv 0 \pmod{p - 1}. \end{cases}$$

---

[26] Dirichlet (1969) vol. 1, pp. 313–342.

It followed that

$$\log(L_0 L_1 \cdots L_{p-2}) = (p-1)\left(\sum \frac{1}{q^s} + \frac{1}{2}\sum \frac{1}{q^{2s}} + \frac{1}{3}\sum \frac{1}{q^{3s}} + \cdots\right), \quad (29.4)$$

where the primes q in the first sum satisfied $q \equiv 1 \pmod{p}$; those in the second sum satisfied $q^2 \equiv 1 \pmod{p}$; those in the third $q^3 \equiv 1 \pmod{p}$; and so on. The second and later sums were convergent for $s \geq 1$; to show that $\sum \frac{1}{q}$ was divergent, Dirichlet had to focus on the behavior of $L_0, L_1, \ldots, L_{p-2}$ as $s \to 1^+$. Dirichlet first expressed the series as an integral; he calculated that for any positive real number k,

$$S = \frac{1}{k^{1+\rho}} + \frac{1}{(k+1)^{1+\rho}} + \frac{1}{(k+2)^{1+\rho}} + \cdots$$

$$= \frac{1}{\Gamma(1+\rho)}\int_0^1 \log^\rho\left(\frac{1}{x}\right) \frac{x^{k-1}}{1-x}\,dx .$$

$$= \frac{1}{\rho} + \frac{1}{\Gamma(1+\rho)}\int_0^1 \left(\frac{x^{k-1}}{1-x} - \frac{1}{\log(\frac{1}{x})}\right)\log^\rho\left(\frac{1}{x}\right)\,dx.$$

He observed that the integral was convergent as $\rho \to 0+$. Note that since the series $L_0(s)$ is given by $\sum \frac{1}{m^s}$ where the sum is over all integers m not divisible by p, Dirichlet could write

$$L_0(1+\rho) = \sum_{m=1}^{p-1}\sum_{l=0}^{\infty} \frac{1}{(m+lp)^{1+\rho}}.$$

Next, he applied the foregoing integral representation for the series to obtain

$$\sum_{l=0}^{\infty}\frac{1}{(m+lp)^{1+\rho}} = \frac{1}{p^{1+\rho}}\sum_{l=0}^{\infty}\frac{1}{(\frac{m}{p}+l)^{1+\rho}} = \frac{1}{p}\cdot\frac{1}{\rho} + \phi(\rho)$$

where $\phi(\rho)$ had a finite limit as $\rho \to 0^+$. So Dirichlet could conclude that

$$L_0(1+\rho) = \frac{p-1}{p}\cdot\frac{1}{\rho} + \phi(\rho),$$

where $\lim_{\rho \to 0^+}\phi(\rho)$ was finite. This implied that $\log L_0(1+\rho)$ behaved like $-\log\rho$ as $\rho \to 0^+$. Dirichlet also showed that the series $L_1(1), L_2(1), \ldots, L_{p-2}(1)$ were convergent. Thus, if $L_j(1) \neq 0$ for $j = 1, 2, \ldots, p-2$, then the product $L_0 L_1 \cdots L_{p-2}$ had to diverge as $\rho \to 0^+$, and the series $\sum \frac{1}{q}$ for $q \equiv 1 \pmod{p}$ would also diverge. This proved that there existed an infinity of primes of the form $pl + 1$.

Dirichlet found a simple proof that for $j \neq \frac{p-1}{2}, L_j(1) \neq 0$. For such a $j, \Omega^{p-1-j} \neq \Omega^j$; Dirichlet therefore considered the product $L_j L_{p-1-j}$. Recall that

$$L_j(s) = \frac{1}{\Gamma(s)}\int_0^1 \frac{\frac{1}{x}f(x)}{1-x^p}\log^{s-1}\left(\frac{1}{x}\right)\,dx = \psi(s) + \chi(s)\sqrt{-1}.$$

Dirichlet noted that $L_j(s)$ was differentiable for $s > 0$, and hence by the mean value theorem he had

$$\psi(1 + \rho) = \psi(1) + \rho\psi'(1 + \delta\rho), \ \chi(1 + \rho) = \chi(1) + \rho\chi'(1 + \epsilon\rho),$$

where $0 < \delta < 1$ and $0 < \epsilon < 1$. Since $L_{p-1-j}(s)$ was the complex conjugate of $L_j(s)$, he got

$$L_{p-1-j}(s)L_j(s) = \psi^2(s) + \chi^2(s).$$

Next, if $L_j(1) = 0$, then $L_{p-1-j}(1) = 0$. This implied that $\psi(1) = 0$ and $\chi(1) = 0$. Thus,

$$\log L_j(1 + \rho)L_{p-1-j}(1 + \rho) = \log \rho^2 \left(\psi'^2(1 + \delta\rho) + \chi'^2(1 + \epsilon\rho)\right)$$

$$= -2\log \frac{1}{\rho} + \log \left(\psi'^2(1 + \delta\rho) + \chi'^2(1 + \epsilon\rho)\right).$$

These calculations implied that if $L_j(1) = 0$ for $j \neq \frac{p-1}{2}$, then

$$\log L_0 L_j L_{p-1-j} = -\log \frac{1}{\rho} + \phi(\rho),$$

and the term on the left-hand side tended to $-\infty$ as $\rho \to 0^+$. Clearly, $\log(L_0 L_1 \cdots L_{p-2})$ was positive from (29.4) so Dirichlet had come to a contradiction. This completed the proof for complex characters.

Dirichlet then dealt with the difficult case in which $j = \frac{p-1}{2}$. In this case, $\omega^{\frac{p-1}{2}} = e^{\pi i} = -1$ and hence the character and the series $L_{\frac{p-1}{2}}$ were real and given by

$$L_{\frac{p-1}{2}}(s) = \sum_{n=1}^{\infty} \left(\frac{n}{p}\right) \frac{1}{p^s}.$$

Recall Dirichlet's results from Chapter 28: When $p \equiv 3 \pmod 4$,

$$\sum \left(\frac{n}{p}\right) \frac{1}{n} = \frac{\pi}{p\sqrt{p}} \left(\sum b - \sum a\right),$$

and when $p \equiv 1 \pmod 4$,

$$\sum \left(\frac{n}{p}\right) \frac{1}{n} = \frac{1}{\sqrt{p}} \log \frac{\prod \sin(\frac{b\pi}{p})}{\prod \sin(\frac{a\pi}{p})},$$

where a and b were quadratic residues and nonresidues, respectively, modulo p. For $p \equiv 3 \pmod 4$, Dirichlet noted that

$$\sum a + \sum b = \sum_{m=1}^{p-1} m = \frac{p(p-1)}{2} = \text{an odd integer.}$$

Hence for $p \equiv 3 \pmod 4$, $\sum b - \sum a$ could not be zero and $L_{\frac{p-1}{2}}(1) \neq 0$. For $p \equiv 1$ (mod 4), Dirichlet used a result Gauss proved in section 357 of his *Disquisitiones*. This important result in cyclotomy stated that

$$2\prod_a \left(x - e^{\frac{2\pi i a}{p}}\right) = Y - Z\sqrt{p}, \quad 2\prod_b \left(x - e^{\frac{2\pi i b}{p}}\right) = Y + Z\sqrt{p}, \qquad (29.5)$$

where Y and Z were polynomials in x with integral coefficients; hence, Gauss had

$$Y^2 - pZ^2 = 4\prod_{k=1}^{p-1} \left(x - e^{\frac{2\pi i k}{p}}\right) = 4\frac{x^p - 1}{x - 1}.$$

Dirichlet set $g = Y(1)$, $h = Z(1)$ so that g and h were integers and $g^2 - ph^2 = 4p$; he could conclude that g was divisible by p. He could then set $g = pk$ to obtain $h^2 - pk^2 = -4$. Since p could not divide 4, he could write that $h \neq 0$. Next, when $x = 1$ in (29.5), he got

$$2\prod_a \left(1 - e^{\frac{2\pi i a}{p}}\right) = 2^{\frac{p+1}{2}}(-1)^{\frac{p-1}{4}} e^{\pi i \sum_a a} \prod_a \frac{e^{\frac{\pi i a}{p}} - e^{-\frac{\pi i a}{p}}}{2i}$$

$$= 2^{\frac{p+1}{2}}(-1)^{\frac{p-1}{4} + \sum_a a} \prod_a \sin\left(\frac{a\pi}{p}\right) = 2^{\frac{p+1}{2}} \prod_a \sin\left(\frac{a\pi}{p}\right).$$

Note that this last equation depends on the fact that when p is of the form $4n + 1$, a and $p - a$ are both quadratic residues, and the residues can be grouped in pairs. There are $\frac{p-1}{4}$ such pairs, and it follows that

$$\sum_a a = \sum_b b = \frac{1}{4}p(p-1).$$

Similarly,

$$2\prod_b \left(1 - e^{\frac{2\pi i b}{p}}\right) = 2^{\frac{p+1}{2}} \prod_b \sin\left(\frac{b\pi}{p}\right),$$

and thus, because $h \neq 0$,

$$\frac{\prod_b \sin\left(\frac{b\pi}{p}\right)}{\prod_a \sin\left(\frac{a\pi}{p}\right)} = \frac{k\sqrt{p} + h}{k\sqrt{p} - h} \neq 1.$$

This proved that $L_{\frac{p-1}{2}}(s)$ did not vanish at $s = 1$ and also that the number of primes $\equiv 1 \pmod p$ was infinite. To show that the number of primes $\equiv m \pmod p$ was infinite, Dirichlet gave a modified argument. He considered the sum

$$\log L_0 + \Omega^{-\gamma_m} \log L_1 + \Omega^{-2\gamma_m} \log L_2 + \cdots + \Omega^{-(p-2)\gamma_m} \log L_{p-2}$$

$$= (p-1)\left(\sum \frac{1}{q^{1+\rho}} + \frac{1}{2} \sum \frac{1}{q^{2+2\rho}} + \frac{1}{3} \sum \frac{1}{q^{3+3\rho}} + \cdots \right),$$

where the primes q in the first sum satisfied $q \equiv m \pmod{p}$ and those in the kth sum satisfied $q^k \equiv m \pmod{p}$. Since he had already proved that $\log L_1$, $\log L_2, \ldots,$ $\log L_{p-2}$ were finite as $\rho \to 0^+$ and that $\log L_0$ behaved like $\log(\frac{1}{\rho})$, Dirichlet could conclude that, when the sum was taken over primes $\equiv m \mod p$, $\sum \frac{1}{q}$ diverged.

29.4   Class Number and $L_\chi(1)$

In a paper of 1838 published in *Crelle's Journal*, "Sur l'usage des séries infinies dans la théorie des nombres,"[27] Dirichlet worked out some particular cases of his class number formula. In this formula, he expressed the class number, a necessarily nonvanishing quantity, in terms of $L_\chi(1)$. In order to give a definition of class number, we first observe that for a, b, c integers, $b^2 - ac$ is called the determinant or discriminant of the quadratic form $ax^2 + 2bxy + cy^2$. Two quadratic forms with the same determinant are in the same class if a linear substitution $x = \alpha x' + \beta y'$ and $y = \gamma x' + \delta y'$ with $\alpha \gamma - \beta \delta = 1$ transforms one quadratic form into the other. This basic definition can be traced to Lagrange. In addition, Lagrange proved that there was a finite number of such classes (called the class number) for a given negative discriminant. Note that Lagrange worked with b instead of $2b$.

In his 1838 paper, Dirichlet considered quadratic forms of determinant $-q$ with q prime. He separated his proof into the two cases $q = 4\nu + 3$ and $q = 4\nu + 1$; we present Dirichlet's proof of the former case. He denoted by f the primes for which the discriminant $-q$ was a quadratic residue and by g the primes for which it was not:

$$\left(-\frac{q}{f} \right) = \left(\frac{f}{q} \right) = 1, \quad \left(-\frac{q}{g} \right) = \left(\frac{g}{q} \right) = -1.$$

He then considered the L-series relations

$$\prod \frac{1}{1 - \frac{1}{f^s}} \prod \frac{1}{1 - \frac{1}{g^s}} = \sum \frac{1}{n^s},$$

$$\prod \frac{1}{1 - \frac{1}{f^s}} \prod \frac{1}{1 + \frac{1}{g^s}} = \sum \left(\frac{n}{q} \right) \frac{1}{n^s},$$

$$\prod \frac{1}{1 - \frac{1}{f^{2s}}} \prod \frac{1}{1 - \frac{1}{g^{2s}}} = \sum \frac{1}{n^{2s}},$$

---

[27] Dirichlet (1969) vol. 1, pp. 357–374.

where the n were odd numbers not divisible by q. He deduced that

$$\frac{\sum \frac{1}{n^s} \cdot \sum \left(\frac{n}{q}\right) \frac{1}{n^s}}{\sum \frac{1}{n^{2s}}} = \prod \frac{1 + \frac{1}{f^s}}{1 - \frac{1}{f^s}} = \prod \left(1 + \frac{2}{f^s} + \frac{2}{f^{2s}} + \frac{2}{f^{3s}} + \cdots\right) = \sum \frac{2^\mu}{m^s},$$

(29.6)

where the summation was over odd integers m divisible only by primes of the type f; μ was the number of distinct primes f by which m was divisible. Dirichlet denoted the inequivalent quadratic forms of determinant $-q$ by

$$ax^2 + 2bxy + cy^2, \ a'x^2 + 2b'xy + c'y^2, \ldots$$

and then observed that articles 180, 155, 156, and 105 of Gauss's *Disquisitiones* implied that

$$2\sum \frac{2^\mu}{m^s} = \sum \frac{1}{(ax^2 + 2bxy + cy^2)^s} + \sum \frac{1}{(a'x^2 + 2b'xy + c'y^2)^s} + \cdots,$$

(29.7)

where the summations on the right-hand side were taken over positive as well as negative values of x and y relatively prime to one another. From this Dirichlet deduced that

$$2\sum \frac{1}{n^s} \cdot \sum \left(\frac{n}{q}\right) \frac{1}{n^s} = \sum \frac{1}{n^{2s}} \cdot \sum \frac{1}{(ax^2 + 2bxy + cy^2)^s} + \cdots. \qquad (29.8)$$

Without giving details, Dirichlet remarked in this paper that "by means of geometric considerations" it could be proved that

$$\sum \frac{1}{n^{2(1+\rho)}} \cdot \sum \frac{1}{(ax^2 + 2bsy + cy^2)^{1+\rho}} \sim \frac{q-1}{2q\sqrt{q}} \cdot \frac{\pi}{\rho}, \quad \rho \to 0^+. \qquad (29.9)$$

Thus, given h different inequivalent forms of determinant $-q$, that is, if h were the class number, then the right-hand side of (29.8) could be expressed as

$$\frac{h(q-1)}{2q\sqrt{q}} \cdot \frac{\pi}{\rho}, \quad \rho \to 0^+. \qquad (29.10)$$

On the other hand, since

$$\sum \frac{1}{n^s} = \left(1 - \frac{1}{2^s}\right)\left(1 - \frac{1}{q^s}\right)\zeta(s),$$

he had

$$\sum \frac{1}{n^{1+\rho}} \sim \frac{q-1}{2q} \cdot \frac{1}{\rho} \quad \text{as } \rho \to 0^+. \qquad (29.11)$$

He applied (29.10) and (29.11) to (29.8) and found a special case of his famous class number formula

$$h = \frac{2\sqrt{q}}{\pi} \sum \left(\frac{n}{q}\right) \frac{1}{n}, \tag{29.12}$$

a formula that expressed the class number in terms of the value of an L-series at $s = 1$. Since the class number had to be at least one, the series had a nonzero value. In 1839, Dirichlet published a proof along similar lines of this result for arbitrary negative determinants. Recall that Dirichlet made liberal use of results from Gauss in his proofs; his contemporaries reported that his copy of the *Disquisitiones* was never kept on the shelf, but on his writing table, and that it always accompanied him on his travels. Through his lectures on number theory, published by Dedekind, Dirichlet made the work of Gauss accessible to all his students.

29.5　Vallée-Poussin's Complex Analytic Proof of $L_\chi(1) \neq 0$

Before he published his famous work on the distribution of primes, Vallée-Poussin published a paper presenting a simpler proof of Dirichlet's theorem on primes in arithmetic progressions, observing that his proof was more natural since it did not depend on the theory of quadratic forms. In this paper, presented to the Belgian Academy in 1896,[28] Vallée-Poussin defined $L_\chi(s)$, where χ was a character modulo an integer M, as a function of a complex variable s. By a simple argument, he showed that for the principal character χ_0, $L_{\chi_0}(s)$ was an analytic function for Re $s > 0$, except for a simple pole at $s = 1$. On the other hand, for any nonprincipal character, the corresponding L-function was analytic for Re $s > 0$ with no exception. Vallée-Poussin's proof that $L_\chi(1) \neq 0$ employed a function similar to one constructed by Dirichlet in his discussion of quadratic forms of negative discriminant. He let χ be a real nonprincipal character; he let q_1 denote primes for which $\chi(q_1) = 1$ and let q_2 denote primes for which $\chi(q_2) = -1$. He set

$$\psi(s) = \frac{L_\chi(s) L_{\chi_0}(s)}{L_{\chi_0}(2s)}.$$

Then, for Re $s > 1$, he observed that

$$\psi(s) = \prod \frac{1 + q_1^{-s}}{1 - q_1^{-s}} = \prod (1 + 2q_1^{-s} + 2q_1^{-2s} + \cdots) = \sum_{n=1}^{\infty} \frac{a_n}{n^s},$$

where $a_n \geq 0$. In addition, since $L_{\chi_0}(2s)$ had a pole at $s = \frac{1}{2}$, he deduced that $\psi(s) = 0$ at $s = \frac{1}{2}$. Vallée-Poussin also observed that there was at least one prime q_1. If not, then $\psi(s) = 1$ for Re $s > 1$ and by analytic continuation $\psi(\frac{1}{2}) = 1$. This contradicted $\psi(\frac{1}{2}) = 0$, and hence $a_n > 0$ for some n in $\sum \frac{a_n}{n^s}$.

---

[28] Vallée-Poussin (1896a).

In order to obtain a proof by contradiction, Vallée-Poussin next assumed that $L_\chi(s) = 0$ at $s = 1$, so that this zero would cancel the pole of $L_{\chi_0}(s)$ at $s = 1$ and $L_\chi(s)L_{\chi_0}(s)$ would be analytic for Re $s > 0$. Again for Re $s > 1$, the derivatives of $\psi(s)$ were given by

$$\psi^{(m)}(s) = (-1)^m \sum_{n=1}^{\infty} \frac{a_n (\log n)^m}{n^s}, \quad m = 1, 2, 3, \ldots .$$

He let $a > 0$ so that $\psi(1 + a + t)$ had radius of convergence greater than $a + \frac{1}{2}$ and

$$\psi(1 + a + t) = \psi(1 + a) + t\psi'(1 + a) + \frac{t^2}{2!}\psi''(1 + a) + \cdots .$$

Denoting $(-1)^m \psi(1 + a)$ by A_m, it was clear that $A_m > 0$ and for $t = -(a + \frac{1}{2})$

$$\psi\left(\frac{1}{2}\right) = \psi(1 + a) + \left(a + \frac{1}{2}\right)A_1 + \frac{(a + \frac{1}{2})^2}{2!}A_2 + \cdots .$$

Since $\psi(\frac{1}{2}) = 0$ and the all the terms on the right were positive, Vallée-Poussin arrived at the necessary contradiction.

29.6　Gelfond and Linnik: Proof of $L_\chi(1) \neq 0$

Gelfond and Linnik's proof that $L_\chi(1) \neq 0$ for any real nonprincipal character χ modulo m was presented in their 1962 book.[29] They made the observation that if ϕ denoted the Euler totient function and $T(n) = \sum_{k=1}^{n} \chi(k)$, then for any positive integer N

$$|T(N) - T(n - 1)| < \phi(m) \quad \text{and} \quad \left|\sum_{k=n}^{N} \frac{\chi(k)}{k}\right| < \frac{2\phi(m)}{n}. \qquad (29.13)$$

Note that the second inequality follows by partial summation. Gelfond and Linnik defined the function

$$U(x) = \sum_{n=1}^{\infty} \frac{\chi(n)x^n}{1 - x^n} = \sum_{n=1}^{\infty}\left(\sum_{d|n} \chi(d)\right) x^n \qquad (29.14)$$

and showed that

$$U(x) > \frac{1}{2\sqrt{1 - x}}. \qquad (29.15)$$

---

[29] Gelfond and Linnik (1966).

They then proved that if $L_\chi(1) = 0$, then

$$U(x) = O\left(\ln\frac{1}{1-x}\right). \tag{29.16}$$

Clearly, (29.15) and (29.16) were in contradiction to one another, indicating that the assumption $L_\chi(1) = 0$ had to be false. Next, to prove (29.15), Gelfond and Linnik made the important observation that

$$f_n \equiv \sum_{d\mid n} \chi(d) = \prod_{k=1}^{s}\left(1 + \chi(p_k) + \cdots + \chi^{v_k}(p_k)\right), \quad \left(n = p_1^{v_1}p_2^{v_2}\cdots p_s^{v_s}\right).$$

Since $\chi(p) = 1, -1$, or 0, it followed that $f_n \geq 0$. Then, if n were a square, all the v_k would be even and each of the s factors of f_n would be ≥ 1. Thus, $f_n \geq 1$ when n was a square. Hence, with $1 > x > \frac{1}{2}, x > x_0$

$$U(x) > \sum_{n=1}^{\infty} x^{n^2} = \int_1^{\infty} x^{t^2}\, dt + O(1)$$

$$= \frac{1}{\sqrt{-\ln x}}\int_0^{\infty} e^{-t^2}\, dt + O(1) = \frac{\sqrt{\pi}}{2\sqrt{-\ln x}} + O(1)$$

$$= \frac{\sqrt{\pi}}{2}\frac{1}{\left(-\ln(1-(1-x))\right)^{\frac{1}{2}}} + O(1) > \frac{1}{2\sqrt{1-x}}.$$

We note that Gelfond and Linnik set $x > x_0$ for some x_0 such that the inequalities would hold. To prove (29.16), they set

$$S_n = \sum_{k=n}^{\infty} \frac{\chi(k)}{k}, \quad S_1 = L_\chi(1);$$

$$R_1(x) = U(x) - \frac{L_\chi(1)}{1-x}$$

$$= \sum_{n=1}^{\infty} \chi(n)\frac{x^n}{1-x^n} - \sum_{n=1}^{\infty}\frac{\chi(n)}{n}\frac{x^n}{1-x^n} + \sum_{n=1}^{\infty}(S_n - S_{n+1})\frac{x^n}{1-x} - \frac{L_\chi(1)}{1-x}$$

$$= \sum_{n=1}^{\infty} \chi(n)\left(\frac{x^n}{1-x^n} - \frac{x^n}{n(1-x)}\right) - \sum_{n=0}^{\infty} S_{n+1}x^n. \tag{29.17}$$

To see how they arrived at the last equation, observe that

$$\frac{1}{1-x}\left(\sum_{n=1}^{\infty}(S_n - S_{n+1})x^n - L_\chi(1)\right)$$

$$= \frac{1}{1-x}\left(S_1 x + \sum_{n=1}^{\infty} S_{n+1}x^n(1-x) - S_1\right) = -\sum_{n=0}^{\infty} S_{n+1}x^n.$$

Next, by (29.13), $S_n = O\left(\frac{1}{n}\right)$, and hence they could write the second sum in (29.17) as

$$\sum_{n=0}^{\infty} S_{n+1} x^n = O\left(\sum_{n=1}^{\infty} \frac{x^n}{n}\right) = O\left(\ln \frac{1}{1-x}\right).$$

By an application of Abel's summation by parts to the first sum in (29.17), they got

$$\left|\sum_{n=1}^{\infty} \chi(n) \left(\frac{x^n}{1-x^n} - \frac{x^n}{n(1-x)}\right)\right|$$

$$= \left|\sum_{n=1}^{\infty} (T(n) - T(n-1)) \left(\frac{x^n}{1-x^n} - \frac{x^n}{n(1-x)}\right)\right|$$

$$= \left|\sum_{n=1}^{\infty} T(n) \left(\frac{x^n}{1-x^n} - \frac{x^n}{n(1-x)} - \frac{x^{n+1}}{1-x^{n+1}} + \frac{x^{n+1}}{(n+1)(1-x)}\right)\right|$$

$$< \frac{\phi(m)}{1-x} \sum_{n=1}^{\infty} \left|\frac{x^n}{1+x+\cdots+x^{n-1}} - \frac{x^{n+1}}{1+x+\cdots+x^n} - \frac{x^n}{n(n+1)} - \frac{(1-x)x^n}{n+1}\right|$$

$$< \frac{\phi(m)}{1-x} \sum_{n=1}^{\infty} \left(\frac{x^n}{1+x+\cdots+x^{n-1}} - \frac{x^{n+1}}{1+x+\cdots+x^n} - \frac{x^n}{n(n+1)}\right)$$

$$+ \phi(m) \sum_{n=1}^{\infty} \frac{x^n}{n+1}$$

$$= 2\phi(m) \sum_{n=1}^{\infty} \frac{x^n}{n+1} = O\left(\ln \frac{1}{1-x}\right).$$

Note that the final inequality was possible because the expression in the first sum was positive; then, since the series was telescoping, it would sum to $\frac{x^n(1-x)}{n+1}$. It then followed from (29.17) that if $L_\chi(1) = 0$, then $U(x) = O\left(\ln \frac{1}{1-x}\right)$; this completed the proof.

29.7 Monsky's Proof That $L_\chi(1) \neq 0$

In 1994, Paul Monsky showed that Gelfond's proof of $L_\chi(1) \neq 0$ could be considerably simplified.[30] Use of the strong result (29.15) turned out to be avoidable. Observe that $\lim_{x \to 1-} U(x) = \infty$, because $U(x) > \sum_{n=1}^{\infty} x^{n^2}$. Monsky demonstrated that if $L_\chi(1)$ vanished, then $U(x)$ was bounded. This contradiction proved the result. His simplification took place in the first sum, $R_1(x)$, in (29.17), where

$$R_1(x) = -\sum_{n=1}^{\infty} \left(\frac{\chi(n)}{n(1-x)} - \frac{\chi(n)x^n}{1-x^n}\right) \equiv -\sum_{n=1}^{\infty} \frac{\chi(n)}{1-x} b_n.$$

---

[30] Monsky (1994).

Monsky first showed that $b_1 \geq b_2 \geq b_3 \geq \cdots$. He noted that

$$b_n - b_{n+1} = \frac{1}{n(n+1)} - \frac{x^n}{(1+x+\cdots+x^{n-1})(1+x+\cdots+x^n)}.$$

Then, by the inequality of the arithmetic and geometric means

$$1 + x + \cdots + x^{n-1} \geq n x^{\frac{n-1}{2}} \geq n x^{\frac{n}{2}}$$

and

$$1 + x + \cdots + x^n \geq (n+1) x^{\frac{n}{2}}.$$

Hence, $b_n \geq b_{n+1}$. Applying Abel's partial summation, Monsky wrote

$$\sum_{n=1}^{\infty} \frac{\chi(x)}{1-x} b_n \leq \frac{\phi(m)b_1}{1-x} = \phi(m). \tag{29.18}$$

He next assumed that $L_\chi(1) = 0$, and this implied $U(x) = R_1(x)$; but (29.18) in turn implied that $\lim_{x \to 1^-} U(x)$ could not be infinite. Thus, he got a contradiction to prove the result.

29.8 Exercises

(1) Investigate Chebyshev's assertion in an 1853 letter to Fuss that

$$\lim_{c \to 0} \left(e^{-3c} - e^{-5c} + e^{-7c} + e^{-11c} - e^{-13c} - e^{-17c} + e^{-19c} + e^{-23c} - \cdots \right)$$

diverges to $+\infty$. See Chebyshev (1899–1907) vol. 1, p. 697. See also Hardy (1966–1979) vol. 2, pp. 42–49, where Hardy and Littlewood derive it from the extended Riemann hypothesis for the series $1^{-s} - 3^{-s} + 5^{-s} - 7^{-s} + \cdots$. Note the editor's comment on this result on p. 98.

(2) Let χ be a real nonprincipal character modulo m, and let $f(n) = \sum \chi(d)$, where the sum is over all divisors d of n. Show that if

$$G(x) = \sum_{n \leq x} \frac{f(n)}{\sqrt{n}}, \quad \text{then} \quad \lim_{x \to \infty} G(x) = \infty.$$

Show also that

$$G(x) = 2\sqrt{x} L(1, \chi) + O(1).$$

Conclude that if $L(1, \chi) = 0$, a contradiction ensues. See Mertens (1895).

(3) Suppose χ is a primitive character mod d, that is, there does not exist a proper divisor m of d such that $\chi(a) = \chi(b)$ whenever $a \equiv b$ and ab is prime to d. Define the Gauss sum $G(\chi)$ by

$$G(\chi) = \sum_{a=1}^{d} \chi(a) e^{\frac{2\pi i a}{d}}.$$

Prove that

$$\overline{\chi}(n)G(\chi) = \sum_{a=1}^{d} \chi(a)e^{\frac{2\pi i n a}{d}}.$$

This result is due to Vallée-Poussin (2000) vol. 1, pp. 358–362; he also defined the concept of a primitive character.

(4) Define the Euler polynomials $E_n(t)$ by the relation

$$\frac{2e^{tx}}{1+e^x} = \sum_{n=0}^{\infty} \frac{E_n(t)}{n!}x^n.$$

Let χ be a primitive character (mod d) and let k be a positive integer such that $\chi(-1) = (-1)^k$. Prove that if q is the greatest integer in $\frac{d-1}{2}$, then

$$\frac{(k-1)!}{(2\pi i)^k}G(\overline{\chi})L(k, \chi) = \frac{1}{2(2^k - \chi(2))} \sum_{a=1}^{q} \overline{\chi}(a)E_{k-1}\left(\frac{2a}{d}\right).$$

This formula is due to Shimura (2007) p. 35; in this book, Shimura observed that most books and papers give only one result on the values of the Dirichlet L-function. Shimura derived several new formulas for these values, including the foregoing example. For a discussion of Shimura's well-known conjecture related to Fermat's theorem, see Gouvêa (1994) and Shimura (2008).

(5) Prove that if the set of positive integers is partitioned into a disjoint union of two nonempty subsets, then at least one of the subsets must contain arbitrarily long arithmetic progressions. This result was conjectured by I. Schur and proved in 1927 by van der Waerden, who studied under E. Noether. The reader may enjoy reading the proof in Khinchin (1998), a book originally written in 1945 as a letter to a soldier recovering from his wounds. In 1927, van der Waerden's theorem was a somewhat isolated result, but it has now become a part of Ramsey theory, an important area of combinatorics. See, Graham, Rothschild, and Spencer (1990). Robert Ellis's algebraic methods in topological dynamics also have applications to this topic. See Ellis, Ellis, and Nerurkar (2000).

(6) Prove that the primes contain arbitrarily long arithmetic progressions. For this result of Ben Green and Terrence Tao, see Green's article in Duke and Tschinkel (2007).

29.9 Notes on the Literature

A historical account of the topic of this chapter was given by Littlewood's student Davenport (1980). This book was very influential because of its treatment of the large sieve, a relatively new topic at the time of first publication in 1967. The extensive notes in each chapter refer to numerous papers and books on this and related topics.

30

Distribution of Primes: Early Results

30.1 Preliminary Remarks

Prime numbers appear to be distributed among the integers in a random way. Mathematicians have searched for a pattern or patterns in the sequence of primes, discovering many interesting features and properties of primes and sequences of primes, but many fundamental questions remain outstanding. In the area of prime number distribution, even apparently very elementary results can be enlightening. For example, in 1737, Euler proved that the series $\sum \frac{1}{p}$, where p is prime, was divergent.[1] He also knew that $\sum_{n=1}^{\infty} \frac{1}{n^2}$ was convergent. By combining these results, one may see that the prime numbers are more numerous than the square numbers. Thus, for large enough x, we expect that $\pi(x)$, the number of primes less than or equal to x, satisfies $\pi(x) > \sqrt{x}$. In fact, extending this type of reasoning, we may expect that

$$\pi(x) > x^{1-\delta} \tag{30.1}$$

for any $\delta > 0$ and x correspondingly large enough. Recall that, in fact, Euler had a fairly definite idea of how the series of prime reciprocals diverged:

$$\sum \frac{1}{p} = \ln(\ln \infty). \tag{30.2}$$

From this, it can easily be shown, by means of a nonrigorous, probabilistic argument, that the density of primes in the interval $(1, x)$ is approximately $\frac{1}{\ln x}$. In 1791 or 1792, when he was about 15 years old, Gauss conjectured just this result.

Gauss never published anything on the distribution of primes, but in 1849 he wrote a letter to the astronomer J. F. Encke giving some insight into his thought in this area.[2] Gauss recounted that he had started making a table of prime numbers from a very young age, noting the number of primes in each chiliad, or interval of a thousand. As

---

[1] Eu. I-14 pp. 217–244. E 72.

[2] For an English translation of this letter, see Goldstein (1973).

a consequence of this work, around 1792 he wrote the following remark in the margin of his copy of J. C. Schulze's mathematical tables:

$$\text{Primzahlen unter } a \ (= \infty) \quad \frac{a}{la}.$$

We may understand this to mean that

$$\lim_{x \to \infty} \frac{\pi(x)}{\frac{x}{\ln x}} = 1. \tag{30.3}$$

This is the prime number theorem. Let us see how Gauss may have come to this conclusion. Consider the following table:

x	$\pi(x)$	$\pi(x)/x$
10	4	0.4
100	25	0.25
1000	168	0.168
10000	1229	0.1229
100000	9592	0.09592
1000000	78498	0.078498

Look at the column for $\frac{\pi(x)}{x}$. If we divide 0.4, the number in the first row, by 2, 3, 4, 5, 6, then we get approximately the numbers in the second, third, fourth, fifth, and sixth rows. The result is even nicer if we change the 0.4 to 0.5 and then do the division. So if we write $\frac{\pi(x)}{x}$ as $\frac{1}{f(x)}$, then $f(x)$ has the property that $f(10^n) = nf(10)$ for $n = 2, 3, 4, 5, 6$. This calculation strongly suggests that $f(x)$ is the logarithmic function. Moreover, $\frac{1}{f(10)} = 0.4$ and $\ln 10 = 2.3$; this may have led Gauss to his conjecture that $f(x) = \ln x$.

In his letter to Encke, Gauss suggested the approximation $\pi(x) \approx \int_2^x \frac{dt}{\ln t}$. In fact, he gave the following table of values for $\pi(x)$ and the corresponding values of the integral:

x	$\pi(x)$	$\int_2^x \frac{dt}{\ln t}$	error
500000	41556	41606.4	+50.4
1000000	78501	79627.5	+126.5
1500000	114112	114263.1	+151.1
2000000	148883	149054.8	+171.8
2500000	183016	183245.0	+229.0
3000000	216745	216970.6	+225.6.

Observe that there are inaccuracies in this table. Gauss made mistakes in his extensive calculations of primes, but the number of his mistakes is surprisingly small. For example, his value of the number of primes less than a million was overestimated by three, while he underestimated those under three million by 72.

In an 1810 letter[3] to the astronomer Olbers, F. W. Bessel (1784–1846) mentioned the logarithmic integral, now defined by

$$\mathrm{li}(x) = \lim_{\epsilon \to 0} \left(\int_0^{1-\epsilon} + \int_{1+\epsilon}^x \right) \frac{dt}{\ln t} = \int_2^x \frac{dt}{\ln t} + 1.0451. \tag{30.4}$$

Gauss had already computed this integral for several values of x and Bessel noted in his letter that he learned from Gauss that $\pi(4,000,000) = 33,859$, while the corresponding value of the logarithmic integral was 33,922.621995.

In his 1798 book on number theory, Legendre made a similar conjecture: that $\frac{x}{A \ln x + B}$ was a good approximation of $\pi(x)$ for suitable A and B. In the second edition of his book, published in 1808, he gave the values $A = 1$ and $B = -1.08366$.[4] Gauss observed in his letter to Encke that as the value of x was made larger, the value of B must likewise increase. However, Gauss was unwilling to conjecture that $B \to -1$ as $x \to \infty$. It is interesting to note that while Gauss was writing these thoughts to Encke, the Russian mathematician Chebyshev was developing his ideas on prime numbers, showing that if B tended to a limit as $x \to \infty$, then the limit had to be -1.[5]

In 1849, the Russian Academy of Sciences published a collection of Euler's papers on number theory. In 1847, the editor, Viktor Bunyakovski, solicited Chebyshev's participation in this project, thereby arousing his interest in number theory. Thus, in 1849 Chebyshev defended his doctoral thesis on theory of congruences, one of whose appendices discussed the number of primes not exceeding a given number. He there expressed doubt about the accuracy of Legendre's formula. He then went on to prove that if n was any fixed nonnegative integer and ρ was a positive real variable, then the sum

$$\sum_{x=2}^{\infty} \left(\pi(x+1) - \pi(x) - \frac{1}{\ln x} \right) \frac{\ln^n x}{x^{1+\rho}}, \tag{30.5}$$

considered as a function of ρ, approached a finite limit as $\rho \to 0$. We mention that Chebyshev wrote $\phi(x)$ for $\pi(x)$. From this theorem, he deduced that $\frac{x}{\pi(x)} - \ln x$ could not have a limit other than -1 as $x \to \infty$. He then observed that this result contradicted Legendre's formula, under which the limit was given as -1.08366.

Chebyshev wrote a second paper on prime numbers in 1850.[6] This important work was apparently motivated by Joseph Bertrand's conjecture that for all integers $n > 3$, there was at least one prime between n and $2n - 2$. In 1845, Bertrand used this conjecture to prove a theorem on symmetric functions.[7] In group theoretic terms, the theorem states that the index of a proper subgroup of the symmetric group S_n is either 2 or $\geq n$. Chebyshev proved Bertrand's conjecture using Stirling's approximation.

---

[3] See Erman (1852) vol. 1, p. 238.
[4] Legendre (1808) p. 394.
[5] Chebyshev (1899–1907) vol. 1, pp. 27–70.
[6] Chebyshev (1850).
[7] Bertrand (1845).

He also showed that the series $\sum_p \frac{1}{p \ln p}$ converged. The results he obtained implied the double inequality

$$0.92129 < \frac{\pi(x)}{\frac{x}{\ln x}} < 1.10555. \tag{30.6}$$

In a paper of 1881,[8] Sylvester used Chebyshev's analysis to give improved bounds, obtaining 0.95695 for the lower bound and 1.04423 for the upper bound. Though Schur and others have succeeded in narrowing the gap between the bounds,[9] it appears that Chebyshev's methods cannot be developed to give a proof of the prime number theorem. We note that, in order to prove Bertrand's conjecture, Chebyshev defined two arithmetical functions of interest even today:

$$\theta(x) = \sum_{p \leq x} \ln p, \tag{30.7}$$

$$\psi(x) = \theta(x) + \theta(x^{\frac{1}{2}}) + \theta(x^{\frac{1}{3}}) + \theta(x^{\frac{1}{4}}) + \cdots. \tag{30.8}$$

In fact, Chebyshev proved the inequalities

$$Ax - \frac{5}{2} \ln x - 1 < \psi(x) < \frac{6}{5} Ax + \frac{5}{4 \ln 6} \ln^2 x + \frac{5}{4} \ln x + 1, \tag{30.9}$$

where

$$A = \ln \left(\frac{2^{\frac{1}{2}} 3^{\frac{1}{3}} 5^{\frac{1}{5}}}{30^{\frac{1}{30}}} \right) = 0.92129202\ldots.$$

He then used inequalities (30.9) to indicate a method for obtaining the result for $\pi(x)$, though he did not give the results explicitly. Chebyshev also proved that if $\lim_{x \to \infty} \frac{\psi(x)}{x}$ existed, its value was 1. Note that this implies that if $\lim_{x \to \infty} \frac{\pi(x)}{\frac{x}{\ln x}}$ exists, then this limit too must be 1.

At the end of his paper, Sylvester noted that for a proof of the prime number theorem, "we shall probably have to wait until some one is born into the world as far surpassing Tchebycheff in insight and penetration as Tchebycheff has proved himself superior in these qualities to the ordinary run of mankind." Chebyshev's elementary but powerful methods formed the basis of a new topic, elementary methods in analytic number theory, and also served as motivation for Alphonse de Polignac (1826–1863) and Franz Mertens (1840–1927) to firmly establish this new subject.[10] In 1874, Mertens showed that Chebyshev's results could be used to obtain asymptotic formulas for the series

$$\sum_{p \leq x} \frac{\ln p}{p} \quad \text{and} \quad \sum_{p \leq x} \frac{1}{p}.$$

[8] Sylvester (1973) vol. 3, pp. 530–545.
[9] Shur (1929).
[10] Polignac (1857) and Mertens (1874a) and (1874b).

Thus, he proved the following refinement of a result of de Polignac:

$$\sum_{p \leq x} \frac{\ln p}{p} = \ln x + O(1), \tag{30.10}$$

where $O(1)$ denoted a quantity bounded as $x \to \infty$. Mertens also gave a more precise formulation of Euler's 1737 result (30.2):

$$\sum_{p \leq x} \frac{1}{p} = \ln \ln x + C + O\left(\frac{1}{\ln x}\right), \tag{30.11}$$

where

$$C = \gamma - \sum_{k=2}^{\infty} \frac{1}{k} \sum_{p} \frac{1}{p^k} \tag{30.12}$$

and γ denoted Euler's constant.

In his famous paper of 1859,[11] Riemann introduced ideas through which the prime number theorem would eventually be proved. Riemann's interest in prime number theory was not surprising, surrounded as he was by great researchers in this field. Riemann began his paper by mentioning Gauss, Euler, and his good friend and teacher Dirichlet, writing that their attention to the subject would surely justify its further study. He did not mention Chebyshev, but he was familiar with the work of Chebyshev to whom he sent a copy of his paper. Also, we know that as a student Riemann studied Legendre's number theory book very carefully. Moreover, Dirichlet stated in a note of 1838[12] that his analytic methods for studying primes could provide a proof of Legendre's conjecture related to the prime number theorem; Dirichlet, however, did not publish any ideas in this direction.[13]

Riemann based his investigation of $\pi(x)$ on Euler's product formula

$$\zeta(s) = \sum_{n=1}^{\infty} \frac{1}{n^s} = \prod_{p} (1 - p^{-s})^{-1}.$$

His innovation here was to take s to be a complex variable with $\mathrm{Re}\, s > 1$. He then defined $\zeta(s)$ as a contour integral, thereby extending its domain to the whole complex plane, except for the pole at $s = 1$. He used the Euler product to show that

$$\frac{\log \zeta(s)}{s} = \int_{1}^{\infty} f(x) x^{-s-1} dx, \quad \mathrm{Re}\, s > 1, \tag{30.13}$$

where

$$f(x) = F(x) + \frac{1}{2} F(x^{\frac{1}{2}}) + \frac{1}{3} F(x^{\frac{1}{3}}) + \cdots. \tag{30.14}$$

[11] Riemann (1859).

[12] Dirichlet (1969) vol. 1, pp. 353–356.

[13] Dirichlet's 1838 note was on asymptotic formulas in number theory; for his work in this area, see Dirichlet (1969) vol. 2, pp. 51–66 and pp. 99–104.

We here write log because complex variables are involved. Riemann defined $F(x)$ as the number of primes less than x when x was not prime; but when x was a prime,

$$F(x) = \frac{F(x+0) + F(x-0)}{2}.$$

Thus, Riemann's $F(x)$ was essentially $\pi(x)$. He obtained the integral representation for $f(x)$ by a method we now call Mellin inversion. Actually, he applied the Fourier inversion to get

$$f(y) = \frac{1}{2\pi i} \int_{a-\infty i}^{a+\infty i} \frac{\log \zeta(s)}{s} y^s ds, \quad a > 1. \tag{30.15}$$

To evaluate this integral, Riemann defined the entire function

$$\xi(s) = (s-1)\Pi\left(\frac{s}{2}\right)\pi^{-\frac{s}{2}}\zeta(s), \tag{30.16}$$

where $\Pi(s) = s\Gamma(s)$. He then obtained an infinite product (or Hadamard product) for $\xi(s)$, given by

$$\xi(s) = \xi(0) \prod_{\rho}\left(1 - \frac{s}{\rho}\right). \tag{30.17}$$

To use this formula effectively in (30.15), one must first understand the distribution of the zeros ρ. It is easy to show that $0 \leq \operatorname{Re}\rho \leq 1$. It follows from the functional equation for $\zeta(s)$ that if ρ is a zero, then so is $1 - \rho$. Riemann then observed that the number of roots of ρ whose imaginary parts lay between 0 and some value T was approximately

$$\frac{T}{2\pi}\log\frac{T}{2\pi} - \frac{T}{2\pi}, \tag{30.18}$$

where the relative error was of the order $\frac{1}{T}$. He sketched a one-sentence proof of this result and added that the estimate for the number of zeros with $\operatorname{Re}\rho = \frac{1}{2}$ was about the same as in (30.18). He remarked that it was very likely, though his passing attempts to prove it had failed, that all the roots had $\operatorname{Re}\rho = \frac{1}{2}$. This is the famous Riemann hypothesis.

By combining (30.15) and (30.17), and assuming the truth of his hypothesis, Riemann derived the formula

$$f(x) = \operatorname{li}(x) - \sum_{\alpha}\left(\operatorname{li}(x^{\frac{1}{2}+\alpha i}) + \operatorname{li}(x^{\frac{1}{2}-\alpha i})\right)$$
$$+ \int_{x}^{\infty}\frac{1}{t^2 - 1}\frac{dt}{t \log t} + \log\xi(0), \tag{30.19}$$

where the sum $\sum_\alpha$ was taken over all positive α such that $\frac{1}{2} + i\alpha$ was a zero of $\xi(s)$. Note that $\xi(0) = \frac{1}{2}$, though due to some confusion in Riemann's notation, he obtained a different value.

In the final remarks in his paper, Riemann noted first that $F(x)$ (or $\pi(x)$) could be obtained from $f(x)$ by the inversion

$$F(x) = \sum_{m=1}^{\infty} \frac{\mu(m)}{m} f(x^{\frac{1}{m}}),$$

where $\mu(m)$ was the Möbius function. We remark that Riemann did not use the brief notation $\mu(m)$. He also noted that the approximation $F(x) = \mathrm{li}(x)$ was correct only to an order of magnitude $x^{\frac{1}{2}}$, yielding a value somewhat too large, while better approximation was given by

$$\mathrm{li}(x) - \frac{1}{2}\mathrm{li}(x^{\frac{1}{2}}) - \frac{1}{3}\mathrm{li}(x^{\frac{1}{3}}) - \frac{1}{5}\mathrm{li}(x^{\frac{1}{5}}) + \frac{1}{6}\mathrm{li}(x^{\frac{1}{6}}) - \cdots. \tag{30.20}$$

Apart from the Riemann hypothesis, the most difficult part of Riemann's paper was his factorization of $\xi(s)$. Indeed, Weierstrass had to develop his theory of product representations of entire functions before even the simpler aspects of $\xi(s)$ could be tackled. Then in 1893, Jacques Hadamard (1865–1963) worked out the theory of factorization of entire functions of a finite order and applied it to $\xi(s)$.[14] That set the stage for his 1896 proof of the prime number theorem.[15] Briefly, Hadamard first proved that $\zeta(s)$ had no zeros on the line $\mathrm{Re}\, s = 1$. Then, using earlier ideas of Cahen and Halphen, he applied Mellin inversion to an integral of a weighted average, say $A(x)$, of Chebyshev's arithmetical function $\theta(x)$. From this inversion, Hadamard derived the asymptotic behavior of $A(x)$ and this in turn yielded the asymptotic behavior of $\theta(x)$, that

$$\lim_{x\to\infty} \frac{\theta(x)}{x} = 1.$$

This proved the prime number theorem. It is interesting that in an 1885 letter to Hermite,[16] Stieltjes claimed to have a proof of the Riemann hypothesis. Aware of this claim, Hadamard remarked that since Stieltjes had not published his proof, he himself would put forward a proof of the simpler result.

Also in 1896, C. J. de la Vallée-Poussin published his own proof of the prime number theorem (PNT), based on similar ideas.[17] After these proofs appeared, research on the prime number theorem centered around efforts to simplify the proof and to understand its logical structure. E. Landau, G. H. Hardy, J. E. Littlewood, and N. Wiener were the main contributors to this endeavor. In 1903, Landau found a new

[14] Hadamard (1893).
[15] Hadamard (1896).
[16] See letter 77 of Baillaud and Bourget (1905).
[17] Vallée-Poussin (1896b).

proof of the prime number theorem,[18] not dependent on Hadamard's theory of entire functions or on the functional relation for the zeta function. Landau required only that $\zeta(s)$ could be continued slightly to the left of Re $s = 1$. This method could be extended to the Dedekind zeta function for number fields and Landau used it to state and prove the prime ideal theorem.

Hardy, Littlewood, and Wiener explicated the key role of Tauberian theorems in prime number theory. It became clear from their work that the prime number theorem was equivalent to the statement that $\zeta(1 + it) \neq 0$ for real t. On the basis of this result, Hardy expected that the zeta function would play a crucial role in any proof of the PNT. But two years after Hardy's death, Atle Selberg and Paul Erdős found an elementary proof of the PNT, obviously without zeta function theory.[19]

30.2 Chebyshev on Legendre's Formula

Recall that Euler proved the divergence of $\sum_p \frac{1}{p}$, where p was prime, by comparing it with $\ln(\sum_{n=1}^\infty \frac{1}{n})$. In his 1848 paper,[20] Chebyshev followed up on this work, proving in his first theorem the existence of the limit

$$\lim_{\rho \to 0} \left(\sum_p \frac{\ln p}{p^{1+\rho}} - \sum_{k=2}^\infty \frac{1}{k^{\rho+1}} \right) \tag{30.21}$$

and, more generally, of the limit after taking derivatives with respect to ρ,

$$\lim_{\rho \to 0} \left(\sum_p \frac{\ln^n p}{p^{1+\rho}} - \sum_{k=2}^\infty \frac{\ln^{n-1} k}{k^{\rho+1}} \right), \quad n = 1, 2, 3, \ldots. \tag{30.22}$$

In order to obtain information about $\pi(x)$, the number of primes less than x, he wrote the series in (30.22) as

$$\sum_{x=2}^\infty \left(\pi(x + 1) - \pi(x) - \frac{1}{\ln x} \right) \frac{\ln^n x}{x^{1+\rho}}. \tag{30.23}$$

Then since

$$\frac{1}{\ln x} - \int_x^{x+1} \frac{dt}{\ln t} = O\left(\frac{1}{x}\right) \quad \text{as} \quad x \to \infty,$$

Chebyshev deduced the important corollary of the existence of the limit

$$\lim_{\rho \to 0} \sum_{x=2}^\infty \left(\pi(x + 1) - \pi(x) - \int_x^{x+1} \frac{dt}{\ln t} \right) \frac{\ln^n x}{x^{1+\rho}}. \tag{30.24}$$

---

[18] Landau (1903).
[19] Selberg (1949) and Erdős (1949).
[20] Chebyshev (1848).

From this, Chebyshev proceeded to derive his second theorem: For any positive real number α, any positive integer n, and for infinitely many integer values of x,

$$\pi(x) > \int_2^x \frac{dt}{\ln t} - \frac{\alpha x}{\ln^n x} \tag{30.25}$$

and with the same conditions on α and n, for infinitely many integer values of x,

$$\pi(x) < \int_2^x \frac{dt}{\ln t} + \frac{\alpha x}{\ln^n x}. \tag{30.26}$$

Using (30.25) and (30.26), Chebyshev could state his remarkable result that if

$$\lim_{x \to \infty} \frac{\pi(x)}{\int_2^x \frac{dt}{\ln t}} \quad \text{or} \quad \lim_{x \to \infty} \frac{\pi(x)}{\frac{x}{\ln x}},$$

existed, then its value had to be 1. Of course, he was unable to show existence here, and that was the essence of the PNT. From (30.25) and (30.26), he also deduced that if

$$\lim_{x \to \infty} \left(\frac{x}{\pi(x)} - \ln x \right)$$

existed, then it had to be -1. He supposed the limit to be L, so that there would exist an N such that for $x > N$,

$$L - \epsilon < \frac{x}{\pi(x)} - \ln x < L + \epsilon.$$

But by (30.25), there would be an infinite number of integers $x > N$ such that

$$\frac{x}{\int_2^x \frac{dt}{\ln t} - \frac{\alpha x}{\ln^n x}} - \ln x > L - \epsilon,$$

or

$$L + 1 < \frac{x - (\ln x - 1)\left(\int_2^x \frac{dt}{\ln t} - \frac{\alpha x}{\ln^n x}\right)}{\int_2^x \frac{dt}{\ln t} - \frac{\alpha x}{\ln^n x}} + \epsilon. \tag{30.27}$$

Similarly, (30.26) implied an inequality in the other direction. At this point, Chebyshev remarked that by a principle of differential calculus (now called l'Hôpital's rule), the expression on the right-hand side of (30.27) could be made arbitrarily small as x became large, so that the result followed. He also remarked that this theorem determined that the limit of $\frac{x}{\pi(x)} - \ln x$ as x went to infinity, was -1, contradicting Legendre, who predicted the limit would be -1.08366.

Chebyshev's proof of (30.26) was similar to his argument for (30.25). He first supposed (30.26) to hold for only a finite number of positive integers x so that there would be an integer a larger than e^n and larger than the largest integer x for which (30.26) would hold. Then for $x > a$,

$$\pi(x) - \int_2^x \frac{dt}{\ln t} \geq \frac{\alpha x}{\ln^n x}, \qquad \frac{n}{\ln x} < 1. \tag{30.28}$$

Chebyshev showed that in this case the series in (30.24) would diverge, a contradiction proving the result. To demonstrate this divergence, he used Abel's summation by parts:

$$\sum_{x=a+1}^{s} u_x(v_{x+1} - v_x) = u_s v_{s+1} - u_a v_{a+1} - \sum_{x=a+1}^{s} v_x(u_x - u_{x-1}).$$

He took

$$v_x = \pi(x) - \int_2^x \frac{dt}{\ln t}, \qquad u_x = \frac{\ln^n s}{x^{1+\rho}},$$

so that

$$\sum_{x=a+1}^{s} \left(\pi(x+1) - \pi(x) - \int_x^{x+1} \frac{dt}{\ln t} \right) \frac{\ln^n x}{x^{1+\rho}}$$

$$= \left(\pi(s+1) - \int_2^{s+1} \frac{dt}{\ln t} \right) \frac{\ln^n s}{s^{1+\rho}} - \left(\pi(a+1) - \int_2^{a+1} \frac{dt}{\ln t} \right) \frac{\ln^n a}{a^{1+\rho}}$$

$$- \sum_{x=a+1}^{s} \left(\pi(x) - \int_2^x \frac{dt}{\ln t} \right) \left(\frac{\ln^n x}{x^{1+\rho}} - \frac{\ln^n(x-1)}{(x-1)^{1+\rho}} \right). \tag{30.29}$$

By the mean value theorem,

$$\frac{\ln^n x}{x^{1+\rho}} - \frac{\ln^n(x-1)}{(x-1)^{1+\rho}} = \left(\frac{n}{\ln(x-\theta)} - (1+\rho) \right) \frac{\ln^n(x-\theta)}{(x-\theta)^{2+\rho}},$$

where $0 < \theta < 1$ and θ depended on x. The sum (30.29) then took the form

$$\sum_{x=a+1}^{s} \left(\pi(x) - \int_2^x \frac{dt}{\ln t} \right) \left(1 + \rho - \frac{n}{\ln(x-\theta)} \right) \frac{\ln^n(x-\theta)}{(x-\theta)^{2+\rho}}. \tag{30.30}$$

Then for $x > a$,

$$1 + \rho - \frac{n}{\ln(x-\theta)} > 1 - \frac{n}{\ln a},$$

and by (30.28)

$$\pi(x) - \int_2^x \frac{dt}{\ln t} \geq \frac{\alpha x}{\ln^n x} \geq \frac{\alpha(x-\theta)}{\ln^n(x-\theta)}.$$

Observe that Chebyshev could derive the last inequality because $\frac{x}{\ln^n x}$ was an increasing function. Therefore, he could see that the sum (30.30) was greater than

$$\alpha \left(1 - \frac{n}{\ln a}\right) \sum_{x=a+1}^{s} \frac{1}{(x-\theta)^{1+\rho}}.$$

Chebyshev thus arrived at a contradiction: when $s \to \infty$, he had the infinite series (30.24) diverging as $\rho \to 0$.

Chebyshev's proof of the existence of the limit (30.21), his first theorem that we are discussing, made use of the formula found in Euler and Abel:

$$\int_0^\infty \frac{e^{-x}}{e^x - 1} x^\rho dx = \int_0^\infty e^{-2x}(1 + e^{-x} + e^{-2x} + \cdots) x^\rho dx$$

$$= \sum_{m=2}^\infty \int_0^\infty e^{-mx} x^\rho dx = \sum_{m=2}^\infty \frac{1}{m^{1+\rho}} \int_0^\infty e^{-x} x^\rho dx. \quad (30.31)$$

To show that the limit (30.21) existed, Chebyshev rewrote the sums contained in it as

$$\frac{d}{d\rho}\left(\sum_p \ln\left(1 - \frac{1}{p^{1+\rho}}\right) + \sum_p \frac{1}{p^{1+\rho}}\right)$$

$$+ \frac{d}{d\rho}\left(\ln\rho - \sum_p \ln\left(1 - \frac{1}{p^{1+\rho}}\right)\right) + \left(\sum_{m=1}^\infty \frac{1}{m^{1+\rho}} - \frac{1}{\rho}\right), \quad (30.32)$$

and proved that each of the three expressions in parentheses was finite as $\rho \to 0$. He proved the more general result for (30.22), by showing that the derivatives of those expressions also had finite limits. Using (30.31), Chebyshev rewrote the third expression in (30.32) as a ratio of two integrals:

$$\sum_{m=2}^\infty \frac{1}{m^{1+\rho}} - \frac{1}{\rho} = \frac{\int_0^\infty \left(\frac{1}{e^x - 1} - \frac{1}{x}\right) e^{-x} x^\rho dx}{\int_0^\infty e^{-x} x^\rho dx}. \quad (30.33)$$

He noted that these integrals converged as $\rho \to 0$ and that the derivatives of (30.33) contained expressions of the form

$$\int_0^\infty \left(\frac{1}{e^x - 1} - \frac{1}{x}\right) e^{-x} x^\rho (\ln x)^k dx \quad \text{or} \quad \int_0^\infty e^{-x} x^\rho (\ln x)^k dx;$$

these integrals also had finite limits as $\rho \to 0$. To show that the middle expression in (30.32),

$$\ln\rho - \sum_p \ln\left(1 - \frac{1}{p^{1+\rho}}\right),$$

was finite as $\rho \to 0$, Chebyshev employed the Euler product

$$\sum_{m=1}^{\infty} \frac{1}{m^{1+\rho}} = \prod_p \left(1 - \frac{1}{p^{1+\rho}}\right)^{-1}.$$

After taking the logarithm of both sides and adding $\ln \rho$ to each side, he had

$$\ln \rho - \sum_p \ln\left(1 - \frac{1}{p^{1+\rho}}\right) = \ln\left(\left(1 + \sum_{m=2}^{\infty} \frac{1}{m^{1+\rho}}\right)\rho\right)$$

$$= \ln\left(1 + \rho + \left(\sum_{m=2}^{\infty} \frac{1}{m^{1+\rho}} - \frac{1}{\rho}\right)\rho\right). \tag{30.34}$$

Noting the expression on the right-hand side, Chebyshev thus proved the existence of the limit of the left-hand side as well as the limits of all its derivatives, as $\rho \to 0$. It is even simpler to show that the first expression in (30.32), and all its derivatives, have a finite limit as $\rho \to 0$. This proves Chebyshev's first theorem.

At the end of his 1849 paper, Chebyshev followed Legendre in assuming the prime number theorem to prove that

$$\frac{1}{2} + \frac{1}{3} + \frac{1}{5} + \cdots + \frac{1}{x} = \ln \ln x + c, \tag{30.35}$$

where x was a very large prime and c was finite. Chebyshev corrected the corresponding formula in Legendre, who had $\ln(\ln x - 0.08366)$ on the right-hand side. Chebyshev also suggested a similar change in Legendre's formula for the product

$$\left(1 - \frac{1}{2}\right)\left(1 - \frac{1}{3}\right)\left(1 - \frac{1}{5}\right) \cdots \left(1 - \frac{1}{x}\right) = \begin{cases} \dfrac{c_0}{\ln x} & \text{in Chebyshev,} \\[2mm] \dfrac{c_0}{(\ln x - 0.08366)} & \text{in Legendre.} \end{cases} \tag{30.36}$$

In 1874, Mertens proved, without assuming the then-unproved PNT, that $c_0 = e^{-\gamma}$, where γ was Euler's constant. Thus, Mertens's result implies

$$\prod_{p \leq x} \left(1 - \frac{1}{p}\right)^{-1} = e^{\gamma} \ln x + O(1). \tag{30.37}$$

An approximate value of e^{γ} is 1.781, whereas, presumably on numerical evidence, Gauss gave the value of the constant to be 1.874.[21]

30.3 Chebyshev's Proof of Bertrand's Conjecture

In his second memoir on prime numbers,[22] Chebyshev proved Bertrand's conjecture, making effective and original use of Stirling's approximation. In the course of his

[21] Gauss (1863–1927) vol. 10, p. 12.
[22] Chebyshev (1850).

discussion of the series for the logarithm of $n!$, he was led to define two related arithmetical functions $\theta(x)$ and $\psi(x)$:

$$\theta(x) = \sum_{p \leq x} \ln p, \qquad \psi(x) = \sum_{p^n \leq x} \ln p, \tag{30.38}$$

where p was prime. Chebyshev immediately noted the clear relation between the two:

$$\psi(x) = \theta(x) + \theta(\sqrt{x}) + \theta(\sqrt[3]{x}) + \theta(\sqrt[4]{x}) + \cdots . \tag{30.39}$$

Keeping in mind the preceding definitions and a result first noted by Legendre,

$$n! = \prod_{p \leq n} p^{\lfloor \frac{n}{p} \rfloor + \lfloor \frac{n}{p^2} \rfloor + \lfloor \frac{n}{p^3} \rfloor + \cdots}, \tag{30.40}$$

Chebyshev observed that if $T(x) = \ln \lfloor x \rfloor!$, then

$$T(x) = \psi(x) + \psi\left(\frac{x}{2}\right) + \psi\left(\frac{x}{3}\right) + \cdots . \tag{30.41}$$

Next, he set $a = \lfloor x \rfloor$ so that, by Stirling's formula,

$$T(x) = \ln a! < \frac{1}{2} \ln 2\pi + a \ln a - a + \frac{1}{2} \ln a + \frac{1}{12a},$$

$$T(x) = \ln(a+1)! - \ln(a+1) > \frac{1}{2} \ln 2\pi + (a+1) \ln(a+1) - (a+1) - \frac{1}{2} \ln(a+1).$$

Thus,

$$\frac{1}{2} \ln 2\pi + x \ln x - x - \frac{1}{2} \ln x < T(x) < \frac{1}{2} \ln 2\pi + x \ln x - x + \frac{1}{2} \ln x + \frac{1}{12x}. \tag{30.42}$$

From these bounds for $T(x)$, Chebyshev obtained bounds for $\psi(x)$. Of course, one might obtain an expression for $\psi(x)$ in terms of $T(x)$ from (30.41) by means of Möbius inversion. However, Chebyshev chose to work with the sum

$$T(x) - T\left(\frac{x}{2}\right) - T\left(\frac{x}{3}\right) - T\left(\frac{x}{5}\right) + T\left(\frac{x}{30}\right). \tag{30.43}$$

He showed that when the value of $T(x)$, taken from (30.41), was substituted in (30.43), the result was the alternating series

$$\psi(x) - \psi\left(\frac{x}{6}\right) + \psi\left(\frac{x}{7}\right) - \psi\left(\frac{x}{10}\right)$$

$$+ \psi\left(\frac{x}{11}\right) - \psi\left(\frac{x}{12}\right) + \psi\left(\frac{x}{13}\right) - \psi\left(\frac{x}{15}\right) + \cdots . \tag{30.44}$$

He observed that, in general, the coefficient of $\psi\left(\frac{x}{n}\right)$ would be

$$+1 \text{ if } n = 30m + k, \; k = 1, 7, 11, 13, 17, 19, 23, 29; \tag{30.45}$$

$$0 \text{ if } n = 30m + k, \; k = 2, 3, 4, 5, 8, 9, 14, 16, 21, 22, 25, 26, 27, 28; \tag{30.46}$$

$$-1 \text{ if } n = 30m + k, \; k = 6, 10, 12, 15, 18, 20, 24; \tag{30.47}$$

$$-1 \text{ if } n = 30m + 30. \tag{30.48}$$

Note here that the series (30.44) was alternating and that the absolute values of terms were nonincreasing, making the sum of the series less than the first term and greater than the sum of the first two terms. Thus, Chebyshev could conclude that

$$\psi(x) - \psi\left(\frac{x}{6}\right) \le T(x) - T\left(\frac{x}{2}\right) - T\left(\frac{x}{3}\right) - T\left(\frac{x}{5}\right) + T\left(\frac{x}{30}\right) \le \psi(x).$$

An application of the two inequalities (30.42) then yielded

$$Ax - \frac{5}{2}\ln x - 1 < T(x) - T\left(\frac{x}{2}\right) - T\left(\frac{x}{3}\right) - T\left(\frac{x}{5}\right) + T\left(\frac{x}{30}\right) < Ax + \frac{5}{2}\ln x,$$

where

$$A = \ln\left(2^{\frac{x}{2}} 3^{\frac{x}{3}} 5^{\frac{x}{5}} 30^{\frac{x}{30}}\right) = 0.92129202\ldots. \tag{30.49}$$

In this way, Chebyshev obtained the two inequalities

$$\psi(x) > Ax - \frac{5}{2}\ln x - 1 \quad \text{and} \quad \psi(x) - \psi\left(\frac{x}{6}\right) < Ax + \frac{5}{2}\ln x. \tag{30.50}$$

The first inequality determined a lower bound for $\psi(x)$; Chebyshev obtained an upper bound from the second inequality by employing an interesting trick. He set

$$f(x) = \frac{6}{5}Ax + \frac{5}{4\ln 6}\ln^2 x + \frac{5}{4}\ln x$$

and by a simple calculation obtained

$$f(x) - f\left(\frac{x}{6}\right) = Ax + \frac{5}{2}\ln x.$$

Therefore, by the second inequality

$$\psi(x) - \psi\left(\frac{x}{6}\right) < f(x) - f\left(\frac{x}{6}\right) \quad \text{or} \quad \psi(x) - f(x) < \psi\left(\frac{x}{6}\right) - f\left(\frac{x}{6}\right).$$

Replacing x by $\frac{x}{6}, \frac{x}{6^2}, \ldots, \frac{x}{6^m}$ successively, he got

$$\psi(x) - f(x) < \psi\left(\frac{x}{6}\right) - f\left(\frac{x}{6}\right) < \psi\left(\frac{x}{6^2}\right) - f\left(\frac{x}{6^2}\right) < \cdots$$

$$< \psi\left(\frac{x}{6^{m+1}}\right) - f\left(\frac{x}{6^{m+1}}\right).$$

Taking m to be the largest integer for which $\frac{x}{6^m} \geq 1$, $\frac{x}{6^{m+1}}$ would have to lie between $\frac{1}{6}$ and 1. Therefore,

$$\psi\left(\frac{x}{6^{m+1}}\right) - f\left(\frac{x}{6^{m+1}}\right) < 1$$

and

$$\psi(x) - f(x) < 1 \quad \text{or} \quad \psi(x) < f(x) + 1.$$

Thus

$$\psi(x) < \frac{6}{5}Ax + \frac{5}{4\ln 6}\ln^2 x + \frac{5}{4}\ln x + 1. \tag{30.51}$$

Chebyshev obtained bounds for $\theta(x)$ from those of $\psi(x)$. He observed that (30.39) implied that

$$\psi(x) - \psi(\sqrt{x}) = \theta(x) + \theta(\sqrt[3]{x}) + \theta(\sqrt[5]{x}) + \cdots$$

$$\psi(x) - 2\psi(\sqrt{x}) < \theta(x) < \psi(x) - \psi(\sqrt{x}). \tag{30.52}$$

He concluded from the bounds for $\psi(x)$ in (30.50) and (30.52) that

$$Ax - \frac{12}{5}Ax^{\frac{1}{2}} - \frac{5}{8\ln 6}\ln^2 x - \frac{15}{4}\ln x - 3 < \theta(x)$$

$$< \frac{6}{5}Ax - Ax^{\frac{1}{2}} + \frac{5}{4\ln 6}\ln^2 x + \frac{5}{2}\ln x + 2. \tag{30.53}$$

With the help of these inequalities, Chebyshev was able to prove Bertrand's conjecture.

Chebyshev argued that if there were exactly m primes between the numbers l and L, then $\theta(L) - \theta(l)$ could be expressed as the sum of the logarithms of these primes and hence

$$m \ln l < \theta(L) - \theta(l) < m \ln L \tag{30.54}$$

or

$$\frac{\theta(L) - \theta(l)}{\ln L} < m < \frac{\theta(L) - \theta(l)}{\ln l}. \tag{30.55}$$

He denoted the upper and lower bounds of $\theta(x)$ in (30.53) by $\theta_I(x)$ and $\theta_{II}(x)$, respectively, and noted that by the last inequality, m was greater than $k = \theta_{II}(L) - \theta_I(l)$. Substituting the values of $\theta_{II}(L)$ and $\theta_I(l)$ and solving for l, Chebyshev obtained

$$l = \frac{5}{6}L - 2L^{\frac{1}{2}} - \frac{25 \ln^2 L}{16A \ln 6} - \frac{5}{6A}\left(\frac{25}{4} + k\right)\ln L - \frac{25}{6A} \tag{30.56}$$

and observed that between l and L there were more than k primes. He then took $k = 0$ and saw that there had to be at least one prime between

$$l = \frac{5}{6}L - 2L^{1/2} - \frac{25 \ln^2 L}{16A \ln 6} - \frac{125}{24A}\ln L - \frac{25}{6A} \quad \text{and} \quad L. \tag{30.57}$$

Finally, he remarked that for $L = 2a - 3$ and $a > 160$, the value of l in (30.57) was larger than a, and hence there was a prime between a and $2a - 3$. Since the conjecture could be confirmed to hold for values of $a \le 160$, this completed the proof. In 1919, Ramanujan published a similar but very brief proof of Bertrand's conjecture. It may also be of interest to note that in 1932, when Paul Erdős was only eighteen years of age, he found a proof quite similar to Ramanujan's.

Chebyshev closed his paper by giving bounds for $\pi(x)$, the number of primes less than x. He derived these bounds as a corollary to an interesting theorem on series: Supposing that for large enough x, $\frac{F(x)}{\ln x}$ was positive and decreasing, the series

$$F(2) + F(3) + F(5) + F(7) + F(11) + F(13) + \cdots$$

converged if and only if the series

$$\frac{F(2)}{\ln 2} + \frac{F(3)}{\ln 3} + \frac{F(4)}{\ln 4} + \frac{F(5)}{\ln 5} + \frac{F(6)}{\ln 6} + \cdots$$

converged. Clearly, this theorem implied the convergence of the series $\sum_p \frac{1}{p \ln p}$. In fact, Chebyshev showed that the sum of the series lay between 1.53 and 1.73. To prove this theorem, Chebyshev took $\alpha, \beta, \gamma, \ldots, \rho$ to be prime numbers between the integers l and L. Then he defined U by

$$S = F(2) + F(3) + F(5) + \cdots + F(\alpha) + F(\beta) + F(\gamma) + \cdots F(\rho)$$

$$= S_0 + F(\alpha) + F(\beta) + F(\gamma) + \cdots + F(\rho) = S_0 + U.$$

Since $\theta(x) - \theta(x-1) = \ln x$ for prime x, and $= 0$ for composite x, Chebyshev could conclude that

$$U = \frac{\theta(l) - \theta(l-1)}{\ln l}F(l) + \frac{\theta(l+1) - \theta(l)}{\ln(l+1)}F(l+1)$$

$$+ \frac{\theta(l+2) - \theta(l+1)}{\ln(l+2)}F(l+2) + \cdots + \frac{\theta(L) - \theta(L-1)}{\ln L}F(L).$$

He then applied summation by parts to obtain

$$U = -\theta(l-1)\frac{F(L)}{\ln l} + \left(\frac{F(l)}{\ln l} - \frac{F(l+1)}{\ln(l+1)}\right)\theta(l) + \left(\frac{F(l+1)}{\ln(l+1)} - \frac{F(l+2)}{\ln(l+2)}\right)\theta(l+1)$$

$$+ \cdots + \left(\frac{F(L)}{\ln L} - \frac{F(L+1)}{\ln(L+1)}\right)\theta(L) + \frac{F(L+1)}{\ln(L+1)}\theta(L). \tag{30.58}$$

He took l large enough that $F(x)\ln x$ was positive and increasing in $l-1 \leq x \leq L+1$ and obtained the inequalities

$$\theta_{II}(l-1)\frac{F(l)}{\ln l} - \theta_I(l-1)\frac{F(l)}{\ln l} + \sum_{x=l}^{L} F(x)\frac{\theta_{ii}(x) - \theta_{II}(x-1)}{\ln x} < U$$

$$< \theta_I(l-1)\frac{F(l)}{\ln l} - \theta_{II}(l-1)\frac{F(L)}{\ln l} + \sum_{x=l}^{L} F(x)\frac{\theta_I(x) - \theta_I(x-1)}{\ln x}. \tag{30.59}$$

Chebyshev noted that

$$\theta_{II}(x) - \theta_{II}(x-1)$$

$$= A - \frac{12}{5}A(\sqrt{x} - \sqrt{x-1}) - \frac{5}{8\ln 6}(\ln^2 x - \ln^2(x-1)) - \frac{15}{4}(\ln x - \ln(x-1))$$

and that this expression was bounded as $x \to \infty$. For example,

$$\sqrt{x} - \sqrt{x-1} = \sqrt{x} - \sqrt{x}\left(1 - \frac{1}{x}\right)^{\frac{1}{2}} \approx \frac{1}{2\sqrt{x}} \to 0, \ x \to \infty,$$

and

$$\ln x - \ln(x-1) = \ln\frac{x-1}{x} = \ln\left(1 - \frac{1}{x}\right) \to 0 \ \text{as} \ x \to \infty.$$

By a similar analysis with $\theta_I(x) - \theta_I(x-1)$, he noted that the two inequalities in (30.59) implied the theorem. Chebyshev obtained the bounds for $\pi(x)$ by taking $F(x) = 1$ and $l = 2$ in (30.59):

$$\frac{\theta_{II}(1)}{\ln 2} - \frac{\theta_I(1)}{\ln 2} + \sum_{x=2}^{L} \frac{\theta_{II}(x) - \theta_{II}'(x-1)}{\ln x} < \pi(x)$$

$$< \frac{\theta_I(1)}{\ln 2} - \frac{\theta_{II}(1)}{\ln 2} + \sum_{x=2}^{L} \frac{\theta_I(x) - \theta_I(x-1)}{\ln x}.$$

30.4   De Polignac's Evaluation of $\sum_{p \le x} \frac{\ln p}{p}$

Inspired by the work of Chebyshev, in the 1850s, Alphonse de Polignac published a number of papers in the *Comptes Rendus* and *Liouville's Journal*. Though his work was largely lacking in rigor, in 1857[23] de Polignac gave a fairly good proof of

$$\sum_{p \le x} \frac{\ln p}{p} = \ln x + \epsilon,$$

where ϵ was a quantity small compared to $\ln x$. In fact, his proof implies that ϵ is bounded. Now by Chebyshev's work,

$$\sum_{p^m \le x} \ln p = \sum_{p \le x} \left(\left\lfloor \frac{x}{p} \right\rfloor + \left\lfloor \frac{x}{p^2} \right\rfloor + \left\lfloor \frac{x}{p^3} \right\rfloor + \cdots \right) \ln p.$$

De Polignac denoted the left-hand side by $\ln F_o(x)$ and $\lfloor \frac{x}{p^k} \rfloor$ by $E \left(\frac{x}{p^k} \right)$. He let n be the largest integer such that $p^n \le x$. Then

$$\sum_{k=1}^{n} \left\lfloor \frac{x}{p^k} \right\rfloor > \sum_{k=1}^{n} \frac{x}{p^k} - \sum_{k=1}^{n} 1,$$

and

$$\sum_{k=1}^{n} \left\lfloor \frac{x}{p^k} \right\rfloor < \sum_{k=1}^{n} \frac{x}{p^k} = \frac{x}{p-1} - \frac{x}{p^n(p-1)}.$$

Therefore,

$$x \sum_{p \le x} \frac{\ln p}{p-1} - x \sum_{p \le x} \frac{\ln p}{p^n(p-1)} > \sum_{p^m \le x} \ln p$$

$$> x \sum_{p \le x} \frac{\ln p}{p-1} - x \sum_{p \le x} \frac{\ln p}{p^n(p-1)} - \ln x.$$

De Polignac then argued that $\sum_{p^m \le x} \ln p = \sum_{n \le x} \ln n = x \ln x +$ terms of smaller order, and $\sum \frac{\ln p}{p-1}$ was of the same order as $\sum \frac{\ln p}{p}$. Moreover, $\sum_{p \le x} \frac{\ln p}{p^n(p-1)}$ was bounded, so that the required result followed from the two inequalities. In 1874, Franz Mertens, aware of de Polignac's work, but motivated by Chebyshev's second paper, proved that

$$\sum_{p \le x} \frac{\ln p}{p} = \ln x + O(1).$$

23 Polignac (1857).

30.5   Mertens's Evaluation of $\prod_{p \leq x} \left(1 - \frac{1}{p}\right)^{-1}$

According to Mertens, his interest in evaluating $\prod_{p \leq x} \left(1 - \frac{1}{p}\right)$ arose from the useful formulas he had seen in the third edition of Legendre's *Théorie des nombres*.[24] Legendre's formulas stated without rigorous proof that for some constants A and C,

$$\sum_{p \leq G} \frac{1}{p} = \ln(\ln G - 0.08366) + C,$$

and

$$\prod_{p \leq G} \left(1 - \frac{1}{p}\right) = \frac{A}{\ln G - 0.08366}.$$

Mertens proved the results: First, that

$$\sum_{p \leq G} \frac{1}{p} = \ln \ln G + \gamma - H + \delta, \tag{30.60}$$

where

$$H = \sum_{k=2}^{\infty} \frac{1}{k} \sum_{p} \frac{1}{p^k}$$

and

$$\delta < \frac{4}{\ln(G + 1)} + \frac{2}{G \ln G}.$$

Secondly,

$$\prod_{p \leq G} \left(1 - \frac{1}{p}\right)^{-1} = e^{\gamma + \delta'} \ln G,$$

where

$$\delta' < \frac{4}{\ln(G + 1)} + \frac{2}{G \ln G} + \frac{1}{2G}.$$

He began by observing that Dirichlet and Chebyshev had shown that for $\rho > 0$,

$$\zeta(1 + \rho) = \frac{1 + (\rho)}{\rho},$$

---

[24] Mertens (1874a) and (1874b).

where (ρ) denoted a quantity tending to 0 as $\rho \to 0$. It followed from this and Euler's product for $\zeta(1 + \rho)$ that

$$\ln \frac{1}{\rho} + (\rho) = \ln \zeta(1 + \rho) = -\sum_p \ln(1 - 1/p^{1+\rho})$$

$$= \sum_p \frac{1}{p^{1+\rho}} + \frac{1}{2} \sum_p \frac{1}{p^{2+2\rho}} + \frac{1}{3} \sum_p \frac{1}{p^{3+3\rho}} + \cdots :$$

Mertens could then easily conclude that

$$\sum_p \frac{1}{p^{1+\rho}} = \ln \frac{1}{\rho} - H + (\rho). \tag{30.61}$$

He proceeded to complete the proof of (30.60) by showing that

$$\sum_{p>G} \frac{1}{p^{1+\rho}} = \ln \frac{1}{\rho} - \ln \ln G - \gamma + \delta + (\rho), \tag{30.62}$$

where

$$|\delta| < \frac{4}{\ln(G + 1)} + \frac{2}{G \ln G}.$$

He set $f(x) = \sum_{p \le x} \frac{\ln p}{p}$. Then summation by parts gave him

$$\sum_{p>G} \frac{1}{p^{1+\rho}} = \sum_{n=G+1}^{\infty} \frac{f(n) - f(n-1)}{n^{\rho} \ln n}$$

$$= -\frac{f(G)}{(G+1)^{\rho} \ln(G+1)} + \sum_{n=G+1}^{\infty} f(n) \left(\frac{1}{n^{\rho} \ln n} - \frac{1}{(n+1)^{\rho} \ln(n+1)} \right). \tag{30.63}$$

Next set $f(n) = \ln n + D_n$. Recall that de Polignac had shown that D_n was small compared to $\ln n$, but Mertens required that D_n be bounded. This was easily achieved. Mertens showed that $D_n < 2$ by computing bounds explicitly, as Chebyshev had done in his work on primes. Now observe that

$$\ln n \left(\frac{1}{n^{\rho} \ln n} - \frac{1}{(n+1)^{\rho} \ln(n+1)} \right) = \frac{1}{n^{\rho}} - \frac{1}{(n+1)^{\rho}} - \frac{\ln \left(1 - \frac{1}{n+1} \right)}{(n+1)^{\rho} \ln(n+1)}$$

$$= \frac{1}{n^{\rho}} - \frac{1}{(n+1)^{\rho}} + \frac{1}{(n+1)^{1+\rho} \ln(n+1)}$$

$$+ \frac{\lambda}{2n(n+1)^{1+\rho} \ln(n+1)},$$

where $0 < \lambda < 1$. Applying this in (30.63) and after cancellation of terms, he had

$$\sum_{p>G} \frac{1}{p^{1+\rho}} = \sum_{n=G+1}^{\infty} \frac{1}{n^{1+\rho} \ln n} + R, \tag{30.64}$$

where

$$R = \frac{\ln(G+1) - f(G)}{(G+1)^{\rho} \ln(G+1)} - \frac{1}{(G+1)^{1+\rho} \ln(G+1)} + \lambda \sum_{n=G+1}^{\infty} \frac{1}{2n(n+1)^{1+\rho} \ln(n+1)}$$

$$+ \sum_{n=G+1}^{\infty} D_n \left(\frac{1}{n^{\rho} \ln n} - \frac{1}{(n+1)^{\rho} \ln(n+1)} \right).$$

It is easy to show that $R = O\left(\frac{1}{\ln G}\right)$ and, in fact, Mertens proved that $|R| < \frac{4}{\ln(G+1)} + \frac{1}{G \ln G}$. To estimate the sum $\sum \frac{1}{n^{1+\rho} \ln n}$, first note that

$$\sum_{n=G+1}^{\infty} \frac{1}{n^{1+t}} = \frac{G^{-t}}{t} - R', \tag{30.65}$$

where

$$R' = \frac{1+t}{2} \sum_{n=G+1}^{\infty} \frac{1}{n^{2+t}} + \frac{(1+t)(2+t)}{2 \cdot 3} \sum_{n=G+1}^{\infty} \frac{1}{n^{3+t}} + \cdots.$$

To prove (30.65), observe that the binomial expansion of $(1 - \frac{1}{n+1})^{-t}$ immediately implies

$$\frac{1}{tn^t} - \frac{1}{t(n+1)^t} = \frac{1}{(n+1)^{1+t}} + \frac{1+t}{2} \frac{1}{(n+1)^{2+t}} + \frac{(1+t)(2+t)}{2 \cdot 3} \frac{1}{(n+1)^{3+t}} + \cdots.$$

The required result followed when this formula was summed from $n = G$ to $n = \infty$. Now integrating (30.65) from ρ to 1, obtain

$$\sum_{n=G+1}^{\infty} \frac{1}{n^{1+\rho} \ln n} - \sum_{n=G+1}^{\infty} \frac{1}{n^2 \ln n} = \int_{\rho}^{1} \frac{G^{-t}}{t} dt - \int_{\rho}^{1} R' dt$$

$$= \int_{\rho \ln G}^{\infty} \frac{dx}{e^x - 1} - \int_{\rho \ln G}^{\infty} \left(\frac{1}{e^x - 1} - \frac{e^{-x}}{x} \right) dx$$

$$- \int_{1}^{\infty} \frac{G^{-x}}{x} dx - \int_{\rho}^{1} R' dt.$$

The first integral in this expression could be written as

$$\int_{\rho \ln G}^{\infty} \frac{e^{-x}}{1 - e^{-x}} dx = \ln(1 - e^{-x}) \Big|_{\rho \ln G}^{\infty} = -\ln(1 - G^{-\rho});$$

the second one, by Gauss's formula for $\psi(1) = \frac{\Gamma'(1)}{\Gamma(1)} = -\gamma$, given in Exercise 17.11, would be

$$\int_0^\infty - \int_0^{\rho \ln G} \left(\frac{1}{e^x - 1} - \frac{e^{-x}}{x} \right) dx = \gamma + (\rho).$$

Also

$$- \ln(1 - G^{-\rho}) = \ln \frac{1}{\rho} - \ln \ln G + (\rho),$$

so that

$$\sum_{n=G+1}^\infty \frac{1}{n^{1+\rho} \ln n} = \ln \frac{1}{\rho} - \ln \ln G - \gamma - \int_1^\infty G^{-x} \frac{dx}{x} + \sum_{n=G+1}^\infty \frac{1}{n^2 \ln n}$$

$$- \int_\rho^1 R' \, dt + (\rho).$$

Next,

$$\int_\rho^1 R' \, dt < \int_0^1 \left(\sum_{n=G+1}^\infty \frac{1}{n^{2+t}} + \sum_{n=G+1}^\infty \frac{1}{n^{3+t}} + \cdots \right) dt$$

$$< \sum_{n=G+1}^\infty \left(\frac{1}{n^2 \ln n} - \frac{1}{n^3 \ln n} \right) + \sum_{n=G+1}^\infty \left(\frac{1}{n^3 \ln n} - \frac{1}{n^4 \ln n} \right) + \cdots$$

$$< \sum_{n=G+1}^\infty \frac{1}{n^2 \ln n} < \sum_{n=G+1}^\infty \left(\frac{1}{(n-1)\ln(n-1)} - \frac{1}{n \ln n} \right) < \frac{1}{G \ln G}$$

and

$$\int_1^\infty \frac{G^{-x}}{x} \, dx < \frac{1}{G \ln G}.$$

Hence,

$$\sum_{n=G+1}^\infty \frac{1}{n^{1+\rho} \ln n} = \ln \frac{1}{\rho} - \ln \ln G - \gamma + \frac{\lambda}{G \ln G} + (\rho).$$

When combined with (30.64), this gave (30.62). Mertens's version of Legendre's formula for $\sum_{p \leq G} \frac{1}{\rho}$ followed from this, (30.62), and (30.61). Mertens's formula for the product was an easy corollary.

In 1926, Hardy commented on Mertens's proof:[25] "The proof is rather difficult to seize or to remember, since it depends on a combination of the method of Tchebycheff

25 Hardy (1966–1979) vol. 2, pp. 210–212, especially p. 210.

on the one hand and the theory of Dirichlet's series on the other, and it may be worth while to give an alternative proof." Hardy himself provided two proofs, the first published in 1927[26] using an integral analog of Littlewood's Tauberian theorem, and the second in 1935[27] using the analog of the simpler Tauber's theorem. Hardy's proofs are quite interesting, but we note that if Mertens's proof is recast in terms of the Stieltjes integral, a very simple proof of (30.62) emerges; note that the latter is the only complex argument in the proof.

To begin this short proof, first observe that

$$\sum_{p \geq G+1} \frac{1}{p^{1+\rho}} = \int_{G+1}^{\infty} x^\rho \ln x \, df(x),$$

where $f(x) = \sum_{p \leq x} \frac{\ln p}{p} = \ln x + \epsilon$, $|\epsilon| \leq 3$. Then integration by parts and an easy calculation produce

$$\sum_{p \geq G+1} \frac{1}{p^{1+\rho}} = (1 + \epsilon\rho) \int_{\ln(G+1)}^{\infty} \frac{e^{-\rho u}}{u} \, du + O\left(\frac{1}{\ln G}\right).$$

Next, again by Gauss's formula for $\Gamma'(1)$ (see Exercise 17.11), this integral can be evaluated as

$$\int_{\rho \ln(G+1)}^{\infty} \frac{e^{-x}}{x} \, dx = -\int_0^{\infty} + \int_0^{\rho \ln(G+1)} \left(\frac{1}{e^x - 1} - \frac{e^{-x}}{x}\right) dx + \int_{\rho \ln(G+1)}^{\infty} \frac{1}{e^x - 1} \, dx$$

$$= \gamma + (\rho) + \ln(1 - (G+1)^\rho)$$

$$= \gamma + (\rho) + \ln \frac{1}{\rho} + \ln \ln(G+1) + (\rho).$$

This proves (30.62) so that the proof of Mertens's formula can now be completed as before.

30.6 Riemann's Formula for $\pi(x)$

Riemann's eight-page paper of 1859,[28] containing his formula for the number of primes less than a given number x, was actually an outline of a research program for the advancement of the theory of distribution of primes. He proved very few statements in this paper, but clearly set forth his conjectures and how some of them they might be verified. It took fifty years of development in complex analysis to prove the first approximation of his formula, the prime number theorem. Almost a century after Riemann's paper appeared, Hardy's student and Oxford professor Edward Titchmarsh wrote, "The memoir in which Riemann first considered the zeta-function has become

[26] ibid.
[27] ibid. pp. 230–233.
[28] Riemann (1859).

famous for the number of ideas it contains which have since proved fruitful, and it is by no means certain that these are even now exhausted."[29]

Recall that Dirichlet and Chebyshev employed Mellin transforms to study prime numbers, but that they limited themselves to real variables. Riemann's great innovation was to employ complex variables. He expressed the Mellin transform of his arithmetic function $f(x)$ defined by (30.14) in terms of the zeta function; then by Mellin inversion, he expressed $f(x)$ as an integral in the complex plane. This made it possible to apply the powerful machinery of complex integration. Riemann observed that if

$$p^{-s} = s \int_p^\infty x^{-s-1}dx, \quad p^{-2s} = s \int_{p^2}^\infty x^{-s-1}dx, \ldots$$

were used in

$$\log \zeta(s) = -\sum_p \log(1 - p^{-s}) = \sum_p p^{-s} + \frac{1}{2}\sum_p p^{-2s} + \frac{1}{3}\sum_p p^{-3s} + \cdots,$$

he got

$$\frac{\log \zeta(s)}{s} = \int_1^\infty f(x)x^{-s-1}dx, \quad \text{Re } s > 1.$$

He then applied the Fourier inversion formula to obtain an integral expression for $f(x)$:

$$f(y) = \frac{1}{2\pi i}\int_{a-\infty i}^{a+\infty i} \frac{\log \zeta(s)}{s}y^s ds, \quad a > 1.$$

Riemann set

$$\xi(s) = \Pi\left(\frac{s}{2}\right)(s-1)\pi^{-\frac{s}{2}}\zeta(s), \tag{30.66}$$

and by using (30.17), he obtained

$$\log \zeta = \frac{s}{2}\log \pi - \log(s-1) - \log \Pi\left(\frac{s}{2}\right) + \sum_\alpha \log\left(1 - \frac{s}{\rho}\right) + \log \xi(0).$$

Riemann noted, however, that when this expression was used in the integral for $f(y)$, the integral became divergent. So he applied integration by parts to get

$$f(s) = -\frac{1}{2\pi i}\frac{1}{\log x}\int_{a-\infty i}^{a+\infty i} \frac{d\left(\frac{\log \zeta(s)}{s}\right)}{ds}x^s ds.$$

[29] Titchmarsh and Heath-Brown (1986) p. 254.

Next he observed that

$$-\log \Pi \left(\frac{s}{2}\right) = \lim_{m \to \infty} \left(\sum_{n=1}^{m} \log \left(1 + \frac{s}{2n}\right) - \frac{s}{2} \log m \right)$$

and therefore

$$-\frac{d}{ds} \frac{\log \Pi(\frac{s}{2})}{s} = \sum_{n=1}^{\infty} \frac{d}{ds} \frac{\log \left(1 + \frac{s}{2n}\right)}{s}.$$

Hence, every term in the expression for $f(s)$, except for the term

$$\frac{1}{2\pi i} \frac{1}{\log x} \int_{a-\infty i}^{a+\infty i} \frac{1}{ss} \log \xi(0) x^s ds = \log \xi(0),$$

took the form

$$\pm \frac{1}{2\pi i} \frac{1}{\log x} \int_{a-\infty i}^{a+\infty i} \frac{d}{ds} \left(\frac{\log \left(1 - \frac{s}{\beta}\right)}{s} \right) x^s ds.$$

To evaluate this integral, Riemann observed that

$$\frac{d}{d\beta} \left(\frac{\log \left(1 - \frac{s}{\beta}\right)}{s} \right) = \frac{1}{(\beta - s)\beta}.$$

Thus, for $\mathrm{Re}(s - \beta) > 0$, he had

$$-\frac{1}{2\pi i} \frac{d}{d\beta} \int_{a-\infty i}^{a+\infty i} \frac{\log(1 - \frac{s}{\beta})}{s} x^s ds = -\frac{1}{2\pi i} \int_{a-\infty i}^{a+\infty i} \frac{x^s ds}{(\beta - s)\beta} = \frac{x^\beta}{\beta}$$

$$= \begin{cases} \int_{\infty}^{x} t^{\beta-1} dt, & \text{when } \mathrm{Re}\ \beta < 0, \\ \int_{0}^{x} t^{\beta-1} dt, & \text{when } \mathrm{Re}\ \beta > 0. \end{cases}$$

Riemann could then conclude that

$$\frac{1}{2\pi i} \frac{1}{\log x} \int_{a-\infty i}^{a+\infty i} \frac{d}{ds} \left(\frac{\log(1 - \frac{s}{\beta})}{s} \right) x^s ds$$

$$= -\frac{1}{2\pi i} \int_{a-\infty i}^{a+\infty i} \frac{\log(1 - \frac{s}{\beta})}{s} x^s ds$$

$$= \begin{cases} \int_{\infty}^{x} \frac{t^{\beta-1}}{\log t} dt, & \text{when } \mathrm{Re}\ \beta < 0, \\ \int_{0}^{x} \frac{t^{\beta-1}}{\log t} dt, & \text{when } \mathrm{Re}\ \beta > 0. \end{cases}$$

Note that it was clear that

$$\int_0^x \frac{t^{\beta-1}}{\log t}dx = \int_0^{x^\beta} \frac{du}{\log u} = \mathrm{li}(x^\beta).$$

By using these results in the expression for $f(x)$, Riemann obtained his famous formula

$$f(x) = \mathrm{li}(x) - \sum_\beta (\mathrm{li}(x^\beta) + \mathrm{li}(x^{1-\beta})) + \int_x^\infty \frac{1}{t^2-1}\frac{dt}{t\log t} + \log \xi(0).$$

In writing this formula, Riemann assumed the truth of the Riemann hypothesis. He wrote $\beta = \frac{1}{2} + i\alpha$, and $1 - \beta = \frac{1}{2} - i\alpha$, so that the expression in the sum appeared as

$$\mathrm{li}\left(x^{\frac{1}{2}+i\alpha}\right) + \mathrm{li}\left(x^{\frac{1}{2}-i\alpha}\right).$$

One may verify that the integral evaluations as sketched by Riemann are indeed correct; one may also consult Harold Edwards's book, offering a detailed discussion of Riemann's paper.[30]

30.7 Exercises

(1) Using Chebyshev's notation, show that for $T(x) = \ln\lfloor x \rfloor! - 2\ln\lfloor \frac{x}{2} \rfloor!$

$$\psi(x) - \psi\left(\frac{x}{2}\right) \le T(x) \le \psi(x) - \psi\left(\frac{x}{2}\right) + \psi\left(\frac{x}{3}\right).$$

Apply Stirling's approximation to prove that $T(x) < \frac{3x}{4}$ for $x > 0$ and $T(x) > \frac{2x}{3}$ for $x > 300$. Use this to show that $\psi(x) < \frac{3x}{2}$. Now show that $\psi(x) - 2\psi(\sqrt{x}) \le \theta(x) \le \psi(x)$ and, therefore,

$$\psi(x) - \psi\left(\frac{x}{2}\right) + \psi\left(\frac{x}{3}\right) \le \theta(x) + 2\psi(\sqrt{x}) - \theta\left(\frac{x}{2}\right) + \psi\left(\frac{x}{3}\right) \quad (30.67)$$

$$< \theta(x) - \theta\left(\frac{x}{2}\right) + \frac{x}{2} + 3\sqrt{x}. \quad (30.68)$$

Apply these results to show that

$$\theta(x) - \theta\left(\frac{x}{2}\right) > \frac{x}{6} - 3\sqrt{x} \quad \text{for} \quad x > 300.$$

Show that this proves Bertrand's conjecture for $x \ge 162$. Finally, show that

$$\pi(x) - \pi\left(\frac{x}{2}\right) > \frac{\frac{x}{6} - 3\sqrt{3}}{\ln x} \quad \text{for} \quad x > 300.$$

See Ramanujan (2000) pp. 208–209.

[30] Edwards (2001).

(2) Let $C(m,n)$ denote the binomial coefficient m choose n, that is, the number of ways of choosing n objects out of m distinct objects. Show that the exponent of a prime p in $C(2n,n)$ is given by

$$\sum_{k=1}^{\infty} \left(\left\lfloor \frac{2n}{p^k} \right\rfloor - 2 \left\lfloor \frac{n}{p^k} \right\rfloor \right).$$

Show that

$$d = \left\lfloor \frac{2n}{p^k} \right\rfloor - 2 \left\lfloor \frac{n}{p^k} \right\rfloor \leq 1,$$

that $d = 1$ for $\sqrt{2n} < p \leq 2n$, and that $d = 0$ for $p > 2n$ and for $\frac{2n}{3} < p < n$. Use these results to conclude that

$$C(2n,n) \leq \prod_{p \leq \sqrt{2n}} (2n) \prod_{\sqrt{2n} < p \leq \frac{2n}{3}} p \prod_{n < p \leq 2n} p.$$

Note that if Bertrand's conjecture is false for some n, then

$$C(2n,n) \leq (2n)^{\sqrt{2n}} \prod_{\sqrt{2n} < p \leq \frac{2n}{3}} p. \qquad (30.69)$$

(3) Show that $2nC(2n,n) > 4^n$. Prove that for $a \geq 5$, $C(2a,a) < 4^{a-1}$ and that $\prod_{a<p<2a} p \leq C(2a,a)$. Use the last two inequalities to show that $\prod_{10<p<n} p < 4^n$. Combine these results with (30.69) to show that if Bertrand's conjecture is false then we get a contradiction. See Erdős (1932). Paul Erdős (1913) founded many aspects of combinatorics and popularized this area of mathematics by continuously traveling all over the world and collaborating with hundreds of mathematicians.

(4) Let $\phi(1) = 1$ and let $\phi(n)$, $n > 1$, be the number of numbers less than n and prime to n. Let $F(t) = \sum_{1 \leq n \leq t} \phi(t)$. Prove that $F(t) = \frac{3t^2}{\pi^2} + O(t \ln t)$. See Mertens (1874b) pp. 290–292.

(5) Show that the integral $\Gamma(s) = \int_0^\infty x^{s-1} e^{-x} \, dx$ has the inversion

$$e^{-x} = \frac{1}{2\pi i} \int_{a-i\infty}^{a+i\infty} \Gamma(x) x^{-s} \, ds, \quad a > 0, \operatorname{Re} x > 0.$$

See Cahen (1894).

(6) Prove Ramanujan's formula

$$\int_0^\infty \frac{\cos \pi x^2}{\sinh \pi x} \sin(2\pi t x) \, dx = \frac{\cosh \pi t - \cos \pi t^2}{2 \sinh \pi t}.$$

Use this formula to show that for $-3 < \operatorname{Re} s < 4$, we have

$$\frac{\Gamma(s)}{2^s}(1 - 2^s)(1 - 2^{1-s})\zeta(s)$$

$$= -\int_0^\infty \left[\left(\frac{\pi}{2}\right)^{s-1}\Gamma(s)\sin\frac{\pi s}{2}x^{-s} + x^{s-1}\right]\frac{\sin^2\left(\frac{x^2}{2\pi}\right)}{\sinh x}\,dx.$$

See Ramanujan (2000) p. 64, for his formula. For the other formula, see Mustafy (1966).

(7) Show that for $\xi(s)$ defined by (30.66) and for $0 < \operatorname{Re} s < 1, \lambda > 0$,

$$\int_0^1 u^{-\frac{1}{2}}k(\lambda, u)\left(u^{s-\frac{1}{2}} + u^{\frac{1}{2}-s}\right)du = \frac{1}{2}B(s)\xi(s)\left(\lambda^{\frac{s}{2}-\frac{1}{4}} + \lambda^{\frac{1}{4}-\frac{s}{2}}\right),$$

$$\text{where} \quad B(s) = \frac{\pi^{-\frac{1}{2}}}{4}\Gamma\left(-\frac{s}{2}\right)\Gamma\left(\frac{s-1}{2}\right),$$

$$\phi(\lambda, u) = \int_0^\infty \left(\lambda^{\frac{1}{4}}e^{-\pi x^2 u\lambda} + \lambda^{-\frac{1}{4}}e^{-\frac{\pi x^2 u}{\lambda}}\right)\frac{x\,dx}{e^{2\pi x} - 1},$$

$$k(\lambda, u) = \frac{1}{4\pi}\left(\lambda^{\frac{1}{4}} + \lambda^{-\frac{1}{4}}\right) - u\phi(\lambda, u^2).$$

Prove also that $k(\lambda, u) > 0$ for $0 \leq u \leq 1$, and that a number $s_0 = \sigma_0 + it_0$ with $0 < \sigma_0 < 1$ is a zero of $\xi(s)$ if and only if for every $\lambda > 0$

$$\int_0^1 u^{-\frac{1}{2}}k(\lambda, u)\left(u^{s_0 - \frac{1}{2}} + u^{\frac{1}{2} - s_0}\right)du = 0.$$

See Mustafy (1972). Ashoke Kumar Mustafy had a thirty-year career in the Indian Administrative Service, including as Vice Chancellor of Lucknow University during 1973–75. In spite of his heavy administrative duties, he worked on mathematics six to seven hours per day and had time to discuss mathematics with a young boy like the author. Mustafy hoped that his result would be useful in proving the Riemann hypothesis; indeed, in this connection he communicated with André Weil who wrote that he found Mustafy's work promising.

30.8   Notes on the Literature

See Smith (1959) pp. 127–148, for an English translation of some parts of Chebyshev's two papers on primes. Delone (2005), an English translation by R. Burns of Delone's Russian original of 1947, gives a detailed commentary on Chebyshev's papers and a discussion of the major contributions to number theory of St. Petersburg mathematicians in the period 1847–1947. Edwards (2001) presents a detailed and fascinating discussion of Riemann's 1859 paper and some of its consequences. Narkiewicz (2000) offers an excellent exposition of the development of the prime number theorem and provides a comprehensive list of references.

Invariant Theory: Cayley and Sylvester

31.1   Preliminary Remarks

The invariant theory of forms, with forms defined as homogeneous polynomials in several variables, was developed extensively in the nineteenth century as an important branch of algebra but with very close connections to algebraic geometry. Several ideas and methods of invariant theory were influential in diverse areas of mathematics: topics as concrete as enumerative combinatorics and the theory of partitions and as general as twentieth-century abstract commutative algebra.

George Boole, the highly original British mathematician, may be taken as the founder of invariant theory, though early examples of the use of invariance can be found in the works of Lagrange, Laplace, and Gauss. Boole had almost no formal training in mathematics, but he carefully studied the work of great mathematicians, including Newton, Lagrange, and Laplace. In a paper on analytic geometry written in 1839,[1] Boole took the first tentative steps toward the idea of invariance, but he gave a clearly formulated definition in his 1841 "Exposition of a General Theory of Linear Transformations."[2] He wrote that he found his inspiration in Lagrange's researches on the rotation of rigid bodies, contained in the 1788 *Mécanique analytique*. Lagrange's result is most economically described in terms of matrices, a concept developed in the 1850s by Cayley. In modern terms, Lagrange's problem was to diagonalize a 3×3 symmetric matrix A; Lagrange expressed this in terms of binary quadratic forms. Given a quadratic form $x^t A x$, with x a three vector, the problem would be to find a matrix P such that $PP^t = I$, the identity matrix, and $P^t A P$ is a diagonal matrix. This means that if x_1, x_2, x_3 are the components of x, y_1, y_2, y_3 of $y = P^t x$, and $\lambda_1, \lambda_2, \lambda_3$ are the diagonal entries in the diagonal matrix, then

$$x^t A x = \lambda_1 y_1^2 + \lambda_2 y_2^2 + \lambda_3 y_3^2, \tag{31.1}$$

$$x_1^2 + x_2^2 + x_3^2 = y_1^2 + y_2^2 + y_3^2. \tag{31.2}$$

---

[1] Boole (1939).
[2] Boole (1941).

It is not surprising that this result of Lagrange also served as the starting point of the spectral theory of matrices. Cauchy, Weierstrass, and Frobenius were the primary developers of this aspect of matrix theory. But Boole took a different turn; he considered a homogeneous polynomial of degree n in m variables and applied a linear transformation to the variables to obtain a new homogeneous polynomial of degree n in m variables. He wished to determine the relations between the coefficients of the two polynomials. Boole's method may perhaps be best understood by studying his simplest example. Let $Q = ax_1^2 + 2bx_1x_2 + cx_2^2$ be a binary quadratic form. Set its two partial derivatives equal to zero and then eliminate the variables x_1 and x_2. Thus,

$$2ax_1 + 2bx_2 = 0 \quad \text{and} \quad 2bx_1 + 2cx_2 = 0. \tag{31.3}$$

Elimination of the variables x_1, x_2 gives

$$\theta(Q) = b^2 - ac = 0. \tag{31.4}$$

Now apply the linear transformation

$$x_1 = py_1 + qy_2 \quad x_2 = ry_1 + sy_2, \tag{31.5}$$

where p, q, r, s are real numbers with $ps - qr \neq 0$, to get a new quadratic form $R = Ay_1^2 + 2By_1y_2 + Cy_2^2$. A calculation similar to the previous one gives $\theta(R) = B^2 - AC$. Boole pointed out that

$$\theta(R) = (ps - qr)^2 \theta(Q); \tag{31.6}$$

the quantity $ps - qr$ is the determinant of the linear transformation (31.5). In addition, the degrees of the homogeneous polynomials $\theta(Q)$ and $\theta(R)$ are defined as equal to the degree of each term, in this case 2.

More generally, Boole showed that, with Q_n a homogeneous polynomial of degree n in m variables, if R_n was the polynomial obtained after the application to Q_n of a linear transformation with determinant E, and if $\theta(Q_n)$ and $\theta(R_n)$ were obtained by the elimination process described earlier, then

$$\theta(R_n) = E^{\frac{\gamma n}{m}} \theta(Q_n). \tag{31.7}$$

Here γ represented the degree of $\theta(R_n)$ and $\theta(Q_n)$. In the 1841 paper, Boole stated but did not prove this theorem, though he gave a few examples to illustrate it. He indicated a proof in a paper appearing four years later.[3] Note that the polynomial $\theta(Q_n)$ is termed an invariant because it satisfies the relation (31.7). Sylvester introduced the term invariant in a long paper on the subject published in 1853, and he coined many other terms used in invariant theory.

At the end of the second part of his 1841 paper, Boole wrote that mathematicians should find invariant theory a fertile area for research and discovery. Indeed, Boole's paper had an immediate impact on Cayley who, upon reading it in 1844, wrote to

---

[3] Boole (1845).

Boole of his enthusiasm for this new area of mathematics.[4] Cayley was then a recent graduate of Cambridge University and had published an 1843 paper on determinants[5] in which he introduced the concept of hyperdeterminants or multidimensional determinants. In a paper of 1845, "On the Theory of Linear Transformations,"[6] Cayley applied these hyperdeterminants to generate new invariants. Cayley's work arose out of his efforts to generalize some well-known results. For example, the invariant $ac - b^2$ for the binary quadratic was known to be the determinant

$$\begin{vmatrix} a & b \\ b & c \end{vmatrix},$$

while the invariant

$$abc + 2fgh - ah^2 - bg^2 - cf^2$$

for the ternary quadratic

$$ax_1^2 + bx_2^2 + cx_3^2 + 2fx_1x_2 + 2gx_1x_3 + 2hx_2x_3$$

was the determinant

$$\begin{vmatrix} a & f & g \\ f & b & h \\ g & h & c \end{vmatrix}. \tag{31.8}$$

The first fact was already contained in Boole; Cayley presented the second invariant in his paper. As an example of the role of hyperdeterminants, so named by Cayley in 1845, he considered the multilinear form

$$\sum \alpha_{ijkl} \, x_i y_j z_k w_l,$$

where the indices i, j, k, l assumed only the values 1 and 2. Each of the four pairs of variables $(x_1, x_2), (y_1, y_2), (z_1, z_2), (w_1, w_2)$ could then be linearly transformed by 2×2 matrices. So the multilinear form corresponded to a $2 \times 2 \times 2 \times 2$ matrix and Cayley used hyperdeterminants to compute an invariant for this form. He then specialized the multilinear form by setting $x_1 = y_1 = z_1 = w_1 = x$ and $x_2 = y_2 = z_2 = w_2 = y$; he then identified the coefficients to get the binary quartic

$$u = ax^4 + 4bx^3y + 6cx^2y^2 + 4dxy^3 + ey^4, \tag{31.9}$$

where $a = \alpha_{1111}, b = \alpha_{2111} = \alpha_{1211} = \alpha_{1121} = \alpha_{1112}$, and so on. By making a similar identification in the invariant for the multilinear form, he obtained the second-degree invariant for the binary quartic:

$$I_1 = ae - 4bd + 3c^2. \tag{31.10}$$

[4] Crilly (2006) p. 86.
[5] Cayley (1843).
[6] Cayley (1845b).

Cayley realized that his result was different from Boole's invariant $\theta(u)$. He communicated his result to Boole, who pointed out that there was also an invariant of the third degree, given by Cayley as:[7]

$$I_2 = ace - b^2 e - ad^2 - c^3 + 2bcd.$$

Cayley in turn showed that his invariant as well as Boole's third-order invariant could most easily be derived by a method Boole had indicated in his very first paper, written in 1839. Boole then informed Cayley of yet another result, obtained by trial and error:

$$\theta(u) = I_1^3 - 27I_2^2, \tag{31.11}$$

showing that the three invariants $\theta(u)$, I_1^3, I_2^2 were linearly dependent. Cayley was intrigued by this result and computed invariants with still greater fervor, though by means of new methods.[8]

Boole soon abandoned invariant theory in favor of analysis and logic; because of the unwieldy computational difficulties of hyperdeterminants, Cayley also gave up using them to find invariants. Perhaps surprisingly, Gelfand, Kapranov, and Zelevinsky rediscovered and promoted the study of hyperdeterminants. In their 1994 *Discriminants, Resultants, and Multidimensional Determinants*,[9] they wrote that although hyperdeterminants had been largely abandoned for 150 years, they found them to be important in their attempt to construct a general theory of hypergeometric functions in several variables.

However, the relation (31.11) suggested to Cayley and Sylvester an important problem, and they began work on it in the 1850s: Determine invariants $I_1, I_2, \ldots, I_s$ of a binary quantic such that all other invariants would be of the form $P(I_1, I_2, \ldots, I_s)$, for some polynomial P. The English mathematicians Arthur Cayley (1821–1895) and J. J. Sylvester were mathematical friends, reminding us of Euler and Goldbach before them and Hardy and Littlewood after them. Cayley and Sylvester met in 1847 as law students; they remained close friends for almost fifty years until Cayley's death, meeting as frequently as possible and exchanging hundreds of letters and hand-delivered notes. Both algebraists, they often worked simultaneously on the same topic. One may ask why they published no joint work. First, Cayley was a reserved and reticent person, while Sylvester was extremely ebullient and volatile. Moreover, Sylvester exhibited a strong need to maintain strict mathematical priority, both for himself and others. For example, in 1882 Sylvester wrote a paper on partitions, divided into a number of distinct sections, each with its own heading and authorship, indicating whether that portion of the argument should be credited to himself or to his student Franklin. In spite of the apparent separateness of their work, Cayley and Sylvester's mutual support and motivation surely led each of them to more progress than they

---

[7] ibid. pp. 205–206.
[8] Cayley (1846) p. 104.
[9] Gelfand, Kapranov, Zelevinsky (1994).

might have achieved separately. E. T. Bell aptly labeled Cayley and Sylvester the invariant twins;[10] we remark that they must have been fraternal twins.

In order to look at the work of Cayley and Sylvester after 1850, we give some definitions in slightly modernized form, largely following Hilbert's notation, presented in his 1897 lectures.[11] Cayley and Sylvester worked primarily on invariants of binary quantics. These are polynomials in two variables, of the form

$$f(x_1, x_2) = a_0 x_1^n + \binom{n}{1} a_1 x_1^{n-1} x_2 + \binom{n}{2} a_2 x_1^{n-2} x_2^2 + \cdots + a_n x_2^n. \qquad (31.12)$$

Suppose the linear transformation (31.5) with determinant $\delta = ps - qr \neq 0$ converts $f(x_1, x_2)$ into

$$A_0 y_1^n + \binom{n}{1} A_1 y_1^{n-1} y_2 + \binom{n}{2} A_2 y_1^{n-2} y_2^2 + \cdots + A_n y_2^n. \qquad (31.13)$$

An invariant I of $f(x_1, x_2)$ is then a polynomial in the coefficients $a_0, a_1, \ldots, a_n$, denoted by $I(a_0, a_1, \ldots, a_n)$, such that for some integer p

$$I(A_0, A_1, \ldots, A_n) = \delta^p I(a_0, a_1, \ldots, a_n), \qquad (31.14)$$

where $A_0, A_1, \ldots, A_n$ are given by (31.13). Next, a covariant of $f(x_1, x_2)$, denoted by $C(a_0, a_1, \ldots, a_n, x_1, x_2)$, is defined as a polynomial in $a_0, a_1, \ldots, a_n$ and in x_1, x_2, such that

$$C(A_0, A_1, \ldots, A_n, y_1, y_2) = \delta^p C(a_0, a_1, \ldots, a_n, x_1, x_2), \qquad (31.15)$$

with $A_0, A_2, \ldots, A_n$ again defined by (31.13).

Within this notation, the invariant of the quadratic form is $a_1^2 - a_0 a_2$; for the quartic form, the invariants mentioned earlier would be

$$I_1 = a_0 a_4 - 4 a_1 a_3 + 3 a_2^2 \quad \text{and} \quad I_2 = a_0 a_2 a_4 - a_1^2 a_4 - a_0 a_3^2 - a_2^3 + 2 a_1 a_2 a_3.$$

Two of these invariants are homogeneous polynomials of degree 2 and the third is of degree 3. If the coefficient a_k is assigned a weight k, then the weight of each term in $a_1^2 - a_0 a_2$ can be given the value 2 by adding the weights in each product. Thus, this invariant is said to be of weight 2. Similarly, the weights of other two invariants are 4 and 6. Note also that the invariant $a_1^2 - a_0 a_2$ is the discriminant of the quadratic form while $I_1^3 - 27 I_2^2$ is the discriminant of the quartic form. In a similar way, the cubic form discriminant given by

$$a_0^2 a_3^2 - 3 a_1^2 a_2^2 + 4 a_1^3 a_3 + 4 a_0 a_2^3 - 6 a_0 a_1 a_2 a_3$$

is an invariant of that form, of degree 4 and weight 6.

---

[10]　Bell (1937), chapter 21.

[11]　For the English translation, see Hilbert (1993).

In his 1854 paper,[12] "An Introductory Memoir upon Quantics," Cayley showed that all invariants of (31.12) were homogeneous polynomials of a given degree, say θ, and weight p, identical to the integer in equation (31.14), and that the relation of these quantities was determined by

$$n\theta = 2p. \tag{31.16}$$

Indeed, this paper included a similar result for covariants. Cayley also found a computationally simpler way of generating invariants by means of differential operators. Interestingly, Cayley later noted that as early as the 1840s, he had observed that

$$\left(a\frac{\partial}{\partial b} + 2b\frac{\partial}{\partial c} \right)(b^2 - ac) = 0,$$

and this then led him to consider such operators even in connection with his researches on hyperdeterminants. So in 1854, Cayley defined the two operators

$$\Omega = a_0\frac{\partial}{\partial a_1} + 2a_1\frac{\partial}{\partial a_2} + 3a_2\frac{\partial}{\partial a_3} + \cdots + na_{n-1}\frac{\partial}{\partial a_n}, \tag{31.17}$$

$$O = na_1\frac{\partial}{\partial a_0} + (n-1)a_2\frac{\partial}{\partial a_1} + (n-2)a_3\frac{\partial}{\partial a_2} + \cdots + a_n\frac{\partial}{\partial a_{n-1}}. \tag{31.18}$$

He showed that an invariant I of (31.12) satisfied the equations

$$\Omega I = 0 \quad \text{and} \quad OI = 0. \tag{31.19}$$

In fact, a seminvariant is defined as a homogeneous and isobaric (each term of the same weight) polynomial S satisfying $\Omega S = 0$.

Sylvester also conceived of the idea of the differential operator and published it before Cayley in an 1852 paper, "On the Principles of the Calculus of Forms."[13] In this paper, Sylvester noted that he had discovered a simple derivation of (31.19) after Cayley had communicated that result to him. Sylvester also remarked that the German mathematician Siegfried Aronhold, "as I collect from private information, was the first to think of the application of this method to the subject."[14]

Cayley and Sylvester each proved that every seminvariant $I(a_0, a_1, \ldots, a_n)$ of degree θ and weight p would be an invariant under the condition $n\theta = 2p$. They generalized this to covariants and it became their favorite method of producing invariants and covariants, of various degrees and weights. Cayley's "Second Memoir upon Quantics"[15] gave a combinatorial method for computing the number of invariants of degree θ and weight p by solving the equation

$$\Omega S(a_0, a_1, \ldots, a_n) = \sum \alpha_{k_0 k_1 \cdots k_n} \Omega a_0^{k_0} a_1^{k_1} \cdots a_n^{k_n} = 0. \tag{31.20}$$

---

[12] Cayley (1854).
[13] Sylvester (1852).
[14] Sylvester (1973) vol. 1, pp. 351–352.
[15] Cayley (1855).

Now the number of terms of the form $a_0^{k_0} a_1^{k_1} \cdots a_n^{k_n}$ that are homogeneous of degree θ and weight p is equal to the number of nonnegative integer solutions of the two equations:

$$k_0 + k_1 + \cdots + k_n = \theta, \tag{31.21}$$

$$k_1 + 2k_2 + \cdots + nk_n = p. \tag{31.22}$$

Let $\omega_n(\theta, p)$ denote this number. When the Ω operator is applied, we get terms of degree θ and weight $p - 1$. So equation (31.20) consists of $\omega_n(\theta, p - 1)$ equations in $\omega_n(\theta, p)$ variables. Cayley conjectured that the equations were independent and proceeded on this certainty. With this assumption, he was able to prove that the number of invariants of degree θ and weight p for a form of $f(x_1, x_2)$ of degree $n = \frac{2p}{\theta}$ would be given by $\omega_n(\theta, p) - \omega_n(\theta, p - 1)$.

Observe that the number of solutions of equation (31.22) is the number of partitions of p, where each part is at most n. This connection between the number of invariants and the number of partitions probably led Cayley and Sylvester in the mid-1850s to investigate partitions; Sylvester gave a course of lectures on the subject in 1857. Interestingly, in 1878 during his later career at Johns Hopkins, Sylvester was able to prove Cayley's conjecture of independence,[16] while he was again working intensely on partitions with his students. This proof implied that $\omega_n(\theta, p)$ was the coefficient of x^p in the Gaussian polynomial

$$\frac{\left(1 - x^{n+1}\right)\left(1 - x^{n+2}\right) \cdots \left(1 - x^{n+\theta}\right)}{(1 - x)(1 - x^2) \cdots (1 - x^\theta)}. \tag{31.23}$$

We observe parenthetically this in turn implies that the Gaussian polynomial is unimodal; recall that a polynomial $a_0 + a_1 x + a_2 x^2 + \cdots + a_n x^n$ is called unimodal if there exists an integer $m \leq n$ such that

$$a_0 \leq a_1 \leq a_2 \leq \cdots \leq a_m \geq a_{m+1} \geq a_{m+2} \geq \cdots \geq a_n. \tag{31.24}$$

In his second memoir, Cayley also considered the problem of determining the fundamental invariants of a binary quantic. He presented the list of such invariants for quadratic through the sextic forms, but due to an error in reasoning he believed that binary forms of order seven and more did not have a finite basis, that is, that there did not exist a finite number of invariants $I_1, I_2, \ldots, I_s$ of a form such that every invariant of that form could be written as a polynomial in these s invariants. This mistake was not corrected until the German mathematician Paul Gordan proved in 1868 that the covariants, and hence also the invariants, of any binary quantic had a finite basis.[17] This was later extended by Hilbert to covariants of m-ary quantics in a very important paper published in 1890.[18] In spite of repeated attempts, Sylvester and Cayley failed to prove Gordan's theorem by their own methods.

---

[16] Sylvester (1878).
[17] Gordan (1868).
[18] Hilbert (1890).

It is very interesting to recognize the origins of German invariant theory in number theory and algebraic geometry. Since Fermat, binary quadratic forms had been studied in number theory. In the 1770s, in order to study such forms, Lagrange applied linear transformations such as in (31.5), except that he took the values p, q, r, and s to be integers. In this context, Gauss mentioned equation (31.6) in article 158 of his 1801 *Disquisitiones*. He did not go beyond the observation that the determinant $\theta(R)$ of the form R divided by the determinant of the form $\theta(Q)$ was a square, $(ps - qr)^2$. Then in 1844, while studying number theoretic properties of the binary cubic, Eisenstein found the invariant[19]

$$a_0^2 a_3^2 - 3a_1^2 a_2^2 + 4a_1^3 a_3 + 4a_0 a_2^3 - 6a_0 a_1 a_2 a_3. \tag{31.25}$$

In the same year, L. O. Hesse (1811–1874), a student of Jacobi, defined the important covariant

$$\frac{\partial^2 Q}{\partial x_1^2} \frac{\partial^2 Q}{\partial x_2^2} - \left(\frac{\partial^2 Q}{\partial x_1 x_2} \right)^2 \tag{31.26}$$

for any binary quantic. He introduced this covariant in order to study critical points of curves. More generally, for any homogeneous polynomial f of degree m in n variables $x_1, x_2, \ldots, x_n$, he defined the determinant

$$\left| \frac{\partial^2 f}{\partial x_i \, \partial x_j} \right|. \tag{31.27}$$

Also note that in 1841, Jacobi published an important paper on functional determinants, defining the Jacobian and drawing attention to this area of study. The determinant (31.27) is now called the Hessian, a name given by Sylvester; in 1949, Hesse's student, Siegfried Aronhold (1819–1894), who also studied with Jacobi and Dirichlet, initiated the symbolic algebraic approach for studying invariants and covariants that characterized German invariant theory until Hilbert took it in a different direction in the 1880s. Clebsch and Gordan made use of Aronhold's approach; Paul Gordan (1837–1912), who wrote his thesis in Berlin under Kummer and had as his only doctoral student the great Emmy Noether, mastered the symbolic method and thereby proved that the invariants and covariants of a binary quantic had a finite basis.

The problem of extending Gordan's result to forms in n variables was very difficult to tackle using the existing algorithmic methods. In 1890, Hilbert introduced new methods and solved the problem. Hilbert's proof depended on his lemma concerning solutions of a system of linear Diophantine equations. He proved the existence of a finite number of solutions of a special kind. This approach lent his theorem a nonconstructive character. Thus, Gordan was said by Max Noether to have commented, "This is not mathematics; this is theology!" Three years later, Hilbert gave a different proof, dependent on what is now known as the Hilbert basis theorem; it has now been reformulated in terms of ideals: If $I \subseteq K(x_1, x_2, \ldots, x_n)$ is any ideal

---

[19] Eisenstein (1975) vol. 1, pp. 1–3.

in the ring of polynomials in n variables with coefficients in the field K, then there exists a finite number of polynomials $f_1, f_2, \ldots, f_m$ in I such that for all f in I

$$f = A_1 f_1 + A_2 f_2 + \cdots + A_m f_m \tag{31.28}$$

for some polynomials $A_1, A_2, \ldots, A_m$ in the ring. With the development of the machinery of Gröbner bases, Hilbert's second method of proof has become computationally quite significant.

Hilbert's work became the foundation for the development of commutative algebra in the twentieth century; it paved the way leading toward the abstract point of view in algebra and to the recent computational methods of ideal theory. In the area of commutative algebra, the chess champion Emanuel Lasker (1868–1941) in 1905 established the main facts behind the primary decomposition of ideals. Another important contributor to the theory of rings of polynomials was F. S. Macaulay (1862–1937), Littlewood's teacher of mathematics at St. Paul's School in London. Although he was an excellent mathematical researcher, Macaulay remained a secondary school teacher throughout his career. In a famous 1921 paper, "Idealtheorie in Ringbereichen," Emmy Noether pioneered the abstract approach to ring theory. It is interesting that while Gordan had an algorithmic and concrete approach to mathematics, his student became one of the founding stars of the abstract and conceptual approach to mathematics.

Although we do not go into detail on the topic, it may be worthwhile to comment briefly on the origins and development of elimination theory. Invariant theorists and mathematicians working with rings of polynomials in several variables found the method of elimination useful in various contexts. Boole made liberal application of elimination theory to produce invariants, and the topic led to the development of several aspects of algebra and algebraic geometry. Remarkably, in the twenty-first century, Eric Feron, an aerospace engineer, saw fit to translate Étienne Bézout's 1779 book on elimination theory as applied to polynomials in several variables, *Théorie géneral des équations algébriques*. In 2006 Feron wrote in the translator's foreword:[20]

Translating Bézout's research centerpiece became necessary to me after attending an illuminating presentation made by Pablo Parrilo at MIT sometime around 2002. His presentation was devoted to polynomially constrained polynomial optimization via sum-of-square arguments. It was illuminating because much of sum-of-square optimization methods rely on (i) using *polynomial multipliers*, and (ii) considering the various monomials appearing in the polynomial expressions as *independent variables*, resulting in interesting algorithmic simplifications. Such was also Bézout's approach when dealing with systems of polynomial equations. I decided I needed to investigate the matter in more detail, by reading Bézout's work and writing the present translation.

Étienne Bézout (1730–1783) became interested in mathematics by studying Euler, and he made many practical mathematical applications, including a six-volume course for the French artillery. His most original investigations involved the analysis of polynomial equations in many variables. Bézout's theorem on the number of intersection points of two plane algebraic curves is a direct consequence of his researches.

---

[20] Bézout (2006).

Even before Bézout, elimination was used in the seventeenth and eighteenth centuries to derive the discriminant of a polynomial or the resultant of two polynomials. In fact, the word resultant was used to signify the result obtained after elimination. The resultant $R(f, g)$ of two polynomials is a polynomial in the coefficients of f and g, and assuming that the coefficients of the highest powers of f and g do not vanish, then $R(f, g) = 0$ if and only if f and g have a common root. In 1665–66, Newton calculated the resultant of two cubics by computing various symmetric functions of the roots, including sums of powers. He wrote out all the thirty-four terms of the resultant.[21] In his published work on algebra, he eliminated the variable from the two equations by a different method.

It is an interesting coincidence that at about this same time, Seki Takakazu (1642–1708) was also thinking about the problem of elimination. Seki was a pioneer in the development of algebra in Japan; he joined with his student, Takebe Katahiro, to lay the foundation of early Japanese mathematics, or Wasan. Around 1670, Seki presented a method of obtaining the resultant by using determinants.[22] Given two polynomial equations of degree n, he first converted them into n equations, each of degree $n - 1$. He then applied a method that amounted to computing the n by n determinant so obtained. He explained the details of his method by taking small values of n, at least up to $n = 4$.[23] And Zhu Shijie investigated resultants in the thirteenth century. As for determinants, Chinese mathematicians had earlier used them to solve simultaneous linear equations and the Japanese mathematicians of the seventeenth and eighteenth centuries were familiar with this aspect of Chinese algebra.[24] Seki's method was rediscovered by Bézout. Consider their method for eliminating the x term from two cubics:

$$f = a_1 x^3 + b_1 x^2 + c_1 x + d_1, \quad g = a_2 x^3 + b_2 x^2 + c_2 x + d_2.$$

The three quadratic polynomials obtained from f and g would be $a_2 f - a_1 g$, $(a_2 x + b_2) f - (a_1 x + b_1) g$, and $(a_2 x^2 + b_2 x + c_2) f - (a_1 x^2 + b_1 x + c_1) g$. The 3×3 determinant formed by the coefficients of the three quadratics would then be the resultant.

Euler and Lagrange and others contributed to elimination theory in the eighteenth century; in the nineteenth century, Sylvester and Cayley were deeply interested in the topic, especially for its connection with invariant theory. The resultant of two binary quantics, for example, was their simultaneous invariant. In 1840, Sylvester published "A Method of Determining by Mere Inspection the Derivatives from Two Equations of Any Degree," giving the modern expression of the resultant of two polynomials of degrees m and n, respectively, as an $m + n$ by $m + n$ determinant. He explained the general rule and illustrated it by computing the 4×4 determinant obtained in the case of two quadratics. Since the computation of determinants is generally tedious, it is interesting to read the remark at the end of Sylvester's paper:[25]

[21] Newton (1967–1981) vol. 1, p. 518.
[22] Mikami (1914, 1974).
[23] Seki (1974).
[24] Mikami (1974).
[25] Sylvester (1973) vol. 1, pp. 54–57, especially p. 57.

Through the well-known ingenuity and kindly proferred help of a distinguished friend, I trust to be able to get a machine made for working Sturm's theorem, and indeed all problems of derivation, after the method here expounded; on which subject I have a great deal more to say, than can be inferred from this or my preceding papers.

The distinguished friend was surely Charles Babbage who at that time was developing his analytical engine to carry out repetitive numerical and algebraic calculations. Babbage was assisted in his endeavor by Ada Lovelace, the daughter of Lord Byron.

Cayley published several papers on elimination theory, reworking and simplifying the methods of earlier writers but also making very original contributions. The comments of Gelfand, Kapranov, and Zelevinsky in this connection are worth noting. In their book,[26] they write that in a short paper of 1848, Cayley "outlined a general method of writing down the resultant of several polynomials in several variables. We were very surprised to find that Cayley introduced in this note several fundamental concepts of homological algebra: complexes, exactness, Koszul complexes, and even the invariant now sometimes called the Whitehead torsion or Reidemeister-Franz torsion of an exact complex. The latter invariant is a natural generalization of the determinant of a square matrix (which itself was a recent discovery back in 1848), so we prefer to call it the determinant of a complex. Using this terminology, Cayley's main result is that the resultant is the determinant of the Koszul complex."

Elimination theory suffered a decline as Emmy Noether's abstract approach came to the forefront. Algebraic algorithms had to be reworked into this new context. Thus, in his 1946 book on the foundations of algebraic geometry,[27] André Weil constructed an abstract device intended to finally make elimination theory superfluous. However, algebraic equations in many variables are also studied by engineers, for whom the abstract approach is not ideal. Moreover, Shreeram Abhyankar, protesting Weil's attempt to eliminate elimination theory,[28] pointed out that some useful mathematical information could be lost in a nonconstructive method. Weil might well have agreed, and this may be indicated by his exposition of Eisenstein and Kronecker's work on the constructive development of elliptic functions. And so elimination theory continues to flourish. A renewed interest in finding efficient algorithms has produced new methods such as Gröbner bases.

31.2 Boole's Derivation of an Invariant

In his two-part paper published in 1841, "Exposition of a General Theory of Linear Transformations," Boole argued that the concept of an invariant could be useful in algebra.[29] He gave a method for the derivation of an invariant of a general form, of degree n and in m variables. Although the method was not of great use in the further development of invariant theory, it is interesting to observe Boole's originality in arriving at this important concept. We follow Boole closely; he supposed h_n and

---

[26] Gelfand, Kapranov, and Zelevinsky (1993) p. 4.
[27] Weil (1946).
[28] Abhyankar (1976).
[29] Boole (1841).

H_n were nth degree homogeneous functions of m variables $x_1, x_2, \ldots, x_m$ expressible linearly in terms of m variables $y_1, y_2, \ldots, y_m$. He also supposed that

$$
\begin{aligned}
h_n(x_1, x_2, \ldots, x_m) &= h'_n(y_1, y_2, \ldots, y_m), \\
H_n(x_1, x_2, \ldots, x_m) &= H'_n(y_1, y_2, \ldots, y_m),
\end{aligned}
\tag{31.29}
$$

where h'_n and H'_n were also homogeneous functions of degree n. In addition, Boole wrote these relations in the simple form

$$
q = r \quad \text{and} \quad Q = R,
\tag{31.30}
$$

respectively. He differentiated both sides of the second equation with respect to $y_1, y_2, \ldots, y_m$, and by means of the chain rule he got

$$
\begin{aligned}
\frac{\partial Q}{\partial x_1}\frac{\partial x_1}{\partial y_1} + \frac{\partial Q}{\partial x_2}\frac{\partial x_2}{\partial y_1} + \cdots + \frac{\partial Q}{\partial x_m}\frac{\partial x_m}{\partial y_1} &= \frac{\partial R}{\partial y_1}, \\
\frac{\partial Q}{\partial x_1}\frac{\partial x_1}{\partial y_2} + \frac{\partial Q}{\partial x_2}\frac{\partial x_2}{\partial y_2} + \cdots + \frac{\partial Q}{\partial x_m}\frac{\partial x_m}{\partial y_2} &= \frac{\partial R}{\partial y_2}, \\
&\vdots \\
\frac{\partial Q}{\partial x_1}\frac{\partial x_1}{\partial y_m} + \frac{\partial Q}{\partial x_2}\frac{\partial x_2}{\partial y_m} + \cdots + \frac{\partial Q}{\partial x_m}\frac{\partial x_m}{\partial y_m} &= \frac{\partial R}{\partial y_m}.
\end{aligned}
\tag{31.31}
$$

We note that Boole did not use the modern partial derivative notation; he wrote $\frac{dQ}{dx_1}$ for $\frac{\partial Q}{\partial x_i}$ and similarly for the other derivatives. Boole then assumed the linear relationship

$$
\begin{aligned}
x_1 &= \lambda_1 y_1 + \lambda_2 y_2 + \cdots + \lambda_m y_m \\
x_2 &= \mu_1 y_1 + \mu_2 y_2 + \cdots + \mu_m y_m \\
&\vdots \\
x_m &= \rho_1 y_1 + \rho_2 y_2 + \cdots + \rho_m y_m,
\end{aligned}
\tag{31.32}
$$

so that he could replace $\frac{\partial x_1}{\partial y_1}, \frac{\partial x_2}{\partial y_1}, \ldots$ by $\lambda_1, \lambda_2, \ldots$. He argued that since the values $\lambda_1, \lambda_2, \ldots, \mu_1, \mu_2, \ldots$ were finite, the equations

$$
\frac{\partial Q}{\partial x_1} = 0, \quad \frac{\partial Q}{\partial x_2} = 0, \cdots, \quad \frac{\partial Q}{\partial x_m} = 0,
\tag{31.33}
$$

implied that

$$
\frac{\partial R}{\partial y_1} = 0, \quad \frac{\partial R}{\partial y_2} = 0, \cdots, \quad \frac{\partial R}{\partial y_m} = 0.
\tag{31.34}
$$

He observed that, since the determinant of the linear transformation (31.32) could be zero, (31.34) did not imply (31.33).

Boole denoted by $\theta(Q)$ the expression obtained when the variables were eliminated from the polynomials

$$\frac{\partial Q}{\partial x_1}, \frac{\partial Q}{\partial x_2}, \ldots, \frac{\partial Q}{\partial x_m}.$$

To eliminate x from two polynomials of degree n, he suggested the Euclidean algorithm. Initially, he had m polynomials in m variables. He could eliminate, for example, x_1 from the first two, and then from the second and third, and so on until there were $m-1$ polynomials in $m-1$ variables. A repetition of this method produced $m-2$ polynomials in $m-2$ variables. Ultimately, he had all the variables eliminated, obtaining an expression $\theta(Q)$ containing only the constants. Thus, if $\frac{\partial Q}{\partial x_i} = 0$, $i = 1, \ldots, n$, he had $\theta(Q) = 0$. Moreover, since (31.33) implied (31.34), he also had $\theta(R) = 0$; and a similar relation of mutual dependence also existed between $\theta(q)$ and $\theta(r)$.

More generally, Boole combined the two relations in (31.30) into one relation of the form $Q + hq = R + hr$. In this case, if h was such that

$$\theta(Q + hq) = 0, \tag{31.35}$$

then an analogous relation

$$\theta(R + hr) = 0 \tag{31.36}$$

would also be satisfied. Next, Boole let v be the number of terms in the homogeneous polynomials q, r, Q, R and denoted the coefficients in these polynomials by $a_1\, a_2,\ldots, a_v, b_1, b_2, \ldots, b_v, A_1, A_2, \ldots, A_v, B_1, B_2, \ldots, B_v$, respectively. Then θ would be a polynomial ϕ in v unknowns, and he could write $\theta(Q) = \phi(A_1, A_2, \ldots, A_v)$. Now from (31.35) and (31.36), Boole reasoned that for any h for which

$$\phi(A_1 + ha_1, A_2 + ha_2, \ldots, A_v + ha_v) = 0, \tag{31.37}$$

he must also have

$$\phi(B_1 + hb_1, B_2 + hb_2, \ldots, B_v + hb_v) = 0. \tag{31.38}$$

The expression on the left-hand side of (31.37) was a polynomial in h where the term independent of h would be $\phi(A_1, A_2, \ldots, A_v) = \theta(Q)$, and the coefficient of the highest power of h would be $\phi(a_1, a_2, \ldots, a_n) = \theta(q)$. If the polynomial was divided across by this coefficient, the resulting monic polynomial would be identical with the monic polynomial obtained after the same procedure was applied to the left-hand side of (31.38). Since the coefficients of the polynomials could also be seen as Taylor coefficients, Boole could deduce that

$$\frac{\theta(Q)}{\theta(q)} = \frac{\theta(R)}{\theta(r)} \tag{31.39}$$

and

$$\frac{\left(a_1 \frac{\partial}{\partial A_1} + a_2 \frac{\partial}{\partial A_2} + \cdots + a_v \frac{\partial}{\partial A_v}\right)^\lambda \theta(Q)}{\theta(q)}$$

$$= \frac{\left(b_1 \frac{\partial}{\partial B_1} + b_2 \frac{\partial}{\partial B_2} + \cdots + b_v \frac{\partial}{\partial B_v}\right)^\lambda \theta(R)}{\theta(r)}. \tag{31.40}$$

As an example, Boole noted the simple case

$$ax^2 + 2bxy + cy^2 = a'x'^2 + 2b'x'y' + c'y'^2,$$
$$Ax^2 + 2Bxy + Cy^2 = A'x'^2 + 2B'x'y' + C'y'^2.$$

The results corresponding to (31.39) and (31.40) were

$$\frac{AC - B^2}{ac - b^2} = \frac{A'C' - B'^2}{a'c' - b'^2}, \tag{31.41}$$

$$\frac{aC - 2bB + cA}{ac - b^2} = \frac{a'C' - 2b'B' + c'A'}{a'c' - b'^2}. \tag{31.42}$$

From (31.39) and at the conclusion of the first part of his paper, Boole arrived at the result that gave rise to algebraic invariant theory:

$$\frac{\theta(Q)}{\theta(R)} = \frac{\theta(q)}{\theta(r)} = E.$$

Boole maintained that E could not depend on the coefficients in Q and R (or q and r). Thus, it must depend only on the coefficients appearing in the linear transformation (31.32). Boole wrote that he had found E to be an appropriate power of the determinant of the linear transformation (31.32), illustrating this by means of the binary quadratic and the cubic. He then went on to state the theorem contained in equation (31.7); this in turn led to the definition of an invariant. In a paper of 1844,[30] Boole gave details of a proof and gave the value of γ in (31.7) as $m(n-1)^{m-1}$.

Near the end of the second part of his 1841 paper, Boole wrote that "Linear transformations have hitherto been chiefly applied to the purpose of taking away from a proposed homogeneous function, those terms which involve the products of the variables. ... [T]he transformations, besides being linear, are understood to represent a geometrical change of axes." He went on to say that linear transformation could be applied to purely algebraic problems without geometric considerations. As an example he posed the problem: "To transform the function, $ax^3 + 3bx^2y + 3cxy^2 + dy^3$, to the form $a'x'^3 + d'y'^3$, a' and d' being given, and the transformation unrestricted by any other condition than that of linearity."

---

[30] Boole (1844a).

After solving this problem by means of his method, he applied it to the solution of a cubic equation with the comment, "The doctrine of linear transformations may be elegantly applied to the solution of algebraic equations." In this connection, Cayley, well aware that a general quintic could not be solved in radicals, corresponded with Boole concerning the solution of a quintic and found that invariants could shed some light on their solution.

It is interesting that in 1930, as a school boy of 16, Mark Kac solved the cubic[31] by an independently discovered method similar to Boole's solution. Kac wrote the cubic as a difference of two cubes, each of which was a linear function of the variable:

$$x^3 + px + q = A(x + m)^3 - B(x + n)^3.$$

By equating the coefficients of x, he found that $A = \frac{n}{n-m}$ and $B = \frac{m}{n-m}$ and that m and n were solutions of the quadratic equation

$$y^2 + \frac{3q}{p} y - \frac{1}{3}p = 0;$$

from this, he was able to derive Cardano's formula. Kac's paper was published in a Polish mathematics journal for students and because of this achievement he went on to become a mathematician. Luckily, the journal's editor had been unaware of Boole's work.

31.3 Differential Operators of Cayley and Sylvester

By the early 1850s, Cayley and Sylvester had discovered several elementary properties of invariants of binary quantics. They knew, for example, that these invariants were homogeneous and isobaric polynomials satisfying certain partial differential equations. They found these equations independently, though Sylvester was the first to publish them in 1852,[32] and they used them as important tools as their work in invariant theory progressed. We present Sylvester's derivation of the partial differential operators, with a slight change in notation, especially in our use of subscripts. Following Sylvester closely, suppose that

$$\phi = a_0 x_1^n + n a_1 x_1^{n-1} x_2 + \frac{1}{2} n(n-1) a_2 x_1^{n-2} x_2^2 + \cdots + a_n x_2^n \qquad (31.43)$$

is a binary quantic and that $I(a_0, a_1, \ldots, a_n)$ is an invariant of ϕ. To derive the differential equation, use the special linear transformation

$$x_1 = y_1 + e y_2, \qquad x_2 = y_2 \qquad (31.44)$$

[31] Kac (1987).
[32] Sylvester (1852).

to obtain the quantic

$$a_0(y_1+ey_2)^n + \binom{n}{1}a_1(y_1+ey_2)^{n-1}y_2 + \cdots + \binom{n}{k}a_k(y_1+ey_2)^{n-k}y_2^k + \cdots + a_n y_2^n$$

$$= A_0 y_1^n + \binom{n}{1}A_1 y_1^{n-1}y_2 + \cdots + \binom{n}{k}A_k y_1^{n-k}y_2^k + \cdots + A_n y_2^n,$$

where $A_0 = a_0$, $A_1 = a_1 + ea_0$, $A_2 = a_2 + 2ea_1 + e^2 a_0$. Note that in general,

$$A_k = a_k + \binom{k}{1}ea_{k-1} + \binom{k}{2}e^2 a_{k-2} + \cdots + e^k a_0. \tag{31.45}$$

Since the determinant of the linear transformation is 1, it follows from the definition (31.14) of an invariant that

$$I(A_0, A_1, \ldots, A_n) = I(a_0, a_1, \ldots, a_n). \tag{31.46}$$

Let $\Delta a_k = A_k - a_k$ and $\Delta I = I(A_0, A_1, \ldots, A_n) - I(a_0, a_1, \ldots, a_n)$. By Taylor's theorem in several variables

$$0 = \Delta I = \sum_{k \geq 1} \frac{1}{k!}\left(\Delta a_0 \frac{\partial}{\partial a_0} + \Delta a_1 \frac{\partial}{\partial a_1} + \Delta a_2 \frac{\partial}{\partial a_2} + \cdots\right)^k I$$

$$= \sum_{k \geq 1} \frac{1}{k!}\left(ea_0 \frac{\partial}{\partial a_1} + (2ea_1 + e^2 a_0)\frac{\partial}{\partial a_2} + \cdots\right)^k I.$$

Since this is true for every value of e, the coefficient of every power of e must be zero. In particular, the coefficient of the first power of e gives

$$\Omega I \equiv \left(a_0 \frac{\partial}{\partial a_1} + 2a_1 \frac{\partial}{\partial a_2} + 3a_2 \frac{\partial}{\partial a_3} + \cdots + na_{n-1}\frac{\partial}{\partial a_n}\right) I = 0. \tag{31.47}$$

As Sylvester pointed out, this differential equation could also be obtained by taking the derivative of (31.46) with respect to e and applying the chain rule:

$$0 = \frac{\partial I}{\partial A_0}\frac{\partial A_0}{\partial e} + \frac{\partial I}{\partial A_1}\frac{\partial A_1}{\partial e} + \cdots + \frac{\partial I}{\partial A_n}\frac{\partial A_n}{\partial e}$$

$$= A_0 \frac{\partial I}{\partial A_1} + 2A_1 \frac{\partial I}{\partial A_2} + \cdots + nA_{n-1}\frac{\partial I}{\partial A_n}.$$

Similarly, apply the transformation

$$x_1 = y_1, \quad x_2 = ey_1 + y_2 \tag{31.48}$$

to get the other differential equation

$$OI \equiv \left(na_1 \frac{\partial}{\partial a_0} + (n-1)a_2 \frac{\partial}{\partial a_1} + \cdots + a_n \frac{\partial}{\partial a_{n-1}} \right) = 0. \tag{31.49}$$

The operators Ω and O defined by (31.47) and (31.49) turned out to be quite important in invariant theory; Cayley and Sylvester made considerable use of them in their researches. The corresponding operators for covariants, defined by (31.15), are given by

$$\left(\Omega - x_2 \frac{\partial}{\partial x_1} \right) C = 0, \tag{31.50}$$

$$\left(O - x_1 \frac{\partial}{\partial x_2} \right) C = 0. \tag{31.51}$$

To prove that invariants are homogeneous and isobaric polynomials, take another special linear transformation

$$x_1 = e_1 y_1, \quad x_2 = e_2 y_2. \tag{31.52}$$

In this case, the quantic is transformed to

$$a_0 e_1^n y_1^n + \binom{n}{1} a_1 e_1^{n-1} e_2 y_1^{n-1} y_2 + \cdots + \binom{n}{k} a_k e_1^{n-k} e_2^k y_1^{n-k} y_2^k + \cdots + a_n e_2^n y_2^n$$

$$= A_0 y_1^n + \binom{n}{1} A_1 y_1^{n-1} y_2 + \cdots + \binom{n}{k} A_k y_1^{n-k} y_2^k + \cdots + A_n y_2^n,$$

where $A_0 = a_0 e_1^n, A_1 = a_1 e_1^{n-1} e_2, \ldots, A_k = a_k e_1^{n-k} e_2^k$.

Now suppose

$$I = (a_0, a_1, \ldots, a_n) = \sum \alpha_{s_0 s_1 \cdots s_n} a_0^{s_0} a_1^{s_1} \cdots a_n^{s_n}.$$

Since the determinant δ of the transformation (31.52) is $e_1 e_2$, we have by definition (31.14)

$$I(A_0, A_1, \ldots, A_n) = I(a_0 e_1^n, a_1 e_1^{n-1} e_2, \ldots, a_n e_2^n)$$

$$= \sum \alpha_{s_0 s_1 \cdots s_n} a_0^{s_0} e_1^{n s_0} a_1^{s_1} e_1^{(n-1)s_1} e_2^{s_2} \cdots a_k^{s_k} e_1^{(n-k)s_k} e_2^{k s_k} \cdots a_n^{s_n} e_n^{n s_n}$$

$$= \sum \alpha_{s_0 s_1 \cdots s_n} a_0^{s_0} a_1^{s_1} \cdots a_n^{s_n} e_1^{n s_0 + (n-1)s_1 + \cdots + (n-k)s_k + \cdots s_{n-1}} e_2^{s_1 + \cdots + k s_k + \cdots + n s_n}$$

$$= \delta^P I(a_0, a_1, \ldots, a_n) = e_1^p e_2^p \sum \alpha_{s_0 s_1 \cdots s_n} a_0^{s_0} a_1^{s_1} \cdots a_n^{s_n}.$$

Equating the coefficients of $a_0^{s_0} a_1^{s_1} \cdots a_n^{s}$ yields

$$ns_0 + (n-1)s_1 + \cdots + (n-k)s_k + \cdots + s_{n-1} = p, \tag{31.53}$$

$$s_1 + 2s_2 + \cdots + k s_k + \cdots + n s_n = p. \tag{31.54}$$

By adding these equations, one obtains

$$n(s_0 + s_1 + \cdots + s_n) = 2p. \tag{31.55}$$

This in turn implies that an invariant I is homogeneous of degree $\theta = s_0 + s_1 + \cdots + s_n$. If we define the weight of a_k to be k, then the weight of $a_0^{s_0} a_1^{s_1} \cdots a_n^{s_n}$ will be given by equation (31.54); this means that I is isobaric of weight p.

In his 1855 memoir, using the same method,[33] Cayley proved a similar result for covariants. Suppose the covariant is given by

$$C(a_0, a_1, \ldots, a_n, x_1, x_2) = C_0 x_1^m + C_1 \binom{m}{1} x_1^{m-1} x_2 + \cdots + C_m x_2^m. \tag{31.56}$$

Following the procedure we have presented, one may conclude that the coefficients $C_0, C_1, \ldots, C_m$ are homogeneous in $a_0, a_1, \ldots, a_n$ and are of the same degree; call the degree θ. Each coefficient is isobaric, and the weights of the coefficients are given by $C_i = p + i, i = 0, 1, \ldots, m$. The weight p of the coefficient C_0 is called the weight of the covariant and the integer m in (31.56) is called the order of the covariant. The argument yielding these results on covariants also shows that

$$m = n\theta - 2p. \tag{31.57}$$

Cayley also used the differential equations (31.50) and (31.51) satisfied by the covariant to derive important relations among coefficients of a covariant. For example, from the equation

$$OC = x_1 \frac{\partial C}{\partial x_2},$$

where C is given by (31.56) and O is the differential operator (31.49), we have

$$OC_0 x_1^m + \binom{m}{1} OC_1 x_1^{m-1} x_2 + \cdots + \binom{m}{k} OC_k x_1^{m-k} x_2^k + \cdots + OC_m x_2^m$$

$$= \binom{m}{1} C_1 x_1^m + 2\binom{m}{2} C_2 x_1^{m-1} x_2 + \cdots$$

$$+ (k+1) \binom{m}{k+1} C_{k+1} x_1^{m-k} x_2^k + \cdots + m C_m x_1 x_2^{m-1}.$$

Equating coefficients produces the relations

$$OC_k = (m - k) C_{k+1}, \quad k = 0, 1, \ldots, m. \tag{31.58}$$

The first m of these equations imply the relations

$$C_k = \frac{1}{m(m-1)\cdots(m-k+1)} O^k C_0, \quad k = 1, 2, \ldots, m. \tag{31.59}$$

---

[33] Cayley (1855).

Since all the coefficients $C_1, C_2, \ldots, C_m$ of the covariant C can be obtained from C_0, Sylvester called C_0 the source of the covariant.

By using the other differential equation for the covariant C, (31.50), Cayley derived the equations

$$\Omega C_k = k C_{k-1}, \quad k = 0, 1, \ldots, m. \tag{31.60}$$

In particular, the source satisfied the differential equation $\Omega C_0 = 0$. Cayley and Sylvester named any homogeneous and isobaric function P of $a_0, a_1, \ldots, a_n$ a semi-invariant, or seminvariant, if it was annihilated by Ω, that is, $\Omega P = 0$. Thus, the source of a covariant turned out to be a seminvariant. Clearly, not all seminvariants are invariants. But Cayley pointed out that if a seminvariant of degree θ and weight p satisfied equation (31.55), that is, $n\theta = 2p$, then the seminvariant would also be an invariant.

31.4   Cayley's Generating Function for the Number of Invariants

Cayley's ambition was to develop an algorithm capable of producing all the invariants of a given binary form. In this pursuit, it was important for him to determine the number of seminvariants of given degree and weight. By 1856, he had discovered a beautiful connection between this problem and Gaussian polynomials. Recall that if I is any seminvariant of degree θ and weight p, and

$$I = \sum \alpha_{s_0 s_1 \cdots s_n} a_0^{s_0} a_1^{s_1} \cdots a_n^{s_n},$$

then

$$s_0 + s_1 + \cdots + s_n = \theta; \quad s_1 + 2s_2 + \cdots + n s_n = p.$$

Next, let $N(n, \theta, p)$ denote the number of seminvariants with given n, θ, and p, and let $\omega_n(\theta, p)$ denote the number of integer solutions of the two previous equations for s_k, with the constraint that $s_k \geq 0$. The differential operator Ω, when applied to I, keeps the degree of each term the same but reduces the weight by one. Note that

$$\Omega I = \sum \alpha_{s_0 s_1 \cdots s_n} \Omega a_0^{s_0} a_1^{s_1} \cdots a_n^{s_n} = 0. \tag{31.61}$$

The number of terms in (31.61) is $\omega_n(\theta, p-1)$, and the coefficient of each of these terms is zero. This implies that there are $\omega_n(\theta, p-1)$ homogeneous linear equations for $\omega_n(\theta, p)$ quantities. In his second memoir, of 1855,[34] Cayley correctly assumed that these equations were independent and concluded that

$$N(n, \theta, p) = \omega_n(\theta, p) - \omega_n(\theta, p-1). \tag{31.62}$$

Cayley was unable to prove his assumption but was so certain of its correctness that he based his invariant theory upon it. Sylvester provided a proof in 1878. Cayley

---

[34] Cauchy (1855).

argued that it was obvious that the number $\omega_n(\theta, p)$ would turn out to be the coefficient of $x^p z^\theta$ in the series expansion of

$$\frac{1}{(1-z)(1-xz)(1-x^2 z)\cdots(1-x^n z)}. \tag{31.63}$$

Indeed, this is not difficult to see if we expand by the geometric series:

$$(1 + z + z^2 + \cdots)(1 + xz + x^2 z^2 + \cdots)\cdots(1 + x^n z + x^{2n} z^2 + \cdots).$$

Clearly, the coefficient of $x^p z^\theta$ will be equal to the number of nonnegative integer solutions of the two equations involving $s_0, s_1, \ldots, s_n$.

Summarizing Cayley's work in obtaining the invariants for forms of low degree, we observe that in 1855 he expanded the generating function (31.63) for $\omega_n(\theta, p)$:

$$\frac{1}{(1-z)\cdots(1-x^n z)} = 1 + G_1(x)z + G_2(x)z^2 + G_3(x)z^3 + \cdots. \tag{31.64}$$

To see the connection with Gaussian polynomials, change z to xz to get

$$\frac{1}{(1-xz)\cdots(1-x^{n+1}z)} = 1 + G_1(x)xz + G_2(x)x^2 z^2 + G_3(x)x^3 z^3 + \cdots. \tag{31.65}$$

The two equations (31.64) and (31.65) imply

$$(1-z)(1 + G_1 z + G_2 z^2 + \cdots + G_m z^m + \cdots)$$
$$= (1 - x^{n+1} z)(1 + G_1 xz + \cdots + G_m x^m z^m + \cdots).$$

Now equate the coefficients of z^m on both sides to get

$$G_m - G_{m-1} = G_m x^m - G_{m-1} x^{m+n}$$

or

$$G_m(x) = \frac{(1 - x^{m+n})}{1 - x^m} G_{m-1}$$
$$= \frac{(1 - x^{m+n})(1 - x^{m+n-1})\cdots(1 - x^{n+1})}{(1 - x^m)(1 - x^{m-1})\cdots(1 - x)}. \tag{31.66}$$

Thus, $G_m(x)$ is a Gaussian polynomial and the coefficient of x^p in the polynomial $G_\theta(x)$ gives $\omega_n(\theta, p)$. Now Cayley realized that the number of seminvariants $N(n, \theta, p)$ could be expressed as the difference between the coefficients of x^p and x^{p-1} in the Gaussian polynomial $G_\theta(x)$, and this difference gave the number of invariants of degree θ and weight p provided that $n\theta = 2p$. Note that $N(n, \theta, p)$ is the coefficient of x^p in

$$\frac{(1 - x^{\theta+1})(1 - x^{\theta+2}) \cdots (1 - x^{\theta+n})}{(1 - x^2)(1 - x^3) \cdots (1 - x^\theta)}. \tag{31.67}$$

Also observe that for invariants with weight $p = \frac{n\theta}{2}$, $\omega_n(\theta, \frac{n\theta}{2})$ and $\omega_\theta(n, \frac{n\theta}{2})$ are equal because they both turn out to be the coefficient of $x^{\frac{n\theta}{2}}$ in

$$\frac{(1 - x)(1 - x^2) \cdots (1 - x^{\theta+n})}{(1 - x) \cdots (1 - x^\theta)(1 - x) \cdots (1 - x^n)}.$$

This immediately implies that

$$N\left(n, \theta, \frac{n\theta}{2}\right) = N\left(\theta, n, \frac{n\theta}{2}\right), \tag{31.68}$$

a result, known as Hermite's reciprocity theorem,[35] established by Hermite in 1852 by a different method. Sylvester noted that this theorem was equivalent to stating that the number of partitions of any number p into at most m parts, with each part at most n, equaled the number of partitions of p into at most n parts, with each part at most m.

Cayley proceeded by applying these results to determine the full invariant systems for forms of degree $n = 2, 3, 4, 5, 6$. For example, by (31.67), when $n = 4$, the number of independent invariants of degree θ would be the coefficient of $x^{2\theta}$ in

$$\frac{(1 - x^{\theta+1})(1 - x^{\theta+2})(1 - x^{\theta+3})(1 - x^{\theta+4})}{(1 - x^2)(1 - x^3)(1 - x^4)}.$$

Observe that in order to find the coefficient of $x^{2\theta}$, we must retain numerator terms of degree 2θ or less; this means that we should determine the coefficient of $x^{2\theta}$ in the power series expansion of

$$\frac{1 - x^{\theta+1}(1 + x + x^2 + x^3)}{(1 - x^2)(1 - x^3)(1 - x^4)} = \frac{1}{(1 - x^2)(1 - x^3)(1 - x^4)} - \frac{x^\theta \cdot x}{(1 - x)(1 - x^2)(1 - x^3)}.$$

This coefficient would be the same as the coefficient of $x^{2\theta}$ in

$$\frac{1}{(1 - x^2)(1 - x^3)(1 - x^4)}$$

minus the coefficient of x^θ in

$$\frac{x}{(1 - x)(1 - x^2)(1 - x^3)}$$

or minus the coefficient of $x^{2\theta}$ in

$$\frac{x^2}{(1 - x^2)(1 - x^4)(1 - x^6)}.$$

---

[35] For Cayley's remarks on Hermite's reciprocity, see Cayley (1854) pp. 256–258.

Thus, we need the coefficient of $x^{2\theta}$ in

$$\frac{1 + x^3 - x^2}{(1 - x^2)(1 - x^4)(1 - x^6)}.$$

We may drop the odd power term x^3; then, we would require the coefficient of $x^{2\theta}$ in

$$\frac{1}{(1 - x^4)(1 - x^6)}$$

or of x^θ in

$$\frac{1}{(1 - x^2)(1 - x^3)} = (1 + x^2 + x^4 + \cdots)(1 + x^3 + x^6 + \cdots). \qquad (31.69)$$

Equation (31.69) implies that the number of independent invariants of degree θ is equal to the number of integer solutions of $2m + 3n = \theta$. Clearly, in each case $\theta = 2$ or $\theta = 3$, there is exactly one invariant, called I_2 or I_3. For nonnegative integers m_1 and n_1, if $2m_1 + 2n_1 = \theta$, then $I_2^{m_1} I_3^{n_1}$ is an invariant of degree θ. It is easy to see that all linearly independent invariants of a given degree can be produced by this method. Hence, I_2 and I_3 generate the full invariant system of a binary form, or quantic, of order 4.

Cayley also showed how the differential operators could be used to determine the invariants I_2 and I_3. For instance, for I_2, since it is of degree 2, it must be of weight 4 by the relation $n\theta = 2p$. The binary form of degree 4 has coefficients a_0, a_1, a_2, a_3, a_4; therefore, the weight 4 and degree 2 monomials are $a_0 a_4$, $a_1 a_3$, and a_2^2. To find an invariant I of degree 2 and weight 4, Cayley could set

$$I = A a_0 a_4 + B a_1 a_3 + C a_2^2$$

and then determine A, B, C by solving the differential equation $\Omega I = 0$, where Ω was defined by (31.47). One may easily check that

$$\Omega a_0 a_4 = 4 a_0 a_3, \quad \Omega a_1 a_3 = a_0 a_3 + 3 a_1 a_2, \quad \Omega a_2^2 = 4 a_1 a_2.$$

Thus, from

$$\Omega I = (4A + B) a_0 a_3 + (3B + 4D) a_1 a_2 = 0,$$

Cayley had $B = -4A$ and $C = 3A$. Hence, there was only one independent invariant in this case, given by

$$I_2 = a_0 a_4 - 4 a_1 a_3 + 3 a_2^2; \qquad (31.70)$$

one may check that the equation $OI = 0$, from (31.49), is also satisfied. A similar calculation would determine the invariant of degree 3 and weight 6:

$$I_3 = a_0 a_2 a_4 - a_0 a_3^2 - a_1^2 a_4 + 2 a_1 a_2 a_3 - a_2^3. \qquad (31.71)$$

Cayley's result for $n = 5$ was that the number of independent invariants of degree θ was the coefficient of x^θ in

$$\frac{1 - x^6 + x^{12}}{(1 - x^4)(1 - x^6)(1 - x^8)},$$

or the coefficient of x^θ in

$$\frac{1 - x^{36}}{(1 - x^4)(1 - x^8)(1 - x^{12})(1 - x^{18})}. \tag{31.72}$$

The result (31.72) allowed Cayley to conclude that there were no invariants of odd degree, but that there was one irreducible invariant of degree 4, one of degree 8, one of 12, and one of 18. However, these were connected by an equation of degree 36, that is, the square of the invariant of degree 18 was a polynomial function of the other three. Sylvester called such a relation a syzygy. Cayley attributed this result to Hermite.[36] In fact, before studying Hermite's work, Cayley had thought that the degrees of the invariants of order 5 binary forms had to be divisible by 4.

In the case of $n = 7$, Cayley made a conceptual error. He stated that the number of independent invariants of degree θ was equal to the coefficient of x^θ in

$$\frac{1 - x^6 + 2x^8 - x^{10} + 5x^{12} + \cdots}{(1 - x^4)(1 - x^6)(1 - x^8)(1 - x^{12})},$$

where the numerator was equal to

$$(1 - x^6)(1 - x^8)^{-2}(1 - x^{10})(1 - x^{12})^{-5}(1 - x^{14})^{-5} \cdots,$$

and where the series of factors did not terminate. Hence, he mistakenly concluded that the invariants did not have a finite basis. Gordan proved this to be incorrect.

31.5   Sylvester's Fundamental Theorem of Invariant Theory

The counting method for finding the fundamental invariants, and Cayley's conjecture in particular, were called into question when Cayley's mistake became evident. But in 1878, Sylvester succeeded in proving this basic result:[37]

$$N(n, \theta, p) = \omega_n(\theta, p) - \omega_n(\theta, p - 1).$$

In his spirited style, his paper began:

I am about to demonstrate a theorem which has been waiting proof for the last quarter of a century and upwards. It is the more necessary that this should be done, because the theorem has been supposed to lead to false conclusions, and its correctness has consequently been impugned . . . but the theorem itself is perfectly true, as I shall show by an argument so irrefragable that it must

[36] Cayley (1854) p. 257.
[37] Sylvester (1878).

be considered for ever hereafter safe from all doubt or cavil. It lies at the basis of the investigations begun by Professor Cayley in his *Second Memoir upon Quantics*, which it has fallen to my lot, with no small labour and contention of mind, to lead to a happy issue, and thereby to advance the standards of the Science of Algebraical Forms to the most advanced point that has hitherto been reached. The stone that was rejected by the builders has become the chief corner-stone of the building.

We follow Sylvester's reasoning very closely, but present it in slightly stream-lined form. The proof depends on Sylvester's lemma that for a seminvariant $F(a_0, a_1, \ldots, a_n)$ of degree θ and weight p,

$$\eta = n\theta - 2p \geq 0. \tag{31.73}$$

To prove the lemma, begin with the observation that if U is any homogeneous, isobaric polynomial of degree θ and weight p, then

$$(\Omega O - O\Omega)U = (n\theta - 2p)U. \tag{31.74}$$

The result (31.74) was well known when Sylvester wrote his paper, but he presented an argument:

$$\Omega O - O\Omega$$

$$= na_0 \frac{\partial}{\partial a_0} + 2(n-1)a_1 \frac{\partial}{\partial a_1} + 3(n-2)a_2 \frac{\partial}{\partial a_2} + \cdots + (n-1)a_{n-1} \frac{\partial}{\partial a_{n-1}}$$

$$- na_1 \frac{\partial}{\partial a_1} - 2(n-1)a_2 \frac{\partial}{\partial a_2} - \cdots - 2(n-1)a_{n-1} \frac{\partial}{\partial a_{n-1}} - na_n \frac{\partial}{\partial a_n}$$

$$= na_0 \frac{\partial}{\partial a_0} + (n-2)a_1 \frac{\partial}{\partial a_1} + (n-4)a_2 \frac{\partial}{\partial a_2} + \cdots$$

$$- (n-2)a_{n-1} \frac{\partial}{\partial a_{n-1}} - na_n \frac{\partial}{\partial a_n}. \tag{31.75}$$

If $\alpha_{s_0 s_1 \cdots s_n} a_0^{s_0} a_1^{s_1} \cdots a_n^{s_n}$ is any monomial in U, then (31.75) implies that

$$(\Omega O - O\Omega)U = n\left(a_0 \frac{\partial}{\partial a_0} + a_1 \frac{\partial}{\partial a_1} + \cdots + a_n \frac{\partial}{\partial a_n}\right)U$$

$$- 2\left(a_1 \frac{\partial}{\partial a_1} + 2a_2 \frac{\partial}{\partial a_2} + \cdots + na_n \frac{\partial}{\partial a_n}\right)U$$

$$= \sum \alpha_{s_0 s_1 \cdots s_n}(n\theta - 2p)a_0^{s_0} a_1^{s_1} \cdots a_n^{s_n}$$

$$= (n\theta - 2p)U = \eta U.$$

So $\Omega O - O\Omega \equiv \eta$. Moreover,

$$\Omega O^2 - O^2\Omega = (\Omega O - O\Omega)O + O(\Omega O - O\Omega).$$

Since the differential operator O raises the weight by 1, we see that

$$(\Omega O - O\Omega)OU = (n\theta - 2(p+1))OU = (\eta - 2)OU. \tag{31.76}$$

Hence,

$$\Omega O^2 - O^2\Omega = (\eta - 2)O + \eta O = 2(\eta - 1)O.$$

By induction one can show that

$$\Omega O^r - O^r\Omega = r(\eta - r + 1)O^{r-1}. \tag{31.77}$$

For a seminvariant F, $\Omega F = 0$; and so

$$\Omega O^r F = r(\eta - r + 1)O^{r-1}F. \tag{31.78}$$

To conclude the proof of Sylvester's lemma, suppose that η is negative. In that case, $|r(\eta - r + 1)|$, for $r = 1, 2, 3, \ldots$ forms an increasing sequence of nonzero integers. Now $O^k F = 0$ for some $k \le n\theta - p + 1$. To understand this statement, note that F, $OF, O^2 F, \ldots$ have weights $p, p + 1, p + 2, \ldots$, but also have the same degree θ, and the greatest possible weight of any homogeneous polynomial of degree θ is $n\theta$, attained by a_n^θ. Thus, $p + k \le n\theta + 1$. So let r be the value of k such that $O^r F = 0$. By (31.77), this implies $O^{r-1}F = 0$. It then follows that $\Omega O^{r-1}F = 0$, and hence $O^{r-2}F = 0$. A repeated application of this procedure gives $\eta F = 0$ or $F = 0$; hence, η cannot be negative, proving the lemma. Sylvester also showed by induction that

$$\Omega^q O^q F = \eta(2\eta - 2)(3\eta - 6) \cdots (q\eta - (q^2 - q))F$$
$$= q!\, (\eta(\eta - 1)(\eta - 2) \cdots (\eta - q + 1))F. \tag{31.79}$$

Now we are equipped to prove Cayley's conjecture. Let $D_n(p, \theta)$ denote the number of linearly independent seminvariants of degree θ and weight p so that the conjecture can be formulated as

$$D_n(p, \theta) = \omega_n(p, \theta) - \omega_n(p - 1, \theta) \equiv \Delta_n(p, \theta). \tag{31.80}$$

Observe that this equality holds if the $\omega_n(\theta, p - 1)$ equations satisfied by $\alpha_{s_0 s_1 \cdots s_n}$ in (31.61) are independent. In any case, we have $D_n(p, \theta) \ge \Delta_n(p, \theta)$. Note that $D_n(0, \theta) = \omega_n(0, \theta)$ since both sides equal 1. It is also clear that

$$D_n(p, \theta) + D_n(p - 1, \theta) + \cdots + D_n(0, \theta)$$
$$\ge \Delta_n(p, \theta) + \Delta_n(p - 1, \theta) + \cdots + \Delta_n(0, \theta) \tag{31.81}$$
$$= \omega_n(p, \theta).$$

If equality holds in this situation, then we see that $D_n(w, \theta) = \Delta_n(w, \theta)$ for all weights $w \le p$. Since, for given n and θ, the weight w satisfies the inequality $n\theta - 2w \ge 0$, we have $w \le \frac{n\theta}{2}$. So the largest value of the weight would be $\frac{n\theta}{2}$ when

$n\theta$ is even, and it would be $\frac{n\theta - 1}{2}$ when $n\theta$ is odd. Let p stand for the maximum weight. Also let $[p]$, $[p-1]$, $[p-2]$, etc., denote semivariants of degree θ in variables $a_0, a_1, \ldots, a_n$ and of weights p, $p-1$, $p-2$, etc., respectively. Then the number of linearly independent $[p]$s would be given by $D_n(p, \theta)$, and the number of linearly independent $[p-1]$s would be $D_n(p-1, \theta)$, and so on. So choose a set of $D_n(p, \theta)$ independent $[p]$s, $D_n(p-1, \theta - 1)$ independent $[p-1]$s, etc. From this set construct a new set S in which all the forms have the same weight p. This can be done by applying the operator O^q to the $D_n(p-q, \theta)$ forms $[p-q]$, since the weights of the forms $O^q[p-q]$ are all p.

To prove that this set S of forms of weight p is linearly independent, we first show that any one set of $O^q[p-q]$ is independent; if not, then the members $O^q[p-q]$ of the set are connected by a linear equation. Apply the operator Ω^q to this equation. By (31.79), $\Omega^q O^q[p-q]$ is a nonzero constant multiple of $[p-q]$. But this contradicts the independence of the $[p-q]$s. Thus, we have shown that the subset consisting of $O^q[p-q]$ is independent. Now suppose that a linear relation holds among/between any number of subsets of the form $O^q[p-q]$ for which m is the largest value of q. Operate on this linear equation by Ω^m. For $q < m$, this operation will introduce quantities of the form $\Omega^{m-q}[p-q]$, but these will in fact vanish because $[p-q]$ is a seminvariant and is hence annihilated by Ω. Thus, only forms of the type $[p-m]$ will remain after the application of Ω^m. This again gives us a contradiction because the seminvariants $[p-m]$ were chosen to be independent. We can therefore conclude that the set S is linearly independent. Therefore, the number of elements in S cannot exceed $\omega_n(p, \theta)$. By construction, the number of elements of S is given by

$$D_n(p, \theta) + D_n(p-1, \theta) + \cdots + D_n(0, \theta).$$

Hence, this sum is less than or equal to $\omega_n(p, \theta)$. Therefore, by (31.81), equality holds and we have proved Cayley's conjecture.

Sylvester's comments on his proof suggest that he may have been a keen student of Kant and valued mathematics as a creative endeavor. He wrote that his proof was accomplished "by aid of a construction drawn from the resources of the Imaginative reason, and founded on the reciprocal properties that have just been exhibited by the famous O and Ω." Later in the paper, he argued that proofs of this type showed that mathematics belonged among the liberal arts. "Whether we look to the advances made in modern geometry, in modern integral calculus, or in modern algebra, in each of these a free handling of the material employed is now possible, and an almost unlimited scope left to the regulated play of the fancy."[38]

31.6 Hilbert's Finite Basis Theorem

David Hilbert (1862–1943) was one of the most influential mathematicians of his time. He is famous for advocating an abstract, structural approach to mathematical problems, though his work on invariant theory had its algorithmic aspect. Hilbert studied

[38] Sylvester (1878) p. 185.

at Königsberg and attended lectures given by the outstanding teacher and number theorist Heinrich Weber (1842–1913). In 1882, Weber and Dedekind collaborated on an important paper in algebraic geometry, in which they presented Riemann surface theory from an algebraic perspective.[39] It is clear that this work influenced Hilbert's later approach to invariant theory. We note parenthetically that Weber wrote a three-volume work on algebra, useful even today.

Hilbert proved his basis theorem for the general situation, beginning with any number of m-ary forms or quantics. In his 1890 paper,[40] Hilbert employed a theorem of Max Noether, father of Emmy, to prove the basic lemma upon which he built his theory: If $F_1, F_2, F_3, \ldots$ is an infinite sequence of forms, that is, homogeneous polynomials in n variables $x_1, x_2, \ldots, x_n$ with coefficients in a field, then there exists an integer m such that every form in the sequence can be expressed as

$$F = A_1 F_1 + A_2 F_2 + \cdots + A_m F_m, \tag{31.82}$$

where $A_1, A_2, \ldots, A_m$ are appropriate forms in the same n variables.

Using this lemma, Hilbert demonstrated that from an arbitrary collection of forms in n variables one can always choose a finite number such that every form in the collection is a linear combination of the chosen forms, as in (31.82). Hilbert proved this by contradiction, assuming the result false. Let $F_1 \neq 0$ be a form in the collection and let F_2 be a form in the collection, but not expressible as $A_1 F_1$. By our assumption, F_2 exists. Now let F_3 be a form not expressible as $A_1 F_1 + A_2 F_2$. Again, F_3 exists by supposition. In this way, we construct a sequence of forms $F_1, F_2, F_3, \ldots$ for which no number m exists to satisfy (31.82). This contradicts Hilbert's lemma. We remark that in more modern books, this theorem is formulated in terms of polynomial ideals.

Hilbert's basis theorem for invariants states that there exists a finite number of invariants $I_1, \ldots, I_m$ of a binary quantic or form Q such that any invariant of Q is some polynomial function of $I_1, \ldots, I_m$. To prove this using Hilbert's reasoning, let S denote the set of all invariants of Q. Though our treatment of this theorem is for only one form, note that Hilbert did not restrict himself to one form Q, but to a finite number of them. His conclusion on the finite basis for the simultaneous invariants is a generalization of the result for one form. Now these invariants are homogeneous and isobaric polynomials in the $n + 1$ variables $a_0, a_1, \ldots, a_n$, the coefficients of the quantic. Hence there exist m invariants $I_1, I_2, \ldots, I_m$ such that every invariant I in S can be written as

$$I = Q_1 I_1 + \cdots + Q_m I_m. \tag{31.83}$$

Now the forms $Q_1, Q_2, \ldots, Q_m$ can be chosen to be isobaric in $a_0, a_1, \ldots, a_n$, but they need not be invariants. To get invariants from $Q_1, Q_2, \ldots, Q_m$, Hilbert constructed an operator using O and Ω:

---

[39] For an English translation of this paper, see Dedekind and Weber (2012).
[40] For an English translation of this paper, see Hilbert (1978) pp. 143–224.

$$L = 1 - \frac{O\,\Omega}{1!\,2!} + \frac{O^2\,\Omega^2}{2!\,3!} - \frac{O^3\,\Omega^3}{3!\,4!} + \cdots. \tag{31.84}$$

This operator has the property that if F is any homogeneous and isobaric polynomial in $a_0, a_1, \ldots, a_n$, of degree θ_1 and weight p_1, such that $n\theta_1 - 2p_1 = 0$, then LF is either zero or an invariant. We shall present the proof of this property of the operator L after we have deduced Hilbert's theorem from it. For this purpose, apply L to (31.83) to get

$$LI = I = (LQ_1)I_1 + \cdots + (LQ_m)I_m, \tag{31.85}$$

a result that follows from the easily proved facts that for any invariant I, $LI = I$ and $L(QI) = (LQ)I$. We must now show that LQ_i is either an invariant or zero. Since $I, I_1, \ldots, I_m$ in (31.85) are invariants, they satisfy the degree and weight condition $n\theta - 2p = 0$, though the $m+1$ invariants may have differing weights and degrees. Thus, the isobaric forms $Q_1, Q_2, \ldots, Q_m$ also satisfy the condition $n\theta - 2p = 0$. Hence $LQ_1, LQ_2, \ldots, LQ_m$ must each be either zero or an invariant. Clearly, all of them cannot be zero for then I would be zero. Thus, the nonzero $LQ_1, LQ_2, \ldots,$ LQ_m are members of the set S of invariants, and can once again be expressed in terms of $I_1, I_2, \ldots, I_m$. However, the LQ_i terms are of lower degree than I and the process will therefore terminate and every invariant I will be a polynomial in $I_1, I_2, \ldots, I_m$.

Hilbert did not bother to write down a proof of the required property of L. In his 1895 book on the algebra of quantics, Edwin Elliott, Sylvester's student at Oxford, gave a simple proof using the Cayley-Sylvester relation (31.74). Let G be a form in $a_0, a_1, \ldots, a_n$ with $\eta = n\theta - 2p \geq 0$. The weights of $\Omega G, \Omega^2 G, \Omega^3 G, \ldots$ are $p - 1$, $p - 2, p - 3, \ldots$, respectively, and hence the quantities corresponding to η become $\eta + 2, \eta + 4, \eta + 6, \ldots$, respectively. Thus, from (31.74) and (31.77), we have the relations

$$\Omega O G - O \Omega G = \eta G,$$

$$\Omega O^2 \Omega G - O^2 \Omega^2 G = 2(\eta + 1)G,$$

$$\vdots$$

$$\Omega O^r \Omega^{r-1} G - O^r \Omega^r G = r(\eta + r - 1)O^{r-1}\Omega^{r-1} G.$$

Multiply the first equation by $\frac{1}{\eta}$, the second by $-\frac{1}{2\eta(\eta+1)}, \ldots$, the rth by

$$\frac{(-1)^{r-1}}{r!\,\eta(\eta + 1)\cdots(\eta + r - 1)},$$

and so on. Add the resulting equations to obtain

$$\Omega O \left\{ \frac{1}{1\cdot\eta} - \frac{1}{2!\,\eta(\eta+1)}O\Omega + \frac{1}{3!\,\eta(\eta+1)(\eta+2)}O^2\Omega^2 - \cdots \right\} G = G. \tag{31.86}$$

This sum is finite since $\Omega^{p+1}G$ vanishes. Now replace G by ΩF where F is an isobaric form in $a_0, a_1, \ldots, a_n$ of weight $p + 1$, and write (31.86) as

$$\Omega \left\{ 1 - \frac{1}{1 \cdot \eta} O\Omega + \frac{1}{1 \cdot 2 \cdot \eta(\eta+1)} O^2\Omega^2 - \frac{1}{3! \, \eta(\eta+1)(\eta+2)} O^3\Omega^3 + \cdots \right\} F = 0.$$

$$(31.87)$$

Now substitute p for $p + 1$, enabling us to write $\eta \geq -2$ and $\eta + 2 \geq 0$. So if F is of weight p, we replace η by $\eta + 2$ in (31.87) to get

$$\Omega \left\{ 1 - \frac{1}{1 \cdot (\eta+2)} O\Omega + \frac{1}{2! \, (\eta+2)(\eta+3)} O^2\Omega^2 - \cdots \right\} F = 0.$$

Therefore, when $\eta = n\theta - 2p = 0$, we have $\Omega(LF) = 0$, and this means that LF is either an invariant or is identically zero.

We note that in his doctoral thesis of 1885, Hilbert introduced the operator L, and other similar operators. He explained that L served as a generalization of transvection, an older method of producing covariants. The subject of Hilbert's dissertation was special binary forms determined by algebraic differential equations and he mainly applied them to spherical functions. He took up this topic at the suggestion of his advisor at Königsberg, Ferdinand Lindemann (1852–1939), who is known for proving the transcendence of π.

31.7   Hilbert's Nullstellensatz

Hilbert's aim in his 1893 paper on invariants[41] was to subsume invariant theory under the general theory of algebraic function fields. This led him to a deeper proof of the basis theorem and to the creation of important new ideas fundamental to the development of twentieth-century commutative algebra and algebraic geometry. This proof of the basis theorem satisfied Gordan's requirement in that it be algorithmic. We briefly discuss one of Hilbert's important results, now known as the Nullstellensatz.

Hilbert proved that for any form or quantic, or system of forms, there existed a finite number of invariants $I_1, I_2, \ldots, I_k$ such that any other invariant I satisfied an algebraic equation

$$I^m + G_1 I^{m-1} + G_2 I^{m-2} + \cdots + G_m = 0, \tag{31.88}$$

where $G_1, G_2, \ldots, G_m$ were integral rational functions of $I_1, I_2, \ldots, I_k$. By homogeneity, the functions $G_1, G_2, \ldots, G_n$ could not have a constant term. With this result in hand, Hilbert considered forms whose coefficients had numerical values such that all the invariants $I_1, I_2, \ldots, I_k$ became zero, meaning that the value of all the invariants was zero, since by (31.88), $I^m = 0$, or $I = 0$. Hilbert called a form null if all its invariants were zero.

---

[41] For an English translation, see Hilbert (1978) pp. 225–301.

The converse of this theorem is of interest. Suppose $I_1, I_2, \ldots, I_k$ are invariants such that their vanishing implies the vanishing of all other invariants of that form or quantic. Hilbert showed that under these conditions, any invariant I of this quantic satisfied an equation of the type (31.88). Hilbert based his proof of this converse on the result now known as the Hilbert Nullstellensatz:

Suppose $f_1, f_2, \ldots, f_m$ are m homogeneous polynomials in $x_1, x_2, \ldots, x_n$, and suppose $F_1, F_2, F_3, \ldots$ are homogeneous polynomials in the same variables, such that they vanish for any values of the variables for which $f_1, \ldots, f_m$ all vanish. Then one can find an integer r such that every product $\Pi^{(r)}$ of r arbitrary functions from the sequence $F_1, F_2, F_3, \ldots$ can be represented in the form

$$\Pi^{(r)} = a_1 f_1 + a_2 f_2 + \cdots + a_m f_m,$$

where $a_1, a_2, \ldots, a_m$ are appropriately chosen homogeneous polynomials in $x_1, x_2, \ldots, x_n$.

31.8 Exercises

(1) Suppose the binary cubic form is

$$q = ax^3 + 3bx^2 y + 3cxy^2 + dy^3.$$

Show that

$$\theta(q) = (ad - bc)^2 - 4(b^2 - ac)(c^2 - bd).$$

See Boole (1841).

(2) Suppose the ternary quadratic form is

$$q = ax^2 + by^2 + cz^2 + 2dyz + 2exz + 2fxy.$$

Show that

$$\theta(q) = abc + 3def - (ad^2 + be^2 + cf^2).$$

See Boole (1841).

(3) Let $q = ax^4 + 4bx^3 y + 6cx^2 y^2 + 4dxy^3 + ey^4$. Show that

$$\begin{aligned}
\theta(q) = {} & a^3 e^3 - 6ab^2 d^2 e - 12a^2 bde^2 - 18a^2 c^2 e^2 - 27a^2 d^4 - 27b^4 e^2 \\
& + 36b^2 c^2 d^2 + 54a^2 cd^2 e + 54ab^2 ce^2 - 54ac^3 d^2 - 54b^2 c^3 e - 64b^3 d^3 \\
& + 81ac^4 e + 108abcd^3 + 108b^3 cde - 180abc^2 de.
\end{aligned}$$

See Boole (1844a).

(4) Prove Sylvester's 1877 generalization of Taylor's theorem: Suppose f is a function of $a, b, c, \ldots$ and f_1 is the same function of

$$a_1 = a, \ b_1 = b + ah, \ c_1 = c + 2bh + ah^2, \ d_1 = d + 3ch + 3bh^2 + ah^3, \ldots,$$

and let Ω represent the operator

$$a\frac{\partial}{\partial b} + 2b\frac{\partial}{\partial c} + 3c\frac{\partial}{\partial d} + \cdots.$$

Then

$$f_1 = f + \Omega.fh + (\Omega.)^2 f\frac{h^2}{1 \cdot 2} + (\Omega.)^3 f\frac{h^3}{1 \cdot 2 \cdot 3} + \cdots.$$

Thus, $f_1 = f$ if and only if $\Omega f = 0$. According to Sylvester, this last statement makes the theorem important in the calculus of invariants. See Sylvester (1973) vol. 3, pp. 88–92.

(5) Find the independent invariants of degrees 4 and 8 for a binary form of order 5. See Cayley (1889–1898) vol. 2, pp. 250–275.

(6) Show that a binary quantic has exactly two linearly independent seminvariants of degree 5 and weight 5. See Elliott (1964) p. 132.

(7) Show that a binary form of order $4n + 2$ has a covariant of the second order and third degree. See Elliott (1964) p. 157. Elliott attributes this result to Hermite.

31.9 Notes on the Literature

See Corry (2004) for the role of invariant theory in the development of the structural method in algebra. He also elaborates on the influence of Dedekind on Hilbert and Emmy Noether. For Kac's very early work on the cubic, see Kalman (2009). K. Parshall's article in Rowe and McCleary (1989), pp. 157–206, gives a history of nineteenth-century invariant theory before Hilbert. Crilly's (2006) biography presents the development of Cayley's mathematical thought with interesting details, especially in connection with invariant theory. The reader may also enjoy Hilbert's (1993) lectures, given in 1897; the first sixty pages cover the work of Cayley and Sylvester. See also Elliott (1964); the first edition of 1895 presented a very readable exposition of nineteenth-century invariant theory in English, but it did not include the symbolic method of the German school. The 1903 book by Grace and Young (1965) filled this need. For recent works on invariant theory incorporating the classical methods of Cayley and Sylvester, see Olver (1999) and Sturmfels (2008).

32

Summability

32.1 Preliminary Remarks

The subject of summability theory encompasses the variety of methods for averaging sequences, series, and integrals; it also includes the relationships among the various methods. This topic originated in the attempts to assign a value to the sum of a divergent series. Guido Grandi (1671–1742) made one of the earliest attempts, giving the sum of the series $1 - 1 + 1 - 1 + \cdots$ to be $\frac{1}{2}$ by setting $x = 1$ in the formula

$$\frac{1}{1+x} = 1 - x + x^2 - x^3 + \cdots.$$

In a letter to Christian Wolf, published in 1713,[1] Leibniz reasoned that since the sum of the first n terms of $1 - 1 + 1 - 1 + \cdots$ would be 0 or 1 depending on whether n was even or odd, the values 0 and 1 would occur with equal frequency, and hence $\frac{1}{2}$ was the most probable value of the sum. This method amounts to taking the limit of the averages of the partial sums assigned to the series $1 - 1 + 1 - 1 + \cdots$ as the number of terms gets larger and larger. Note also that $1 - x + x^2 - x^3 + \cdots$ may be seen as a type of weighted average of $1 - 1 + 1 - 1 + \cdots$. Newton also dealt with divergent series, although in unpublished work. A significant example is his transformation formula, now named after Euler,

$$\sum_{n=0}^{\infty} A_n x^{n+1} = \sum_{n=0}^{\infty} y^{n+1} \Delta^n A_0,$$

where $y = \frac{x}{1-x}$. Newton discovered this transformation in 1684, but it unfortunately remained unpublished for almost three centuries.[2] He used it to evaluate the alternating series for $\ln(1+x)$ and for $\arctan x$, taking the absolute value of x to be greater than 1. Newton explained that this transformation could be applied to convert an alternating divergent series to a convergent one; then, the value of the divergent series would be

[1] Leibniz (1713).
[2] Newton (1967–1981) vol. 4, pp. 604–611.

given by the corresponding value of the convergent series. From this we can see that Newton's ideas on divergent series were groundbreaking.

Between 1720 and 1740, de Moivre, Stirling, Euler, and Maclaurin gained significant, though partial, insights into divergent asymptotic series. Their method, based on the Euler–Maclaurin summation formula, was to begin with a finite series and convert it to an infinite asymptotic series, yielding an excellent numerical approximation of the finite series. Interestingly, in the twentieth century, Ramanujan also used the Euler–Maclaurin formula in an attempt to construct a theory for summing divergent series.[3]

Euler and Lagrange also made considerable use of divergent series in their work, though Euler's work was clearly more incisive. In 1749, Euler gave a brilliant application of summability by defining

$$1^n - 2^n + 3^n - 4^n + \cdots = \lim_{x \to 1^-} (1^n x - 2^n x^2 + 3^n x^3 - \cdots), \qquad (32.1)$$

an equation he used to discover the functional relation for the zeta function.[4] Recall that Euler's initial motivation may have been to study the series on the left, hoping it would illuminate the problem of summing the zeta value $\zeta(2n + 1)$ where n was a positive integer. By generalizing (32.1), we may say that Euler defined the sum of the series $\sum_{n=0}^{\infty} a_n$ by the equation

$$\sum_{n=0}^{\infty} a_n = \lim_{x \to 1^-} \sum_{n=0}^{\infty} a_n x^n. \qquad (32.2)$$

As discussed in Chapter 4, in 1826 Abel proved that if $\sum a_n$ was convergent, then (32.2) would hold. For this reason we say that if the value of the limit in (32.2) is taken to be L, then the series $\sum a_n$ is Abel-summable or A-summable to L. Expressed in another way, the Abel mean of $\sum a_n$ is L. Although Euler defined this summability method, it is named after Abel. As a matter of fact, when n is a positive even integer, then the value of the series on the left-hand side of (32.1) sums to 0. As we have mentioned, Abel ironically called this situation "horrible" and in a letter to Humboe, quoted Horace: "Risum teneatis, amici." [Restrain your laughter, friends.][5]

Interestingly, in the 1820s, Poisson applied Abel summability to the convergence of Fourier series.[6] Recall that Fourier claimed in his famous 1807 memoir and other works that an arbitrary function could be expanded as a Fourier series; though he presented several ingenious arguments in favor of this proposition, he did not provide a real proof. In a paper published in 1820, Poisson attempted to demonstrate that the Fourier series of a continuous function converged to that function by showing that

$$\lim_{r \to 1^-} \left(\frac{1}{2} a_0 + \sum_{n=1}^{\infty} (a_n \cos n\theta + b_n \sin n\theta) r^n \right) = f(\theta), \qquad (32.3)$$

---

[3] See Hardy (1949) pp. 346–347.
[4] Eu. I-15 pp. 70–90. E 352.
[5] See Ore (1974) p. 97 or Stubhaug (2000) pp. 343–344.
[6] Poisson (1826).

where a_n and b_n were the Fourier coefficients of a continuous function $f(\theta)$. Poisson showed that the expression within parentheses in (32.3) could be expressed as

$$P(r,\theta) = \frac{1}{2\pi} \int_0^{2\pi} \frac{1 - r^2}{1 - 2r\cos(\theta - \phi) + r^2} f(\phi) \, d\phi,$$

now called the Poisson integral. He then gave an argument that as r approached 1, the integral approached $f(\theta)$, but this argument was full of gaps. But even had Poisson's proof of (32.3) been complete, he undermined it from the beginning by falsely assuming the converse of Abel's theorem. Recall that Cauchy made a similar error at around the same time. Tauber and Littlewood later established that the converse of Abel's theorem required a growth condition on the coefficients. Cauchy, Abel, and others concluded that divergent series had no sum, effectively banishing this topic for nearly fifty years. It was only after the theory of convergent series was established on a sound footing, through the efforts of Gauss, Cauchy, Abel, Dirichlet, and Weierstrass, that mathematicians could confidently address the summability of divergent series.

The German mathematician Ferdinand Georg Frobenius (1849–1917) initiated the modern theory of summability by proving the first theorem establishing a relation between two different methods of summation. In a short paper of 1880[7] he showed that if $s_n = \sum_{k=0}^n a_k$ and

$$\frac{s_0 + s_1 + \cdots + s_n}{n+1} \to S \text{ as } n \to \infty, \tag{32.4}$$

then

$$\lim_{x \to 1^-} \sum_{n=0}^{\infty} a_n x^n = S.$$

This theorem explained why Grandi and Leibniz obtained the same value for the sum of the series $1 - 1 + 1 - 1 + \cdots$. Frobenius was a student of Weierstrass, and he initially worked in differential equations and their series solutions. He branched out into number theory and algebra with particular emphasis on groups. In answering a question of Dedekind on group determinants, Frobenius created and developed the topic for which he is best known, group representation theory. Two years after Frobenius's important paper, Otto Hölder (1859–1937), who also studied with Weierstrass, extended that work. He defined[8]

$$H_n^{(r+1)} = \frac{H_0^{(r)} + H_1^{(r)} + \cdots + H_n^{(r)}}{n+1}, \quad r = 0, 1, 2, \ldots, \tag{32.5}$$

where $H_k^{(0)} = s_k$. He pointed out that there were sequences $s_0, s_1, s_2, \ldots$ for which the limit (32.4) did not exist but such that there was an integer r for which the

---

[7] Frobenius (1880).
[8] Hölder (1882).

$\lim_{n\to\infty} H_n^{(r)}$ existed. Thus, such a series $\sum a_n$ is said to be (H,r) summable. Moreover, Hölder proved that if $\lim_{n\to\infty} H_n^{(r)} = S$, then $\lim_{x\to 1^-} \sum_{n=0}^{\infty} a_n x^n = S$.

The Italian mathematician Ernesto Cesàro (1859–1906), in spite of financial and other challenges, managed to learn mathematics from a number of good teachers, obtain positions in Italian universities, and publish prolifically in differential geometry and number theory. He studied under Eugène Catalan in Liège and spent a year in Paris attending lectures by Hermite and Gaston Darboux. He had wide interests, including mathematical physics. In 1890,[9] Cesàro gave an important application of the summability method (32.4), shedding light on a classical question on products of infinite series: Suppose that $\sum a_n = A$ and $\sum b_n = B$, and let the Cauchy product of these two series be $\sum c_n$, where $c_n = a_0 b_n + a_1 b_{n-1} + \cdots + a_n b_0$. When does the Cauchy product converge? Cesàro proved that even when the product did not converge, the limit of the arithmetic means of the partial sums of the product would converge to AB. In other words, if $C_n = c_0 + c_1 + \cdots + c_n$ then

$$\frac{C_0 + C_1 + \cdots + C_n}{n} \to AB \text{ as } n \to \infty. \tag{32.6}$$

Note that Cesàro's theorem generalized Abel's theorem that if $\sum a_n = A$, $\sum b_n = B$, and $\sum c_n = C$, then $AB = C$. In today's terminology, we would say that a series $\sum a_n$ is Cesàro summable or $(C, 1)$ summable to S if (32.4) holds true. We may also say that the Cauchy product of two series converging to A and B is Cesàro summable to AB. Cesàro next extended his result to not necessarily convergent series: If

$$A_n = \sum_{k=0}^{n} a_k, \quad B_n = \sum_{k=0}^{n} b_k, \quad C_n = \sum_{k=0}^{n} c_k$$

and

$$\frac{\sum_{k=0}^{n} A_k}{n+1} \to A, \quad \frac{\sum_{k=0}^{n} B_k}{n+1} \to B \quad \text{as } n \to \infty,$$

then

$$\frac{\sum_{k=0}^{n} C_k}{n+1} \to AB \quad \text{as } n \to \infty.$$

Cesàro also defined a more general form of convergence, starting with

$$\binom{k+n}{n} A_{n,k} = a_n + \binom{k+1}{1} a_{n-1} + \cdots + \binom{k+n}{n} a_0. \tag{32.7}$$

He defined a series $\sum a_n$ as summable (today we say (C, k) summable) to A if there was a k such that $\lim_{n\to\infty} A_{n,k} = A$. Note that it is not necessary for k to be a nonnegative integer. Since we can write

---

[9] Cesàro (1890).

$$\binom{k+j}{j} = \frac{(k+1)(k+2)\cdots(k+j)}{j!},$$

we may take k to be a real number > -1.

In 1900, the Hungarian mathematician Lipót Fejér (1880–1959) delivered a big boost to the Cesàro summability method by proving that the Fourier series of any continuous function was $(C,1)$ summable to the function.[10] It is interesting that Fejér's result arose out of an earlier attempt to solve the Dirichlet problem for the unit circle: For a continuous function $f(\theta)$ on the unit circle, determine a harmonic function $\phi(x,y) = \Phi(r,\lambda)$ inside the unit disk such that $\Phi(r,\lambda)$ tends to $f(\theta)$ as $re^{i\lambda}$ approaches $e^{i\theta}$ from inside the unit disk. Note that $\phi(x,y)$ would be harmonic if it satisfied Laplace's equation

$$\frac{\partial^2\phi}{\partial x^2} + \frac{\partial^2\phi}{\partial y^2} = 0. \tag{32.8}$$

In 1870, Carl Neumann (1832–1925), son of Franz Neumann and one of the founders of the *Mathematische Annalen*, made an attempt at solving this problem by means of the harmonic function determined by the Poisson integral $P(r,\theta)$. He used Poisson's result, that $P(r,\theta)$ tended to $f(\theta)$ as $r \to 1^-$. Recall, however, that the proof given by Poisson was incomplete; this in turn undermined Neumann's proof. As a third year student at the Technical University in Hungary, Fejér spent 1899–1900 in Berlin, attending lectures by L. Fuchs, Schwarz, and Frobenius, all students of Weierstrass. Fejér learned of Neumann's attempt from Schwarz, who in 1871 had solved the Dirichlet problem by an alternative method. Examining the gap in Neumann's proof, Fejér proved the $(C,1)$ summability of the Fourier series of a continuous function. This result, combined with Frobenius's theorem that $(C,1)$ summability implied Abel summability, mended Poisson's proof. As a corollary, Fejér obtained the theorem that a continuous function could be uniformly approximated by trigonometric polynomials on a closed interval. Since the sine and cosine functions could be approximated by their Taylor polynomials, he further deduced Weierstrass's theorem on the uniform approximation of continuous functions by polynomials.

Similar to his first mathematical efforts, a number of Fejér's later papers presented elegant solutions to interesting but circumscribed problems, where both the problems and the solutions had significant implications in several areas. While a professor at Budapest, he had a broad influence on the development of mathematics in Hungary. His mathematical style, his outgoing personality, and his wide-ranging cultural interests attracted many good students, including Erdős, Pólya, Szegő, Turán, and von Neumann.

The Austrian mathematician Alfred Tauber (1866–1942) was a professor of mathematics at Vienna and an accomplished actuary; he served as chief mathematician for the Phönix Insurance Company. He and Georg Pick died in the Theresienstadt concentration camp at about the same time. Tauber gave a new direction to summability

---

[10] Fejér (1900).

theory with a result on a converse of Abel's theorem on series.[11] He proved that if $\sum a_n$ was Abel-summable to A and $na_n \to 0$, or $a_n = o(\frac{1}{n})$, as $n \to \infty$, then $\sum a_n = A$. Tauber also proved an Abel summable series $\sum a_n$ to be convergent if and only if

$$\frac{a_1 + 2a_2 + \cdots + na_n}{n} \to 0 \text{ as } n \to \infty. \tag{32.9}$$

In a paper of 1907, the German analytic number theorist Edmund Landau (1877–1938), extended Tauber's theorem to series of the form $\sum_{n=1}^{\infty} a_n e^{-\lambda_n x}$, where $\lambda_1 < \lambda_2 < \cdots$ and $\lambda_n \to \infty$ as $n \to \infty$. Note that this covers power series as well as Dirichlet series.[12] Landau also proved an integral analog of Tauber's theorem: If

$$J(x) = \int_1^{\infty} f(t) t^{-x} \, dt \to A \quad \text{as} \quad x \to 0 \tag{32.10}$$

and

$$f(t) = o\left(\frac{1}{t \ln t}\right) \quad \text{as} \quad t \to \infty, \tag{32.11}$$

then

$$J(0) = \int_1^{\infty} f(t) \, dt = A. \tag{32.12}$$

Recall that in 1749, Euler attempted to use Abel summability to prove the functional equation for the zeta function. In 1906, Landau vindicated Euler's efforts by proving that the Abel sum of the series $\sum_{n=1}^{\infty} \frac{(-1)^{n-1}}{n^s}$ yielded the value $(1-2^{1-s})\zeta(s)$, obtained by the analytic continuation of the zeta function.[13] In this work, Landau employed an 1898 result of the Finnish mathematician Hjalmar Mellin. Landau, a student of Frobenius, also introduced the one-sided Tauberian condition on the coefficients of series, especially applicable in number theory. In 1903, Landau derived the prime number theorem[14] without using Hadamard's theory of entire functions of finite order; in 1907, he obtained an important generalization of Picard's theorem on entire functions.[15] Concerning Landau's 1927 *Vorlesungen über Zahlentheorie*, Hardy and Heilbronn wrote in an obituary notice for Landau, "This remarkable work is complete in itself; he does not assume ... even a little knowledge of number-theory or algebra. It stretches from the very beginning to the limits of knowledge, in 1927, of the 'additive,' 'analytic,' and 'geometric' theories."[16]

The preliminary summability results of Frobenius, Cesàro, Fejér, Tauber, and Landau laid the foundation for a cohesive theory of summability with wide

---

[11] Tauber (1897).
[12] Landau (1907b).
[13] Landau (1906).
[14] Landau (1903).
[15] Landau (1907a).
[16] Hardy and Heilbronn (1938).

applicability. The British mathematicians G. H. Hardy (1877–1947) and J. E. Littlewood (1885–1977) were the first to fully understand the potential and scope of this mathematical theory. Hardy's many mathematical contributions included the circle method, discovered jointly with Ramanujan in their work on the asymptotic theory of partitions; and the concept of maximal functions, developed in collaboration with Littlewood. His influence was felt as much through his teaching as in his research. He helped raise British standards of teaching in analysis by publishing his 1908 *A Course of Pure Mathematics*, still in print today. In his preface to the 1937 edition of this book, Hardy remarked that if he were to rewrite the book, "I should not write (to use Prof. Littlewood's simile) like 'a missionary talking to cannibals,' but with decent terseness and restraint."[17] Hardy enjoyed mathematical collaboration, and his association with Littlewood was one of the most productive in the history of mathematics. They published almost one hundred joint papers in analysis and analytic number theory. According to Harald Bohr's birthday lecture of 1953, reprinted by Bollobás in his foreword to Littlewood's *Miscellany*, the Hardy–Littlewood collaboration was based on four rules:[18]

(1) When one wrote to the other, it was completely indifferent whether what they wrote was right or wrong.
(2) When one received a letter from the other, he was under no obligation whatsoever to read it, let alone answer it.
(3) Although it did not really matter if they both simultaneously thought about the same detail, still, it was preferable that they should not do so.
(4) It was quite indifferent if one of them had not contributed the least bit to the contents of a paper under their common name.

Although both Hardy and Littlewood lived on the Trinity College grounds, within one or two hundred yards of one another, and ate their meals in the same dining hall, their rules suggest that most of their communications were via the written word. Littlewood, unlike Hardy, had an interest in applied mathematics. In collaboration with Mary Cartwright (1900–1998), he also made important contributions to nonlinear differential equations and topological dynamics. Concerning Littlewood, V. I. Arnold wrote, "In mathematics he was a direct successor of Newton and Poincaré, doing research even on artillery ballistics. I was surprised to discover his estimates of the time of preservation of an adiabatic invariant in a Hamiltonian system." It is even more surprising that the 'theory of chaos' in dynamical systems, including 'Smale's horseshoe,' had been already developed and published by Littlewood."[19]

In a 1909 paper, Hardy showed that if $\sum a_n$ was $(C, 1)$ summable to S and $a_n = O(\frac{1}{n})$ then $\sum a_n$ converged to S. He noted that by combining this result with Fejér's $(C, 1)$ summability of the Fourier series of a continuous function $f(x)$, one obtained Dirichlet's theorem on Fourier series. Take $f(x)$ to be monotonic, and apply the second mean value theorem

---

[17] Hardy (1937).
[18] Littlewood (1986) pp. 10–11.
[19] Arnold (2007) pp. 115–116.

$$\int_0^{2\pi} f(x)\cos nx\,dx = f(0)\int_0^{\xi} \cos nx\,dx + f(2\pi)\int_{\xi}^{2\pi'} \cos nx\,dx$$

to see that the Fourier coefficients are $O(\frac{1}{n})$. Hardy was not successful in his attempt to prove the more general result that if $\sum a_n$ was Abel summable to S and $a_n = O(\frac{1}{n})$, then $\sum a_n = S$. In fact, he thought the result could well be false. He suggested the problem to his former student Littlewood, who succeeded in solving it in the affirmative. In his "A Mathematical Education," Littlewood gave an account of his discovery of the proof.[20] Surprisingly, as he grappled with the problem, he forgot that Hardy had already proved the Cesàro-Tauber theorem. In 1911, during his attempt to reprove this, he discovered the derivatives theorem.[21] Note that in intuitive terms, the derivatives theorem states that the orders of magnitude of two derivatives of a function restrict the order of magnitude of the intermediate derivatives. Hardy and Littlewood made considerable use of this concept in their early work. But, as they mentioned in a paper of 1914,[22] Hadamard had already proved the derivatives theorem and had published it in an 1897 paper on waves. Indeed, A. Kneser also independently obtained the theorem in the same year. Littlewood stated his Abel-Tauber theorem in the general form that included Dirichlet series:

If $0 < \lambda_1 < \lambda_2 < \cdots < \lambda_n \to \infty$ as $n \to \infty$,

$$\lim_{n\to\infty} \sum_{n=1}^{\infty} a_n e^{-x\lambda_n} = S, \tag{32.13}$$

and

$$|a_n| < K\left(\lambda_n - \frac{\lambda_{n-1}}{\lambda_n}\right) \quad \text{for a constant } K, \tag{32.14}$$

then

$$\sum_{n=1}^{\infty} a_n = S.$$

Observe that when $\lambda_n = n$, the sum in (32.13) reduces to a power series, whereas when $\lambda_n = \ln n$, one gets a Dirichlet series. Littlewood's condition $|a_n| = O(\frac{1}{n})$ is quite natural, since is it is easy to see that under this condition, if $\sum a_n x^n$ oscillates finitely as $x \to 1^-$, then so does the sequence $\sum_{k=1}^{n} a_k$ as $n \to \infty$. In fact, Littlewood pointed out that the condition $|a_n| = o(\frac{1}{n})$ implied the much stronger result: that the limits of oscillation of $\sum a_n x^n$ as $x \to 1^-$ and of $\sum_{k=1}^{n} a_k$ as $n \to \infty$ were the same. These results must have suggested to him that in order for Abel summability to imply convergence, a weaker condition would suffice.

---

[20] Littlewood (1986) pp. 80–93.
[21] Littlewood (1911).
[22] Hardy and Heilbronn (1914).

In an interesting 1910 paper,[23] Landau proved Hardy's $(C, 1)$ summability theorem with a weaker one-sided Tauberian condition $na_n \geq -K$, where K was a constant. He mentioned that one-sided Tauberian arguments had been used by Hadamard and Vallée-Poussin in their proofs of the prime number theorem (PNT). In a 1913 paper,[24] Hardy and Littlewood proved a one-sided extension of Littlewood's theorem: If $a_n \geq 0$, $\alpha > 0$, and

$$\lim_{x \to 1^-} (1 - x)^\alpha \sum_{n=0}^\infty a_n x^n = A, \tag{32.15}$$

then

$$\lim_{n \to \infty} \frac{\sum_{k=0}^n a_k}{n^\alpha} = \frac{A}{\Gamma(1 + \alpha)}. \tag{32.16}$$

Hardy and Littlewood soon saw that this theorem had important implications in prime number theory. They showed that[25]

$$\lim_{\xi \to 0^+} \xi \sum_{n=1}^\infty \Lambda(n) e^{-n\xi} = 1, \tag{32.17}$$

and since $\Lambda(n) \geq 0$, the hypothesis of their theorem, given by (32.15), was true with $\alpha = 1$, $x = e^{-\xi}$, $a_n = \Lambda(n)$, and $A = 1$. Recall that when n is a positive integer power of a prime p, $\Lambda(n) = \ln p$; otherwise, it is 0.

Next, by (32.16)

$$\lim_{n \to \infty} \frac{1}{n} \sum_{k=1}^n \Lambda(k) = 1. \tag{32.18}$$

It was well known that (32.18) was equivalent to the PNT, and to prove (32.17) Hardy and Littlewood needed the fact that with $s = \sigma + it$, $\zeta(s)$ had no zeros on $\sigma = 1$ and satisfied a very mild growth condition for large t and $1 \leq \sigma \leq 2$. This growth condition was so weak that they concluded that there should be a proof of the PNT requiring only $\zeta(1 + it) \neq 0$ for real t. In looking for such a proof, they investigated the Lambert summability method. Note that a series $\sum a_n$ is Lambert summable to S if

$$\lim_{x \to 1^-} (1 - x) \sum_{n=1}^\infty n a_n \frac{x^n}{1 - x^n} = S. \tag{32.19}$$

In a paper written in 1919, Hardy and Littlewood proved that Lambert summability implied Abel summability.[26] From this theorem they could easily derive a result

[23] Landau (1910).
[24] Hardy and Littlewood (1913).
[25] Hardy and Littlewood (1918).
[26] Hardy and Littlewood (1921).

equivalent to the PNT. Unfortunately, this did not give a new proof of the PNT because to prove their Lambert summability theorem, they had used the fact that, with μ the Möbius function,

$$g(n) = \sum_{m=1}^{n} \frac{\mu(m)}{m} = O\left(\frac{1}{(\ln n)^2}\right).$$

(32.20)

Thus, they relied on a result a little deeper than the PNT, since the PNT is equivalent to $g(n) = o(1)$ as $n \to \infty$. Though they failed to offer another proof of the PNT, their work set the stage for Wiener. In 1928, Norbert Wiener (1894–1964) found a method to directly handle Lambert summability. Wiener received his doctoral degree from Harvard University at the age of 18 with a thesis in logic. He then spent a part of 1913 at Cambridge University to study under Bertrand Russell who advised him to study mathematics and physics, especially the papers of Einstein and Niels Bohr on relativity, Brownian motion and quantum theory. Wiener was greatly impressed and influenced by Hardy's course on real and complex variables and all of this bore fruit about a decade later. Wiener was a professor at M.I.T. from 1919 to his death in 1964. He interacted vigorously with his engineering colleagues. The electrical engineering department requested that he provide a rigorous basis for Heaviside's operational methods. This work led Wiener to a very fruitful study of a generalized harmonic analysis. He encountered a technical problem in his harmonic analysis research: Show that for a class of nonnegative functions $f(t)$

$$\lim_{T \to \infty} \frac{1}{T} \int_0^T f(t)\,dt = \lim_{\epsilon \to 0} \frac{2}{\pi \epsilon} \int_0^\infty f(t) \frac{\sin^2 \epsilon t}{t^2}\,dt.$$

(32.21)

At this point in his researches, in 1926, Wiener was visiting Göttingen, as was his friend, the English mathematician A. E. Ingham. Wiener learned from Ingham that his problem was Tauberian in nature and that Hardy and Littlewood had worked on similar problems. Wiener corresponded with Hardy on this question but finally decided to follow his own approach, using Fourier transforms. In his autobiography, *I Am a Mathematician*, Wiener wrote that he also consulted Toeplitz's student R. Schmidt, who had published an important paper on Tauberian theory in 1925. Wiener had hoped to collaborate with Schmidt on this problem, for there was a connection in their approaches, but this collaboration did not work out. However, Schmidt suggested that, since his own method had failed for Lambert summability and the PNT, Wiener might test his own approach in those cases. Wiener was soon able to discover a comprehensive method, covering all known Tauberian results.

To get a sense of Wiener's work, begin by writing the Abel sum of $\sum a_n$ in the form

$$A = \lim_{r \to 1^-} (1 - r) \sum_{n=0}^{\infty} s_n r^n = \lim_{x \to \infty} \frac{1}{x} \sum_{n=0}^{\infty} s_n e^{-\frac{n}{x}}.$$

(32.22)

With this, we have another form of the Hardy–Littlewood theorem: If

$$\lim_{x \to \infty} \frac{1}{x} \sum_{n=0}^{\infty} s_n e^{-\frac{n}{x}} = A \quad \text{and} \quad s_n = O(1), \quad \text{then} \quad \lim_{x \to \infty} \frac{1}{x} \sum_{n \le x} s_n = A. \quad (32.23)$$

Now we can write the integral analog: If $F(t)$ is bounded and

$$\lim_{x \to \infty} \frac{1}{x} \int_0^{\infty} e^{-\frac{t}{x}} F(t) \, dt = A \quad \text{then} \quad \lim_{x \to \infty} \frac{1}{x} \int_0^x F(t) \, dt = A. \quad (32.24)$$

Note that the first limit in (32.24) is a weighted average of the function $F(t)$ where the weight function is given by $e^{-\frac{t}{x}}$. More generally, let the weight function be expressed as $G(\frac{t}{x})$ so that the integral takes the form

$$\int_0^{\infty} G\left(\frac{t}{x}\right) F(t) \, dt = \int_{-\infty}^{\infty} e^{u-y} G(e^{u-y}) F(e^y) \, dy = \int_{-\infty}^{\infty} K_1(u - y) f(u) \, du, \quad (32.25)$$

after applying the change of variables $t = e^u$, $x = e^y$, $F(e^u) = f(u)$, $e^{u-y} G(u - y) = K_1(u - y)$. Wiener could then pose the very general question: Given a bounded function $f(u)$ and kernel K_1 integrable over $(-\infty, \infty)$, under what conditions does the equation

$$\lim_{y \to \infty} \int_{-\infty}^{\infty} K_1(u - y) f(u) \, du = A \int_{-\infty}^{\infty} K_1(u) \, du \quad (32.26)$$

imply

$$\lim_{y \to \infty} \int_{-\infty}^{\infty} K_2(u - y) f(u) \, du = A \int_{-\infty}^{\infty} K_2(u) \, du \quad (32.27)$$

for a different integrable kernel K_2? To determine a simple condition on K_1, Wiener assumed that K_2 was a convolution of K_1 with an integrable function R, that is

$$K_2(y) = \int_{-\infty}^{\infty} K_1(y - u) R(u) \, du. \quad (32.28)$$

Now note that the Fourier transform converts a convolution of two functions to the ordinary product of the transforms of the two functions. So, where $\widehat{K}$ denotes the Fourier transform of K,

$$\widehat{K}_2 = \widehat{K}_1 \cdot \widehat{R}. \quad (32.29)$$

The beauty of relation (32.29) is that it allows us to determine $\widehat{R}$ at all points if for all x

$$\widehat{K}_1(x) = \int_{-\infty}^{\infty} e^{-ixt} K_1(t) \, dt \ne 0. \quad (32.30)$$

This was Wiener's now-famous condition, that the existence of the first average would imply the existence of the second. In his 1932 paper "Tauberian Theorems," Wiener stated two forms of this theorem.[27] Note that he wrote L_p for L^p. The first version of Wiener's Tauberian theorem: Let $f(x)$ be a bounded measurable function, defined over $(-\infty, \infty)$. Let $K_1(x)$ be a function in L_1, and let

$$\frac{1}{\sqrt{2\pi}} \int_{-\infty}^{\infty} K_1(x) e^{-iux} \, dx \neq 0 \tag{32.31}$$

for all real u. Let

$$\lim_{x \to \infty} \int_{-\infty}^{\infty} f(\xi) K_1(\xi - x) \, d\xi = A \int_{-\infty}^{\infty} K_1(\xi) \, d\xi. \tag{32.32}$$

Then if $K_2(x)$ is any function in L_1,

$$\lim_{x \to \infty} \int_{-\infty}^{\infty} f(\xi) K_2(\xi - x) \, d\xi = A \int_{-\infty}^{\infty} K_2(\xi) \, d\xi. \tag{32.33}$$

Conversely, let $K_1(\xi)$ be a function of L_1, and let $\int_{-\infty}^{\infty} K_1(\xi) \, d\xi \neq 0$. Let (32.32) imply (32.33) whenever $K_2(x)$ belongs to L_1 and $f(x)$ is bounded. Then (32.31) holds.

In his initial 1928 form of the theorem,[28] Wiener required a growth condition $O(\frac{1}{\xi^2})$ at $\pm\infty$ for the kernels $K_1(\xi)$ and $K_2(\xi)$. In the 1932 version, he refined his theory by means of his well-known theorem on absolutely convergent Fourier series: If a nonvanishing function f has an absolutely convergent Fourier series, then $\frac{1}{f}$ has an absolutely convergent Fourier series. Although this was a difficult result, it emerged less than a decade later as a corollary of I. M. Gelfand's work on commutative Banach algebras. Wiener stated a second general theorem, directly applicable to infinite series, involving Stieltjes integrals; he derived a form of the PNT from this result. Thus, Wiener got his second Tauberian theorem: Let $f(x)$ be a function of limited total variation over every finite range, and let

$$\int_{y}^{y+1} |df(x)| \tag{32.34}$$

be bounded in y. Let $K_1(x)$ be a continuous function in L_1, and let

$$\sum_{k=-\infty}^{\infty} \max_{k \leq x \leq k+1} |K_1(x)| \tag{32.35}$$

converge. Now assume

$$\frac{1}{\sqrt{2\pi}} \int_{-\infty}^{\infty} K_1(x) e^{iux} \, dx \neq 0 \quad (-\infty < u < \infty) \tag{32.36}$$

---

[27] Wiener (1932).
[28] Wiener (1928).

and

$$\lim_{x\to\infty} \int_{-\infty}^{\infty} K_1(\xi - x)\, df(\xi) = A \int_{-\infty}^{\infty} K_1(\xi)\, d\xi. \qquad (32.37)$$

If $K_2(x)$ is a continuous function in L_1 satisfying the condition (32.35), then

$$\lim_{x\to\infty} \int_{-\infty}^{\infty} K_2(\xi - x)\, df(\xi) = A \int_{-\infty}^{\infty} K_2(\xi)\, d\xi. \qquad (32.38)$$

Note that Wiener also stated a converse of this theorem. Then in 1938, H. R. Pitt (1914–2005) formulated a simple theorem containing both Wiener theorems as corollaries. Pitt took undergraduate courses from Hardy and Littlewood at Cambridge in the 1930s. After graduation in 1936, he studied under Wiener at M.I.T. In his 1938 paper "General Tauberian Theorems,"[29] Pitt proved: Suppose $K(x) \in L_1(-\infty,\infty)$ and its Fourier transform $\widehat{K}(t)$ does not vanish for any real t. If $f(x)$ is bounded, slowly oscillating, that is

$$f(y) - f(x) \to 0 \text{ when } y > x, x \to \infty, y - x \to 0, \qquad (32.39)$$

and

$$\int_{-\infty}^{\infty} K(x - t) f(t)\, dt \to A \int_{-\infty}^{\infty} K(x)\, dx,$$

then

$$f(x) \to A \text{ as } x \to \infty.$$

The Serbian mathematician Jovan Karamata (1902–1967) also made an important contribution to Tauberian theory. In 1930, he published a two-page proof of the Hardy–Littlewood theorem, that Abel summability with a one-sided condition implied Cesàro summability.[30] Karamata's proof used only the Weierstrass approximation theorem to prove his main result that if $a_n \geq 0$ and $\sum a_n$ was Abel summable to s, then for every Riemann integrable function $g(x)$,

$$\lim_{x\to 1^-} (1 - x) \sum_{n=0}^{\infty} a_n x^n g(x^n) = s \int_0^1 g(t)\, dt. \qquad (32.40)$$

This elegant proof took researchers in Tauberian theory completely by surprise, since up to that time all the proofs of the Hardy–Littlewood theorem had required a fair amount of machinery. Karamata graduated from the University of Belgrade in 1925, where he came under the influence of Mihailo Petrović (1868–1943) who had studied at the École Normale in Paris under Hermite, Poincaré, and Picard. Petrović brought to Serbia the spirit of scientific research he learned in France. By the time he

[29] Pitt (1938).
[30] Karamata (1930).

met Karamata, he had ceased to do mathematical research but he advised Karamata to study the latest mathematical discoveries. Karamata regarded himself as self-taught and would say that his teacher in classical analysis was Pólya and Szegő's *Aufgaben und Lehrsätze aus der Analysis*, published in 1925. In fact, the topic of Karamata's doctoral thesis was the development of Weyl's work on the uniform distribution of sequences $x_1, x_2, x_3, \ldots$ in the interval $(0, 1)$. We observe that Weyl's theorems were given as a set of five problems in Pólya and Szegő's book. The first of these problems was to show that a sequence $x_1, x_2, x_3, \ldots$ in $(0, 1)$ was uniformly distributed if and only if for every Riemann integrable function f

$$\lim_{n \to \infty} \frac{f(x_1) + f(x_2) + \cdots + f(x_n)}{n} = \int_0^1 f(x) \, dx. \qquad (32.41)$$

One may compare this with Karamata's theorem. Again, it is interesting to note that, following their section on uniform distribution, Pólya and Szegő's book posed a problem requiring the use of Weyl's formula as well as Frobenius's theorem on summability. Karamata also introduced the important concept of a regularly varying function.

32.2 Fejér: Summability of Fourier Series

In 1900, L. Fejér made an application of $(C, 1)$ summability to Fourier series by proving that the Fourier series of f was $(C, 1)$ summable to $\frac{f(x+0)+f(x-0)}{2}$ at every point where $f(x \pm 0)$ existed.[31] He assumed that f was bounded and integrable on $[0, 2\pi]$. Recall that the Fourier coefficients are given by

$$a_n = \frac{1}{\pi} \int_0^{2\pi} f(t) \cos nt \, dt, \ b_n = \frac{1}{\pi} \int_0^{2\pi} f(t) \sin nt \, dt$$

and that the nth partial sum of a Fourier series is given by

$$s_n(x) = \frac{1}{2}a_0 + \sum_{k=1}^n (a_k \cos kx + b_k \sin kx)$$

$$= \frac{1}{2\pi} \int_0^{2\pi} f(t)dt + \sum_{k=1}^n \frac{1}{\pi} \int_0^{2\pi} f(t) \cos k(t - x) \, dt.$$

Fejér began his proof with the observation that

$$\sigma_{n-1} = \frac{1}{2} + \cos \theta + \cdots + \cos (n-1)\theta = \frac{1}{2} \frac{\cos (n-1)\theta - \cos n\theta}{1 - \cos \theta};$$

---

[31] Fejér republished his paper in the *Math Annalen* in 1904. See Fejér (1904).

hence,

$$\frac{\sigma_0 + \sigma_1 + \cdots + \sigma_{n-1}}{n} = \frac{1}{2n} \frac{1 - \cos n\theta}{1 - \cos \theta} = \frac{1}{2n} \left(\frac{\sin \left(\frac{n\theta}{2} \right)}{\sin \left(\frac{\theta}{2} \right)} \right)^2.$$

Thus, for the arithmetic mean of the partial sums, he had

$$S_n(x) = \frac{s_0(x) + s_1(x) + s_2(x) + \cdots + s_{n-1}(x)}{n}$$

$$= \frac{1}{n\pi} \int_{-\frac{x}{2}}^{\pi - \frac{x}{2}} f(x + 2u) \left(\frac{\sin nu}{\sin u} \right)^2 du.$$

Fejér immediately perceived that this integral was simpler than the one found by Dirichlet for the partial sum $s_n(x)$ because the kernel $\frac{\sin^2 v}{\sin^2 u}$ was always nonnegative, unlike the corresponding kernel in Dirichlet's integral $\frac{\sin(2n-1)u}{\sin u}$. Fejér first considered the case where f was continuous at x. He let $\epsilon > 0$, so that there existed a $\delta > 0$ such that

$$|f(x + h) - f(x)| < \epsilon \text{ for } |h| \le \delta.$$

We note that Fejér's notation interchanged ϵ and δ. He next wrote the integral for $S_n(x)$ in three parts:

$$S_n(x) = \frac{1}{2n\pi} \int_0^{x-\delta} \frac{1 - \cos n(t - x)}{1 - \cos (t - x)} f(t) \, dt$$

$$+ \frac{1}{2n\pi} \int_{x-\delta}^{x+\delta} \frac{1 - \cos n(t - x)}{1 - \cos(t - x)} f(t) \, dt$$

$$+ \frac{1}{2n\pi} \int_{x+\delta}^{2\pi} \frac{1 - \cos n(t - x)}{1 - \cos (t - x)} f(t) \, dt.$$

He assumed $|f(t)| \le M$ in $[0, 2\pi]$. Then the absolute values of the first and third integrals were bounded by $\frac{2M}{n(1 - \cos \delta)}$. For the second integral, the positivity of the term multiplying $f(t)$ implied that

$$\int_{x-\delta}^{x+\delta} \frac{1 - \cos n(t - x)}{1 - \cos (t - x)} f(t) \, dt = (f(x) + \eta) \int_{x-\delta}^{x+\delta} \frac{1 - \cos n(t - x)}{1 - \cos (t - x)} dt,$$

where $|\eta| < \epsilon$. Fejér then noted that

$$\frac{1}{2n\pi} \int_0^{2\pi} \frac{1 - \cos n(t - x)}{1 - \cos (t - x)} dt = 1.$$

Thus,

$$\frac{1}{2n\pi} \int_{x-\delta}^{x+\delta} \frac{1 - \cos n(t - x)}{1 - \cos (t - x)} dt$$

$$= 1 - \left(\frac{1}{2n\pi} \int_{0}^{x-\delta} \frac{1 - \cos n(t - x)}{1 - \cos (t - x)} dt + \frac{1}{2n\pi} \int_{x+\delta}^{2\pi} \frac{1 - \cos n(t - x)}{1 - \cos (t - x)} dt \right).$$

He observed that each of the last two integrals was less that $\frac{2}{n(1-\cos \delta)}$. With all this information, he could conclude that for n large enough

$$|S_n(x) - f(x)| < 2\epsilon.$$

This proved Fejér's theorem for the case in which f was continuous at x. Assuming only the existence of the limits $f(x - 0)$ and $f(x + 0)$, Fejér broke the integral for $S_n(x)$ into two parts:

$$I_1(x) = \frac{1}{2n\pi} \int_{0}^{x} \frac{1 - \cos n(t - x)}{1 - \cos (t - x)} f(t) dt,$$

$$I_2(x) = \frac{1}{2n\pi} \int_{x}^{2\pi} \frac{1 - \cos n(t - x)}{1 - \cos (t - x)} f(t) dt.$$

Then by a similar argument

$$\lim_{n \to \infty} I_1(x) = \frac{1}{2} f(x - 0), \quad \lim_{n \to \infty} I_2(x) = \frac{1}{2} f(x + 0).$$

Fejér went on to observe that if $f(x)$ was everywhere continuous, then $S_n(x)$ converged uniformly to $f(x)$. He also noted the following immediate corollaries of his theorem:

- If the Fourier series converges at a point of continuity of a function, then its sum is the value of the function at that point.
- A continuous function on a closed interval is a uniform limit of a sequence of polynomials. This is Weierstrass's approximation theorem.
- Poisson's integral yields a solution for Dirichlet's problem for the circle.

Hermann A. Schwarz was the first to prove the third result. He felt that a proof by Fourier series was probably not possible. As noted before, Fejér's motivation in the discovery of his theorem was to provide a proof using Fourier series.

Hardy recognized that his Tauberian theorem on $(C, 1)$ summability, combined with Fejér's theorem, immediately yielded a result on Fourier series: If the Fourier coefficients of a continuous function f are $a_n = O(\frac{1}{n})$ and $b_n = O(\frac{1}{n})$, the Fourier series of f at x converges to $f(x)$. Hardy then reasoned that since the Fourier coefficients of a periodic function f of bounded variation satisfied $a_n = O(\frac{1}{n})$, $b_n = O(\frac{1}{n})$, then the Fourier series of a such a function converged to $\frac{1}{2}(f(x + 0) + f(x - 0))$. In fact, this is the classical Dirichlet–Jordan theorem. Furthermore, observe that since Cesàro summability implies Abel summability, it follows that for f as in Fejér's theorem, we have

$$\lim_{r \to 1^-} \left(\frac{1}{2}a_0 + (a_1 \cos x + b_1 \sin x)r + (a_2 \cos 2x + b_2 \sin 2x)r^2 + \cdots \right)$$

$$= \frac{1}{2}\big(f(x+0) + f(x-0)\big).$$

This equation simplifies to

$$\lim_{r \to 1^-} \left(\frac{1}{2\pi} \int_0^{2\pi} \frac{1 - r^2}{1 - 2r \cos(x - t) + r^2} f(t)\,dt \right) = \frac{1}{2}\big(f(x+0) + f(x-0)\big).$$

When Hilbert saw Fejér's work, he requested Fejér to attempt a proof of a similar theorem for the Laplace series where a function $f(\theta, \phi)$ was expanded in terms of surface harmonics. Fejér was unsuccessful in this effort for some years. Finally, while looking at a book on Bessel functions, he saw F. G. Mehler's integral formula for Legendre polynomials:

$$P_n(\cos \theta) = \frac{2}{\pi} \int_\theta^\pi \frac{\sin(2n+1)(\frac{t}{2})}{\sqrt{2(\cos \theta - \cos t)}}\,dt, \quad 0 < \theta < \pi.$$

With the help of this result, in 1908 Fejér was able to prove that the Laplace series of a bounded integrable function was $(C, 2)$ summable to the function at any point of continuity.[32] In 1913, H. Gronwall proved that $(C, 2)$ could be replaced by $(C, 1)$.[33]

32.3 Karamata's Proof of the Hardy–Littlewood Theorem

Karamata's short proof[34] of Littlewood's theorem and of the more general Hardy–Littlewood theorem relied on Weierstrass's approximation theorem. Karamata used it in the following form: For any Riemann integrable function $g(x)$ on $(0, 1)$ and every $\epsilon > 0$ there exist two polynomials $p(t)$ and $P(t)$ such that

$$p(t) \le g(t) \le P(t) \quad \text{for} \quad 0 \le t \le 1, \tag{32.42}$$

$$\int_0^1 \big(P(t) - p(t)\big)\,dt \le \epsilon. \tag{32.43}$$

Karamata did not give the details of the proof of this result. It can be proved, however, by first taking $g(t)$ to be a continuous function. By Weierstrass's theorem, there are polynomials $p(t)$ and $P(t)$ differing by at most $\frac{\epsilon}{4}$ from $g(t) - \frac{\epsilon}{4}$ and $g(t) + \frac{\epsilon}{4}$, respectively, for all $t \in [0, 1]$. Clearly, the required result follows for $g(t)$ continuous. We next take $g(t)$ to be piecewise continuous, and the result follows because $g(t)$ can

[32] Fejér (1908).
[33] Gronwall (1913).
[34] Karamata (1930).

be approximated by continuous functions. Finally, for any Riemann integrable function $g(t)$, there are step functions $m(t)$ and $M(t)$ such that $m(t) \leq f(t) \leq M(t)$ and

$$\int_0^1 \left(M(t) - m(t) \right) dt < \frac{\epsilon}{2}.$$

Karamata's theorem: If $a_n \geq -K$, with $K \geq 0$ independent of n and

$$(1 - x) \sum_{n=0}^{\infty} a_n x^n \to A \quad \text{as} \quad x \to 1^-,$$

then

$$(1 - x) \sum_{n=0}^{\infty} a_n g(x^n) x^n \to A \int_0^1 g(t) \, dt$$

for every Riemann integrable function $g(t)$.

In Karamata's proof, it was obviously sufficient to take $K = 0$, for he could replace a_n by $a_n + K$. Karamata then supposed $g(x) = x^{\alpha}$, $\alpha \geq 0$. Then he had

$$(1 - x) \sum_{n=0}^{\infty} a_n g(x^n) x^n$$

$$= (1 - x) \sum_{n=0}^{\infty} a_n x^{(\alpha+1)n}$$

$$= \frac{(1 - x)}{1 - x^{\alpha+1}} (1 - x^{\alpha+1}) \sum_{n=0}^{\infty} a_n x^{(\alpha+1)n} \to \frac{A}{\alpha + 1} = A \int_0^1 t^{\alpha} \, dt,$$

as $x \to 1^-$. It followed by linearity that for every polynomial $P(x)$,

$$(1 - x) \sum_{n=0}^{\infty} a_n P(x^n) x^n \to A \int_0^1 P(t) \, dt.$$

He could next apply (32.42) and (32.43) because a_n was positive; Karamata's theorem followed.

To derive the Hardy–Littlewood theorem, Karamata set $x = e^{-\frac{1}{n}}$ and let $g(t)$ be the piecewise continuous function

$$g(t) = \begin{cases} 0 & 0 \leq t < \frac{1}{e}, \\ \frac{1}{t} & \frac{1}{e} \leq t \leq 1. \end{cases}$$

He then arrived at $g(x^m) = 0$ for $m > n$, $g(x^m) x^m = 1$ for $m \leq n$, and $\int_0^1 g(t) \, dt = 1$, thereby reducing his theorem to the Hardy–Littlewood theorem. In other words, given the one-sided Tauberian condition $a_n \geq -K$, if the Abel sum of $\sum_{n=0}^{\infty} a_n x^n$ was A, then the Cesàro sum of $\sum a_n$ was also A.

32.4 Wiener's Proof of Littlewood's Theorem

Littlewood's Tauberian theorem of 1910 was the first difficult and deep Tauberian result to be proved. It is therefore interesting to see how Wiener derived this theorem from his general theorem.[35] We restate Littlewood's result:

$$\text{If } \lim_{y \to 1^-} \sum_{n=0}^{\infty} a_n y^n = s \text{ and } n|a_n| < K, \text{ then } \sum_{n=0}^{\infty} a_n = s.$$

The first step in Wiener's proof was to express $\sum a_n y^n$ as an integral. For that purpose he showed that $s(x) = \sum_{n \leq x} a_n$ was bounded for $0 \leq x < \infty$. By hypothesis, $\sum_{n=0}^{\infty} a_n e^{-\frac{n}{x}}$ was bounded for $0 \leq x < \infty$ and (using $n|a_n| < K$)

$$\left| s(x) - \sum_{n=0}^{\infty} a_n e^{-\frac{n}{x}} \right| = \left| \sum_{n \leq x} a_n (1 - e^{-\frac{n}{x}}) - \sum_{n > x} a_n e^{-\frac{n}{x}} \right|$$

$$\leq \sum_{n \leq x} \frac{K}{n} \cdot \frac{n}{x} + \sum_{n > x} \frac{K}{n} e^{-\frac{n}{x}}$$

$$\leq 2K + K \int_x^{\infty} e^{-\frac{u}{x}} \frac{du}{u}$$

$$\leq 3K + K \int_1^{\infty} e^{-u} \frac{du}{u} = \text{constant}.$$

This showed that $s(x)$ was bounded so that he had

$$\sum_{n=0}^{\infty} a_n e^{-nx} = \int_{0^-}^{\infty} e^{-ux} \, ds(u) = \int_0^{\infty} x e^{-ux} s(u) \, du.$$

Hence,

$$s = \lim_{x \to 0^+} \int_0^{\infty} x e^{-ux} s(u) \, du = \lim_{\xi \to \infty} \int_{-\infty}^{\infty} e^{-\xi} e^{-e^{\eta - \xi}} s(e^{\eta}) e^{\eta} \, d\eta.$$

So Wiener set $K_1(\xi) = e^{-\xi} e^{-e^{-\xi}}$ and observed that

$$\int_{-\infty}^{\infty} K_1(\xi) \, d\xi = \int_{-\infty}^{\infty} e^{-\xi} e^{-e^{-\xi}} \, d\xi = \int_0^{\infty} e^{-x} \, dx = 1.$$

Thus,

$$\lim_{\xi \to \infty} \int_{-\infty}^{\infty} K_1(\xi - \eta) s(e^{\eta}) \, d\eta = s \int_{-\infty}^{\infty} K_1(\xi) \, d\xi$$

[35] Wiener (1958) pp. 104–106.

and

$$\frac{1}{\sqrt{2\pi}} \int_{-\infty}^{\infty} K_1(\xi) e^{-iu\xi}\, d\xi = \frac{1}{\sqrt{2\pi}} \int_0^{\infty} x^{iu} e^{-x}\, dx = \frac{1}{\sqrt{2\pi}} \Gamma(1 + iu) \neq 0.$$

Therefore, $K_1(\xi)$ satisfied the hypotheses, (32.31) and (32.32), of his first Tauberian theorem. Wiener then chose $K_2(\xi)$ in such a manner that he obtained the $(C, 1)$ summability of $\sum a_n$ to s. He set

$$K_2(\xi) = \begin{cases} 0 & \xi < 0, \\ e^{-\xi} & \xi > 0, \end{cases}$$

so that by his first Tauberian theorem,

$$s = s \int_0^{\infty} e^{-\xi}\, d\xi = s \int_{-\infty}^{\infty} K_2(\xi)\, d\xi = \lim_{\xi \to \infty} \int_{-\infty}^{\infty} K_2(\xi - \eta) s(e^{\eta})\, d\eta$$

$$= \lim_{\xi \to \infty} \int_{-\infty}^{\xi} e^{\eta - \xi} s(e^{\eta})\, d\eta = \lim_{x \to \infty} \frac{1}{x} \int_0^x s(y)\, dy.$$

Note that by applying Hardy's theorem that $(C, 1)$ summability together with $a_n = O(\frac{1}{n})$ implies convergence, the Hardy–Littlewood theorem follows. However, Wiener included a simple argument to prove Hardy's theorem: For $\lambda > 0$,

$$s = \frac{(1 + \lambda)s - s}{\lambda} = \lim_{x \to \infty} \frac{1}{\lambda x} \left(\int_0^{(1+\lambda)x} s(y)\, dy - \int_0^x s(y)\, dy \right)$$

$$= \lim_{x \to \infty} \frac{1}{\lambda x} \int_x^{(1+\lambda)x} s(y)\, dy = \lim_{x \to \infty} \left(s(x) + \frac{1}{\lambda x} \int_x^{(1+\lambda)x} (s(y) - s(x))\, dy \right).$$

The condition $a_n = O(\frac{1}{n})$ then implied the necessary result:

$$\left| \frac{1}{\lambda x} \int_x^{(1+\lambda)x} (s(y) - s(x))\, dy \right| \le \frac{1}{\lambda x} \int_x^{(1+\lambda)x} \sum_{x < n < y} \frac{K}{n}\, dy$$

$$\le \sum_{\lfloor x \rfloor + 1}^{\lfloor (1+\lambda)x \rfloor} \frac{K}{\lfloor x \rfloor} \le \frac{\lfloor \lambda x \rfloor K}{\lfloor x \rfloor} < 2\lambda K,$$

for sufficiently large x. Hence $\overline{\lim}_{x \to \infty} |s(x) - s| < 2\lambda K$; or, because λ was an arbitrary positive number, $\lim_{x \to \infty} |s(x) - s| = 0$. This completed Wiener's proof of Littlewood's theorem.

32.5 Hardy and Littlewood: The Prime Number Theorem

In their 1921 paper, "On a Tauberian Theorem for Lambert Series," Hardy and Littlewood gave a very simple proof of the PNT based on the result that Lambert summability implied Abel summability.[36] As we mentioned before, their proof of the Lambert summability theorem employed a result, due to Landau, stronger than the PNT. Thus, although they did not produce a new proof, their derivation of the PNT insightfully reveals its Tauberian character. In this derivation, Hardy and Littlewood employed a number-theoretic result describing the average behavior of the arithmetic function $d(n)$, the number of divisors of n. Dirichlet first proved this result by his ingenious hyperbola method in an 1849 paper on the average behavior of arithmetic functions.[37]

Hardy and Littlewood first showed that the series

$$\sum_{n=1}^{\infty} \frac{\Lambda(n) - 1}{n}$$

was Lambert summable to -2γ, where γ was Euler's constant. Note here that the Lambert series could be written as

$$f(y) = y \sum_{n=1}^{\infty} \frac{(\Lambda(n) - 1)e^{-ny}}{1 - e^{-ny}} = y \sum_{n=1}^{\infty} (\Lambda(n) - 1)e^{-ny}(1 + e^{-ny} + e^{-2ny} + \cdots)$$

$$= y \sum_{n=1}^{\infty} c_n e^{-ny}, \text{ where } c_n = \sum_{d|n} (\Lambda(d) - 1) = \ln n - d(n).$$

Next, they observed that

$$\sum_{i=1}^{n} c_i = \ln n! - \sum_{i=1}^{n} d(i).$$

To estimate the logarithmic term, they applied Stirling's formula and to estimate the second term they used the Dirichlet divisor theorem:

$$\sum_{i=1}^{n} d(i) = n \ln n + (2\gamma - 1)n + O(\sqrt{n}).$$

These calculations gave them

$$\frac{1}{n} \sum_{i=1}^{n} c_i \sim -2\gamma \quad \text{as } n \to \infty.$$

By Frobenius's theorem, the last result implied that

$$\lim_{y \to \infty} f(y) = \lim_{y \to \infty} y \sum_{n=1}^{\infty} c_n e^{-ny} = -2\gamma.$$

[36] Hardy and Littlewood (1921).
[37] Dirichlet (1969) vol. II, pp. 50–66.

This proved the Lambert summability of $\sum_{n=1}^{\infty} \frac{\Lambda(n)-1}{n}$. Hence, by their theorem that Lambert summability implies Abel summability, the series was Abel summable to -2γ. It was also clear that $\frac{\Lambda(n)-1}{n} \geq -1$. Moreover, Hardy and Littlewood had earlier extended Littlewood's theorem and this extension showed that this one-sided Tauberian condition was sufficient to obtain the ordinary convergence of $\sum_{n=1}^{\infty} \frac{\Lambda(n)-1}{n}$. Recall that by (32.9), the convergence of $\sum a_n$ implied that

$$\frac{a_1 + 2a_2 + \cdots + na_n}{n} \to 0 \quad \text{as } n \to \infty.$$

For $a_k = \frac{\Lambda(k)-1}{k}$, the last condition translated to

$$\frac{\Lambda(1) - 1 + \Lambda(2) - 1 + \cdots + \Lambda(n) - 1}{n} \to 0 \quad \text{as } n \to \infty$$

or

$$\lim_{N \to \infty} \frac{1}{N} \sum_{n=1}^{N} \Lambda(n) = 1,$$

and this was equivalent to the prime number theorem.

In his 1971 paper, "The Quickest Proof of the Prime Number Theorem," Littlewood observed that in 1918 he and Hardy proved (32.17).[38] He pointed out that though they had earlier proved the Tauberian theorem (with the one-sided condition mentioned above) necessary to deduce the quickest proof, they did not mention the PNT in their 1918 paper. We note that it was this Tauberian theorem for which Karamata gave his nice proof, described by Littlewood as "highly sophisticated."

32.6　Wiener's Proof of the PNT

In his work on the Tauberian theorem, one of Wiener's fundamental aims was to prove the prime number theorem by means of Lambert summability.[39] Thus, he wished to determine the behavior of $\sum_{n \leq x} \Lambda(n)$ as $x \to \infty$ from the behavior of

$$\sum_{n=1}^{\infty} \Lambda(n) \frac{x^n}{1 - x^n} \quad \text{as } x \to 1^-.$$

First, Wiener observed that

$$\sum_{n=1}^{\infty} \Lambda(n) \frac{x^n}{1 - x^n} = \sum_{n=1}^{\infty} x^n \sum_{m|n} \Lambda(m) = \sum_{n=1}^{\infty} x^n \ln n$$

$$= \sum_{n=1}^{\infty} \ln n \frac{x^n - x^{n+1}}{1 - x} = \sum_{n=1}^{\infty} \frac{x^{n+1}}{1 - x}(\ln(n+1) - \ln n)$$

[38] Littlewood (1982) vol. 1, pp. 951–955.
[39] Wiener (1958) pp. 112–124.

$$= \frac{x}{1-x} \sum_{n=1}^{\infty} \ln\left(1 + \frac{1}{n}\right) = \frac{x}{1-x} \sum_{n=1}^{\infty} \left(\frac{1}{n} + O\left(\frac{1}{n^2}\right)\right) x^n$$

$$= \frac{x}{1-x} \left(\ln\frac{1}{1-x} + \sum_{n=1}^{\infty} O\left(\frac{1}{n^2}\right) x^n\right).$$

Note that the second line used summation by parts. Wiener next set $x = e^{-\xi}$ and multiplied by $-\xi$ to obtain

$$\sum_{n=1}^{\infty} \Lambda(n) \frac{\xi e^{-n\xi}}{e^{-n\xi} - 1} = \frac{\xi e^{-\xi}}{1 - e^{-\xi}} \left(\ln(1 - e^{-\xi}) - \sum_{n=1}^{\infty} O\left(\frac{1}{n^2}\right) e^{-n\xi}\right).$$

It followed from the right-hand side that as $\xi \to 0^+$, the series behaved like $\ln \xi$. Wiener therefore worked with the differentiated series. Upon differentiating the last equation, he arrived at

$$\sum_{n=1}^{\infty} \Lambda(n) \frac{d}{dn\xi} \frac{n\xi e^{-n\xi}}{e^{-n\xi} - 1} = \sum_{n=1}^{\infty} \Lambda(n) \frac{e^{-2n\xi} - e^{-n\xi} + n\xi e^{-n\xi}}{(e^{-n\xi} - 1)^2}$$

$$= \frac{e^{-\xi} - e^{-2\xi} - \xi e^{-\xi}}{(1 - e^{-\xi})^2} \left(\ln(1 - e^{-\xi}) + \sum_{n=1}^{\infty} O\left(\frac{1}{n^2}\right) e^{-n\xi}\right)$$

$$+ \frac{\xi e^{-\xi}}{1 - e^{-\xi}} \left(\frac{e^{-\xi}}{1 - e^{-\xi}} + \sum_{n=1}^{\infty} O\left(\frac{1}{n}\right) e^{-n\xi}\right)$$

$$= O(1)\big(O(\ln \xi) + O(1)\big)$$

$$+ \big(1 + O(\xi)\big)\left(\frac{1}{\xi} + O(1) + O(\ln \xi)\right)$$

$$= \frac{1}{\xi} + O(\ln \xi),$$

as $\xi \to 0^+$.

Thus, he had

$$\lim_{\xi \to 0^+} \xi \sum_{n=1}^{\infty} \Lambda(n) \frac{e^{-2n\xi} - e^{-n\xi} + n\xi e^{-n\xi}}{(e^{-n\xi} - 1)^2} = 1.$$

Wiener wrote the sum as a Stieltjes integral so that he could apply his second Tauberian theorem. Toward that end, he set

$$g(y) = \sum_{n=1}^{\lfloor e^y \rfloor} \frac{\Lambda(n)}{n},$$

so that the previous equation containing the limit took the form

$$1 = \lim_{\xi \to 0^+} \int_0^\infty \eta\xi \frac{e^{-2n\xi} - e^{-n\xi} + \eta\xi e^{-n\xi}}{(e^{-n\xi} - 1)^2} \, dg(\ln \eta)$$

$$= \lim_{x \to \infty} \int_{-\infty}^\infty \frac{e^{y-x}(e^{-2e^{y-x}} - e^{-e^{y-x}} + e^{y-x}e^{-e^{y-x}})}{(e^{-e^{y-x}} - 1)^2} \, dg(y).$$

To understand the next step, compare the last expression with the corresponding expression in Wiener's theorem. On this basis, we can see how Wiener next wrote

$$K_1(x) = \frac{e^{-x}(e^{-2e^{-x}} - e^{-e^{-x}} + e^{-x}e^{-e^{-x}})}{(e^{-e^{-x}} - 1)^2}$$

and then

$$\int_{-\infty}^\infty K_1(x) \, dx = \int_0^\infty \frac{e^{-2\xi} - e^{-\xi} + \xi e^{-\xi}}{(e^{-\xi} - 1)^2} \, d\xi = \int_0^\infty \frac{d}{d\xi} \frac{\xi e^{-\xi}}{e^{-\xi} - 1} \, d\xi$$

$$= \lim_{\xi \to 0^+} \frac{\xi e^{-\xi}}{1 - e^{-\xi}} = 1.$$

Thus, he had $A = 1$ in the hypothesis of his theorem. One may check that $K_1(x)$ satisfies (32.35) and since $g(y)$ is monotomic,

$$\int_n^{n+1} |dg(x)| = \int_n^{n+1} dg(x).$$

Moreover, the latter expression is bounded for $-\infty < n < \infty$. Finally, Wiener had only to check that the Fourier transform of $K_1(x)$ did not vanish. So he computed

$$\frac{1}{\sqrt{2\pi}} \int_{-\infty}^\infty K_1(x)e^{-iux} \, dx = \frac{1}{\sqrt{2\pi}} \int_0^\infty \frac{d}{d\xi}\left(\frac{\xi e^{-\xi}}{e^{-\xi} - 1}\right) \xi^{iu} \, d\xi$$

$$= \lim_{\lambda \to 0^+} \frac{1}{\sqrt{2\pi}} \int_0^\infty \frac{d}{d\xi}\left(\frac{\xi e^{-\xi}}{e^{-\xi} - 1}\right) \xi^{iu+\lambda} \, d\xi.$$

Integration by parts converted the last expression to

$$\lim_{\lambda \to 0^+} \frac{iu + \lambda}{\sqrt{2\pi}} \int_0^\infty \frac{\xi^{iu+\lambda}e^{-\xi}}{1 - e^{-\xi}} \, d\xi = \lim_{\lambda \to 0^+} \frac{iu + \lambda}{\sqrt{2\pi}} \int_0^\infty \xi^{iu+\lambda} \sum_{n=1}^\infty e^{-n\xi} \, d\xi$$

$$= \lim_{\lambda \to 0^+} \frac{\lambda + iu}{\sqrt{2\pi}} \sum_{n=1}^\infty \frac{\Gamma(\lambda + 1 + iu)}{n^{\lambda+1+iu}}$$

$$= \lim_{\lambda \to 0^+} \frac{\lambda + iu}{\sqrt{2\pi}} \zeta(\lambda + 1 + iu)\Gamma(\lambda + 1 + iu)$$

$$= iu\zeta(1 + iu)\Gamma(1 + iu).$$

The work of Hadamard and Vallée-Poussin showed that $\zeta(1 + iu)$ did not vanish for any real u, and hence the Fourier transform of $K_1(x)$ did not vanish.

Finally, Wiener had to choose $K_2(x)$ appropriately so that he got the PNT in the form $\lim_{N \to \infty} \frac{1}{N} \sum_{n=1}^{N} \Lambda(n) = 1$. Note that in this application, $K_2(x)$ had to be continuous; it could not be the piecewise continuous function

$$K_2(x) = \begin{cases} 0, & x < 0, \\ e^{-x}, & x > 0, \end{cases}$$

although, if allowed, this would have yielded the result immediately. So Wiener defined two continuous functions

$$K_{21}(x) = \begin{cases} 0, & x < -\epsilon, \\ \dfrac{x + \epsilon}{\epsilon}, & -\epsilon \le x < 0, \\ e^{-x}, & 0 \le x, \end{cases}$$

$$K_{22}(x) = \begin{cases} 0, & x < 0, \\ \dfrac{x}{\epsilon} e^{-\epsilon}, & 0 \le x < \epsilon, \\ e^{-x}, & \epsilon \le x. \end{cases}$$

Here he verified that

$$\int_{-\infty}^{\infty} K_{21}(x)\, dx = 1 + \frac{\epsilon}{2} \quad \text{and} \quad \int_{-\infty}^{\infty} K_{22}(x)\, dx = e^{-\epsilon}\left(1 + \frac{\epsilon}{2}\right).$$

Wiener's second Tauberian theorem then implied

$$
\begin{aligned}
1 + \frac{\epsilon}{2} &= \lim_{x \to \infty} \int_{-\infty}^{\infty} K_{21}(x - y)\, dg(y) \\
&= \lim_{x \to \infty} \left(\int_{-\infty}^{\infty} e^{y-x}\, dg(y) + \int_{x}^{x+\epsilon} \frac{\epsilon - y + x}{\epsilon}\, dg(y) \right) \\
&\ge \varlimsup_{x \to \infty} \int_{-\infty}^{x} e^{y-x}\, dg(y) = \varlimsup_{N \to \infty} \frac{1}{N} \int_{0}^{N} \eta\, dg(\ln \eta) \\
&= \varlimsup_{N \to \infty} \frac{1}{N} \sum_{n=1}^{N} \Lambda(n), \qquad\qquad (32.44)
\end{aligned}
$$

and

$$e^{-\epsilon}\left(1 + \frac{\epsilon}{2}\right) = \lim_{x \to \infty} \int_{-\infty}^{\infty} K_{22}(x - y)\, dg(y)$$

$$= \lim_{x \to \infty} \left(\int_{\infty}^{x-\epsilon} e^{y-x}\, dg(y) + \int_{x-\epsilon}^{x} \frac{x - y}{\epsilon} e^{-\epsilon}\, dg(y) \right)$$

$$= \lim_{x \to \infty} \left(\int_{-\infty}^{x} e^{y-x}\, dg(y) - \int_{x-\epsilon}^{x} \left(e^{y-x} - \frac{x - y}{\epsilon} e^{-\epsilon} \right) dg(y) \right)$$

$$\leq \lim_{x \to \infty} \int_{\infty}^{x} e^{y-x}\, dg(y) = \lim_{N \to \infty} \frac{1}{N} \sum_{n=1}^{N} \Lambda(n). \qquad (32.45)$$

Note that in the above calculation, one may use the fact that

$$e^{y-x} - \frac{x - y}{\epsilon} e^{-\epsilon} \geq 0 \text{ for } x - \epsilon \leq y \leq x.$$

Wiener let $\epsilon \to 0$ in the inequalities (32.44) and (32.45) to get

$$1 \geq \overline{\lim_{N \to \infty}} \frac{1}{N} \sum_{n=1}^{N} \Lambda(n) \text{ and } 1 \leq \overline{\lim_{N \to \infty}} \frac{1}{N} \sum_{n=1}^{N} \Lambda(n).$$

These inequalities implied that $\lim_{N \to \infty} \frac{1}{N} \sum_{n=1}^{N} \Lambda(n)$ existed and was equal to 1. This proof of the PNT used only one property of the zeta function: that it did not vanish on the line consisting of points with real part equal to 1.

32.7 Kac's Proof of Wiener's Theorem

The basic principle behind Wiener's Tauberian theorem is simple but penetrating. Mark Kac illustrated this insight by producing a short proof of the 1928 form of Wiener's theorem.[40] This proof uses only Fubini's theorem and the uniqueness of Fourier transforms; like Wiener's 1928 theorem, it is powerful enough to produce the PNT as a consequence.

Kac's theorem: Suppose

$$K_1(x) \in L^1(-\infty, \infty), \quad x^2 K_1(x) \in L^1(-\infty, \infty)$$

and

$$k_1(\xi) = \int_{-\infty}^{\infty} K_1(x) e^{i\xi x}\, dx \neq 0, \quad -\infty < \xi < \infty.$$

---

[40] Kac (1965).

If $m(y)$ is a bounded measurable function such that for all x,

$$\int_{-\infty}^{\infty} K_1(x - y)m(y)\,dy = 0,$$

then $m(y) = 0$ almost everywhere.

In proving this theorem, Kac realized that the condition $x^2 K_1(x) \in L^1$ implied that $k_1(\xi)$ was twice continuously differentiable. Thus, let Φ be the set of all twice continuously differentiable functions with compact support. Since $k_1(\xi) \neq 0$ for all ξ, it follows that every $\phi \in \Phi$ is of the form $k_1 \psi$ for some $\psi \in \Phi$. In short, $k_1 \Phi = \Phi$. Let $\phi \in \Phi$ and let F be the Fourier transform of ϕ; that is, let

$$F(x) = \int_{-\infty}^{\infty} \phi(\xi)e^{ix\xi}\,d\xi.$$

Because ϕ has compact support, $F(x)$ is defined for all $x \in \mathbb{C}$ and $F'(x)$ exists. Hence, F is an entire function, and ϕ can be chosen such that F is not identically zero. Thus, F has only a countable number of zeros. Since $F \in L^1(-\infty, \infty)$ and $|F(x)||k_1(x - y)||m(y)|$ is integrable as a function of two variables (x, y), we can apply Fubini's theorem and change the order of integration:

$$\begin{aligned}
0 &= \int_{-\infty}^{\infty} F(x) \left(\int_{-\infty}^{\infty} K_1(x - y)m(y)\,dy \right) dx \\
&= \int_{-\infty}^{\infty} m(y) \left(\int_{-\infty}^{\infty} K_1(x - y)F(x)\,dx \right) dy \\
&= \int_{-\infty}^{\infty} m(y) \left(\int_{-\infty}^{\infty} k_1(\xi)\phi(\xi)e^{i\xi y}\,d\xi \right) dy.
\end{aligned}$$

Because $k_1\Phi = \Phi$, we can conclude that for all $\phi \in \Phi$, we have

$$0 = \int_{-\infty}^{\infty} m(y) \left(\int_{-\infty}^{\infty} \phi(\xi)e^{i\xi y}\,d\xi \right) dy.$$

Now Φ is closed under translation so that we can replace $\phi(\xi)$ by $\phi(\xi - \alpha)$ and change variables to arrive at

$$0 = \int_{-\infty}^{\infty} m(y) \left(\int_{-\infty}^{\infty} \phi(\xi)e^{i\xi y}\,d\xi \right) e^{i\alpha y}\,dy,$$

for all real α. By the definition of F, this gives

$$0 = \int_{-\infty}^{\infty} m(y)F(y)e^{i\alpha y}\,dy$$

for all real α; and the uniqueness of Fourier transforms implies $m(y)F(y) = 0$ for almost all y. Since F can be chosen to have countably many zeros, we conclude that $m(y) = 0$ almost everywhere.

32.8   Gelfand: Normed Rings

Wiener derived the final form of his Tauberian theorem by means of his famous
theorem on nonvanishing and absolutely convergent Fourier series. About ten years
later, in 1941, Izrail Gelfand provided a short and elegant derivation of this theorem,
based on his theory of normed rings.[41] In this effort, Gelfand utilized an abstract
formulation of the Fourier transform, now known as the Gelfand transform. In a short
note published in 1939, Gelfand developed the elements of the theory of commutative
Banach algebras. These are algebras B over the complex numbers, containing a
multiplicative identity e; such algebras are complete with respect to a norm $|| \; ||$, such
that $||e|| = 1$ and $||xy|| \leq ||x|| \cdot ||y||$ for x and y in B; thus, Gelfand named them
normed rings. In two further short notes published in the same year, Gelfand gave
applications of his theory to absolutely convergent Fourier series and integrals and to
the ring of almost periodic functions. Gelfand used the work of Wiener and Pitt on
absolutely convergent Fourier series and integrals as a springboard in his construction
of Banach algebras. He succeeded in obtaining short proofs of the Wiener-Pitt results
by revealing their essentially algebraic character.

In his two-page fundamental paper of 1939, Gelfand denoted a normed ring, or
commutative Banach algebra, by R and observed as his first theorem that any maximal
ideal M was closed in R. His third theorem, now called the Gelfand-Mazur theorem,
stated that R/M was isomorphic to the field of complex numbers. This theorem
originated with the 1918 result of Alexander Ostrowski (1893–1986), student of
Landau and Klein, that a complete Archimedean field is isomorphic to either the field
of real numbers or the field of complex numbers. In 1938 this was generalized by
Stanislaw Mazur (1905–1981), student of Stefan Banach, who proved that a normed
associative real division algebra was isomorphic to the field of real numbers, or to
the field of complex numbers, or to the noncommutative field of quarternions. In his
1941 paper "Normierte Ringe," Gelfand gave a beautiful proof of the particular case of
Mazur's theorem he needed. This proof employed Liouville's theorem that a bounded
entire function is a constant.

Gelfand was then able to associate with each $x \in R$ a complex number $x(M)$, to
obtain a complex valued function on the set of all maximal ideals of R. He defined
a topology on the set of maximal ideals to make the set into a compact Hausdorff
space and the functions $x(M)$ continuous. In his 1941 paper, Gelfand also noted the
easily proved result that $x(M) \leq ||x||$. This depended on the lemma that for the
multiplicative identity e and any $y \in R$, if $||e + y|| < 1$, then y was invertible. In
fact, it is easily verified that

$$y^{-1} = -(e + (e + y) + (e + y)^2 + \cdots). \qquad (32.46)$$

Now observe that if $x(M) = \lambda \in \mathbf{C}$, then $x = \lambda e + z$ where $z \in M$. Assume $\lambda \neq 0$,
because if $\lambda = 0$, then the inequality $|x(M)| \leq ||x||$ is obvious. Then for $y = \frac{z}{\lambda}$,
we have

$$\frac{|x(M)|}{||x||} = \frac{\lambda}{||\lambda e + z||} = \frac{1}{||e + y||} \leq 1,$$

because if $||e + y|| < 1$, then y would be invertible and not be in M.

We now have the result needed to understand Gelfand's simple proof of Wiener's theorem on nonvanishing absolutely convergent Fourier series. He let R be the set of all functions $f(t) = \sum_{n=-\infty}^{\infty} a_n e^{int}$ such that $\sum_{n=-\infty}^{\infty} |a_n| < \infty$; he then let $||f|| = \sum_{n=-\infty}^{\infty} |a_n|$. This gave R the structure of a commutative Banach algebra. Gelfand argued that for any maximal ideal M and $e^{it} \in R$, $e^{it}(M)$ was some complex number a. He then obtained

$$|e^{it}(M)| = |a| \leq ||e^{it}|| = 1, \quad \text{and} \quad \frac{1}{|a|} \leq ||e^{-it}|| = 1,$$

and hence $a = e^{it_0}$ for some real number t_0. This meant that any trigonometric polynomial $\sum_{n=-N}^{N} a_n e^{int}$ corresponded to the number $\sum_{n=-N}^{N} a_n e^{int_0}$. And since the mapping $R \to R/M$ was continuous, to every function $f(t) \in R$, there corresponded a number $f(t_0)$. Therefore, the maximal ideal M consisted of all function $f(t)$ such that $f(t_0) = 0$. It followed that if a function $f(t)$ did not vanish at any point, then $f(t)$ was not a member of any maximal ideal of R. Wiener could then conclude that $f(t)$ had an inverse in R, proving the theorem. Gelfand's proof of the Gelfand-Mazur theorem, that $R' = R/M$ is isomorphic to $\mathbb{C}$, began by supposing that $x \in R'$ and $x \neq \lambda e$ for any complex number λ. Then $(x - \lambda e)^{-1}$ exists for every λ. Moreover,

$$\lim_{h \to 0} \frac{(x - (\lambda + h)e)^{-1} - (x - \lambda e)^{-1}}{h} = -(x - \lambda e)^{-2},$$

and

$$|\lambda^{-1}| \left|\left|\left(e - \frac{x}{\lambda}\right)^{-1}\right|\right| \to 0 \text{ as } \lambda \to \infty, \tag{32.47}$$

since, for $|\lambda| > ||x||$, equation (32.46) implies that $||(e - \frac{x}{\lambda})^{-1}|| \leq \frac{1}{1 - ||\frac{x}{\lambda}||}$. Hence for any multiplicative linear functional $\phi : R \to \mathbb{C}$, that is, $\phi(xy) = \phi(x)\phi(y)$, the function $\phi((x - \lambda e)^{-1})$ is a bounded entire function and therefore a constant. By (32.47), this constant must be zero. It follows that $(x - \lambda e)^{-1}$ is zero and the theorem is proved by the contradiction:

$$e = (x - \lambda e)^{-1}(x - \lambda e) = 0.$$

F. Riesz, S. Mazur, and others studied normed rings before Gelfand, but Gelfand's concept of the space of maximal ideals unified several isolated earlier results and opened up new avenues for further research. In fact, the space of maximal ideals became important in algebraic geometry also, though in that area Alexander Grothendieck showed that the space of prime ideals produced better results.

In 1930, Gelfand (1913–2009) moved to Moscow from Odessa without completing his secondary education. He had studied mathematics on his own from an early age; his lack of books spurred him to great creativity. At the age of 15, for example, he discovered the Euler–Maclaurin formula. He studied a textbook on differential calculus and Taylor series, but he had no book on the integral calculus. While investigating the problem of the area under $y = x^n$, he was led to consider the sums $1^n + 2^n + \cdots + m^n$. He soon found the Euler–Maclaurin formula and the generating function for the Bernoulli numbers by means of the Taylor series. In a similar way, he discovered Newton's formula for the sums of powers in the theory of symmetric functions. In Moscow, he worked in odd jobs such as doorkeeper at the Lenin Library, while also teaching mathematics. The great Russian mathematician A. N. Kolmogorov (1903–1987) took an interest in Gelfand, who very soon found himself lecturing at the Moscow State University and studying with Kolmogorov who directed him to problems in functional analysis. This resulted in Gelfand's 1935 thesis, "Abstract Functions and Linear Operators." The theory of commutative normed rings was the subject of his 1938 doctoral thesis.

Gelfand made major contributions to several areas of mathematics such as representation theory, differential equations, computational mathematics, and biocybernetics. At the age of 80, in collaboration with M. Kapranov and A. Zelevinsky, he was starting to develop a theory of hypergeometric functions of many variables. Though he was unable to bring this work to perfection, he had seen the importance of such a theory when he was much younger. For example, in his 1956 lecture "On Some Problems of Functional Analysis," he gave his thoughts on the matter:

> It is known that almost all the special functions of one variable to be met with in mathematical physics may be obtained from the general hypergeometric function of Gauss by a suitable choice of parameters. These same functions appear as elements of representations of the simplest classical groups, namely the groups of rotations of the sphere and of the Lobacevskii plane. This connection lies in the nature of the matter, since the special functions make their appearance by way of considerations connected with this or that invariance of a problem under transformations of a space. Hence it is natural to construct the theory of hypergeometric functions of several variables, relying on results and methods of the theory of the representations of compact or locally compact Lie groups. It is thus necessary so to construct the theory of hypergeometric functions that it should contain the theory of general spherical functions, connected with the representations of semi-simple groups.

32.9 Exercises

(1) Prove that if $A_1, A_2, A_3, \ldots, A_n, \ldots$ is a sequence such that the difference $A_{n+1} - A_n$ converges to a limit A as $n \to \infty$, then $\frac{A_n}{n}$ converges to the same limit. Cauchy stated and proved this result in his *Analyse algébrique*; see Cauchy (1989) or Bradley and Sandifer (2009) pp. 35 and 42. Show that this result implies that if a series $\sum a_n$ converges to A, then it converges $(C, 1)$ to A.

(2) Prove the theorem of Frobenius that Cesàro summability implies Abel summability. Observe that

$$\sum_{n=0}^{\infty} a_n x^n = (1 - x)^2 \sum_{n=0}^{\infty} (A_0 + \cdots + A_n) x^n,$$

where $A_n = \sum_{k=0}^{n} a_k$. See Frobenius (1880).

(3) Prove Borel's theorem that ordinary convergence implies Borel summability, that is, if $a_n \to A$ as $n \to \infty$, then

$$e^{-x} \sum_{n=0}^{\infty} \frac{a_n x^n}{n!} \to A \text{ as } x \to \infty.$$

See Hardy (1949) p. 80.

(4) Show that the condition in the second theorem of Tauber given by (32.9) is implied by the condition $na_n \to 0$ as $n \to \infty$. See Tauber (1897).

(5) Prove Cesàro's theorem that if $\sum a_n = A$ and $\sum b_n = B$ and $C_n = c_0 + c_1 + c_2 + \cdots + c_n$, where $c_n = a_0 b_n + a_1 b_{n-1} + \cdots + a_n b_0$, then

$$\frac{C_0 + C_1 + \cdots + C_n}{n+1} \to AB \text{ as } n \to \infty.$$

See Cesàro (1890).

(6) Suppose $a(x)$ and $b(x)$ are continuous functions. Prove that if

$$\int_0^\infty a(x)\, dx = A, \qquad \int_0^\infty b(x)\, dx = B,$$

then

$$\lim_{x\to\infty} \frac{1}{x} \int_0^x dt \int_0^t du \int_0^u a(w) b(u-w)\, dw = AB.$$

Next, deduce that if $\int_0^\infty dx \int_0^x a(t) b(x-t)\, dt$ is convergent, then its value is AB.

(7) If $\int_0^\infty a(x)\, dx = A$, $\int_0^\infty b(x)\, dx = B$, $|xa(x)| < K$, and $|xb(x)| < K$,

then

$$\int_0^\infty dx \int_0^x a(t) b(x-t)\, dt = AB.$$

The theorems in this and the previous exercise are due to Hardy. See Hardy (1966–1979) vol. 6, pp. 210–212.

(8) In 1971, at the age of 86 and in memory of his student Harold Davenport (1907–1969), Littlewood gave a short proof of the PNT depending on the following known results:

- The Hardy–Littlewood theorem of which Karamata gave a two-page proof;
- The functional equation of the zeta function;
- The Cahen–Mellin integral for e^{-y} in terms of the gamma function;
- The Dirichlet series for $-\frac{\zeta'(s)}{\zeta(s)}$ when $\operatorname{Re} s > 1$;
- The complex zeros ρ of $\zeta(s)$ have a real part between 0 and 1;
- $\frac{\zeta'(s)}{\zeta(s)} = O(\log t) + s \sum_\rho \frac{1}{\rho(s-\rho)}$;
- For $s = -1 + it$, $\frac{\zeta'(s)}{\zeta(s)} = O(t^A)$, where A is a positive absolute constant; Littlewood remarked that A would not necessarily have the same value from one occurrence to the next;

- If $N(T)$ denotes the number of zeros of $\rho = \beta + i\gamma$ with $0 \leq \gamma \leq T$, then $N(T) = O(T^A)$.

Note that the last result is extremely weak and far more is and was known about $N(T)$, but Littlewood did not require a stronger result. First prove Littlewood's first lemma: Given a large positive T_0, there is a T, with $AT_0 < T < AT_0$, such that $\frac{\zeta'(s)}{\zeta(s)} = O(T^A)$ for $s = \sigma + iT$, $-1 \leq \sigma \leq 2$. The corresponding result for $s = \sigma - iT$ follows by symmetry. Next, demonstrate Littlewood's second lemma that for

$$y > 0, \quad -2\pi i \sum \Lambda(n)e^{-ny} = \int_{2-i\infty}^{2+i\infty} \Gamma(s)\frac{\zeta'(s)}{\zeta(s)}y^{-s}\,ds.$$

From this, Littlewood deduced the PNT in one page. Show how this can be done. See Littlewood (1982) vol. 2, pp. 951–955.

(9) Show that $\zeta(s)$ has no zeros when the real part of s is one. The original proofs of Hadamard and Vallée-Poussin were slightly more complicated than later proofs. See Titchmarsh and Heath-Brown (1986) p. 48.

32.10 Notes on the Literature

Hardy (1966–1979) vol. 6 contains all his papers on summability or Tauberian theory, as well as his joint papers with Littlewood on the topic. Littlewood (1982) vol. 1 includes his papers on this subject that were not joint with Hardy.

All of Wiener's work on Tauberian theory can be found in Wiener (1976–1985) vol. 2. Most of Wiener's work discussed in this book was taken from Wiener (1958), a course of lectures on Fourier transforms and their applications, given by Wiener at Cambridge University in 1932. Norman Levinson, a student of Wiener, has written an interesting article explaining Wiener's progression from harmonic analysis to Tauberian theory. This appeared in the *Bulletin of the AMS*, vol. 72, 1, part II and this volume also includes very helpful accounts by experts on Wiener's many contributions to mathematics. Masani (1990) is a comprehensive biography of Wiener, discussing his mathematical work, its myriad applications, the development of his thought, and giving good references.

In his original paper on the short proof of Littlewood's theorem, Karamata also introduced the concept of majorizability to obtain a new condition for the convergence of Abel summable series. Although this idea sheds light on Karamata's proof, the portion of the paper dealing with majorizability was removed by Landau and the paper was reduced to Karamata (1930). See Nikolić (2009).

Tucciarone (1973) gives a helpful history of summable series, from their origins through the 1920s; it includes a good bibliography. Korevaar (2004) is an encyclopedic treatment of Tauberian theory, covering a century of developments, with numerous historical comments and references.

Elliptic Functions: Eighteenth Century

33.1 Preliminary Remarks

In 1847, Jacobi wrote to Fuss that Euler had been motivated to found elliptic function theory by reading Count Fagnano's *Produzioni Matematiche*.[1] Indeed, in the work of the then unknown Fagnano, Euler discovered the key to the apparently intractable elliptic integral. Giulio Carlo Fagnano (1682–1766) studied theology and philosophy in Rome but avoided mathematics, though he was encouraged to study it. Many years later, after reading Malebranche's *Concerning the Search for Truth*, he taught himself mathematics with great devotion, and from 1714 to 1720, he published some interesting papers on integrals in little-known Italian journals. In 1718 he published his now-famous paper on dividing the lemniscate into several equal parts. His results were not noted at first, but were brought to light by an interesting chain of events. In the early 1740s, Fagnano was consulted concerning the possible instability of the dome of St. Peter's. In 1750, in compensation for his help, Fagnano's collected papers were published at the order of Pope Benedict XIV. Fagnano then applied for membership in the Berlin Academy. Euler was assigned the task of evaluating the quality of the mathematical portion of Fagnano's papers. Euler was intrigued by Fagnano's results on the lemniscatic integral $\int \frac{dx}{\sqrt{1-x^4}}$; these results inspired some of Euler's most brilliant work on integral calculus, laying the foundation for the theory of elliptic integrals and functions. It goes without saying that Fagnano was admitted to the Academy.

The early work of Jakob and Johann Bernoulli[2] on the lemniscatic integral $\int \frac{dx}{\sqrt{1-x^4}}$ led Fagnano to investigate the topic. The equation of the lemniscate is given in cartesian coordinates by

$$(x^2 + y^2)^2 = a^2(x^2 - y^2) \tag{33.1}$$

---

[1] Fagnano (1911).
[2] Bernoulli (1744) vol. 1, pp. 601–612 and Bernoulli (1742) vol. 1, pp. 119–122.

and in polar coordinates by

$$r^2 = a^2 \cos 2\theta, \tag{33.2}$$

where a is a constant. The graph of the lemniscate resembles the symbol for infinity, the diameter of one side given by $r = a$ when $\theta = 0$. For convenience, take $a = 1$. The cartesian coordinates x and y are then given by

$$2x^2 = r^2 + r^4, \tag{33.3}$$

$$2y^2 = r^2 - r^4, \tag{33.4}$$

where $0 \leq r \leq 1$. A simple calculation shows that if s denotes the arc length of the lemniscate, then

$$ds = \frac{dr}{\sqrt{1 - r^4}},$$

or

$$s(r) = \int_0^r \frac{dt}{\sqrt{1 - t^4}}. \tag{33.5}$$

The lemniscate appeared in Jakob Bernoulli's solution to his own 1691 problem on the shape of an elastic band constrained by its own weight. Johann Bernoulli later encountered the same curve when he asked how to find a curve such that the time taken to traverse it was proportional to the distance from a fixed point. Jakob Bernoulli opined that the lemniscatic integral could not be evaluated in terms of the inverse trigonometric, logarithmic, or rational functions. However, he offered the series expansion

$$\int_0^1 \frac{dt}{\sqrt{1 - t^4}} = \sum_{n=0}^{\infty} \frac{1 \cdot 3 \cdot 5 \cdots (2n - 1)}{n! \, 2^n (4n + 1)}, \tag{33.6}$$

obtained by expanding the denominator by the binomial theorem and integrating term by term. The Bernoullis also investigated the problem of bisecting the arcs of curves such as the parabolic spiral.

Fagnano's most famous accomplishment was to bisect an arc of a lemniscate and trisect and quinsect the full arc from $r = 0$ to $r = 1$.[3] His methods were such that these procedures could actually be accomplished using a straight edge and compass. His proofs were based on obtaining appropriate changes of variables and on transforming lemniscatic integrals into other lemniscatic integrals. For example, he found that if

$$t = \frac{1}{r} \sqrt{1 \pm \sqrt{1 - r^4}}, \tag{33.7}$$

[3] Fagnano (1911) vol. 2, pp. 293–297 and 304–313.

then

$$\frac{dr}{\sqrt{1-r^4}} = \frac{\sqrt{2}\,dt}{\sqrt{1+t^4}};$$ (33.8)

and if

$$\frac{u\sqrt{2}}{\sqrt{1-u^4}} = \frac{1}{r}\sqrt{1 - \sqrt{1-r^4}},$$ (33.9)

then

$$\frac{dt}{\sqrt{1-r^4}} = \frac{2\,du}{\sqrt{1-u^4}}.$$ (33.10)

In order to better understand (33.7), write it as

$$r^2 = \frac{2t^2}{1+t^4}.$$ (33.11)

Fagnano may have made this substitution on the basis of a similar transformation

$$r^2 = \frac{2t}{1+t^2},$$ (33.12)

used to rationalize the integrand for arscine: $\arcsin x = \int_0^x \frac{dr}{\sqrt{1-r^2}}$. Note that from the substitution given by (33.12), we have

$$\frac{dr}{\sqrt{1-r^2}} = \frac{2\,dt}{1+t^2}.$$ (33.13)

It was thus natural for Fagnano to consider (33.11), even though it did not rationalize the integrand, so that he instead obtained (33.8). To understand (33.9), compare it with (33.7) to get

$$t^2 = \frac{2u^2}{1-u^4}.$$ (33.14)

Making this substitution is only reasonable, since it produces

$$\frac{dt}{\sqrt{1+t^4}} = \frac{\sqrt{2}\,du}{\sqrt{1-u^4}}.$$ (33.15)

This means that if substitutions (33.11) and (33.14) are applied successively, the result is (33.10). Moreover, the relationship between r and u can be expressed by

$$r^2 = \frac{4u^2(1-u^4)}{(1+u^4)^2}.$$ (33.16)

Thus, any arc in the first quadrant of a lemnisate, with an endpoint at the origin, can be bisected using straight edge and compass. See this by observing that the arc length of the lemniscate is given by (33.5) and that the arc length corresponding to the radius vector r is double the arc length given by u, where r and u are related by (33.16). One may check that r and u can be obtained from one another by solving only quadratic equations. Also recall that one may use (33.3) and (33.4) to obtain the coordinates of the points from the radius vector. This shows that we have geometric constructibility.

Fagnano's use of $\frac{\sqrt{2}\,dt}{\sqrt{1+t^4}}$ in (33.8) may seem peculiar, since this expression does not appear to take the form of an arc length of a lemniscate. Watson and Siegel have both explained this very nicely[4] in terms of later ideas due to Gauss and Abel. Set $t = e^{\frac{i\pi}{4}} v$ in (33.11) so that

$$ r^2 = \frac{2iv^2}{1-v^4} $$

and

$$ \frac{dr}{\sqrt{1-r^4}} = \frac{(1+i)\,dv}{\sqrt{1-v^4}}. \tag{33.17} $$

Moreover, by (33.14)

$$ v^2 = \frac{-2iu^2}{1-u^4} \tag{33.18} $$

and

$$ \frac{dv}{\sqrt{1-v^4}} = \frac{(1-i)\,du}{\sqrt{1-u^4}}. \tag{33.19} $$

Now note that these transformations produce points on the lemniscate, but they are imaginary points. Thus, Siegel points out that (33.17) and (33.19) are examples of "complex multiplication" of the lemniscatic integral and, when applied successively, produce the bisection. Indeed, Fagnano was familiar with the use of complex numbers in integrals; he discovered that

$$ \int \frac{dt}{1+t^2} = \log\left(\frac{1+it}{1-it}\right)^{\frac{1}{2i}}, $$

a result also published by Johann Bernoulli in 1702, as mentioned in Chapter 12. Fagnano noted the amusing particular case

$$ \pi = 2i \log\left(\frac{1-i}{1+i}\right). $$

---

[4] Watson (1933) and Siegel (1969) vol. 1, pp. 1–7.

Upon reading Fagnano, Euler perceived that the doubling of the arc length of the lemniscate corresponded to the double angle formula for the sine function. This in turn was a particular case of the addition formula for the sine function. Thus, he gradually understood that Fagnano's transformation formulas might be particular cases of an addition formula for elliptic integrals. Euler's earlier efforts to evaluate these integrals in terms of elementary functions having reached a dead end, he sensed in the work of Fagnano an innovative and productive direction for the theory of elliptic integrals.

Consider Euler's state of mind when he began reading Fagnano. As a student of Johann Bernoulli, he knew that the integral $\int \frac{dt}{\sqrt{1-t^4}}$ could probably not be evaluated in terms of logarithms or inverse trigonometric functions. Then in 1738, he reproved Fermat's theorem that the equation $z^2 = x^4 - y^4$ had no nontrivial integer solutions. Note that this was one result in number theory for which Fermat wrote down a proof! Now the substitution (33.12), rationalizing $\frac{dt}{\sqrt{1-t^2}}$, also provided the rational solutions of $z^2 = 1-x^2$ or the integer solutions of $z^2 = y^2-x^2$. Euler realized that $\frac{dt}{\sqrt{1-t^4}}$ could not be rationalized by substitution, since that would imply that Fermat's equation could have integer solutions, a contradiction.

Here note that Euler was well aware of the connection between Diophantine equations of the form $y^2 = ax^2 + bx + c$ and the integration of expressions of the form $\sqrt{ax^2 + bx + c}$. In a 1723 letter to Goldbach, Daniel Bernoulli made specific mention of this connection and so did Johann Bernoulli in his integral calculus lectures, published in the 1740s, long after he delivered them. See Section 12.8 in this connection. Thus, it is safe to assume that Euler was aware that the elliptic integral could not be evaluated in terms of elementary functions. He was searching for a new path, and found it in Fagnano, upon whom he heaped praise. Within a few weeks of receiving Fagnano's *Produzioni Matematiche*, Euler gave a favorable report to the Berlin Academy, including some of his own reflections. He soon wrote a paper reworking and generalizing Fagnano's results and then went on to publish several more papers, which now fill two volumes of his *Opera Omnia*.

Euler's papers and letters to Goldbach indicate that he saw a close connection between $\int \frac{dt}{\sqrt{1-t^2}}$ and $\int \frac{dt}{\sqrt{1-t^4}}$. In fact, in his letter of May 30, 1752,[5] Euler mentioned to Goldbach that

$$\frac{dx}{\sqrt{1 - xx}} = \frac{dy}{\sqrt{1 - yy}} \tag{33.20}$$

had the complete integral

$$yy + xx = cc + 2xy\sqrt{(1 - cc)}, \tag{33.21}$$

while

$$\frac{dx}{\sqrt{1 - x^4}} = \frac{dy}{\sqrt{1 - y^4}} \tag{33.22}$$

[5] Fuss (1968) vol. 1, pp. 564–568.

had the complete integral

$$yy + xx = cc + 2xy\sqrt{(1 - c^4)} - ccxxyy. \qquad (33.23)$$

Now from (33.16) we see that

$$\int_0^r \frac{dt}{\sqrt{1 - t^4}} = 2 \int_0^u \frac{dt}{\sqrt{1 - t^4}} \qquad (33.24)$$

when

$$r = \frac{2u\sqrt{1 - u^4}}{1 + u^4}. \qquad (33.25)$$

The corresponding result for the arcsine function is

$$\int_0^r \frac{dt}{\sqrt{1' - t^2}} = 2 \int_0^u \frac{dt}{\sqrt{1 - t^2}} \qquad (33.26)$$

when

$$r = 2u\sqrt{1 - u^2}. \qquad (33.27)$$

These two relations are equivalent to the double angle formula for $\sin x$, that is, $\sin 2x = 2 \sin x \cos x$. And this is in turn a particular case of the addition formula

$$\sin(x + y) = \sin x \cos y + \cos x \sin y.$$

Next write this in terms of integrals as

$$\int_0^u \frac{dt}{\sqrt{1 - t^2}} + \int_0^v \frac{dt}{\sqrt{1 - t^2}} = \int_0^z \frac{dt}{\sqrt{1 - t^2}} \quad \text{for } z = u\sqrt{1 - v^2} + v\sqrt{1 - u^2}.$$

$$(33.28)$$

Recall that Euler thought that he could view Fagnano's bisection of the lemniscatic arc as a particular case of a possible addition formula for the lemniscatic function. In 1753, Euler found the required addition formula:[6]

$$s(u) + s(v) = s(r), \quad r = \frac{u\sqrt{1 - v^4} + v\sqrt{1 - u^4}}{1 + u^2 v^2}, \qquad (33.29)$$

where $s(u)$ was the lemniscatic integral defined by (33.5). To understand the method by which Euler obtained (33.29), note first that upon integrating (33.20), one obtains

$$\arcsin x = \arcsin y \pm \arcsin c = \arcsin(y\sqrt{1 - c^2} \pm c\sqrt{1 - y^2}),$$

---

[6] Eu. I-20 pp. 58–79, especially pp. 65–66. E 251.

implying that the complete integral of (33.20) is

$$x = y\sqrt{1 - c^2} \pm c\sqrt{1 - y^2}.$$

This is actually the addition formula for the sine function, also given by (33.28), and it is equivalent to the complete integral (33.21) from Euler's letter. In a similar manner, Euler derived the addition formula for the lemniscatic function (33.29) by solving equation (33.23) for x or for y. None of his papers on this topic give an account of how he found the complete integral; they merely verify that the differential $\dfrac{dx}{\sqrt{1-x^4}}$ remained invariant under the transformation obtained by solving (33.23) for y in terms of x.

Euler also extended (33.29) to the more general quartic[7] $P(x) = 1 + mx^2 + nx^4$. He proved that the complete integral of

$$\frac{dx}{\sqrt{P(x)}} = \frac{dy}{\sqrt{P(y)}}$$

turned out to be the equation

$$-nc^2x^2y^2 + x^2 + y^2 = c^2 + 2xy\sqrt{1 + mc^2 + nc^4},$$

where c was an arbitrary constant. Upon solving for y, one obtained

$$y = \frac{x\sqrt{P(c)} \pm c\sqrt{P(x)}}{1 - nc^2x^2}.$$

Euler obtained the addition formula for the case in which $P(x)$ was the general quartic $A + 2Bx + Cx^2 + 2Dx^3 + Ex^4$. By means of a fractional linear transformation, he reduced the general quartic to the particular case $1 + mx^2 + nx^4$. The slight drawback in Euler's technique was that it introduced complex coefficients, whereas he intended to use only real coefficients. This lacuna was filled by Legendre in a paper of 1792.[8] Euler proved these results on the addition formula during the 1750s; during the next twenty years, he went on to prove similar results for elliptic integrals of the second and third kinds, to use terminology introduced by Legendre.

This body of Euler's results brought the theory of elliptic integrals to prominence, not only in the context of the integral calculus and Diophantine equations, but also in areas of applied mathematics such as elasticity and dynamics, where numerical evaluations were paramount. Since elliptic integrals could not be evaluated in terms of elementary functions, numerical methods were sought. The early work of Jakob Bernoulli showed this to be a tough problem. His 1694 paper discussed the elastic curve defined by

$$f(x) = \int_0^x \frac{t^2\, dt}{\sqrt{1 - t^4}} \tag{33.30}$$

---

[7] ibid. pp. 63–67.
[8] Legendre (1792) pp. 9–10.

with arc length $s(x)$ given by (33.5). Bernoulli determined the intervals within which the values of $f(1)$ and $s(1)$ would have to fall. In a 1704 paper, Bernoulli was able use series methods to specify these values within shorter intervals: $1.3088173 < s(1) < 1.3152635$ and $0.5983546 < f(1) < 0.6004034$. But Bernoulli's hypergeometric series (33.6) did not converge rapidly enough, so that these values were not too accurate. Then, in his *Methodus Differentialis*, James Stirling produced a vastly better evaluation, correct to fifteen decimal places.

Recall that Stirling's book contained several methods for transforming hypergeometric and other series into more rapidly convergent series. In proposition 11 of his book,[9] he applied a specific method to Bernoulli's hypergeometric series, obtaining:

$$\int_0^1 \frac{dx}{\sqrt{1-x^4}} = 1.31102877714605987, \tag{33.31}$$

$$\int_0^1 \frac{x^2\, dx}{\sqrt{1-x^4}} = 0.59907011736779611. \tag{33.32}$$

Incidentally, it was a source of pride to Euler when he found, through his 1737 work on the elastic curve, that the product of these two integrals was exactly $\frac{\pi}{4}$. To get this result today, one would evaluate the beta integrals in terms of the gamma function. And it is interesting that soon after 1737, Euler found this method of evaluating in terms of the gamma function. Euler also used the series method to obtain numerical approximations of some elliptic integrals. Again, these results were not accurate to many decimal places because the series did not converge rapidly enough. Using the transformation of integrals, Lagrange, Legendre, and Gauss found better methods, also of great theoretical significance. But these methods owed a debt to the work of John Landen.

In 1771, John Landen presented a fundamental transformation of elliptic integrals; he elaborated on this result in another paper published four years later.[10] He stated his problem in geometric terms, expressing the length of the arc of any hyperbola in terms of two elliptic arcs. The Landen transformation can be stated, as formulated by Legendre,[11] as the theorem: If $\sin(2\phi - \theta) = k \sin\theta$, then

$$(1+k) \int_0^\theta (1 - k^2 \sin^2 u)^{-\frac{1}{2}}\, du = 2 \int_0^\phi \left(1 - \frac{4k}{(1+k)^2} \sin^2 u\right)^{-\frac{1}{2}} du. \tag{33.33}$$

We note the important particular case

$$(1+k) \int_0^\pi (1 - k^2 \sin^2 \theta)^{-\frac{1}{2}}\, d\theta = 2 \int_0^{\frac{\pi}{2}} \left(1 - \frac{4k}{(1+k)^2} \sin^2 \theta\right)^{-\frac{1}{2}} d\theta. \tag{33.34}$$

[9] Stirling and Tweddle (2003) pp. 74–75.
[10] Landen (1771) and (1775).
[11] Legendre (1811–1817) vol. 1, p. 81.

In Legendre's notation

$$K(k) = \int_0^{\frac{\pi}{2}} (1 - k^2 \sin^2 \theta)^{-\frac{1}{2}} \, d\theta \tag{33.35}$$

so that (33.34) can be expressed as

$$K\left(\frac{2\sqrt{k}}{1+k}\right) = (1+k)K(k). \tag{33.36}$$

Note that (33.36) is also referred to as Landen's (quadratic) transformation. This and Euler's addition formula were the two pillars on which the early theory of elliptic integrals and functions was constructed. Mittag-Leffler insightfully wrote, "Landen does not seem, however, to have fully understood the value of his discovery."[12]

But Lagrange quickly grasped the applicability of Landen's transformation to the numerical approximation of elliptic integrals, and in this connection, he discovered the concept of the arithmetic-geometric mean of two numbers. In 1784–1785, Lagrange presented these ideas in the Turin Academy journal under the title "Sur une nouvelle méthode de calcul intégral."[13] Lagrange was then a member of the Berlin Academy, but was born in Turin and was a founder of its academy. Consequently, he was quite interested in the growth of the Turin Academy and published several papers in its journal. In his paper, Lagrange expressed Landen's transformation in the elegant form: If $p > q > 0$,

$$p' = p + \sqrt{p^2 - q^2}, \quad q' = p - \sqrt{p^2 - q^2} \tag{33.37}$$

and

$$R(p,q,y) = \sqrt{(1 \pm p^2 y^2)(1 \pm q^2 y^2)}, \quad y' = \frac{yR}{1 \pm q^2 y^2}, \tag{33.38}$$

then

$$\frac{dy}{R} = \frac{dy'}{R'}, \tag{33.39}$$

where $R' = R(p', q', y')$.

He also observed that $p = \frac{p'+q'}{2}$, the arithmetic mean of p' and q', and that $q = \sqrt{p'q'}$, the geometric mean of p' and q'. He used these relations to define two sequences. Let $p_0 = p, q_0 = q$ with $p > q$ and for any positive or negative integer n set

$$p_n = p_{n-1} + \sqrt{p_{n-1}^2 - q_{n-1}^2}, \quad q_n = p_{n-1} - \sqrt{p_{n-1}^2 - q_{n-1}^2}, \tag{33.40}$$

---

[12] Mittag-Leffler (1923).
[13] Lagrange (1867–1892) vol. 2, pp. 281–312.

or

$$p_n = \frac{p_{n+1} + q_{n+1}}{2}, \quad q_n = \sqrt{p_{n+1}q_{n+1}}. \tag{33.41}$$

The relations (33.40) and (33.41) define the bilateral sequence

$$\cdots p_{-1}, q_{-1}, p_0, q_0, p_1, q_1, \ldots.$$

In the positive direction, for increasing n, the q_n terms tend to zero and p_n terms tend to infinity. So for the purpose of approximate evaluation of the integral, whose form did not change by (33.39), Lagrange took $q_n = 0$ for large enough n. This reduced the elliptic integral to $\int \frac{dx}{\sqrt{1 \pm p_n^2 x^2}}$, an exactly computable integral. In the negative direction, Lagrange observed that the arithmetic means p_{-n} and the geometric means q_{-n} had the same limit, because $p_{-n} - q_{-n} \to 0$ as $n \to \infty$. Thus, for sufficiently large n, the elliptic integral could be approximately evaluated from the exactly computable integral $\int \frac{dx}{1 \pm p_{-n}^2 x^2}$.

Apparently Lagrange did not enjoy numerical calculation as much as Newton, Stirling, and Euler did, so he did not actually apply his method to find approximate values for elliptic integrals. This was left to Legendre, who effectively used iterated forms of (33.33) and (33.34) to construct numerical tables of such integrals.

We mention in passing that Euler used results obtainable from the addition formula to study Diophantine equations of the form $y^2 = p(x)$ where $p(x)$ was a polynomial of degree four with integer coefficients. Since Euler did not note the connection with elliptic integrals, we are not certain that he was aware of it. In 1834, Jacobi reviewed some of these papers of Euler and on that basis, he concluded that Euler knew of this relationship. In a similar way, Euler as well as Lagrange employed quadratic (or second-order) transformations (isogenies) to study the special Diophantine equations $z^2 = x^4 \pm y^4$ and $z^2 = 2x^4 - y^4$. This may be one reason that Lagrange did not refer to Landen in his 1784–85 paper on elliptic integrals. Another reason, of course, is that mathematicians in the eighteenth century were not in the habit of giving an exhaustive list of references. Thus, Mittag-Leffler assumed that Landen and Lagrange had independently discovered the Landen transform. In fact, on January 3, 1777, Lagrange wrote to his friend Condorcet that he had seen Landen's 1775 paper containing the theorem reducing the problem of the rectification of arcs of ellipses to a problem of hyperbolic arcs.[14] Lagrange wrote that he found this a singular result and that he had not yet verified it. Apparently, he found the time to study Landen and went on to discover the arithmetic-geometric mean and its use in numerical evaluation of integrals. It is remarkable that he did not do more with it. Perhaps he was already beginning to lose interest in mathematical research. After 1785, he produced no further original mathematical results, though he did publish important and influential books, including *Mécanique analytique* of 1788 and *Fonctions analytiques* of 1797.

---

[14] Lagrange (1867–1892) vol. 14, p. 41.

33.2 Fagnano Divides the Lemniscate

In 1691, Jakob Bernoulli observed that the arc length of the parabolic spiral $(a-r)^2 = 2ab\theta$ was given by

$$s = \int \sqrt{1 + \frac{r^2(a-r)^2}{a^2b^2}}\, dr,$$

and that the integrand was an even function of $\frac{1}{2}a - r$.

Therefore, he had

$$\int_{\frac{1}{2}a-c}^{\frac{1}{2}a} = \int_{\frac{1}{2}a}^{\frac{1}{2}a+c}.$$

He could then conclude that the length of the arc of the spiral joining the points corresponding to $r = \frac{1}{2}a - c$ and $r = \frac{1}{2}a$ equaled the length of the arc from $r = \frac{1}{2}a$ to $r = \frac{1}{2}a + c$, the two arcs were incongruent.

Fagnano extended this and other of Bernoulli's results to arcs of other curves, including the lemniscate, given by

$$(x^2 + y^2)^2 = x^2 - y^2 \quad \text{or} \quad r^2 = \cos 2\theta.$$

In 1718, Fagnano published his two-part work on the division of the lemniscate.[15] In the first part he stated that if

$$u = \frac{\sqrt{1-z^2}}{\sqrt{1+z^2}}, \tag{33.42}$$

then

$$\int \frac{dz}{\sqrt{1-z^4}} = \int -\frac{du}{\sqrt{1-y^4}}. \tag{33.43}$$

Fagnano's statement of this result took an apparently more general form, but we can obtain it from this one by replacing u and z by $\frac{u}{a}$ and $\frac{z}{a}$, where a is a constant. He observed that the result could be proved by differentiating (33.42) and substituting in (33.43). As an immediate consequence of this theorem, it was clear that

$$u^2z^2 + u^2 + z^2 - 1 = 0 \tag{33.44}$$

was an integral of the equation

$$\frac{dz}{\sqrt{1-z^4}} = \pm \frac{du}{\sqrt{1-u^4}}. \tag{33.45}$$

---

[15] Fagnano's results discussed here can be found in Fagnano (1911) pp. 293–297, pp. 304–313.

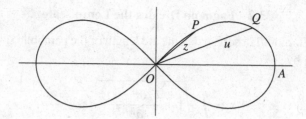

Figure 33.1 Fagnano's lemniscate.

As mentioned earlier, Euler noticed this fact very quickly. We can write the theorem in modern form as

$$\int_0^z \frac{dt}{\sqrt{1-t^4}} = \int_u^1 \frac{dt}{\sqrt{1-t^4}}, \qquad (33.46)$$

when u and z are related by (33.42) or (33.44). This means that in Figure 33.1, if $O, P, Q,$ and A denote points on the lemniscate corresponding to the values $0, z, u$ and 1 of the radius vector, then arc OP = arc QA in length.

In the last section of part 1 of his paper, Fagnano observed that the full lemniscatic arc OA would be bisected if the points P and Q coincided. This would happen when $z = u$ and then (33.44) would imply

$$z^4 + 2z^2 - 1 = 0, \quad \text{or} \quad z = u = \sqrt{\sqrt{2}-1}.$$

Thus, this constructible number would express the distance between the point of bisection and the origin.

Fagnano started the second part of the paper on the division of the lemniscate with the theorem: If

$$x = \frac{\sqrt{1 \mp \sqrt{1-z^4}}}{z}, \qquad (33.47)$$

then

$$\frac{\pm dz}{\sqrt{1-z^4}} = \frac{dx\sqrt{2}}{\sqrt{1+x^4}}. \qquad (33.48)$$

His proof consisted of the observations that

$$dx = \frac{\pm dz\sqrt{1 \mp \sqrt{1-z^4}}}{z^2\sqrt{1-z^4}},$$

and

$$\frac{\sqrt{1+x^4}}{\sqrt{2}} = \frac{\sqrt{1 \mp \sqrt{1-z^4}}}{z^2}.$$

Note that in (33.48) the differential on the right-hand side is apparently not a lemniscatic differential. So Fagnano stated another theorem: If

$$x = \frac{u\sqrt{2}}{\sqrt{1-u^4}},$$ (33.49)

then

$$\frac{du}{\sqrt{1-u^4}} = \frac{1}{\sqrt{2}} \times \frac{dx}{\sqrt{1+x^4}}.$$ (33.50)

Once again, his proof simply noted that differentiating (33.49) resulted in

$$\frac{dx}{\sqrt{2}} = \frac{du}{\sqrt{1-u^4}} \times \frac{1+u^4}{1-u^4},$$

and that (33.49) also implied

$$\sqrt{1+x^4} = \frac{1+u^4}{1-u^4}.$$

By combining these theorems, he obtained the result on the duplication of the lemniscatic arc starting at the origin: If

$$\frac{u\sqrt{2}}{\sqrt{1-u^4}} = \frac{1}{z}\sqrt{1-\sqrt{1-z^4}},$$ (33.51)

then

$$\frac{dz}{\sqrt{1-z^4}} = \frac{2\,du}{\sqrt{1-u^4}}.$$ (33.52)

Note that if P corresponds to z and Q to u, and if z and u are related by (33.51), then (33.52) shows that arc $OP = 2$ arc OQ. This means that if the value of z is given, then u can be obtained by taking square roots and conversely. Hence, duplication and bisection can be done by straight edge and compass. Fagnano made the observation that (33.51) was equivalent to the relation

$$z = \frac{2u\sqrt{1-u^4}}{1+u^4}.$$ (33.53)

Recall that Euler saw this result as the extension of the double angle formula for arcsine, and it was perhaps this result that led him to the addition formula for the lemniscatic integral.

Fagnano trisected the full arc OA of the lemniscate by combining (33.42) and (33.43) with (33.51) and (33.52). To obtain the trisection in a simpler form, he presented another transformation: If

$$\frac{\sqrt{1-t^4}}{t\sqrt{2}} = \frac{1}{z}\sqrt{1 - \sqrt{1-z^4}}, \tag{33.54}$$

then

$$\frac{dz}{\sqrt{1-z^4}} = -\frac{2\,dt}{\sqrt{1-t^4}}. \tag{33.55}$$

He then noted that he could obtain a point of trisection by setting $t = z$ and that the trisection point would be given by $t = \sqrt[4]{2\sqrt{3} - 3}$. One may check that for $t = z$, (33.54) simplifies to the equation $t^8 + 6t^4 - 3 = 0$, and that $2\sqrt{3} - 3$ is a solution of $x^2 + 6x - 3 = 0$.

Fagnano went on to work out how the arc OA could be divided into five equal parts. He did not write down the details, but, based on his trisection method, the method would probably begin by taking points on the lemniscate for which the distances from the origin are $t, z, v,$ and u such that arc $Ot = 2$ arc Oz, arc $Oz = 2$ arc Ov, and arc $Ov = $ arc uA. Fagnano's formulas give the relations connecting t with z, z with v, and v with u. Finally, if we take $t = u$, then we get arc $Ot = \frac{4}{5}$ arc uA and the equation for t reduces to

$$t^{24} + 50t^{20} - 125t^{16} + 300t^{12} - 105t^8 - 62t^4 + 5 = 0.$$

Although Fagnano did not publish this equation, it is likely that he obtained and solved it. Gauss derived the equation, and it is explicitly given in his collected works.[16] The twenty-fourth degree polynomial has factors

$$t^8 - 2t^4 + 5 \quad \text{and} \quad t^8 + (26 \pm 12\sqrt{5})t^4 + 9 \pm 4\sqrt{5}, \tag{33.56}$$

where we choose either both plus signs or both minus signs. Note that the first polynomial in (33.56) has only complex roots, but the real roots of the other two polynomials can be expressed in terms of square roots and are therefore constructible. The quinsection may be obtained by solving the polynomial with both negative signs. Fagnano stated the corollary that the quadrant of the leminscate could be divided algebraically into a number of equal parts if that number were of the form $2 \times 2^m, 3 \times 2^m, 5 \times 2^m$ for any positive integer m. He wrote that this was a "new and singular property" of his curve.

33.3 Euler: Addition Formula

Although Euler quickly perceived the importance of Fagnano's work on the lemniscatic integral, he could not at first locate any fundamental guiding principle among the large number of apparently ad hoc transformations applied somewhat randomly.

---

[16] Gauss (1863–1927) vol. 3, pp. 404–405 and vol. 10, p. 162.

It took him a little while to discover the required unifying ideas: the addition formula and the complete integral for the equation[17]

$$\frac{m\,dx}{\sqrt{1-x^4}} = \frac{n\,dy}{\sqrt{1-y^4}}. \tag{33.57}$$

It appears that Euler spotted a hint: Fagnano's result that

$$(x^2+1)(y^2+1) = 2 \quad \text{or} \quad x^2y^2 + x^2 + y^2 - 1 = 0$$

actually gave a special integral of the preceding differential equation, when $m = n = 1$. In his first paper on this topic, presented to the Academy in 1753, Euler gave some preliminary results on this hint.[18] But soon after this, as his letter to Goldbach indicated, he discovered the general algebraic integral and published it in his 1753 paper. In the first theorem of this paper, Euler took $m = n = 1$ and stated that the differential equation

$$\frac{dx}{\sqrt{1-x^4}} = \frac{dy}{\sqrt{1-y^4}} \tag{33.58}$$

had the complete integral

$$xx + yy + ccxxyy = cc + 2xy\sqrt{(1-c^4)}. \tag{33.59}$$

Here note that by taking $c = 1$, one obtains Fagnano's result (33.44).
Euler argued that taking the differential of (33.59) gave

$$x\,dx + y\,dy + ccxy(x\,dy + y\,dx) = (x\,dy + y\,dx)\sqrt{(1-c^4)},$$

and hence

$$dx(x + ccxyy - y\sqrt{(1-c^4)}) + dy(y + ccxxy - x\sqrt{(1-c^4)}) = 0. \tag{33.60}$$

He solved (33.59) as a quadratic in y, choosing the signs of the square roots so that $y = c$ when $x = 0$. He similarly solved it as a quadratic in x. Thus he got

$$y = \frac{x\sqrt{(1-c^4)} + c\sqrt{(1-x^4)}}{1 + ccxx} \quad \text{and} \quad x = \frac{y\sqrt{(1-c^4)} - c\sqrt{(1-y^4)}}{1 + ccyy}.$$

These equations implied that

$$x + ccxyy = y\sqrt{(1-c^4)} - c\sqrt{(1-y^4)},$$

$$y + ccxxy - x\sqrt{(1-c^4)} = c\sqrt{(1-x^4)}.$$

---

[17] Eu. I-20 pp. 58–79. E 251.
[18] ibid. pp. 80–107. E 252.

Euler substituted these relations in (33.60) to obtain

$$-c\,dx\sqrt{(1-y^4)} + c\,dy\sqrt{(1-x^4)} = 0.$$

This was equivalent to (33.58), and the theorem was proved. Euler then noted that this theorem was equivalent to the formula

$$\int_0^u \frac{dt}{\sqrt{1-t^4}} + \int_0^c \frac{dt}{\sqrt{1-t^4}} = \int_0^x \frac{dt}{\sqrt{1-t^4}} \qquad (33.61)$$

with

$$x = \frac{u\sqrt{(1-c^4)} + c\sqrt{(1-u^4)}}{1+c^2u^2}. \qquad (33.62)$$

Note that (33.61) and (33.61) represent the famous addition formula for the lemniscatic integral and it generalized Fagnano's duplication formula obtained by taking $u = c$. Thus, Euler saw that the transformation that left the differential $\frac{dx}{\sqrt{1-x^4}}$ invariant also provided the addition formula.

Euler then considered the more general differential $\frac{dx}{\sqrt{1+mx^2+nx^4}}$ and proved in a similar way that it remained invariant under the transformation

$$cc - xx - yy + nccxxyy + 2xy\sqrt{(1+mcc+nc^4)} = 0.$$

This in turn yielded an appropriate addition formula for this more general elliptic integral.

33.4 Cayley on Landen's Transformation

Landen's exposition of his transformation is not easy to read; in fact, G. N. Watson aptly described it as "clumsy." However, Cayley's 1876 text (reprinted in 1895) on elliptic functions, described in his preface as "founded upon Legendre's *Traité des fonctions elliptiques* and upon Jacobi's *Fundamenta Nova*, and Memoirs by him in *Crelle's Journal*," presents Landen's work in more felicitous notation and in such a manner as to outline its geometric underpinnings and make clear its essential and useful features.[19]

Summarizing Cayley, with reference to Figure 33.2, we begin by taking a point P on the circle with center O and another point Q, on the diameter AB. Set

$$QA = a, \quad QB = b, \quad A\hat{Q}P = \phi_1, \quad A\hat{B}P = \phi$$

so that $A\hat{O}P = 2\phi$. Now let

$$a_1 = \frac{a+b}{2}, \quad b_1 = \sqrt{ab}, \quad c_1 = \frac{a-b}{2}.$$

[19] Cayley (1895) pp. 327–328.

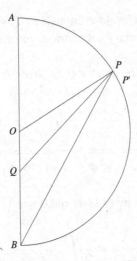

Figure 33.2 Cayley's diagram for Landen's transformation.

Then

$$OA = OB = OP = a_1, \quad OQ = a_1 - b = \frac{a-b}{2} = c_1,$$

$$QP \sin \phi_1 = a_1 \sin 2\phi,$$
$$QP \cos \phi_1 = c_1 + a_1 \cos 2\phi,$$
$$QP^2 = c_1^2 + 2c_1 a_1 \cos 2\phi + a_1^2$$
$$= \frac{1}{2}(a^2 + b^2)(\cos^2 \phi + \sin^2 \phi) + \frac{1}{2}(a^2 - b^2)(\cos^2 \phi - \sin^2 \phi)$$
$$= a^2 \cos^2 \phi + b^2 \sin^2 \phi.$$

Therefore,

$$\sin \phi_1 = \frac{a_1 \sin 2\phi}{\sqrt{a^2 \cos^2 \phi + b^2 \sin^2 \phi}} \quad \text{and} \quad \cos \phi_1 = \frac{c_1 + a_1 \cos 2\phi}{\sqrt{a^2 \cos^2 \phi + b^2 \sin^2 \phi}},$$

$$\tag{33.63}$$

$$a_1^2 \cos^2 \phi_1 + b_1^2 \sin^2 \phi_1 = \frac{a_1^2 (a \cos^2 \phi + b \sin^2 \phi)^2}{a^2 \cos^2 \phi + b^2 \sin^2 \phi}.$$

A simple calculation produces

$$\sin(2\phi - \phi_1) = \frac{1}{2} \cdot \frac{(a-b) \sin 2\phi}{\sqrt{a^2 \cos^2 \phi + b^2 \sin^2 \phi}}, \tag{33.64}$$

$$\cos(2\phi - \phi_1) = \frac{1}{a_1} \cdot \sqrt{a_1^2 \cos^2 \phi + b_1^2 \sin^2 \phi}.$$

Take a point P' on the circle close enough to P so that we can regard PP' as tangent to the circle. An elementary geometric argument shows that

$$PQd\phi_1 = PP' \sin P'\hat{P}Q = 2a_1 d\phi \cos(2\phi - \phi_1).$$

This is equivalent to

$$\frac{2d\phi}{\sqrt{a^2 \cos^2 \phi + b^2 \sin^2 \phi}} = \frac{d\phi_1}{\sqrt{a_1^2 \cos^2 \phi_1 + b_1^2 \sin^2 \phi_1}}. \qquad (33.65)$$

Thus, if ϕ and ϕ_1 are related by (33.64) and $0 \le \phi \le \frac{\pi}{2}$, then

$$\int_0^\phi \frac{dt}{\sqrt{a^2 \cos^2 t + b^2 \sin^2 t}} = \frac{1}{2} \int_0^{\phi_1} \frac{dt}{\sqrt{a_1^2 \cos^2 t + b_1^2 \sin^2 t}}. \qquad (33.66)$$

In particular,

$$\int_0^{\frac{\pi}{2}} \frac{dt}{\sqrt{a^2 \cos^2 t + b^2 \sin^2 t}} = \frac{1}{2} \int_0^\pi \frac{dt}{\sqrt{a_1^2 \cos^2 t + b_1^2 \sin^2 t}}$$

$$= \int_0^{\frac{\pi}{2}} \frac{dt}{\sqrt{a_1^2 \cos^2 t + b_1^2 \sin^2 t}}. \qquad (33.67)$$

Using the notation of Legendre, we set

$$k^2 = 1 - \frac{b^2}{a^2}, \quad k' = \frac{b}{a} \quad \text{and} \quad k_1^2 = 1 - \frac{b_1^2}{a_1^2}.$$

Then (33.66) can be written as

$$\int_0^\phi \frac{dt}{\sqrt{1 - k^2 \sin^2 t}} = \frac{1}{2} \frac{a}{a_1} \int_0^{\phi_1} \frac{dt}{\sqrt{1 - k_1^2 \sin^2 t}}$$

$$= \frac{1}{2}(1 + k_1) \int_0^{\phi_1} \frac{dt}{\sqrt{1 - k_1^2 \sin^2 t}}; \qquad (33.68)$$

note that this is in fact Landen's transformation (33.33). To see this, compare the first equations in (33.63) and (33.64) to get

$$\sin(2\phi - \phi_1) = k_1 \sin \phi_1. \qquad (33.69)$$

Now (33.33) follows from (33.68) by noting that $\frac{4k_1}{(1+k_1)^2} = k^2$.

Finally, we observe that if we write $y = \sin\phi_1$ and $x = \sin\phi$, then (33.69) can be expressed as the quadratic transformation

$$y = \frac{(1+k')x\sqrt{1-x^2}}{\sqrt{1-k^2x^2}},\qquad(33.70)$$

and (33.65) takes the form

$$\frac{(1+k_1)\,dy}{\sqrt{(1-y^2)(1-k_1^2y^2)}} = \frac{2\,dx}{\sqrt{(1-x^2)(1-k^2x^2)}}.\qquad(33.71)$$

33.5 Lagrange, Gauss, Ivory on the agM

Lagrange was the first mathematician to observe the connection between the arithmetic geometric mean (agM) and elliptic integrals. His 1784–1785 result (33.39) essentially expressed this connection. He made this discovery as he pursued a numerical method for evaluating elliptic integrals, and he did not further investigate the concept of the agM. In his 1818 paper on astronomy,[20] Gauss gave a formula relating the agM of two positive real numbers with an elliptic integral and he worked out an extensive theory on this topic, but published very little of it. Gauss denoted the agM of two positive numbers a and b, with $a \geq b$, as $M(a,b)$. He considered the sequence

$$a_1 = \frac{a+b}{2},\quad b_1 = \sqrt{ab},\quad a_2 = \frac{a_1+b_1}{2},\quad b_2 = \sqrt{a_1b_1},\quad\text{etc.,}\qquad(33.72)$$

and noted that

$$b \leq b_1 \leq b_2 \leq \cdots \leq b_n \leq \cdots \leq a_n \leq \cdots \leq a_1 \leq a.$$

He observed that if $a = b$, then $a_n = b_n$ for all n. On the other hand, if $a > b$, then

$$\frac{a_n - b_n}{a_{n-1} - b_{n-1}} = \frac{a_{n-1} - b_{n-1}}{4(a_n + b_n)} = \frac{a_{n-1} - b_{n-1}}{2(a_{n-1} + b_{n-1}) + 4b_n}$$

and hence

$$a_n - b_n < \frac{a_{n-1} - b_{n-1}}{2} < \frac{a - b}{2^n}.\qquad(33.73)$$

Consequently, the increasing sequence b_n and the decreasing sequence a_n converged to the same number denoted by $M(a,b)$.

Gauss also noted the simple properties of $M(a,b)$ given by the equations

$$M(a,b) = M(a_1,b_1)\quad\text{and}\quad M(na,nb) = nM(a,b)\quad\text{for any real } n > 0.\quad(33.74)$$

---

[20] Gauss (1863–1927) vol. 3, pp. 331–355.

From these equations he deduced a number of relations; for instance, for $x = \frac{2t}{1+t^2}$, he had

$$M(1+x, 1-x) = M\left(1, \frac{1-t^2}{1+t^2}\right) = \frac{1}{1+t^2} M(1-t^2, 1+t^2). \qquad (33.75)$$

Gauss's theorem, published in 1818 but proved much earlier, stated that

$$\frac{1}{M(a,b)} = \frac{2}{\pi} \int_0^{\frac{\pi}{2}} \frac{d\theta}{\sqrt{a^2 \cos^2 \theta + b^2 \sin^2 \theta}}. \qquad (33.76)$$

To see how Lagrange's transformation implies this theorem of Gauss, denote the integral in (33.76) by $I(a,b)$ and set $x = \frac{\cot \theta}{b}$ to get

$$I(a,b) = \frac{2}{\pi} \int_0^\infty \frac{dx}{\sqrt{(1+a^2 x^2)(1+b^2 x^2)}}. \qquad (33.77)$$

Recall Lagrange's result (33.39), that if

$$x = y \sqrt{\frac{1 + a_1^2 y^2}{1 + b_1^2 y^2}}, \qquad (33.78)$$

then

$$\frac{dx}{\sqrt{(1+a^2 x^2)(1+b^2 x^2)}} = \frac{dy}{\sqrt{(1+a_1^2 y^2)(1+b_1^2 y^2)}}. \qquad (33.79)$$

When this result is applied to (33.77), we see that

$$I(a,b) = I(a_1, b_1). \qquad (33.80)$$

Upon iteration, we conclude that if $c = M(a,b)$, then

$$I(a,b) = I(a_n, b_n) = I(c,c) = \frac{2}{\pi} \int_0^\infty \frac{dx}{1 + c^2 x^2} = \frac{1}{c}.$$

This proves Gauss's theorem. Note that (33.67) is identical to (33.80).

Gauss derived (33.76) by means of a different transformation. He set

$$\sin \theta = \frac{2a \sin \theta'}{(a+b) \cos^2 \theta' + 2a \sin^2 \theta'} = \frac{2a \sin \theta'}{a + b + (a-b) \sin^2 \theta'}$$

and observed that (33.76) would follow. Jacobi provided more details on this transformation in section 38 of his *Fundamenta Nova*, published a decade after Gauss's paper. Since Jacobi was pursuing other threads, his presentation is not as direct as that in Cayley's 1876 treatise. Following Cayley, replace $\sin\theta$ and $\sin\theta'$ by y and x, respectively, to write Gauss's substitution as

$$y = \frac{(1+k)x}{1+kx^2}, \quad k = \frac{a-b}{a+b}. \tag{33.81}$$

This is the form in which Gauss's transformation is often presented, particularly in connection with the transformation theory of elliptic functions. One then perceives that proving the relation $I(a,b) = I(a_1,b_1)$ is equivalent to showing that

$$\frac{dy}{\sqrt{(1-y^2)(1-\lambda^2 y^2)}} = \frac{(1+k)\,dx}{\sqrt{(1-x^2)(1-k^2 x^2)}}, \tag{33.82}$$

where $\lambda = \frac{2\sqrt{k}}{1+k}$.

To prove (33.82), let $D = 1 + kx^2$, the denominator of y in (33.81). Then

$$1 - y = \frac{(1-x)(1-kx)}{D},$$

$$1 + y = \frac{(1+x)(1+kx)}{D},$$

$$1 - \lambda x = \frac{1 - \sqrt{k}x}{D},$$

$$1 + \lambda x = \frac{1 + \sqrt{k}x}{D}.$$

Consequently,

$$\sqrt{(1-y^2)(1-\lambda^2 y^2)} = (1-kx^2)\sqrt{(1-x^2)(1-k^2 x^2)}D^2 \tag{33.83}$$

$$dy = \frac{(1+k)(1-kx^2)\,dx}{D^2}. \tag{33.84}$$

Note that formulas (33.83) and (33.84) imply (33.82), giving us another proof of Gauss's theorem.

Gauss gave yet another proof, by means of power series, though he did not publish it.[21] We reproduce Gauss's proof, but we use subscript and factorial symbols where Gauss did not. In this derivation, he assumed that $M(1+x, 1-x)$ had a series expansion so that he could write

$$\frac{1}{M(1+x, 1-x)} = A_0 + A_1 x^2 + A_2 x^4 + A_3 x^6 + \cdots, \quad A_0 = 1.$$

---

[21] Gauss (1863–1927) vol. 3, pp. 361–403, especially pp. 367–369.

Using (33.75), he had

$$\frac{2t}{1+t^2} + A_1 \left(\frac{2t}{1+t^2}\right)^3 + A_2 \left(\frac{2t}{1+t^2}\right)^5 + \cdots = 2t(A_0 + A_1 t^4 + A_2 t^8 + \cdots).$$

He equated the coefficients of powers of t to obtain the relations

$$A_0 = 1,$$
$$0 = 1 - 4A_1,$$
$$A_1 = 1 - 12A_1 + 16A_2,$$
$$0 = 1 - 24A_1 + 80A_2 - 64A_3,$$
$$A_2 = 1 - 40A_1 + 240A_2 - 448A_3 + 256A_4, \quad \text{etc.}$$

Unlike earlier mathematicians, he also presented the general nth relation

$$M = 1 - 4A_1 \frac{n(n-1)}{2!} + 16A_2 \frac{(n+1)n(n-1)(n-2)}{4!}$$
$$- 64A_3 \frac{(n+2)(n+1)n(n-1)(n-2)(n-3)}{6!} + \cdots,$$

with the remark that $M = 0$ when n was even and $M = A_{\frac{n-1}{2}}$ when n was odd; in other words, M was the $\frac{n+1}{2}$th term of the series $A_0, A_1, A_2, \ldots$. Taking $0 = 0$ as the 0th equation, and abbreviating and labeling the equations as $[0], [1], [2], \ldots$, he wrote down the equations

$$1^2[2] - 0^2[0], \ 2^2[3] - 1^2[1], \ 3^2[4] - 2^2[2], \ 4^2[5] - 3^2[3], \ 5^2[6] - 4^2[4], \ldots.$$

In general, for the nth equation he had

$$n^2 N - (n-1)^2 L = (2n-1)\left(1 - 2^2 A_1 \frac{3n^2 - 3n + 2}{2!} + 2^4 A_2 \frac{n(n-1)(5n^2 - 5n + 6)}{4!}\right)$$
$$- (2n-1)\left(2^6 A_3 \frac{(n+1)n(n-1)(n-2)(7n^2 - 7n + 12)}{6!} + \cdots\right). \tag{33.85}$$

A editorial note to Gauss's paper observes that L and N were equal to $A_{\frac{n-2}{2}}$ and $A_{\frac{n}{2}}$, respectively, when n was even, and zero when n was odd. In another footnote, it is pointed out that the derivation connected with the forms $n^2 N - (n-1)^2 L$ was explained in article 162 of the *Disquisitiones Arithmeticae*. It may be of interest to note that in that article, Gauss discussed the problem of determining all transformations of the form

$$X = \alpha' x + \beta' y, \ Y = \gamma' x + \sigma' y,$$

given one known transformation

$$X = \alpha x + \beta y, \ Y = \gamma x + \sigma y$$

of

$$AX^2 + 2BXY + CY^2 \quad \text{to} \quad ax^2 + 2bxy + cy^2.$$

Continuing the proof, let $k = 2l - 1$. Gauss wrote the lth term of the expression in (33.85) (without the factor $2n - 1$) as

$$\pm 2^{k-1} A_{\frac{k-1}{2}} \frac{\left(n + \frac{k-5}{2}\right)\left(n + \frac{k-7}{2}\right)\cdots\left(n - \frac{k-3}{2}\right)\left(kn^2 - kn + \frac{k^2-1}{4}\right)}{(k-1)!}. \qquad (33.86)$$

We note that the sign is plus when $l - 1$ is even and minus when $l - 1$ is odd. Gauss then divided each such term into two parts and added the second part of one term to the first part of the succeeding term. As a first step in this process, he observed that

$$kn^2 - kn + \frac{k^2 - 1}{4} = k\left(n - \frac{k-1}{2}\right)\left(n + \frac{k-3}{2}\right) + \frac{(k-1)^3}{4}.$$

Using this, he could express the term (33.86) as a sum of two terms

$$\pm\left(2^{k-1}k^2 A_{\frac{k-1}{2}} \frac{\left(n + \frac{k-3}{2}\right)\left(n + \frac{k-1}{2}\right)\cdots\left(n - \frac{k-1}{2}\right)}{k!}\right.$$

$$\left. + \left(2^{k-1}A_{\frac{k-1}{2}} \frac{n + \frac{k-5}{2}\cdots n - \frac{k-3}{2}}{(k-2)!} \cdot \left(\frac{k-1}{2}\right)^2\right)\right).$$

When he added the second part of this expression to the first part of the succeeding expression, he obtained

$$2^{k-1}\frac{\left(n + \frac{k-3}{2}\right)\left(n + \frac{k-5}{2}\right)\cdots\left(n - \frac{k-1}{2}\right)}{k!}\left((k+1)^2 A_{\frac{k-1}{2}} - k^2 A_{\frac{k-1}{2}}\right).$$

Thus, he could express (33.85) as

$$n^2 N - (n-1)^2 L = (2n-1)\left((A_0 - 2^2 A_1) - 4\frac{n(n-1)}{2!\,3}(3^2 A_1 - 4^2 A_2)\right)$$

$$+ (2n-1)\left(4^2\frac{(n+1)n(n-1)(n-2)}{4!\,5}(5^2 A_2 - 6^2 A_3)\right)$$

$$-(2n-1)\left(4^3\frac{(n+2)(n+1)\cdots(n-2)(n-3)}{6!\,7}(7^2 A_3 - 6^2 A_2) + \cdots\right).$$

To clarify the general result implied by this procedure, Gauss wrote the first few special cases:

$$0 = 1 - 4A_1,$$
$$4A_1 - 1 = 3(1 - 4A_1) - 4(9A_1 - 16A_2),$$
$$0 = 5(1 - 4A_1) - 20(9A_1 - 16A_2) + 16(25A_2 - 36A_3),$$
$$16A_2 - 9A_1 = 7(1 - 4A_1) - 56(9A_1 - 16A_2) + 112(25A_2 - 36A_3)$$
$$- 65(49A_3 - 64A_4) \quad \text{etc.}$$

Thus, he found that

$$A_1 = \frac{1^2}{2^2}, \quad A_2 = \frac{3^2}{4^2}A_1 = \frac{1^2 \cdot 3^2}{2^2 \cdot 4^2}, \quad A_3 = \frac{1^2 \cdot 3^2 \cdot 5^2}{2^2 \cdot 4^2 \cdot 6^2}, \dots,$$

and in general

$$A_n = \frac{1^2 \cdot 3^2 \cdots (2n-1)^2}{2^2 \cdot 4^2 \cdots (2n)^2}.$$

Gauss then related $\frac{1}{M(1+x,1-x)}$ with the elliptic integral by evaluating the following integral as a series:

$$\frac{1}{\pi} \int_0^\pi (1 - x^2 \cos^2 \theta)^{-\frac{1}{2}} \, d\theta = \frac{1}{\pi} \int_0^\pi \left(1 + \frac{1}{2}x^2 \cos^2 \theta + \frac{1}{2} \cdot \frac{3}{4}x^4 \cos^4 \theta + \cdots \right) d\theta$$

$$= 1 + \frac{1^2}{2^2}x^2 + \frac{1^2 \cdot 3^2}{2^2 \cdot 4^2}x^4 + \cdots . \tag{33.87}$$

Since the series for the integral and the agM were the same, he concluded that

$$\frac{1}{M(1+x, 1-x)} = \frac{2}{\pi} \int_0^{\frac{\pi}{2}} \frac{d\theta}{\sqrt{1 - x^2 \sin^2 \theta}}.$$

Finally, one may see that by taking $x = \sqrt{\frac{1-b^2}{a^2}}$, we have Gauss's formula (33.76).

In a 1796 paper,[22] James Ivory gave an interesting new method to prove the formula

$$K\left(\frac{2\sqrt{x}}{1+x}\right) = (1+x)K(x), \tag{33.88}$$

where

$$K(x) = \frac{2}{\pi} \int_0^{\frac{\pi}{2}} \frac{d\theta}{\sqrt{1 - x^2 \cos^2 \theta}}.$$

Legendre was the first to prove this result, for the purpose of numerically evaluating complete elliptic integrals; his proof used the Landen transformation. In his paper,

[22] Ivory (1796).

Ivory did not mention the agM and, indeed, he may not have been aware of its significance even if he had noticed it in Lagrange's 1785 paper. In the cover letter accompanying his paper, Ivory explained that his aim was to present a simple method for obtaining the expansion

$$(a^2 + b^2 - 2ab \cos \phi)^n = A + B \cos \phi + C \cos 2\phi + \cdots . \tag{33.89}$$

Ivory started with the relation $\sin(\psi - \phi) = c \sin \psi$, took its fluxion (derivative), and simplified to get

$$\dot\phi = \frac{\sqrt{1 - c^2 \sin^2 \psi} - c \cos \psi}{\sqrt{1 - c^2 \sin^2 \psi}} \dot\psi .$$

He performed an elementary calculation to show that the numerator could be expressed as $\sqrt{1 + c^2 - 2c \cos \phi}$. This led him to the equation

$$\frac{\dot\phi}{\sqrt{1 + c^2 - 2c \cos \phi}} = \frac{\dot\psi}{\sqrt{1 - c^2 \sin^2 \psi}} . \tag{33.90}$$

He then set

$$c' = \frac{1 - \sqrt{1 - c^2}}{1 + \sqrt{1 - c^2}} \tag{33.91}$$

to find that

$$\sqrt{1 - c^2 \sin^2 \psi} = \frac{\sqrt{1 + c'^2 + 2c' \cos 2\psi}}{1 + c'} .$$

Thus, he could express (33.90) in the form

$$\frac{\dot\phi}{\sqrt{1 + c^2 - 2c \cos \phi}} = \frac{(1 + c')\dot\psi}{\sqrt{1 + c'^2 + 2c' \cos 2\psi}} . \tag{33.92}$$

Ivory's contribution was a new method for evaluating the integrals

$$\int_0^\pi \frac{d\phi}{\sqrt{1 + c^2 - 2c \cos \phi}} = \int_0^\pi \frac{(1 + c')\, d\psi}{\sqrt{1 + c'^2 + 2c' \cos 2\psi}} . \tag{33.93}$$

First, he observed that by the binomial theorem

$$(1 + c^2 - 2c \cos \phi)^{-\frac{1}{2}} = (1 - ce^{i\phi})^{-\frac{1}{2}}(1 - ce^{-i\phi})^{-\frac{1}{2}}$$

$$= \left(1 + \frac{1}{2}ce^{i\phi} + \frac{1 \cdot 3}{2 \cdot 4}c^2 e^{2i\phi} + \cdots\right)\left(1 + \frac{1}{2}ce^{-i\phi} + \frac{1 \cdot 3}{2 \cdot 4}c^2 e^{-2i\phi} + \cdots\right).$$

$$\tag{33.94}$$

Ivory then noted that multiplying these two series gave him the cosine expansion when $n = -\frac{1}{2}$ in (33.89) . This showed that the value of the integral on the left-hand side of (33.93) was simply the constant A. Since the constant term A was easy to evaluate from the product of the two series, he could express equation (33.93) as

$$\int_0^\pi \frac{d\phi}{\sqrt{1+c^2-2c\cos\phi}} = 1 + \frac{1^2}{2^2}c^2 + \frac{1^2\cdot 3^2}{2^2\cdot 4^2}c^4 + \cdots$$

$$= (1+c')\left(1 + \frac{1^2}{2^2}c'^2 + \frac{1^2\cdot 3^2}{2^2\cdot 4^2}c'^4 + \cdots\right). \quad (33.95)$$

Observe that this proves (33.88).

Ivory remarked that c' was smaller than c; as an example, he pointed out that when $c = \frac{4}{5}$, then $c' = \frac{1}{4}$. Thus, (33.95) was very useful for computational purposes. Ivory noted, as Legendre had done before him, that the formula's computational effectiveness was greatly improved by iteration. Now note that (33.95) can be expressed as a quadratic transformation of hypergeometric functions:

$$F\left(\frac{1}{2},\frac{1}{2},1,c^2\right) = (1+c')F\left(\frac{1}{2},\frac{1}{2},1,c'^2\right),$$

where c' is given by (33.91). Gauss may have been motivated to study quadratic transformations of hypergeometric functions because of this and similar results.

We have discussed two different quadratic transformations of elliptic integrals. We can rewrite Landen's transformation (33.33) in the form: If

$$\lambda = \frac{2\sqrt{k}}{1+k}, \quad \lambda^2 + \lambda'^2 = 1, \quad z = (1+\lambda')y\frac{\sqrt{(1-y^2)}}{\sqrt{(1-\lambda^2 y^2)}},$$

then

$$\frac{(1+k)\,dz}{\sqrt{(1-z^2)(1-k^2 z^2)}} = \frac{2\,dy}{\sqrt{(1-y^2)(1-k^2 y^2)}}. \quad (33.96)$$

Note: (33.96) gives one quadratic transformation and the other is Gauss's transformation given by the equations (33.81) and (33.82). It is easy to check that if these transformations are applied one after the other, we get the duplication of the elliptic integral:

$$\frac{dz}{\sqrt{(1-z^2)(1-k^2 z^2)}} = \frac{2\,dx}{\sqrt{(1-x^2)(1-k^2 x^2)}},$$

$$z = 2x\frac{\sqrt{1-x^2}\sqrt{1-k^2 x^2}}{1-k^2 x^4}.$$

33.6 Remarks on Gauss and Elliptic Functions

Gauss wrote in an 1816 letter to his friend Schumacher that he rediscovered the arithmetic-geometric mean in 1791 at the age of 14.[23] From 1791 until 1800, Gauss made a series of discoveries advancing the theory of elliptic integrals and functions to new and extraordinary heights. Among his great achievements in this area were the inversion of the elliptic integral and the consequent discovery of double periodicity and the development of elliptic functions as power series and as double products, leading to series and product expansions in terms of trigonometric functions. This in turn brought him to the discovery of the theta functions and the triple product identity. The initial motivation behind Gauss's work on elliptic functions was the problem of the division of the lemniscate. Gauss solved the problem by means of complex multiplication of elliptic functions. Finally, in 1800, he extended his youthful work on the arithmetic-geometric mean by considering the agM of two complex numbers. He found that the agM in this case was countably many-valued, and his attempts to find a relation among the values led him deeper into the theory of theta and modular functions. In this connection, Gauss discovered an important transformation of theta functions:

$$\sum_{k=-\infty}^{\infty} e^{-\alpha(k+w)^2} = \sqrt{\frac{\pi}{\alpha}} \sum_{k=-\infty}^{\infty} e^{-\frac{n^2\pi^2}{\alpha} + n\omega\pi i}.$$

It is remarkable that Gauss published very little of these groundbreaking theories, although they surely rank among his greatest discoveries in pure mathematics. Perhaps he wished to first develop a coherent theory of functions of complex variables. Consider the fact that, though he initially discovered double periodicity through a formal use of complex numbers, his 1800 work defined an elliptic function by means of a ratio of two theta functions. His early definition of an elliptic function by means of the inversion of an elliptic integral would require the concept and careful use of analytic continuation, not then developed. We note that Gauss's 1811 letter to Bessel shows that he was making inroads into the mysteries of complex variables. The unpublished portion of Gauss's 1813 paper on hypergeometric series also gives some indication of his understanding of analytic continuation. However, it seems that after 1805–06, Gauss never found the time to completely develop his ideas in number theory, elliptic functions, or complex variables. From 1801 onward, he researched applied topics such as astronomy, geodesy, telegraphy, magnetism, crystallography, and optics. Of course, mathematical problems in these areas led him to interesting and important discoveries such as the method of least squares, trigonometric interpolation, numerical integration, the technique of fast Fourier transforms, and the theory of curved surfaces.

---

[23] Peters (1860–1865) vol. 1, p. 125.

Thus, Gauss never wrote up a detailed account of his researches on elliptic functions, though he wrote a substantial amount on theta functions. In his fragmentary notes on elliptic functions, one may see some of the main results but usually there are no details of the methods he employed. However, in a letter to Schumacher,[24] Gauss wrote that Abel's first paper on the theory of elliptic functions followed the same path he himself had trod in 1798 and that Abel's work relieved him of the burden of publishing that part of his work. He wrote a similar letter to Crelle and the entries in Gauss's diary from the period 1797–1800 bear out these assertions.

Gauss's investigations relating to elliptic functions, the agM, and theta functions took place during 1791–1800, the critical period when he was maturing into a most formidable mathematical mind. Gauss's early interest in mathematics was kindled by his association with Johann Bartels (1769–1836), a teacher's assistant at the school Gauss attended. Gauss was 11 years old when he and Bartels studied infinite series and the binomial theorem. Bartels later became professor of mathematics at the University of Kazan where he taught the great Russian mathematician N. I. Lobachevsky (1793–1856), one of the discoverers of non-Euclidean geometry. Having become known as a promising student, in 1791 Gauss met the Duke of Braunschweig and was presented with a table of logarithms by the minister of state. Greatly impressed with the genius of Gauss, the Duke provided financial support for Gauss to attend the Collegium Carolinum in Braunschweig (Brunswick). Entering the Collegium in 1792, Gauss became accomplished in languages and began studying the works of Newton, Euler, and Lagrange. He was impressed by Euler's pentagonal number theorem:

$$\prod_{n=1}^{\infty}(1 - x^n) = \sum_{n=-\infty}^{\infty} (-1)^n x^{\frac{n(3n-1)}{2}}.$$

This result led him to investigate series whose exponents were square or triangular numbers. Gauss is reported to have said that in 1794 he knew the connection between such series and the agM. This is very likely because the series identities needed for this question could be easily proved by methods Gauss had seen in the works of Euler. To state the identities, set $A(x) = \sum x^{n^2}$ and $B(x) = \sum (-1)^n x^{n^2}$ where the sums are over all integers. Then

$$A(x) + B(x) = 2A(x^4), \tag{33.97}$$

$$A^2(x) + B^2(x) = 2A^2(x^2), \tag{33.98}$$

$$A(x)B(x) = B^2(x^2). \tag{33.99}$$

The first identity, (33.97), is almost obvious; the second, (33.98), can be proved by first observing that the coefficient of x^n in $A^2(x)$ is the number of ways n can be expressed as a sum of two integer squares and then noting that this is the same as the number of ways $2n$ can be expressed as a sum of two integer squares. The third identity, (33.99), is a consequence of the first two because

---

[24] Gauss (1863–1927) vol. 10_1, p. 248.

$$A(x)B(x) = \frac{(A(x) + B(x))^2}{2} - \frac{A^2(x) + B^2(x)}{2}$$

$$= 2A^2(x^4) - A^2(x^2) = B^2(x^2).$$

It follows from (33.97) and (33.99) that the arithmetic mean and geometric mean of $A^2(x)$ and $B^2(x)$ are $A^2(x^2)$ and $B^2(x^2)$. This kind of reasoning must have been very familiar to Gauss in 1794, both from his investigations in collaboration with Bartels and from his study of Euler's *Introductio*. Gauss wrote up his results connecting series with the agM at a later date.[25] In that manuscript, he derived the properties of the series by means of their product representations obtained from the triple product identity.

In 1794–1795, Gauss does not seem to have been aware of the connection of these series or the agM with elliptic integrals. Then in October 1795, with the continued support of the Duke, Gauss registered as a student at the University of Göttingen where he had access to an excellent library. For example, in early 1796 he borrowed many volumes of the *Mémories de l'Académie de Berlin* from the library. The volumes contained several works of Lagrange on number theory, algebra, and other mathematical topics. Gauss first mentioned elliptic integrals in his mathematical diary on September 9, 1796.[26] He gave the power series expansion of the inverse of the elliptic integral $\int (1 - x^3)^{-\frac{1}{2}} dx$; he found it by Newton's reversion of series method. A few days later, he noted the series for the inverse of the more general integral $\int (1 - x^n)^{-\frac{1}{2}} dx$. In January 1797, his interest in elliptic integrals became more serious. His notes indicate that he had already studied Stirling and Euler on this topic. He noted in his January 7 entry in the diary that

$$\int \sqrt{\sin x}\, dx = 2 \int \frac{yy\, dy}{\sqrt{1 - y^4}}, \quad \int \sqrt{\frac{1}{\sin x}}\, dx = 2 \int \frac{dy}{\sqrt{1 - y^4}}, \quad yy = \frac{\sin}{\cos} x,$$

and a day later he recorded that he had started investigating the elastic curve depending on $\int (1 - x^4)^{-\frac{1}{2}} dx$. Later he crossed out the words "elastic curve" and replaced them with "lemniscate." In order to understand the reason for this change in point of view, we note that Gauss started his mathematical diary in March 1796 when he discovered the principles underlying the problem of dividing the circle into n equal parts. In particular, this problem required a study of the polynomials obtained when $\sin(nx)$ and $\cos(nx)$ were expressed in terms of $\sin x$ and $\cos x$. Note that the sine or cosine function can be defined in terms of the inverse of the integral $\int \frac{dx}{\sqrt{1-x^2}}$. Around March 1797, Gauss found that in order to divide the lemniscate into n equal parts, he had to study the properties of the lemniscatic function, defined as the inverse of the integral $\int \frac{dx}{\sqrt{1-x^4}}$.

On March 19, Gauss observed in his diary that the division of the lemniscate into n parts led to an algebraic equation of degree n^2. In fact, this follows from the addition formula for the lemniscatic integral. However, it appears from his September 1796

[25] ibid. vol. 3, pp. 466–469.
[26] See Dunnington (2004) pp. 469–484.

note that he was already thinking in terms of the inverse of the integral $\int \frac{dx}{\sqrt{1-x^4}}$. If we denote the inverse by sl x, then we see that he had discovered that sl(nx) could be expressed as a rational function of sl x and that the numerator was a polynomial of degree n^2 in sl x. Since only n solutions correspond to real division points, this discovery showed him that a majority of the solutions of the equation of degree n^2 had to be complex. It is possible that this led him to make an imaginary substitution in the integral $\int \frac{dx}{\sqrt{1-x^4}}$, and this in turn led him to the discovery of the double periodicity of the lemniscatic function. In his undated diary entry between March 19 and March 21, Gauss noted that the lemniscate was geometrically divisible into five equal parts. This is a remarkable statement; it shows that Gauss had not only found double periodicity but had also found an example of complex multiplication of elliptic functions. We note that in general, an elliptic function $\phi(x)$ has two fundamental periods whose ratio must be a complex number. Moreover, for any integer n, the addition formula for elliptic functions shows that $\phi(nx)$ is a rational function of $\phi(x)$. However, in the case where the ratio of the fundamental periods is a root of a quadratic with rational coefficients, there exists a complex number α such that $\phi(\alpha x)$ can be expressed as a rational function of $\phi(x)$. In this situation, we say that $\phi(x)$ permits complex multiplication by α. Apparently, in 1828 Abel was the first to study this phenomenon. It is not clear to what extent Gauss had investigated complex multiplication, but he certainly used it in connection with dividing the lemniscate into n equal parts, at least when $n = 5$. To be able to show that a fifth part of the lemniscatic curve could be obtained by geometric construction, Gauss had to solve two appropriate quadratic equations. The surviving fragments of Gauss's work on this problem do not contain these equations, but they can be found in Abel's first paper on elliptic functions.[27]

Abel noted that $5 = (2+i)(2-i)$, and that for $y = \phi(x) \equiv \text{sl}(x)$,

$$\phi((2+i)x) = yi\frac{1 - 2i - y^4}{1 - (1 - 2i)y^4} \equiv z, \tag{33.100}$$

$$\phi(5x) = \phi((2-i)(2+i)x) = -zi\frac{1 + 2i - z^4}{1 - (1 + 2i)z^4}. \tag{33.101}$$

We note that Abel here used complex multiplication of ϕ by $2 \pm i$. Next, using (33.101) to solve the equation $\phi(5x) = 0$, he had to first solve $z^4 = 1 + 2i$, and then solve

$$yi\frac{1 - 2i - y^4}{1 - (1 - 2i)y^4} = (1 + 2i)^{\frac{1}{4}}. \tag{33.102}$$

Solve the latter by dividing the previous equation by the conjugate equation

$$-yi\frac{1 + 2i - y^4}{1 - (1 + 2i)y^4} = (1 - 2i)^{\frac{1}{4}} \tag{33.103}$$

---

[27] Abel (2007) p. 248.

to obtain a quadratic in y^4. Note that all the equations can be solved by appropriate quadratic equations.

In his notes from this period,[28] Gauss defined the lemniscatic sine and cosine functions by the equations

$$\text{sin lemn} \left(\int_0^x \frac{dt}{\sqrt{1 - t^4}} \right) = x, \quad \text{cos lemn} \left(\frac{1}{2}\omega - \int_0^x \frac{dt}{\sqrt{1 - t^4}} \right) = x,$$

where $\omega = \int_0^1 (1 - t^4)^{-\frac{1}{2}} dt$. Gauss sometimes abbreviated sine lemn and cos lemn as s and c. We use the more common sl and cl. By an application of Euler's addition formula for elliptic integrals, Gauss found the addition formulas for the elliptic functions sl and cl:

$$1 = ss + cc + sscc, \tag{33.104}$$

$$\text{sl}(a \pm b) = \frac{sc' \mp ss'}{1 \mp scs'c'}, \tag{33.105}$$

$$\text{cl}(a \pm b) = \frac{cc' \mp ss'}{1 \pm ss'cc'}, \tag{33.106}$$

where $s = \text{sl}(a), s' = \text{sl}(b)$ and c, c' are similarly defined. He employed these formulas to express $\text{sl}(n\phi)$ and $\text{cl}(n\phi)$ in terms of $\text{sl}(\phi)$ and $\text{cl}(\phi)$. By a formal change of variables $t = iu$, Gauss obtained

$$i \int_0^x (1 - t^4)^{-\frac{1}{2}} dt = \int_0^{ix} (1 - u^4)^{-\frac{1}{2}} du,$$

or, in terms of the lemniscatic functions:

$$\text{sl}(iy) = i\,\text{sl}(y), \quad \text{cl}(iy) = \frac{1}{\text{cl}(y)}.$$

Thus, with (33.105), he had the formula for complex arguments

$$\text{sl}(a + ib) = \frac{\text{sl}(a) + i\,\text{sl}(b)\text{cl}(a)\text{cl}(b)}{\text{cl}(b) - i\,\text{sl}(a)\text{sl}(b)\text{cl}(a)}. \tag{33.107}$$

Gauss used these formulas to determine that the periods of sl were 2ω and $2i\omega$. The ratio of the periods would then be $i = \sqrt{-1}$, a root of the quadratic equation $x^2 + 1 = 0$, so that complex multiplication by $\sqrt{-1}$ was possible. Gauss also found that the zeros and poles of $\text{sl}(\phi)$ were of the form $(m + in)\omega$ and $((m + \frac{1}{2}) + i(n + \frac{1}{2}))\omega$, where $m, n \in \mathbb{Z}$. These results allowed him to express the lemniscatic function as a quotient of two entire functions

$$\text{sl}(\phi) = \frac{M(\phi)}{N(\phi)}, \tag{33.108}$$

---

[28] Gauss (1863–1927) vol. 10_1, pp. 147–154.

where M and N were double infinite products. Gauss's diary entry, for March 29, 1797, confirms that by that time he was aware of all these results. In the same entry, he gave numerical evidence that $\log N(\omega)$ agreed with $\frac{\pi}{2}$ to six decimal places and noted that a proof would be an important advance in analysis. Gauss's fragmentary notes from this period also show that he had found the significant formula

$$M(\phi)^4 + N(\phi)^4 = N(2\phi).\tag{33.109}$$

He also wrote, perhaps based on numerical evidence, that

$$M\left(\frac{\omega}{2}\right) = N\left(\frac{\omega}{2}\right).\tag{33.110}$$

Subsequent entries in Gauss's diary suggest that he abandoned his intensive study of elliptic functions for a year. The ninety-second entry, written July 1798, concerned the lemniscatic function; he noted that he had "found out the most elegant things exceeding all expectations and that by methods which open to us a whole new field ahead." According to his notes, his result was[29]

$$\mathrm{sl}(\phi) = \frac{P(\phi)}{Q(\phi)},\tag{33.111}$$

where

$$P(\phi) = \frac{\omega}{\pi}s\left(1 + \frac{4ss}{(e^\pi - e^{-\pi})^2}\right)\left(1 + \frac{4ss}{(e^{2\pi} - e^{-2\pi})^2}\right)\left(1 + \frac{4ss}{(e^{3\pi} - e^{-3\pi})^2}\right)\cdots,\tag{33.112}$$

$$Q(\phi) = \left(1 - \frac{4ss}{(e^{\frac{\pi}{2}} - e^{-\frac{\pi}{2}})^2}\right)\left(1 - \frac{4ss}{(e^{\frac{3\pi}{2}} - e^{-\frac{3\pi}{2}})^2}\right)$$
$$\cdot\left(1 - \frac{4ss}{(e^{\frac{5\pi}{2}} - e^{-\frac{5\pi}{2}})^2}\right)\cdots.\tag{33.113}$$

From Abel, we may surmise that Gauss used the product formula for $\sin x$ to transform the products M and N into new products expressed in terms of the variable $s = \sin(\frac{\pi\phi}{\omega})$.

Gauss also gave the equations connecting M and N with P and Q:

$$M(\psi\omega) = e^{\frac{\pi}{2}\psi\psi}P(\psi\omega),\quad N(\psi\omega) = e^{\frac{\pi}{2}\psi\psi}Q(\psi\omega),\tag{33.114}$$

particular cases of Weierstrass's relations connecting his sigma function with the theta function. Note that when $\psi = 1, s = \sin(\pi\psi) = \sin\pi = 0$ and $Q(\omega) = 1$. Therefore, $N(\omega) = e^{\frac{\pi}{2}}$; thus, Gauss resolved the questions he raised the year before.

[29] ibid. vol. 3, pp. 415–416.

In the summer of 1798, Gauss discovered another important set of relations, the significance of which is better understood by observing that for $s = \sin \psi \pi$,

$$1 + \frac{4s^2}{(e^{n\pi} - e^{-n\pi})^2} = 1 + \frac{4s^2 e^{-2n\pi}}{(1 - e^{-2n\pi})^2}$$

$$= \frac{(1 - 2e^{-2n\pi} \cos 2\psi\pi + e^{-4n\pi})}{(1 - e^{-2n\pi})^2}$$

$$= \frac{(1 - e^{-2n\pi} e^{2i\psi\pi})(1 - e^{-2n\pi} e^{-2i\psi\pi})}{(1 - e^{-2n\pi})^2},$$

an equation that converts the previous product for P (33.112) to

$$P(\psi\omega) = \frac{\omega}{\pi} \sin \psi\pi \prod_{n=1}^{\infty} \frac{(1 - e^{-2n\pi} e^{2i\psi\pi})(1 - e^{-2n\pi} e^{-2i\psi\pi})}{(1 - e^{-2n\pi})^2}. \tag{33.115}$$

Similarly, the product for Q, as in (33.113), can be rewritten as

$$Q(\psi\omega) = \prod_{n=1}^{\infty} \frac{(1 + e^{-(2n-1)\pi} e^{2i\psi\pi})(1 + e^{-(2n-1)\pi} e^{-2i\psi\pi})}{(1 - e^{-(2n-1)\pi})^2}. \tag{33.116}$$

From these results, Gauss found the Fourier series expansion of $\text{sl}(\phi)$:[30]

$$\text{sl}(\psi\omega) = \frac{\pi}{\omega} \frac{4}{e^{\frac{\pi}{2}} + e^{-\frac{\pi}{2}}} \sin \psi\pi - \frac{\pi}{\omega} \frac{4}{e^{\frac{3\pi}{2}} + e^{-\frac{3\pi}{2}}} \sin 3\psi\pi + \cdots. \tag{33.117}$$

He also found Fourier series for $\log Q(\psi\omega)$, $\log P(\psi\omega)$, $\log \text{sl}(\psi\omega)$, etc. For example,

$$\log Q(\psi\omega) = -\frac{1}{2} \log 2 + \frac{\pi}{12} + \frac{2}{e^\pi - e^{-\pi}} \cos 2\psi\pi - \frac{1}{2} \frac{2}{e^{2\pi} - e^{-2\pi}} \cos 4\psi\pi + \cdots. \tag{33.118}$$

Note that products (33.115) and (33.116) are theta products revealing the form of the product representation for a general theta function.

It appears that Gauss had reached this point in his researches on the lemniscatic function by the end of summer 1798. And on September 28, he completed his studies at Göttingen and departed for Braunschweig. From a letter to his great friend Wolfgang Bolyai, we learn that Gauss was uncertain of his financial future. Note that Bolyai's son was the noted János Bolyai, discoverer of non-Euclidean geometry. Gauss's financial uncertainty remained until the end of the year when the Duke guaranteed him further support, suggesting that Gauss earn a doctoral degree in mathematics. Gauss accomplished this by submitting a thesis to the University of Helmstedt on the fundamental theorem of algebra, work he had completed a year earlier. He noted

[30] ibid. p. 417.

in his diary, "Proved by a valid method that equations have imaginary roots." In a later addendum to this diary entry, from October 1798, Gauss wrote that this method was published August 1799 as his dissertation; he received his degree in July 1799 on the recommendation of Johann Friedrich Pfaff (1765–1825). Gauss was then free to continue his mathematical researches with the Duke's financial assistance. This increased after Gauss turned down an offer from St. Petersburg in 1802 and continued until he was appointed director of the observatory at the University of Göttingen where he remained to the end of his life.

In spite of his financial insecurity during the fall of 1798, Gauss's creativity did not abate. In October 1798, Gauss noted, "New things in the field of analysis opened up to us, namely, investigation of a function etc." Gauss had earlier found the Fourier expansion of $\frac{P}{Q}$, but he was excited now to discover the Fourier expansions of the functions P and Q themselves:[31]

$$P(\psi\omega) = 2^{\frac{3}{4}}\sqrt{\frac{\pi}{\omega}}(e^{-\frac{\pi}{4}}\sin\psi\pi - e^{-\frac{9\pi}{4}}\sin 3\psi\pi + e^{-\frac{25\pi}{4}}\sin 5\psi\pi - \cdots) \quad (33.119)$$

and

$$Q(\psi\omega) = 2^{-\frac{1}{4}}\sqrt{\frac{\pi}{\omega}}(1 + 2e^{-\pi}\cos 2\psi\pi + 2e^{-4\pi}\cos 4\psi\pi + \cdots). \quad (33.120)$$

As consequences, he noted

$$1 - 2e^{-\pi} + 2e^{-4\pi} - 2e^{-9\pi} + \cdots = \sqrt{\frac{\omega}{\pi}}, \quad (33.121)$$

$$e^{-\frac{\pi}{4}} + e^{-\frac{9\pi}{4}} + e^{-\frac{25\pi}{4}} + \cdots = \frac{1}{2}\sqrt{\frac{\omega}{\pi}}, \quad (33.122)$$

$$\sqrt{\frac{\omega}{\pi}} = 0.9135791381561168214072425 9.$$

He found this value by computing $2e^{-\frac{\pi}{4}}$ to thirty-nine decimal places. Note that (33.121) in fact gives the period of the lemniscatic function as a theta series value.

By comparing the products for P and Q in (33.115) and (33.116) with the series for P and Q in (33.119) and (33.120), we see that by 1798 Gauss knew the triple product identity. In fact, to derive the series from the product, one requires not only the triple product identity but also an additional formula. Gauss could have derived this from what he already knew. First consider how the factor $e^{-\frac{\pi}{4}}$ arises in the series for P. The addition formula (33.107) implies Gauss's observation that

$$\text{sl}\left(\omega\psi + \frac{\omega}{2} + i\frac{\omega}{2}\right) = \frac{-i}{\text{sl}(\omega\psi)}. \quad (33.123)$$

---

[31] ibid. vol. 10₁, pp. 536–537.

Here we mention that Jacobi used a similar formula in his *Fundamenta Nova*. We also note that

$$e^{i\pi(\psi+\frac{1}{2}+\frac{i}{2})} = ie^{-\frac{\pi}{2}}e^{i\psi\pi}.$$

When these two relations are applied to the formulas (33.111), (33.115), and (33.116), the result after a simple calculation for $q = e^{-\pi}$ is that

$$\frac{4\pi e^{-\frac{\pi}{2}}}{\omega} \cdot \prod_{n=1}^{\infty}\left(\frac{1-q^{2n}}{1+q^{2n-1}}\right)^2 = \frac{\omega}{\pi}\prod_{n=1}^{\infty}\left(\frac{1+q^{2n-1}}{1-q^{2n}}\right)^2.$$

This simplifies to

$$\frac{2\pi}{\omega}e^{-\frac{\pi}{4}} = \prod_{n=1}^{\infty}\left(\frac{1+q^{2n-1}}{1-q^{2n}}\right)^2.$$

When the value $\frac{\pi}{\omega}$ from this equation is substituted in (33.115), we arrive at

$$P(\psi\omega) = 2e^{-\frac{\pi}{4}}\sin\psi\pi\prod_{n=1}^{\infty}\frac{(1-e^{-2n\pi}e^{2i\psi\pi})(1-e^{-2n\pi}e^{-2i\psi\pi})}{(1+e^{-(2n-1)\pi})^2}.$$

Here observe that the factor $e^{-\frac{\pi}{4}}$ is accounted for. After applying the triple product identity to this equation, we obtain the series

$$P(\psi\omega) = \frac{2e^{-\frac{\pi}{4}}\sum_{n=0}^{\infty}(-1)^n e^{-n(n+1)\pi}\sin(2n+1)\psi}{\prod_{n=1}^{\infty}(1-e^{-2n\pi})(1+e^{-(2n-1)\pi})}. \tag{33.124}$$

Since $n(n+1) + \frac{1}{4} = (2n+1)^2$, we see that this is Gauss's series except for the infinite product in the denominator. To eliminate this term, we apply Gauss's relations (33.109), (33.110), and (33.114) to get

$$P\left(\frac{\omega}{2}\right) = Q\left(\frac{\omega}{2}\right) = 2^{-\frac{1}{4}}.$$

Setting $\psi = \frac{1}{2}$ and using the last relation in (33.115) and (33.116), we find

$$2^{-\frac{1}{4}} = \frac{\pi}{\omega}\prod_{n=1}^{\infty}\frac{(1+e^{-2n\pi})^2}{(1+e^{-(2n-1)\pi})^2},$$

$$2^{-\frac{1}{4}} = \prod_{n=1}^{\infty}\frac{(1-e^{-(2n-1)\pi})^2}{(1+e^{-(2n-1)\pi})^2}.$$

From these two equations, a few lines of calculation yield

$$2^{\frac{1}{4}}\sqrt{\frac{\omega}{\pi}} = \prod_{n=1}^{\infty} \frac{1}{(1 - e^{-2n\pi})(1 + e^{-(2n-1)\pi})}.$$

Ultimately, we obtain Gauss's series (33.119) when we apply this equation to (33.124). Obtain series (33.120) similarly.

Although the triple product identity is difficult to prove ab initio, Gauss gave at least two documented proofs of this kind, thought to date from approximately 1808. However, it is possible that Gauss could have proved the triple product identity in 1798 by assuming Euler's pentagonal number theorem and in that case the proof would have been straightforward, the necessary technique having been established by Euler. Consider the product in the numerator of $Q(\psi\omega)$ in (33.116). For convenience, set $q = e^{-\pi}$ and $x = e^{2i\psi\pi}$, so that the product becomes

$$f(x) = \prod_{n=1}^{\infty}(1 + q^{2n-1}x)\left(1 + \frac{q^{2n-1}}{x}\right).$$

Then

$$f(q^2x) = \prod_{n=1}^{\infty}(1 + q^{2n+1}x)\left(1 + \frac{q^{2n-3}}{x}\right)$$

$$= \frac{1 + \frac{1}{qx}}{1 + qx}f(x) = \frac{1}{qx}f(x).$$

Now let $f(x) = \sum_{-\infty}^{\infty} a_n x^n$ so the previous equation becomes

$$\sum_{n=-\infty}^{\infty} a_n x^n = qx \sum_{n=-\infty}^{\infty} a_n q^{2n} x^n = \sum_{n=-\infty}^{\infty} a_n q^{2n+1} x^{n+1}.$$

By equating the coefficients of x^n, we see that

$$a_n(q) = a_{n-1}(q)q^{2n-1} = a_{n-2}(q)q^{2n-1+2n-3} = a_0(q)q^{n^2}.$$

Hence

$$\prod_{n=1}^{\infty}(1 + q^{2n-1}x)\left(1 + \frac{q^{2n-1}}{x}\right) = a_0(q) \sum_{n=-\infty}^{\infty} q^{n^2} x^n. \tag{33.125}$$

These simple calculations appear in Gauss's notes of 1799. Now to employ Euler's pentagonal number theorem

$$\prod_{n=1}^{\infty}(1 - p^n) = \sum_{n=-\infty}^{\infty} (-1)^n p^{\frac{n(3n+1)}{2}}, \tag{33.126}$$

we set $q = p^{\frac{3}{2}}$ and $x = -p^{\frac{1}{2}}$ in (33.125) to get

$$\prod_{n=1}^{\infty}(1 - p^{3n-1})(1 - p^{3n-2}) = a_0(q) \sum_{n=-\infty}^{\infty} (-1)^n p^{\frac{n(3n+1)}{2}}. \qquad (33.127)$$

Comparing (33.126) and (33.127), we arrive at

$$a_0(q) = \frac{1}{\prod_{n=1}^{\infty}(1 - p^{3n})} = \frac{1}{\prod_{n=1}^{\infty}(1 - q^{2n})}$$

and this proves the triple product identity.

After his remarkable work of 1798, Gauss's journal entry of May 30, 1799, connected the agM with the lemniscatic integral: "We have proved that the arithmetic-geometric mean of 1 and $\sqrt{2}$ is $\frac{\pi}{\omega}$ to 11 places, which thing being proved a new field will certainly be opened up."

We derive the agM of $\sqrt{2}$ and 1 from some previously mentioned formulas of Gauss. By taking $\psi = 1$ in (33.120), we get

$$2^{\frac{1}{2}}\sqrt{\frac{\omega}{\pi}} = 1 + 2e^{-\pi} + 2e^{-4\pi} + 2e^{-9\pi} + \cdots.$$

In terms of the functions A and B in (33.97) and (33.99), the previous formula and (33.121) imply that

$$A^2(e^{-\pi}) = 2^{\frac{1}{2}}\frac{\omega}{\pi} \quad \text{and} \quad B^2(e^{-\pi}) = \frac{\omega}{\pi}.$$

When this is combined with the fact that $A^2(x^2)$ and $B^2(x^2)$ are the arithmetic and geometric means, respectively, of $A^2(x)$ and $B^2(x)$, it follows that the agM of $\sqrt{2}$ and 1 is $\frac{\pi}{\omega}$. This means that, if in 1795 Gauss knew the connection of the two series $\sum x^{n^2}$ and $\sum (-1)^n x^{n^2}$ with the agM, then in 1799 he had a proof of the result quoted from the diary. Since he enjoyed numerical computation, he also verified this result to eleven places. Felix Klein and Ludwig Schlesinger, the editors of Gauss's mathematical diary, have remarked that the May 30 entry could represent a conclusion or a conjecture. It is very likely that it was a conclusion and that when he spoke of a new field, Gauss had in mind a generalization to any two real numbers a and b instead of the pair 1, $\sqrt{2}$. As we have seen, Lagrange had already found this generalization in 1785. Gauss published his work in an astronomical paper of 1818,[32] where he wrote that he discovered the result before he saw the paper of Lagrange.

It appears that up to 1798, Gauss did not investigate elliptic functions beyond the lemniscatic function, but with his discovery of the connection between the agM and the elliptic integral $\int_0^{\frac{\pi}{2}} \frac{d\theta}{\sqrt{a^2\cos^2\theta + b^2\sin^2\theta}}$, he began to explore the inversion of more

---

[32] ibid. vol. 3, pp. 331–355.

general elliptic integrals. This culminated with his journal entry of May 6, 1800: "We have led the theory of transcendental quantities:

$$\int \frac{dx}{\sqrt{(1 - \alpha xx)(1 - \beta xx)}}$$

to the summit of universality." Gauss's notes show that he used the agM to define two theta functions whose ratio he demonstrated to be the elliptic function inverting the integral. Gauss's approach to elliptic functions as ratios of theta functions was the same as the point of view taken by Jacobi in his 1836 Königsberg lectures. Gauss started with the integral[33]

$$\int \frac{du}{\sqrt{(1 + \mu\mu \sin^2 u)}} = \phi = \psi\omega$$

and set

$$\mu = \tan v, \quad \frac{\pi}{M\sqrt{(1 + \mu\mu)}} = \frac{\pi \cos v}{M \cos v} = \omega, \quad \frac{\pi}{\mu M \sqrt{(1 + \frac{1}{\mu\mu})}} = \frac{\pi \cos v}{M \sin v} = \omega'.$$

Note that Gauss denoted the agM of $\sqrt{(1 + \mu^2)}$ and 1 by

$$M(\sqrt{(1 + \mu^2)}, 1) \equiv M\sqrt{(1 + \mu\mu)},$$

so that

$$M \cos v = M(1, \cos v) \quad \text{and} \quad M \sin v = M(1, \sin v).$$

Note also that the Lagrange–Gauss agM theorem implied that

$$\frac{\omega}{2} = \int_0^{\frac{\pi}{2}} \frac{du}{\sqrt{(1 + \mu^2 \sin^2 u)}} = \int_0^1 \frac{dx}{\sqrt{((1 - x^2)(1 + \mu^2 x^2))}},$$

$$\frac{\omega'}{2} = \frac{1}{\mu} \int_0^{\frac{\pi}{2}} \frac{du}{\sqrt{(1 + \frac{1}{\mu^2} \sin^2 u)}} = \frac{1}{\mu} \int_0^1 \frac{dx}{\sqrt{(1 - x^2)(1 + \frac{1}{\mu^2}x^2)}}.$$

Gauss then wrote the elliptic function as

$$S(\psi\omega) = \frac{\pi}{\mu\omega} \left(\frac{4 \sin \psi\pi}{e^{\frac{\omega'}{2\omega}\pi} + e^{-\frac{\omega'}{2\omega}\pi}} - \frac{4 \sin 3\psi\pi}{e^{\frac{3\omega'}{2\omega}\pi} + e^{-\frac{3\omega'}{2\omega}\pi}} + \cdots \right)$$

$$= \frac{T(\psi\omega)}{W(\psi\omega)},$$

[33] ibid. vol. 10_1, pp. 194–198.

where the theta functions T and W were defined by the series

$$W(\psi\omega) = \sqrt{M} \cos v(1 + 2e^{-\frac{\omega'\pi}{\omega}} \cos 2\psi\pi + 2e^{-\frac{4\omega'\pi}{\omega}} \cos 4\psi\pi + \cdots),$$

$$T(\psi\omega) = \sqrt{\cot v}\sqrt{M} \cos v(2e^{-\frac{\omega'\pi}{4\omega}} \sin \psi\pi - 2e^{-\frac{9\omega'\pi}{4\omega}} \sin 3\psi\pi + \cdots).$$

To demonstrate that his elliptic function was actually the inversion of his original elliptic integral, he effectively showed that if

$$\int_0^u \frac{dt}{\sqrt{(1 + \mu\mu \sin^2 t)}} = \phi,$$

then $s(\phi) = \sin u$. Without giving details, he next wrote down the zeros of W and T and extended to this elliptic function all the results he had obtained for the lemniscatic function.

33.7 Exercises

(1) Show that if $t = \frac{\sqrt{1-u^2}}{\sqrt{1+u^2}}$, then

$$\int -dt\, \frac{\sqrt{1+t^2}}{\sqrt{1-t^2}} - \int du\, \frac{\sqrt{1+u^2}}{\sqrt{1-u^2}} = t^3 \frac{\sqrt{1-t^4}}{1+t^4} + u^3 \frac{\sqrt{1-u^4}}{1+u^4}.$$

See Fagnano (1911) vol. 2, p. 453.

(2) Show that the complete integral of $\dfrac{dx}{\sqrt{(f+gx^3)}} = \dfrac{dy}{\sqrt{(f+gy^3)}}$ is given by

$$f(x^2 + y^2) + \frac{g^2c^2x^2y^2}{4f} - gcxy(x+y) - 2fxy - gc^2(x+y) - 2fc = 0.$$

See Eu. I-20 p. 78.

(3) Let $(x, y) = \prod_{n=1}^\infty (1 + x^{2n-1}y)\left(1 + \frac{x^{2n-1}}{y}\right)$ and $[x] = \prod_{n=1}^\infty (1 - x^n)$. Show that

$$(x, \alpha y) \cdot \left(x, \frac{y}{\alpha}\right) = ((x^2, \alpha^2) \cdot (x^2, y^2) + x\alpha y(x^2, \alpha^2 x^2)(x^2, x^2 y^2)) \cdot \frac{[x^4]^2}{[x^2]^2}.$$

See Gauss (1863–1927) vol. 3, p. 458.

(4) Let

$$\prod_{n=1}^\infty (1 + x^{2n-1}y)\left(1 + \frac{x^{2n-1}}{y}\right) = Fx \sum_{m=-\infty}^\infty x^{m^2}(y^m + y^{-m}),$$

and let $[x]$ be defined as in the problem above. Show that

(a)

$$Fx = \frac{(1-x)^2(1-x^2)^2(1-x^3)^2\cdots}{1-2x+2x^4-2x^9+\cdots} = \frac{[x]^2}{[x^2]^2}\cdot\frac{1}{1-2x+2x^4-2x^9+\cdots},$$

(b)

$$Fx = \frac{(1+x^2)(1+x^6)(1+x^{10})\cdots}{1-2x^4+2x^{16}-2x^{36}+\cdots} = \frac{[x^4]^2}{[x^2][x^8]}\cdot\frac{1}{1-2x^4+2x^{16}-\cdots},$$

(c)

$$[x^2]Fx = [x^8]Fx^4 = [x^{32}]Fx^{16} = [x^{128}]Fx^{64} = \text{etc.} = 1,$$

(d)

$$1-2x+2x^4-\cdots = \frac{[x]^2}{[x^2]} = \frac{1-x}{1+x}\cdot\frac{1-x^2}{1+x^2}\cdot\frac{1-x^3}{1+x^3}\cdots,$$

(e)

$$1+2x+2x^4+\cdots = \frac{[x^2]^5}{[x]^2[x^4]^2} = \frac{1+x}{1-x}\cdot\frac{1-x^2}{1+x^2}\cdot\frac{1+x^3}{1-x^3}\cdots.$$

See Gauss (1863–1927) vol. 3, pp. 446–447. Observe that this was one of Gauss's proofs of the triple product identity.

(5) Set

$$Px = 1+2\sum_{n=1}^{\infty}x^{n^2}, \quad Qx = 1+2\sum_{n=1}^{\infty}(-1)^nx^{n^2}, \quad Rx = 2\sum_{n=1}^{\infty}x^{\frac{(2n-1)^2}{4}}.$$

Note that we would write $Px = P(x)$, etc. Show that

(a) $Rx = 2x^{\frac{1}{4}}\frac{[x^4]^2}{[x^2]}$,

(b) $Px\cdot Qx = (Qxx)^2$; $\quad Px\cdot Rx = \frac{(R\sqrt{x})^2}{2}$,

(c) $Px+Qx = 2P(x^4)$; $\quad Px-Qx = 2R(x^4)$; $\quad (Px)^2-(Qx)^2 = 2(Rxx)^2$,

(d) $Px+iQx = (1+i)Q(ix)$; $\quad Px-iQx = (1-i)P(ix)$,

(e) $(Px)^2+(Qx)^2 = 2(Pxx)^2$; $\quad (Px)^4-(Qx)^4 = (Rx)^4$,

(f) $(Px)^2$, $(Qx)^2$ have an arithmetic geometric mean that is always 1,

(g)

$$\int_0^{2\pi}\frac{d\theta}{\sqrt{((Px)^4\cos^2\theta+(Qx)^4\sin^2\theta)}} = 2\pi.$$

We note that Gauss wrote $\cos\theta^2$ for $\cos^2\theta$, etc. See Gauss (1863–1927) vol. 3, pp. 465–467.

(6) Show that

$$(1 + 2x + 2x^4 + 2x^9 + \cdots)^4 = 1 + \frac{8x}{1-x} + \frac{16xx}{1+xx} + \frac{24x^3}{1-x^3} + \frac{32x^4}{1+x^4} + \cdots$$

$$= 1 + \frac{8x}{(1-x)^2} + \frac{8xx}{(1+xx)^2} + \frac{8x^3}{(1-x^3)^2} + \frac{8x^4}{(1+x^4)^2} + \cdots .$$

See Gauss (1863–1927) vol. 3, p. 445.

(7) Show that

$$1 + \left(\frac{1}{2}\right)^3 + \left(\frac{1 \cdot 3}{2 \cdot 4}\right)^3 + \left(\frac{1 \cdot 3 \cdot 5}{2 \cdot 4 \cdot 6}\right)^3 + \cdots = 2 \left(\frac{\tilde{\omega}}{\pi}\right)^2,$$

where

$$\frac{\tilde{\omega}}{2} = \int_0^1 \frac{dz}{\sqrt{(1-z^4)}}.$$

See Gauss (1863–1927) vol. 3, p. 425.

(8) This exercise gives a proof of the transformation formula for a theta function, first published by Cauchy and Poisson. See Section 34.11. Gauss worked out the details given here in an unpublished paper.

(a) Expand $T = \sum_{k=-\infty}^{\infty} e^{-\alpha(k+\omega)^2}$ as a Fourier series

$$T = A_0 + 2 \sum_{n=1}^{\infty} A_n \cos n\omega P, \quad \text{where} \quad A_n = \int_0^1 T \cos n\omega P \, d\omega \quad \text{and} \quad P = 2\pi.$$

(b) Show that

$$A_n = \int_{-\infty}^{\infty} e^{-\alpha\omega\omega} \cos n\omega P \, d\omega = e^{-\frac{nn\pi i}{\alpha}} \sqrt{\frac{\pi}{\alpha}}.$$

(c) Conclude that

$$\sum_{k=-\infty}^{\infty} e^{-\alpha(k+\omega)^2} = \sqrt{\frac{\pi}{\alpha}} \cdot e^{-\alpha\omega\omega} \cdot \sum_{k=-\infty}^{\infty} e^{-\frac{\pi\pi}{\alpha}(k+\frac{\alpha\omega i}{\pi})^2}.$$

See Gauss (1863–1927) vol. 3, pp. 436–437.

33.8 Notes on the Literature

Fagnano (1911) is a reprint of his *Produzioni Matematiche*. See pp. 293–297, 304–313 of vol. 2 for material on the lemniscate. This volume contains several more articles by Fagnano on the integral calculus and on the lemniscatic calculus.

Volumes 20–21 of series 1 of Euler's *Opera Omnia* contain the papers of Euler providing the foundation for the theory of algebraic functions and their integrals.

Of course, Euler dealt only with the integrals arising from the algebraic equation $y^2 = p(x)$, where $p(x)$ was of degree 4.

See Landen (1771) and (1775) for his original contributions to elliptic integrals. Our exposition is based largely on Cayley (1895), pp. 327–330. Cayley's book contains a good account of Jacobi's *Fundamenta Nova*, elaborating the transformation theory of elliptic integrals of Landen, Legendre, and Gauss.

Gauss's extensive work on the agM and his work on elliptic functions in general can be found in Gauss (1863–1927), vols. 3 and 10. Pieper (1998) suggests that Gauss discovered the triple product identity between April and June 1800. He has also pointed out that this identity can be proved easily by applying Euler's pentagonal number theorem. Berggren, Borwein, and Borwein (1997) contains a number of interesting papers on the agM and its application to the computation of π.

There are several interesting historical accounts of the theory of elliptic functions and integrals. Weil (1983) deals with Euler's work on this topic and its relation to Diophantine equations. Varadarajan (2006) gives a brief analysis of Euler in terms of Riemann surfaces of genus one. Watson (1933) gives a very entertaining and detailed mathematical exposition of Fagnano, Landen, and Ivory's contributions to elliptic integrals. Watson also wrote, without giving a reference, that Jacobi called December 23, 1751, the birthday of elliptic functions. Later, André Weil observed, "According to Jacobi, the theory of elliptic functions was born between the twenty-third of December 1751, and the twenty-seventh of January 1752." See Weil (1983) p. 1.

Ozhigova (2007), first published in 1988, (also reprinted in Bogolyubov, Mikhaïlov, and Yushkevich (2007)) refers to Jacobi 's 1847 letter to Fuss, saying on p. 55 that Euler's study of Fagnano inaugurated the subject of elliptic functions. We observe that in his letter, of October 24, 1847, to Euler's great-grandson P. H. Fuss, Jacobi strongly recommended the publication of Euler's papers, arguing that they were very important to the advancement of science. As further support for his point, Jacobi mentioned that by reading of the minutes of the Berlin Academy, he discovered a critical date in the history of mathematics: when the Academy assigned Euler the task of refereeing Count Fagnano's mathematical work. Jacobi then stated that Euler's evaluation of these papers served to found the theory of elliptic functions. See Stäckel and Ahrens (1908) p. 23.

Cox (1984) contains a fascinating and enlightening resumé of Gauss's remarkable work on the agM of two complex numbers. He shows that Gauss may have had significant ideas on the modular group and some of its subgroups and their fundamental domains. The reader may wish to read this paper before reading Gauss's somewhat fragmentary original papers on the topic. Mittag-Leffler (1923) and Almkvist and Berndt (1988) are both interesting papers. The first is an insightful account of work on elliptic functions and integrals from 1718 to 1870; the second focuses on topics related to the quadratic transformation and the agM. Mittag-Leffler (1923) is in fact an English translation of an 1876 paper published in Swedish.

The first chapter of Siegel (1969) vol. 1 contains perceptive remarks concerning Fagnano and Euler on the addition formula. Bühler (1981) and Dunnington (2004) are well-written biographies of Gauss. Bühler has more mathematical exposition, but the value of Dunnington is enhanced by the inclusion of an English translation with commentary by J. J. Gray of Gauss's diary; we have made use of this translation in the text.

34

Elliptic Functions: Nineteenth Century

34.1 Preliminary Remarks

The eighteenth century saw two major new results in elliptic functions: the addition formula of Euler and the second-order transformation of Landen and Lagrange. Gauss discovered yet another quadratic transformation, in connection with his proof that the agM of two positive numbers could be represented by an elliptic integral. These transformations changed the parameters in the elliptic integrals, without alterinsing their basic form. In fact, Gauss went well beyond this elementary transformation theory, and before the end of the eighteenth century he had greatly refined elliptic function theory. He did not publish his work; it was rediscovered by Abel and Jacobi in the 1820s.

Adrien-Marie Legendre was the main contributor to elliptic integrals in the period between Lagrange and Abel. In his first major work, *Exercices de calcul intégral* of 1811–1817, he reduced any elliptic integral $\int A(x) \frac{dx}{\sqrt{R(x)}}$, where $R(x)$ was a fourth-degree polynomial in x and $A(x)$ was a rational function in x and $R(x)$, to integrals of three kinds:[1]

$$F(k,x) = \int_0^x \frac{dt}{\sqrt{(1-t^2)(1-k^2t^2)}}, \tag{34.1}$$

$$E(k,x) = \int_0^x \frac{\sqrt{1-k^2t^2}}{\sqrt{1-t^2}} \, dt, \tag{34.2}$$

$$\Pi(n,k,x) = \int_0^x \frac{dt}{(1+nx^2)\sqrt{(1-x^2)(1-k^2x^2)}}. \tag{34.3}$$

Legendre's second major work was the three-volume *Traité des fonctions elliptiques* of 1825–1828;[2] the first volume presented the received eighteenth-century theory of elliptic integrals with some improvements and additions; the second volume gave

---

[1] Legendre (1811–1817) vol. 1, p. 19.
[2] Legendre (1825–1828).

extensive and long-useful numerical tables, constructed by Legendre's own methods. At the age of seventy-five, upon learning of the more advanced results of Abel and Jacobi, Legendre did his best to give a flattering exposition of their work, and this was the topic of the third volume.

Legendre (1752–1833) studied at the College Mazarin in Paris, where he received an excellent education. Apparently, Legendre wished to be remembered for his works alone and not much is known of his personal life. In fact, it was only recently discovered that the portrait by which he had been identified for a century was actually that of an unrelated politician named Louis Legendre. Thus, the only portrait now available is a sketched caricature made by Julien-Léopold Boilly. Legendre's research on his two favorite subjects, number theory and elliptic functions, was immediately superseded after his books appeared. Nevertheless, Legendre's name became permanently associated with several mathematical objects, including the Legendre polynomials, the Legendre symbol, and the Legendre differential equation. Though he studied elliptic functions for almost forty years, Legendre apparently never considered inverting the integral. Abel was the first to publish this idea, inaugurating a great advance in this topic.

The mathematical career of Niels Henrik Abel began in 1821 with his attempt to solve the general quintic equation. His mathematics professors at the University of Christiania could find no errors in Abel's solution and communicated it to Ferdinand Degen in Copenhagen. Though Degen could not find the mistake, he made two suggestions: that Abel apply his method to specific examples, since that could reveal hidden errors; and that he abandon the sterile subject of algebraic equations to exercise his brilliance in the more fruitful subject of elliptic integrals. Degen's advice led Abel to find the mistake in his work and eventually to prove the impossibility of solving the quintic in radicals. He also began to work on elliptic integrals, and it is fairly certain that by 1823 he had inverted the elliptic integral to rediscover elliptic functions.[3] We recall that Gauss had already done this without publishing it. Moreover, the problem of the division of elliptic functions carried Abel deeper into the theory of algebraic equations and ultimately to his famous theorem on solvable equations. In this manner, Abel found an extremely productive connection between elliptic functions and the theory of algebraic equations.

Abel's first work on elliptic functions, the first part of "Recherches sur les fonctions elliptiques," appeared in *Crelle's Journal* in September 1827.[4] In this paper he defined the elliptic function $\phi\alpha = x$ when

$$\alpha = \int_0^x \frac{dt}{\sqrt{(1 - c^2 t^2)(1 + e^2 t^2)}}.$$ (34.4)

He showed that $\phi\alpha$ was a meromorphic function with two independent periods, $2w$ and $2i\tilde{w}$, given by

---

[3] Abel (1965) vol. 2, p. 254.
[4] For an English translation of this paper, see Abel (2007) pp. 145–245.

$$\omega = 2 \int_0^{\frac{1}{c}} \frac{dx}{\sqrt{(1 - c^2 x^2)(1 + e^2 x^2)}}, \qquad \tilde{\omega} = 2 \int_0^{\frac{1}{e}} \frac{dx}{\sqrt{(1 - e^2 x^2)(1 + c^2 x^2)}}. \tag{34.5}$$

He then gave a new proof for the addition formula for elliptic functions and used it to express $\phi(n\alpha)$, where n was an integer, as a rational function of $\phi\alpha$, $f\alpha = \sqrt{(1 - c^2\phi^2\alpha)}$, and $F\alpha = \sqrt{(1 + e^2\phi^2\alpha)}$. This was analogous to expressing $\sin(nx)$ as a polynomial in $\sin x$ and $\cos x = \sqrt{(1 - \sin^2 x)}$. With n an odd integer, he noted that the rational function could be written as $\frac{xp(x)}{q(x)}$, where p and q were polynomials in $x = \phi\alpha$ and of degree $n^2 - 1$. Abel next showed that the solution of the equation $p(x) = 0$, whose roots were

$$\phi\left(\frac{a\omega + ib\tilde{\omega}}{n}\right),$$

for integers a and b, depended on an equation of degree $n + 1$ which could be solved algebraically only in particular cases.

In the second part of his "Recherches," Abel applied his theory to the division of the lemniscate.[5] Recall that Fagnano divided the full arc of the lemniscate in the first quadrant into two, three, and five equal parts. In his *Disquisitiones*, without reference to Fagnano, Gauss stated that the theory he had constructed for the division of a circle into n equal parts could be extended to the lemniscate, but he never gave details on this.[6] In the case of the circle, Gauss was able to simplify the problem, so that he had only to prove that $\cos\left(\frac{2\pi}{n}\right)$ could be expressed in terms of square roots, when n was a prime of the form $2^k + 1$. He did this by showing that $\cos\left(\frac{2\pi}{n}\right)$ satisfied an appropriate algebraic equation of degree $\frac{n-1}{2} = 2^{k-1}$.

Thus, in order to extend Gauss's theory to the lemniscate, Abel had to find the division point by working with $\mathrm{sl}\left(\frac{\omega}{n}\right)$, where 2ω was the period of the lemniscatic function. However, note that $\mathrm{sl}\left(\frac{\omega}{n}\right)$ satisfies an equation of degree $n^2 - 1$, and this cannot be a power of 2, except when $n = 3$. This drawback would apparently suggest that Gauss's theory for the circle could not be extended to the case of the lemniscate. But Abel found a resolution to this roadblock by discovering complex multiplication of elliptic functions.[7]

The primes expressible as $2^k + 1$, except for 3, take the form $4m + 1$ and can be written as sums of two squares, $4m + 1 = a^2 + b^2$, where $a + b$ is odd. Abel used this fact to solve the problem of dividing the lemniscate into $n = 4m + 1$ parts. He showed that the complex number $\mathrm{sl}\left(\frac{\omega}{a+ib}\right)$ was the solution to an equation of degree $n - 1 = 4m$ with coefficients of the form $c + id$, where c and d were rational numbers. To prove this, he used the addition formula for $\mathrm{sl}\,\alpha$ to first prove that $\mathrm{sl}((a + ib)\alpha)$ could be expressed as a rational function $\frac{xp(x)}{q(x)}$, $x = \mathrm{sl}\,\alpha$, where $p(x)$ and $q(x)$ were polynomials of degree $n - 1$. Next, he employed the Lagrange resolvent, just as Gauss had done for the cylotomic case, to show that $\mathrm{sl}\left(\frac{w}{a+ib}\right)$ could be evaluated by means of

[5] Abel (2007) pp. 245–283.
[6] Gauss (1965) p. 407.
[7] Abel (2007) pp. 245–255.

only square roots, providing n was of the form $2^k + 1$. Abel pointed out that the value of $\mathrm{sl}(\frac{\omega}{n})$ could then be found by means of square roots and so this value was constructible by straight edge and compass. The second part of the "Recherches" also dealt with the transformation of elliptic functions, but on this topic Jacobi had published earlier than Abel.

Carl Gustav Jacob Jacobi (1804–1851) studied at the University of Berlin, though he largely preferred to study on his own, especially Euler's works. His interest in elliptic integrals was aroused by the quadratic transformations in Legendre's *Exercices de calcul intégral*. In June 1827, Jacobi communicated a short note to the *Astronomische Nachrichten* giving two cubic transformations and two fifth-order transformations of elliptic integrals.

In fact, in 1825 Legendre had already discovered this cubic transformation, though it was published in the second volume of his *Traité des fonctions elliptiques*. Heinrich Schumacher, editor of the *Astronomiche Nachrichtèn*, noted that Jacobi did not refer to Legendre's book, though this was not surprising, since Jacobi had not seen Legendre's work at the time. In any case, Jacobi also had the new result on the fifth-order transformation. Then in August 1827, Jacobi communicated to Schumacher a general odd-order transformation, allowing the division of an elliptic integral into an arbitrary odd number of parts. Unfortunately, Jacobi included no proof, and so Schumacher consulted his friend Gauss about the correctness of the results. Gauss replied that the results were correct but asked Schumacher not to communicate with him further on this topic. Gauss himself was planning to publish his twenty-five-year-old results on elliptic functions and wished to avoid priority disputes.

Schumacher published Jacobi's notes but urged him to supply the proofs as soon as possible. Legendre saw the paper and was eager to see the proofs. Jacobi told Legendre that he had only guessed the theorem for odd-order transformations; in November 1827 he was able to derive a proof by means of the inversion of elliptic integrals. Meanwhile, in September, the first part of Abel's "Recherches" had appeared. It is curious that Jacobi did not refer to Abel's paper and avoided the question of whether he had borrowed any idea from Abel. In 1828, the second part of the "Recherches" was published, in which Abel added an appendix explaining how his own results could prove Jacobi's theorem. Jacobi's proof was published after Abel had written the second part of his paper. Christoffer Hansteen reported that when Abel saw Jacobi's inversion of the elliptic integral without reference to him, he was visibly shocked. In fact, Abel wrote in a letter to Bernt Holmboe that he published his "Transformations des fonctions elliptiques" in order to supercede Jacobi; he called the paper his "knockout" of Jacobi. In 1828, Gauss wrote Schumacher that Abel's "Recherches" had relieved him of the duty of writing up a third of his investigations on elliptic functions. The other two thirds consisted of the arithmetic-geometric mean and the elliptic modular and theta functions.

Abel's early and tragic death in 1829 cut short the rivalry between Abel and Jacobi. In that same year, Jacobi published the results of two years' labor on elliptic functions, in his *Fundamenta Nova*.[8] This work presented an extensive

---

[8] See Jacobi (1969) vol. 1, pp. 49–239.

development of transformation theory and applied it to the derivation of series and product representations of elliptic functions, their moduli, and periods. The problem of converting products into series led Jacobi to the discovery of the triple product identity, though Gauss had anticipated him. In fact, these series and products were theta functions; thus, Jacobi had discovered that elliptic functions could be expressed as quotients of theta functions.

Jacobi earned his doctoral degree from Berlin with a thesis on partial fractions in 1825 and a year later he took a position at Königsberg. Because of his sharp wit and tongue, Jacobi might have faced obstacles to advancement. However, he gained quick recognition from French mathematicians and Legendre in particular, who had presented Jacobi's work to the French Academy in 1827. Finally, with the publication of the *Fundamenta Nova*, Jacobi became known as one of the most outstanding mathematicians in Europe. In a paper of 1834,[9] Jacobi proved two important theorems on functions of one variable: First, he showed that such a function could not have two fundamental periods whose ratio was real; secondly, he showed that such a function could have only two fundamental periods whose ratio was complex. He argued that the functions would otherwise have arbitrarily small periods, a condition he assumed to be absurd. Jacobi had not yet conceived of an analytic function, but when Weierstrass and Cauchy later confirmed his assumption, the proof relied on the fact that the zeros of analytic functions were isolated.

Using a suggestion from Hermite that he use Fourier series, Joseph Liouville (1809–1882) in 1844 reproved Jacobi's first theorem.[10] This work initiated Liouville's definitive theory of elliptic functions. According to Weierstrass, this work was very important, though Liouville published little of it; Weierstrass also criticized Briot and Bouquet for publishing Liouville's ideas without giving him sufficient credit. Liouville's innovation was to define elliptic functions as doubly-periodic functions, rather than as inverses of integrals. He showed that doubly-periodic functions could not be bounded and, in fact, had to have at least two simple poles. Except for two short notes, he did not publish these results, but in 1847 he began a series of lectures on this topic. These lectures were first published in 1880[11] by the longtime editor of *Crelle's Journal*, Carl Borchardt (1817–1880), who in 1847 had attended the lectures. Reportedly, Borchardt also showed the notes to Jacobi and informed Liouville that Jacobi was extremely impressed. A typeset manuscript of these lectures, said by Weierstrass to have been taken from the notes of Borchardt, was found among Dirichlet's papers after his death in 1859. Apparently, Liouville had intended to publish the notes in his own journal, but had perhaps asked his friend Dirichlet to review the proofs. Why did Liouville not see to it that the proofs were published? This sequence of events remains a mystery, even after Jesper Lützen's comprehensive and detailed book on Liouville, published in 1990. It is interesting to note that the book by Liouville's students, Briot and Bouquet, started with Liouville's approach,

[9] Jacobi (1969) vol. 2, pp. 23–50.
[10] See Lützen (1990) pp. 535–540.
[11] Liouville (1880).

but proved the results by using the complex analytic methods of Cauchy and Laurent. Many standard textbooks of today make use of these methods.

At about the same time as Liouville, Gotthold Eisenstein (1823–1852) provided yet another important approach to elliptic functions. Eisenstein was dissatisfied with the inversion of the elliptic integral in Abel and Jacobi. He observed that, since integrals defined single-valued functions, the periodicity of their inverses must be problematic. Eisenstein was a number theorist of extraordinary vision. He viewed the theory of periodic functions as inseparable from number theory. In fact, in 1847 he published a 120-page treatise in *Crelle's Journal*, developing a new basis for elliptic function theory, using double series and double products; this approach was well suited for number theoretic applications.[12] The Weierstrass elliptic function $\wp(z)$ first appeared in this work. Although this paper was soon republished in a collection of Eisenstein's papers, with a foreword by no less a personage than Gauss, it unfortunately did not receive recognition in the nineteenth century. In the preface to his 1975 book *Elliptic Functions According to Eisenstein and Kronecker*, André Weil brought this paper to the attention of the mathematical community. He wrote,[13]

> It is not merely out of an antiquarian interest that the attempt will be made here to resurrect them [Eisenstein's ideas]. Not only do they provide the best introduction to the work of Hecke; but we hope to show that they can be applied quite profitably to some current problems, particularly if they are used in conjunction with Kronecker's late work which is their natural continuation.

Weil's treatment of Eisenstein is thorough and insightful as well as easily available. Thus, the reader may profitably consult Weil for Eisenstein's 1847 work.[14]

Eisenstein's objection to the inversion of the elliptic integral was addressed by Cauchy in the 1840s and then by Riemann in the 1850s.[15] Cauchy had been vigorously developing the theory of complex integration since 1814; this work provided him with the tools necessary to address this problem. Riemann was familiar with Cauchy's work, but he added his original idea of a Riemann surface to study Abelian, and in particular elliptic, functions.

34.2   Abel: Elliptic Functions

Abel's great paper of 1827, "Recherches sur les fonctions elliptiques," was published in two parts in volumes two and three of *Crelle's Journal*. In this paper, Abel defined an elliptic function as the inverse of the elliptic integral[16]

$$\alpha = \int_0^x \frac{dx}{\sqrt{(1 - c^2 x^2)(1 + e^2 x^2)}}, \quad 0 \le x \le \frac{1}{c}. \tag{34.6}$$

---

[12] Eisenstein (1975) vol. 1, pp. 357–478.
[13] Weil (1976) p. 4.
[14] Also see Roy (2017) chapter 4.
[15] For Riemann's work on elliptic functions, see his lectures: Riemann (1899).
[16] Abel (2007) pp. 145–156.

He expressed x as a function of α and set $x = \phi\alpha$. He noted that α was positive and increasing as x moved from 0 to $\frac{1}{c}$, and set

$$\frac{\omega}{2} = \int_0^{\frac{1}{c}} \frac{dx}{\sqrt{(1 - c^2x^2)(1 + e^2x^2)}}. \tag{34.7}$$

Thus, $\phi\alpha$ was positive and increasing in $0 \le \alpha \le \frac{\omega}{2}$ and

$$\phi(0) = 0, \quad \phi\left(\frac{\omega}{2}\right) = \frac{1}{c}. $$

Moreover, since α changed sign when x was changed to $-x$, he had $\phi(-\alpha) = -\phi(\alpha)$. Abel then formally changed x to ix without a rigorous justification, just as Euler, Laplace, and Poisson had done earlier. Now in an 1814 paper published in 1827[17] and in papers published as early as 1825,[18] Cauchy discussed functions of complex variables in a more systematic manner. Abel could have employed Cauchy's ideas to give a more rigorous foundation of his theory of elliptic functions. It is possible that Abel was not aware of this aspect of Cauchy's work. In any case, with the above change of variables, Abel set

$$xi = \phi(\beta i), \quad \text{where} \quad \beta = \int_0^x \frac{dx}{\sqrt{(1 + c^2x^2)(1 - e^2x^2)}}, \tag{34.8}$$

and observed that β was real and positive for $0 \le x \le \frac{1}{e}$. He then set

$$\frac{\tilde{\omega}}{2} = \int_0^{\frac{1}{e}} \frac{dx}{\sqrt{(1 - e^2x^2)(1 + c^2x^2)}}, \tag{34.9}$$

so that $-i\phi(\beta i)$ was positive for $0 \le \beta \le \frac{\tilde{\omega}}{2}$; he also had

$$\phi\left(\frac{\tilde{\omega}i}{2}\right) = \frac{1}{e}. \tag{34.10}$$

Abel then defined two auxiliary functions

$$f\alpha = \sqrt{1 - c^2\phi^2\alpha}, \tag{34.11}$$

$$F\alpha = \sqrt{1 + e^2\phi^2\alpha}, \tag{34.12}$$

and noted that when c and e were interchanged, $f(\alpha i)$ and $F(\alpha i)$ were transformed into $F(\alpha)$ and $f(\alpha)$, respectively.

---

[17] Cauchy (1882–1974) Ser. 1, vol. 1, pp. 329–506.
[18] ibid. Ser. 2, vol. 15, pp. 41–89.

At this point, Abel observed that $\phi(\alpha)$ was already defined for $-\frac{\omega}{2} \leq \alpha \leq \frac{\omega}{2}$, and $\phi(\beta i)$ for $-\frac{\tilde{\omega}}{2} \leq \beta \leq \frac{\tilde{\omega}}{2}$; next he wished to define ϕ for all complex numbers. In order to achieve this, Abel employed the addition formula for ϕ:

$$\phi(\alpha + \beta) = \frac{\phi\alpha \cdot f\beta \cdot F\beta + \phi\beta \cdot f\alpha \cdot F\alpha}{1 + e^2 c^2 \phi^2 \alpha \cdot \phi^2 \beta}. \tag{34.13}$$

He also stated the addition formulas for the auxiliary functions $f\alpha$ and $F\alpha$, remarking that these formulas could be deduced from the results in Legendre's *Exercices* but he wanted to give an alternative derivation. He first deduced the easily proved formulas for the derivatives:

$$\phi'\alpha = f\alpha \cdot F\alpha, \quad f'\alpha = -c^2 \phi\alpha \cdot F\alpha, \quad F'\alpha = e^2 \phi\alpha \cdot f\alpha.$$

Abel then let r designate the right-hand side of (34.13) and showed that

$$\frac{dr}{d\alpha} = \frac{(1 - e^2 c^2 \phi^2 \alpha \phi^2 \beta)[(e^2 - c^2)\phi\alpha\phi\beta + f\alpha f\beta F\alpha F\beta] - 2e^2 c^2 \phi\alpha\phi\beta(\phi^2\alpha + \phi^2\beta)}{(1 + e^2 c^2 \phi^2 \alpha \phi^2 \beta)^2}.$$

By symmetry in α and β, Abel concluded that

$$\frac{dr}{d\alpha} = \frac{dr}{d\beta}.$$

He observed that this partial differential equation implied that $r = \psi(\alpha + \beta)$ for some function ψ. Moreover, since $\phi(0) = 0$, $f(0) = 1$, $F(0) = 1$, he set $\beta = 0$ in the expression for r on the right-hand side of (34.13) and found that $r = \phi\alpha$. But $r = \psi\alpha$ when $\beta = 0$. So he had $\phi\alpha = \psi\alpha$ or $\phi = \psi$. This proved the addition formula.

Abel deduced the periodicity of ϕ from the addition formula. He first set $\beta = \pm\frac{\omega}{2}$ and $\beta = \pm\frac{\tilde{\omega}i}{2}$ in (34.13). Observing that $f(\pm\frac{\omega}{2}) = 0$, and $F(\pm\frac{\tilde{\omega}i}{2}) = 0$, he then obtained the formulas

$$\phi\left(\alpha \pm \frac{\omega}{2}\right) = \pm\phi\left(\frac{\omega}{2}\right)\frac{f\alpha}{F\alpha} = \pm\frac{f\alpha}{cF\alpha},$$

$$\phi\left(\alpha \pm \frac{\tilde{\omega}}{2}\right) = \pm\phi\left(\frac{\tilde{\omega}i}{2}\right)\frac{F\alpha}{f\alpha} = \pm\frac{iF\alpha}{ef\alpha}.$$

These results implied that

$$\phi\left(\frac{\omega}{2} + \alpha\right) = \phi\left(\frac{\omega}{2} - \alpha\right), \tag{34.14}$$

$$\phi\left(\frac{\tilde{\omega}}{2}i + \alpha\right) = \phi\left(\frac{\tilde{\omega}}{2}i - \alpha\right), \tag{34.15}$$

$$\phi\left(\alpha \pm \frac{\omega}{2}\right)\phi\left(\alpha + \frac{\tilde{\omega}}{2}i\right) = \pm\frac{i}{ce}. \tag{34.16}$$

Replacing α by $\alpha + \frac{\omega}{2}$ in (34.14), and α by $\alpha + \frac{\tilde{\omega}i}{2}$ in (34.15), Abel found

$$\phi(\alpha + \omega) = \phi(-\alpha) = -\phi\alpha, \tag{34.17}$$

$$\phi(\alpha + \tilde{\omega}i) = -\phi\alpha. \tag{34.18}$$

By means of these formulas, he defined $\phi\alpha$ and $\phi(\alpha i)$ for all real α and then by the addition formula (34.13) he obtained $\phi(\alpha + \beta i)$ for any complex value $\alpha + \beta i$. Moreover, from (34.17) and (34.18), it followed that ϕ was doubly-periodic with periods 2ω and $2\tilde{\omega}i$:

$$\phi(2\omega + \alpha) = -\phi(\omega + \alpha) = \phi\alpha,$$
$$\phi(2\tilde{\omega}i + \alpha) = -\phi(\tilde{\omega}i + \alpha) = \phi\alpha.$$

Abel also determined the zeros and poles of ϕ. For example, from (34.16) he obtained

$$\phi\left(\frac{\omega}{2} + \frac{\tilde{\omega}i}{2}\right) = \frac{1}{0}.$$

Then by (34.17) and (34.18),

$$\phi\left[\left(m + \frac{1}{2}\right)\omega + \left(n + \frac{1}{2}\right)\tilde{\omega}i\right] = \frac{1}{0},$$

when m and n were integers. Then with a little more work, Abel showed that $(m + \frac{1}{2})\omega + (n + \frac{1}{2})\tilde{\omega}i$ were all the poles of ϕ. Similarly, he showed that $m\omega + n\tilde{\omega}i$ were all the zeros of ϕ.

34.3 Abel: Infinite Products

Recall that one way of deriving the infinite product for $\sin x$ is to express $\sin(2n+1)x$ by means of the addition theorem as a polynomial of degree $2n + 1$ in $\sin x$, factorize this polynomial, and then take the limit as n tends to infinity. Abel applied a similar procedure to obtain the product for ϕx. Abel deduced from the addition formula that for a positive integer n,

$$\phi(n + 1)\beta = -\phi(n - 1)\beta + \frac{2\phi(n\beta)f\beta \cdot F\beta}{1 + c^2 e^2 \phi^2(n\beta)\phi^2\beta}.$$

After some further calculation, he proved by induction that

$$\phi(2n\beta) = \phi\beta.f\beta.F\beta.T, \qquad \phi(2n + 1)\beta = \phi\beta.T_1,$$

where T and T_1 were rational functions of $(\phi\beta)^2$. He then wrote

$$\phi(2n + 1)\beta = \frac{P_{2n+1}}{Q_{2n+1}}, \tag{34.19}$$

where P_{2n+1} and Q_{2n+1} were polynomials of degree $(2n + 1)^2$ and $4n(n + 1)$, respectively. He noted that the roots of $P_{2n+1} = 0$ were clearly given by

$$x = (-1)^{m+\mu}\phi\left(\beta + \frac{m}{2n + 1}\omega + \frac{\mu}{2n + 1}\tilde{\omega}i\right), \tag{34.20}$$

for $-n \le m, \mu \le n$; by setting $\beta = \frac{\alpha}{2n+1}$, the roots were

$$x = (-1)^{m+\mu}\phi\left(\frac{\alpha}{2n + 1} + \frac{m\omega + \mu\tilde{\omega}i}{2n + 1}\right).$$

Abel next expressed $\phi(2n + 1)\beta$ as a sum and as a product of terms of the form (34.20). His method was similar to Euler's derivation in the *Introductio in Analysin Infinitorum*, where Euler expressed $\sin(2n + 1)x$ as a product of terms of the form $\sin\left(x + \frac{m\pi}{2n+1}\right)$. See Section 15.4 in this connection. Abel wrote

$$P_{2n+1} = Ax^{(2n+1)^2} + \cdots + Bx,$$
$$Q_{2n+1} = Cx^{(2n+1)^2-1} + \cdots + D,$$

so that by (34.19), he had

$$(Ax^{(2n+1)^2} + \cdots + Bx) = \phi(2n + 1)\beta.(Cx^{(2n+1)^2-1} + \cdots + D).$$

He observed that the highest-power term had coefficient A, the second highest term had coefficient $-\phi(2n + 1)\beta.C$, and the last term was $-\phi(2n + 1)\beta.D$. Then, since the roots of the equation were given by (34.20), the sum of the roots could be obtained from the coefficient of the second highest term and the product of the roots from the last term. Thus, he had the equations

$$\phi(2n + 1)\beta = \frac{A}{C}\sum_{m=-n}^{n}\sum_{\mu=-n}^{n}(-1)^{m+\mu}\phi\left(\beta + \frac{m\omega + \mu\tilde{\omega}i}{2n + 1}\right) \tag{34.21}$$

$$= \frac{A}{D}\prod_{m=-n}^{n}\prod_{\mu=-n}^{n}\phi\left(\beta + \frac{m\omega + \mu\tilde{\omega}i}{2n + 1}\right). \tag{34.22}$$

Abel set $\beta = \frac{\omega}{2} + \frac{\tilde{\omega}}{2}i + \alpha$, and let $\alpha \to 0$ to determine

$$\frac{A}{C} = \frac{1}{2n + 1}. \tag{34.23}$$

He then let $\beta \to 0$, to obtain

$$(2n + 1) = \frac{A}{D}\prod_{m=1}^{n}\phi^2\left(\frac{m\omega}{2n + 1}\right)\prod_{\mu=1}^{n}\phi^2\left(\frac{\mu\tilde{\omega}i}{2n + 1}\right)$$

$$\times \prod_{m=1}^{n}\prod_{\mu=1}^{n}\phi^2\left(\frac{m\omega + \mu\tilde{\omega}i}{2n + 1}\right)\phi^2\left(\frac{m\omega - \mu\tilde{\omega}i}{2n + 1}\right). \tag{34.24}$$

This gave him an expression for $\frac{A}{D}$ and he substituted it back in (34.22). To simplify the resulting product, he applied a consequence of the addition formula:

$$\frac{\phi(\beta+\alpha)\phi(\beta-\alpha)}{\phi^2\alpha} = -\frac{1-\frac{\phi^2\beta}{\phi^2\alpha}}{1-\frac{\phi^2\beta}{\phi^2\left(\alpha+\frac{\omega}{2}+\frac{\tilde\omega}{2}i\right)}}.$$

Thus, Abel obtained

$$\phi(2n+1)\beta = (2n+1)\phi\beta \prod_{m=1}^{n}\frac{N_{m,0}}{R_{m,0}} \prod_{\mu=1}^{n}\frac{N_{0,\mu}}{R_{0,\mu}} \prod_{m=1}^{n}\prod_{\mu=1}^{n}\frac{N_{m,\mu}}{R_{m,\mu}}\cdot\frac{\overline{N}_{m,\mu}}{\overline{R}_{m,\mu}}, \qquad (34.25)$$

where

$$N_{m,\mu} = 1 - \frac{\phi^2\beta}{\phi^2\left(\frac{m\omega+\mu\tilde\omega i}{2n+1}\right)},$$

$$\overline{N}_{m,\mu} = 1 - \frac{\phi^2\beta}{\phi^2\left(\frac{m\omega-\mu\tilde\omega i}{2n+1}\right)},$$

$$R_{m,\mu} = 1 - \frac{\phi^2\beta}{\phi^2\left(\frac{\omega}{2}+\frac{\tilde\omega}{2}i+\frac{m\omega+\mu\tilde\omega i}{2n+1}\right)},$$

$$\overline{R}_{m,\mu} = 1 - \frac{\phi^2\beta}{\phi^2\left(\frac{\omega}{2}+\frac{\tilde\omega}{2}i+\frac{m\omega-\mu\tilde\omega e}{2n+1}\right)}.$$

He then set $\beta = \frac{\alpha}{2n+1}$, let $n \to \infty$, and used the formula

$$\lim_{n\to\infty}\frac{\phi^2\left(\frac{\alpha}{2n+1}\right)}{\phi^2\left(\frac{\lambda}{2n+1}\right)} = \frac{\alpha^2}{\lambda^2}$$

to obtain an infinite product for $\phi\alpha$. Abel carried out several pages of calculations to show that the limiting procedure was valid and that the product converged to $\phi\alpha$. It is not clear that Abel's justification was complete. Anyhow, Abel obtained the formula[19]

$$\phi\alpha = \alpha \prod_{m=1}^{\infty}\left(1-\frac{\alpha^2}{(m\omega)^2}\right) \cdot \prod_{\mu=1}^{\infty}\left(1+\frac{\alpha^2}{(\mu\tilde\omega)^2}\right)$$

$$\times \prod_{m=1}^{\infty}\left(\prod_{\mu=1}^{\infty}\left(\frac{1-\frac{\alpha^2}{(m\omega+\mu\tilde\omega i)^2}}{1-\frac{\alpha^2}{\left(\left(m-\frac{1}{2}\right)\omega+\left(\mu-\frac{1}{2}\right)\tilde\omega i\right)^2}}\right) \prod_{\mu=1}^{\infty}\left(\frac{1-\frac{\alpha^2}{(m\omega-\mu\tilde\omega i)^2}}{1-\frac{\alpha^2}{\left(\left(m-\frac{1}{2}\right)\omega-\left(\mu-\frac{1}{2}\right)\tilde\omega i\right)^2}}\right)\right).$$

$$(34.26)$$

---

[19] Abel (2007) p. 236.

Recall that in 1797 Gauss obtained a similar formula for the particular case of the lemniscatic function. Abel then expressed (34.26) in terms of sines, just as Gauss had done in 1798. This and other similarities in their work led Gauss to remark that Abel followed the same steps as he did in 1797. Abel next rewrote the double product in (34.26):

$$\prod_{m=1}^{\infty} \prod_{\mu=1}^{\infty} \frac{1 + \frac{(\alpha+m\omega)^2}{\mu^2\tilde{\omega}^2}}{1 + \frac{\left(\alpha+\left(m-\frac{1}{2}\right)\omega\right)^2}{\left(\mu-\frac{1}{2}\right)^2\tilde{\omega}^2}} \cdot \frac{1 + \frac{(\alpha-m\omega)^2}{\tilde{\mu}^2\tilde{\omega}^2}}{1 + \frac{\left(\alpha-\left(m-\frac{1}{2}\right)\omega\right)^2}{\left(\mu-\frac{1}{2}\right)^2\tilde{\omega}^2}} \cdot \left(\frac{1 + \frac{\left(m-\frac{1}{2}\right)^2\omega^2}{\left(\mu-\frac{1}{2}\right)^2\tilde{\omega}^2}}{1 + \frac{m^2\omega^2}{\mu^2\tilde{\omega}^2}} \right)^2 \cdot$$

Then by means of the product for $\sin x$ given by

$$\sin x = x \prod_{\mu=1}^{\infty} \left(1 - \frac{x^2}{\mu^2\pi^2} \right),$$

and using the addition formula for sine given by

$$\sin(a - b) \cdot \sin(a + b) = \sin^2 a - \sin^2 b,$$

he obtained

$$\phi\alpha = \frac{\tilde{\omega}}{\pi} \frac{s}{i} \prod_{m=1}^{\infty} \frac{1 - \frac{s^2}{A_m^2}}{1 - \frac{s^2}{B_m^2}}$$

where

$$s = \sin\left(\frac{\alpha\pi i}{\tilde{\omega}}\right), \quad A_m = \sin\left(\frac{m\omega\pi i}{\tilde{\omega}}\right), \quad B_m = \cos\left(\left(m - \frac{1}{2}\right)\frac{\omega\pi i}{\tilde{\omega}}\right).$$

Finally, by the use of $\phi(i\alpha) = i\phi\alpha$, he obtained his product formula:

$$\phi\alpha = \frac{\omega}{\pi} \sin\frac{\alpha\pi}{\omega} \prod_{m=1}^{\infty} \frac{1 + \frac{4\sin^2\left(\frac{\alpha\pi}{\omega}\right)}{\left(e^{\frac{m\tilde{\omega}\pi}{\omega}} - e^{-\frac{m\tilde{\omega}\pi}{\omega}}\right)^2}}{1 - \frac{4\sin^2\left(\frac{\alpha\pi}{\omega}\right)}{\left(e^{\frac{(2m-1)\tilde{\omega}\pi}{(2\omega)}} + e^{-\frac{(2m-1)\tilde{\omega}\pi}{(2\omega)}}\right)^2}}. \tag{34.27}$$

Abel also used the series (34.21) to obtain various other formulas, including[20]

$$\phi\left(\frac{\alpha\omega}{2}\right) = \frac{4\pi}{\omega} \left(\frac{e^{\frac{\pi}{2}}}{1+e^{\pi}} \sin\frac{\alpha\pi}{2} - \frac{e^{\frac{3\pi}{2}}}{1+e^{3\pi}} \sin\frac{3\alpha\pi}{2} + \frac{e^{\frac{5\pi}{2}}}{1+e^{5\pi}} \sin\frac{5\alpha\pi}{2} - \cdots \right).$$

[20] ibid. p. 244.

34.4 Abel: Division of Elliptic Functions and Algebraic Equations

In his 1827 paper, Abel considered Gauss's algebraic theory on the division of periodic functions and extended it to the division of doubly-periodic functions. To understand Abel's motivation, recall that from a study of Viète, Newton determined that for an odd number n, $\sin nx$ could be expressed as a polynomial of degree n in $\sin x$. Note that a similar result holds for $\cos nx$. Gauss proved that these polynomials could be solved algebraically; Euler and Vandermonde had earlier done this for values of n up to eleven. Abel determined from the addition theorem that for a positive integer n, $\phi(2n+1)\alpha$ was a rational function of $\phi\alpha$ such that the numerator took the form $xR(x^2)$, where R was a polynomial of degree $n(2n+2)$ and $x = \phi\alpha$. His problem was to find out whether R could be solved algebraically, that is, by radicals. He discovered that he could employ Lagrange resolvents, an idea due to Waring, Vandermonde, and Lagrange, as Gauss had also done. However, Abel's problem was more complicated than Gauss's and took him deeper into the theory of equations.

As a mathematical aside, we briefly discuss Abel's related contributions to the theory of equations. His work in elliptic function theory gave him glimpses into the nature of algebraically solvable equations. In particular, he sought to determine solvability in terms of the structure of the roots of the equation. In an 1826 letter to Crelle, Abel stated a result on the form of the roots of a solvable quintic. He later generalized this result to irreducible equations of prime degree, published posthumously in the first edition of his collected papers of 1837. This paper contained the remarkable theorem that an irreducible equation of prime degree was solvable by radicals if and only if all its roots were rational functions of any two of the roots. Galois rediscovered this theorem a few years later, but his work arose out of a study of those permutations of the roots preserving algebraic relations among the roots. Because the group theory of algebraic equations, developed by Galois, gained recognition before Abel's theory, based on structure of roots, Abel's theorems have now become recast and known in terms of groups. It might be fruitful to make a parallel study of the two approaches.

Recall that Abel proved that $\phi(2n+1)\beta$ was a rational function of $x = \phi\beta$ whose numerator took the form $xR(x^2)$ where R was a polynomial of degree $n(2n+2)$. Abel then proved the important theorem that the solutions of $R = 0$ depended on the solutions of a certain equation of degree $2n+2$ with coefficients that were rational functions of c and e. He proceded to demonstrate that if the latter equation could be solved by radicals, then so could $R = 0$. He went on to observe that, in general, this equation was not solvable by radicals but could be solved in particular cases, such as for $e = c$, $e = \sqrt{3}c$, $e = (2 \pm \sqrt{3})c$, etc. The case $e = c$ corresponded to the lemniscatic function and had already been discussed in Gauss's unpublished work, at least in special cases.

Abel's proof of this theorem was lengthy. We present a brief summary, using his notation. First note that by (34.25) and the fact that the zeros of ϕ occur at $mw + in\tilde{w}$, it follows that the solutions of $R = 0$ must be given by

$$r = \phi^2 \left(\frac{m\omega \pm i\mu\tilde{\omega}i}{2n+1} \right).$$

By periodicity of ϕ, the number of different values of r can be reduced to the $n(2n+2)$ values given by

$$r_\nu = \phi^2\left(\frac{\nu\omega}{2n+1}\right), \quad r_{\nu,m} = \phi^2\left(\nu\frac{m\omega + i\tilde{\omega}}{2n+1}\right), \tag{34.28}$$

where $1 \le \nu \le n$, $0 \le m \le 2n$. Now let ω' denote any quantity of the form $m\omega + i\mu\tilde{\omega}$ and define ψ by the equation

$$\psi\left(\phi^2\left(\frac{\omega'}{2n+1}\right)\right) = \theta\left(\phi^2\left(\frac{\omega'}{2n+1}\right), \phi^2\left(\frac{2\omega'}{2n+1}\right), \cdots, \phi^2\left(\frac{n\omega'}{2n+1}\right)\right), \tag{34.29}$$

where θ is a rational symmetric function of the n quantities. It is clear from the definition of ψ that

$$\psi\left(\phi^2\left(\frac{\nu\omega'}{2n+1}\right)\right) = \psi\left(\phi^2\left(\frac{\omega'}{2n+1}\right)\right), \quad 1 \le \nu \le n. \tag{34.30}$$

In particular,

$$\psi r_\nu = \psi r_1; \quad \psi r_{\nu,m} = \psi r_{1,m}, \quad 1 \le \nu \le n. \tag{34.31}$$

The aforementioned equation of degree $2n+2$ can be given by

$$(p - \psi r_1)(p - \psi r_{1,0})(p - \psi r_{1,1})\cdots(p - \psi r_{1,2n})$$
$$= q_0 + q_1 p + q_2 p^2 + \cdots + q_{2n+1}p^{2n+1} + p^{2n+2}. \tag{34.32}$$

It is easy to see that $q_0, q_1, \ldots, q_{2n+1}$ are rational functions of c and e. Note that the sum of the kth powers of the roots of (34.32) are symmetric functions of the $n(2n+2)$ roots r_ν and $r_{\nu,m}$ of $R = 0$, where r_ν and $r_{\nu,m}$ are given by (34.28). To see this, observe that

$$(\psi r_1)^k = \frac{1}{n}[(\psi r_1)^k + (\psi r_2)^k + \cdots + (\psi r_n)^k],$$

$$(\psi r_{1,m})^k = \frac{1}{n}[(\psi r_{1,m})^k + (\psi r_{2,m})^k + \cdots + (\psi r_{n,m})^k], \quad 0 \le m \le 2n,$$

and

$$(\psi r_1)^k + (\psi r_{1,0})^k + (\psi r_{1,1})^k + \cdots + (\psi r_{1,2n})^k$$

$$= \frac{1}{n}\left[(\psi r_1)^k + (\psi r_2)^k + \cdots + (\psi r_n)^k\right]$$

$$+ \frac{1}{n}\left[(\psi r_{1,0})^k + (\psi r_{2,0})^k + \cdots + (\psi r_{n,0})^k\right]$$

$$\cdots\cdots\cdots$$

$$+ \frac{1}{n}\left[(\psi r_{1,2n})^k + (\psi r_{2,2n})^k + \cdots + (\psi r_{n,2n})^k\right].$$

Since the coefficients of the polynomial R are rational functions of c and e, we may now conclude that each sum of the kth powers of the roots of the polynomial (34.32) is a rational function of c and e. Since the power sum symmetric functions form a basis for the symmetric functions, it follows that $q_0, q_1, \ldots, q_{2n+1}$ are rational functions of c and e.

Next, we show that if $p = \psi r_1$ and $q = \theta r_1$ are rational symmetric functions of $r_1, r_2, \ldots, r_n$, then q can be determined in terms of p. Note that a similar result holds for $\psi r_{1,m}$ and $\theta r_{1,m}$. For $k = 0, 1, \ldots, 2n + 1$, set

$$s_k = (\psi r_1)^k \theta r_1 + (\psi r_{1,0})^k \theta r_{1,0} + \cdots + (\psi r_{1,2n})^k \theta r_{1,2n}. \tag{34.33}$$

We prove that s_k can be expressed as a rational function of c and e. Note that

$$(\psi r_1)^k \theta r_1 = (\psi r_\nu)^k \theta r_\nu = \frac{1}{n}[(\psi r_1)^k \theta r_1 + (\psi r_2)^k \theta r_2 + \cdots + (\psi r_n)^k \theta r_n];$$

$$(\psi r_{1,m})^k \theta r_{1,m} = (\psi r_{\nu,m})^k \theta r_{\nu,m} = \frac{1}{n}[(\psi r_{1,m})^k \theta r_{1,m} + \cdots + (\psi r_{n,m})^k \theta r_{n,m}].$$

When these values are substituted in (34.33), we observe that s_k is a symmetric rational function of the roots of $R = 0$; therefore, s_k, $k = 0, 1, \ldots, 2n + 1$, are rational functions of c and e. We can apply Cramer's rule to solve these equations for $\theta r_1, \theta r_{1,0}, \ldots, \theta r_{1,2n}$ in terms of rational functions of $\psi r_1, \ldots, \psi r_{1,2n}$. This result in turn implies that the coefficients of the equation

$$(r - r_1)(r - r_2) \cdots (r - r_n) = r^n + p_{n-1}f r^{n-1} + p_{n-2}r^{n-2} + \cdots + p_1 r + p_0 \tag{34.34}$$

can be determined by the equation (34.32). There are $2n + 1$ additional equations of degree n with roots $r_{1,\nu}, \ldots, r_{n,\nu}$ for $0 \le \nu \le 2n$; the coefficients of these equations are also determined by (34.32).

In this way, Abel reduced the problem of solving the equation $R = 0$ of degree $n(2n \pm 2)$ to that of solving $2n + 2$ equations of the form (34.34). We demonstrate by means of the Lagrange resolvent (Gauss's method for solving the cyclotomic equation)[21] that the solutions of these equations can be expressed in terms of the solutions to (34.32). Let

$$\phi^2\left(\frac{\omega'}{2n+1}\right), \quad \phi^2\left(\frac{2\omega'}{2n+1}\right), \quad \ldots, \quad \phi^2\left(\frac{n\omega'}{2n+1}\right)$$

denote the solutions of (34.34), where ω' stands for ω or $m\omega + i\tilde{\omega}$. By a theorem of Gauss, there exists a number α generating the numbers $1, 2, \ldots, 2n$ (modulo $2n + 1$). Then by the periodicity of ϕ, the set

$$\phi^2(\epsilon), \phi^2(\alpha\epsilon), \phi^2(\alpha^2\epsilon), \ldots, \phi^2(\alpha^{n-1}\epsilon),$$

where $\epsilon = \frac{\omega'}{2n+1}$, represents all the solutions of (34.34). We omit Abel's straightforward proof of this result.

---

[21] See Neumann (2007b).

Now let θ denote any imaginary root of $\theta^n - 1 = 0$, and define the Lagrange resolvent

$$\psi(\epsilon) = \phi^2(\epsilon) + \phi^2(\alpha\epsilon)\theta + \phi^2(\alpha^2\epsilon) + \cdots + \phi^2(\alpha^{n-1}\epsilon)\theta^{n-1}. \qquad (34.35)$$

It is clear that $\psi(\epsilon)$ is a rational function of $\phi^2(\epsilon)$, expressible as $\psi(\epsilon) = \chi\left(\phi^2(\epsilon)\right)$. By a simple calculation involving roots of unity, we can show that

$$\psi(\alpha^m\epsilon) = \theta^{-m}\psi(\epsilon) \quad \text{or} \quad \psi(\epsilon) = \theta^m\chi(\phi^2(\alpha^m\epsilon)),$$

implying that $(\psi\epsilon)^n = [\chi(\phi^2(\alpha^m\epsilon))]^n$. Taking $m = 0, 1, \ldots, n - 1$ and adding we arrive at

$$n(\psi\epsilon)^n = \left[\chi\left(\phi^2(\epsilon)\right)\right]^n + \left[\chi\left(\phi^2(\alpha\epsilon)\right)\right]^n + \cdots + \left[\chi\left(\phi^2(\alpha^{n-1}\epsilon)\right)\right]^n. \qquad (34.36)$$

The expression on the right-hand side of (34.36) is a rational symmetric function of

$$\phi^2\epsilon, \ \phi^2(\alpha\epsilon), \ \ldots, \phi^2(\alpha^{n-1}\epsilon).$$

That is, it is a rational symmetric function of the roots of (34.34). Therefore, $(\phi\epsilon)^n = v$ is a rational function of $p_0, p_1, \ldots, p_{n-1}$ and

$$\sqrt[n]{v} = \phi^2\epsilon + \theta\phi^2(\alpha\epsilon) + \theta^2\phi^2(\alpha^2\epsilon) + \cdots + \theta^{n-1}\phi^2(\alpha^{n-1}\epsilon). \qquad (34.37)$$

Note also that v is a rational function of the roots of (34.32); so if (34.32) can be solved by radicals, then v can be expressed in terms of radicals. By changing θ to $\theta^2, \theta^3, \ldots, \theta^{n-1}$ and denoting the corresponding values of v by $v_2, v_3, \ldots, v_{n-1}$, we have

$$\sqrt[n]{v_k} = \phi^2(\epsilon) + \theta^k\phi^2(\alpha\epsilon) + \cdots + \theta^{k(n-1)}\phi^2(\alpha^{n-1}\epsilon), \quad k = 1, 2, \ldots, n - 1. \qquad (34.38)$$

When these $n - 1$ equations are combined with the equation

$$-p_{n-1} = \phi^2(\epsilon) + \phi^2(\alpha\epsilon) + \cdots + \phi^2(\alpha^{n-1}\epsilon),$$

we can easily solve these n linear equations to get

$$\phi^2(\alpha^m\epsilon) = \frac{1}{n}(-p_{n-1} + \theta^{-m}\sqrt[n]{v_1} + \theta^{-2m}\sqrt[n]{v_2} + \cdots + \theta^{-(n-1)m}\sqrt[n]{v_{n-1}}), \qquad (34.39)$$

for $m = 0, 1, \ldots, n - 1$.

It can also be shown that

$$s_k = \frac{\sqrt[n]{v_k}}{(\sqrt[n]{v_1})^k}$$

is a rational function of $p_0, p_1, \ldots, p_{n-1}$. For this purpose, it is sufficient to check that s_k is unchanged by $\epsilon \to \alpha^m \epsilon$. This gives us Abel's final formula for $\phi^2(\alpha^m \epsilon)$:

$$\phi^2(\alpha^m \epsilon) = \frac{1}{n}\left(-p_{n-1} + \theta^{-m} v^{\frac{1}{n}} + s_2 \theta^{-2m} v^{\frac{2}{n}} + \cdots + s_{n-1}\theta^{-(n-1)m} v^{\frac{n-1}{m}}\right),$$

(34.40)

for $m = 0, 1, \ldots, n-1$. This implies that if v can be expressed in terms of radicals, then $R = 0$ can be solved by radicals.

34.5 Abel: Division of the Lemniscate

Recall Abel's remark that in the case $\frac{e}{c} = 1$, the division points $\phi^2\left(\frac{k\omega'}{2n+1}\right)$ could be obtained by solving an algebraic equation by radicals. When $e = c = 1$, Abel's integral (34.6) is reduced to

$$\alpha = \int_0^x \frac{dx}{\sqrt{1-x^4}}, \quad \text{or} \quad x = \phi\alpha. \tag{34.41}$$

It is easy to check that

$$\phi(\alpha i) = i\,\phi\alpha \tag{34.42}$$

and

$$\frac{\omega}{2} = \frac{\tilde{\omega}}{2} = \int_0^1 \frac{dx}{\sqrt{1-x^4}}. \tag{34.43}$$

Abel applied the addition formula to show that for $m + \mu$ odd and $x = \phi\delta$,

$$\phi(m + \mu i)\delta = x\psi(x^2), \tag{34.44}$$

for some rational function ψ. Then by changing δ to $i\delta$ and using (34.42), he obtained $\phi(m + \mu i)\delta = x\psi(-x^2)$, or $\psi(-x^2) = \psi(x^2)$. He therefore concluded that[22]

$$\phi(m + \mu i)\delta = x \cdot \frac{T}{S}, \tag{34.45}$$

where T and S were polynomials in powers of x^4. This very significant result showed that the elliptic function $\phi\delta$ permitted complex multiplication, that is, $\phi(m + \mu i)\delta$ could be expressed as a rational function of $\phi\delta$. As an example, he noted that

$$\phi(2 + i)\delta = ix \cdot \frac{1 - 2i - x^4}{1 - (1 - 2i)x^4}, \tag{34.46}$$

---

[22] ibid. p. 248.

a result proved by Gauss in an unpublished work, wherein he also divided the lemniscate into $5 = (2 + i)(2 - i)$ parts.

Abel showed how (34.45) could be applied to the problem of dividing the lemniscate into $4\nu + 1$ parts. By Fermat's theorem on sums of two squares, Abel could write

$$\alpha^2 + \beta^2 = 4\nu + 1 = (\alpha + i\beta)(\alpha - i\beta),$$

where $\alpha + \beta$ was odd. With $m = \alpha, \mu = \beta$, and $\delta = \frac{\omega}{\alpha + i\beta}$, he could use (34.45) to obtain $x = \phi(\delta)$ as a root of $T = 0$. By using the periodicity of ϕ and the addition formula, Abel proved that

$$\pm\phi\left(\frac{\omega}{\alpha + i\beta}\right), \ \pm\phi\left(\frac{2\omega}{\alpha + i\beta}\right), \ \ldots, \ \pm\phi\left(\frac{\alpha^2 + \beta^2 - 1}{2} \cdot \frac{\omega}{\alpha + \beta i}\right)$$

comprised all the roots of the polynomial T. By setting $T(x) = R(x^2)$, he obtained

$$\phi^2(\delta), \ \phi^2(2\delta), \ \phi^2(3\delta), \ldots, \phi^2(2\nu\delta) \tag{34.47}$$

as all the roots of $R = 0$. Next, Abel once again applied Gauss's method. He first showed that for a primitive root ϵ modulo $4\nu + 1 = \alpha^2 + \beta^2$, the set $\phi^2(\epsilon^m\delta)$, $m = 0, 1, \ldots, 2\nu - 1$, was equal to the set given in (34.47). He then referred to the method of Lagrange resolvents to conclude that[23]

$$\phi^2(\epsilon^m\delta) = \frac{1}{2\nu}\left(A + \theta^{-m} \cdot v^{\frac{1}{2\nu}} + s_2\theta^{-2m} \cdot v^{\frac{2}{2\nu}} + \cdots + s_{2\nu-1}\theta^{-(2\nu-1)m} \cdot v^{\frac{2\nu-1}{2\nu}}\right),$$
$$\tag{34.48}$$

where θ was an imaginary root of $\theta^{2\nu} - 1 = 0$, and v, s_k were determined by the expressions

$$v = \left[\phi^2(\delta) + \theta \cdot \phi^2(\epsilon\delta) + \theta^2 \cdot \phi^2(\epsilon^2\delta) + \cdots + \theta^{2\nu-1} \cdot \phi^2(\epsilon^{2\nu-1}\delta)\right]^{2\nu}, \tag{34.49}$$

$$s_k = \frac{\phi^2(\delta) + \theta^k \cdot \phi^2(\epsilon\delta) + \cdots + \theta^{(2\nu-1)k} \cdot \phi^2(\epsilon^{2\nu-1}\delta)}{[\phi^2(\delta) + \theta \cdot \phi^2(\epsilon\delta) + \cdots + \theta^{2\nu-1} \cdot \phi^2(\epsilon^{2\nu-1}\delta)]^k}, \tag{34.50}$$

$$A = \phi^2(\delta) + \phi^2(\epsilon\delta) + \cdots + \phi^2(\epsilon^{2\nu-1}\delta). \tag{34.51}$$

Moreover, the expressions (34.49), (34.50), and (34.51) could be written as rational functions of the coefficients of $R = 0$. Recall that the coefficients of $R = 0$ took the form $a + bi$ with a, b rational. Thus, v, s_k and A were of the form $c + id$, with c and d rational.

Abel then noted that if $4\nu + 1 = 1 + 2^n$, then $2\nu = 2^{n-1}$ and the values in (34.48) could be computed by repeatedly taking square roots.[24] Thus, the values of $\phi\left(\frac{m\omega}{\alpha+i\beta}\right)$

---

[23] ibid. p. 253.
[24] ibid. p. 255.

could be evaluated by taking square roots; hence, by applying the addition formula, the value of $\phi(\frac{\omega}{4\nu+1})$ could be so determined. This proved the result that the lemniscate could be geometrically divided into 2^n+1 parts, when this was a prime number. Recall that in his *Disquisitiones*, Gauss had stated that this was true.

34.6 Jacobi's Elliptic Functions

In his *Fundamenta Nova*, Jacobi presented a detailed account of his theory of elliptic functions. He inverted the elliptic integral[25]

$$u = \int_0^\phi \frac{d\phi}{\sqrt{1 - k^2 \sin^2 \phi}} = \int_0^x \frac{dx}{\sqrt{(1-x^2)(1-k^2x^2)}}$$

by defining the function $x = \sin \operatorname{am} u$, where $\phi = \operatorname{am} u$, calling ϕ the amplitude of u. He noted that, in general, any trigonometric function of ϕ, such as $\cos \phi = \cos \operatorname{am} u$, $\tan \phi = \tan \operatorname{am} u$, could be defined in this manner. Jacobi worked mainly with the functions $\sin \phi$, $\cos \phi$, and

$$\Delta \operatorname{am} u = \sqrt{1 - k^2 \sin^2 \operatorname{am} u} = \frac{d \operatorname{am} u}{du}.$$

Following Gudermann, we employ modern notation for these functions: $\operatorname{sn} u$, $\operatorname{cn} u$, and $\operatorname{dn} u$. When we emphasize dependence on modulus k, we write $\operatorname{sn}(u,k)$, $\operatorname{cn}(u,k)$, and $\operatorname{dn}(u,k)$. The complementary modulus k', defined by $k^2 + k'^2 = 1$, is also important. Legendre denoted by K the complete elliptic integral obtained by taking $x = 1$ in the preceding integral; he denoted the corresponding complete integral for the modulus k' by K'.

Jacobi listed the addition theorems and related identities, results he obtained directly from those of Euler and Legendre:

$$\operatorname{sn}(u + v) = \frac{\operatorname{sn} u \operatorname{cn} v \operatorname{dn} v + \operatorname{sn} v \operatorname{cn} u \operatorname{dn} u}{D},$$

$$\operatorname{cn}(u + v) = \frac{\operatorname{cn} u \operatorname{cn} v - \operatorname{sn} u \operatorname{dn} u \operatorname{sn} v \operatorname{dn} v}{D},$$

$$\operatorname{dn}(u + v) = \frac{\operatorname{dn} u \operatorname{dn} v - k^2 \operatorname{sn} u \operatorname{cn} u \operatorname{sn} v \operatorname{cn} v}{D},$$

$$\operatorname{sn}(u + v)\operatorname{sn}(u - v) = \frac{\operatorname{sn}^2 u - \operatorname{sn}^2 v}{D},$$

where

$$D = 1 - k^2 \operatorname{sn}^2 u \operatorname{sn}^2 v.$$

[25] Jacobi (1969) vol. 1, pp. 81–87.

Jacobi then extended the domain of the elliptic functions by applying the transformation, later called Jacobi's imaginary transformation:

$$\sin \phi = i \tan \psi. \tag{34.52}$$

This implied $\cos \phi = \sec \psi$ and $d\phi = \frac{i \, d\psi}{\cos \psi}$ and

$$\frac{d\phi}{\sqrt{1 - k^2 \sin^2 \phi}} = \frac{i \, d\psi}{\sqrt{\cos^2 \psi + k^2 \sin^2 \psi}} = \frac{i \, d\psi}{\sqrt{1 - k'k' \sin^2 \psi}}.$$

Jacobi used these this to write

$$\sin \operatorname{am} (iu, k) = i \tan \operatorname{am} (u, k'),$$

$$\cos \operatorname{am} (iu, k) = \sec \operatorname{am} (u, k'),$$

$$\tan \operatorname{am} (iu, k) = i \sin \operatorname{am} (u, k'),$$

and other similar formulas. From these results, Jacobi deduced that $\operatorname{sn}(u,k)$ had periods $4K$ and $2iK'$; $\operatorname{cn}(u,k)$ had periods $4K$ and $2K + 2iK'$; and $\operatorname{dn}(u,k)$ had periods $2K$ and $4iK'$. Moreover, in a period parallelogram, $\operatorname{sn} u$ had zeros at $u = 0$ and at $u = 2K$ and had poles at iK' and $2K + iK'$. Jacobi had similar results for $\operatorname{cn} u$ and $\operatorname{dn} u$.

We note an application of Jacobi's imaginary transformation to the quadratic transformations discussed earlier. This will provide an introduction to the higher-order transformations appearing in the next two sections. Recall that Landen's quadratic transformation

$$y = \frac{(1 + k')x\sqrt{1 - x^2}}{\sqrt{1 - k^2 x^2}} \tag{34.53}$$

produces the differential relation

$$\frac{dy}{\sqrt{(1 - y^2)(1 - \lambda^2 y^2)}} = \frac{(1 + k') \, dx}{\sqrt{(1 - x^2)(1 - k^2 x^2)}}, \tag{34.54}$$

where $\lambda = \frac{1-k'}{1+k'}$ or, in other words, $k = \frac{2\sqrt{\lambda}}{1+\lambda}$. This algebraic relation between the moduli λ and k is called a modular equation. By means of this relation, we may write (34.54) as

$$\frac{(1 + \lambda) \, dy}{\sqrt{(1 - y^2)(1 - \lambda^2 y^2)}} = \frac{2 \, dx}{\sqrt{(1 - x^2)(1 - k^2 x^2)}}. \tag{34.55}$$

If we integrate the differential on the right-hand side of (34.55) from 0 to 1, we get $2K$. However, as x increases from 0 to $\frac{1}{\sqrt{1+k'}}$, y increases from 0 to 1; and as x

continues to increase to 1, y decreases from 1 to 0. Thus, if Λ denotes the complete integral corresponding to the modulus λ, we get the equation

$$2(1+\lambda)\Lambda = 2K \quad \text{or} \quad K = (1+\lambda)\Lambda. \tag{34.56}$$

Now note that the second quadratic transformation of Gauss

$$z = \frac{(1+\lambda)y}{1+\lambda y^2} \tag{34.57}$$

produces the differential relation

$$\frac{dz}{\sqrt{(1-z^2)(1-\gamma^2 z^2)}} = \frac{(1+\lambda)\,dy}{\sqrt{(1-y^2)(1-\lambda^2 y^2)}}, \tag{34.58}$$

where $\gamma = \frac{2\sqrt{\lambda}}{1+\lambda}$. We can therefore take $\gamma = k$ and apply (34.55) followed by (34.58) to obtain duplication:

$$\frac{dz}{\sqrt{(1-z^2)(1-k^2 z^2)}} = \frac{2\,dx}{\sqrt{(1-x^2)(1-k^2 x^2)}}.$$

One of Jacobi's earliest discoveries was that there were, similarly, two cubic transformations, and when these were applied consecutively, they produced triplication. He then extended this to general odd order transformations.

Jacobi's imaginary transformation (34.52) when written in terms of x and y amounts to setting

$$x = \frac{iX}{\sqrt{1-X^2}} \quad \text{and} \quad y = \frac{iY}{\sqrt{1-Y^2}}. \tag{34.59}$$

When these expressions for x and y are substituted in Landen's transformation (34.53), we obtain, after simplification, Gauss's form of the transformation:

$$\frac{Y}{\sqrt{1-Y^2}} = \frac{(1+k')X}{\sqrt{1-X^2)(1-k^2 X^2)}}$$

or

$$Y = \frac{(1+k')X}{1+k'X^2}.$$

Moreover, the differential relation (34.55) converts to

$$\frac{(1+\lambda)\,dY}{\sqrt{(1-Y^2)(1-\lambda^2 Y^2)}} = \frac{2\,dX}{\sqrt{(1-X^2)(1-k'^2 X^2)}}.$$

Observe that, since X and Y increase simultaneously from 0 to 1, this relation can be written in terms of complete integrals:

$$2K' = (1 + \lambda)\Lambda'.$$

Dividing this equation by (34.56) gives another form of the modular relation, also used by Legendre:

$$\frac{2K'}{K} = \frac{\Lambda'}{\Lambda}. \tag{34.60}$$

As one might expect, when Jacobi's imaginary transformation is applied to Gauss's transformation, one obtains Landen's transformation, except that k and λ are converted to their complements k' and λ'. These results also carry over to general transformations.

34.7　Jacobi: Cubic and Quintic Transformations

In a letter of June 13, 1827,[26] Jacobi communicated to Schumacher, editor of the *Astronomische Nachrichten*, two cubic and two quintic transformations. Jacobi's first result stated: If we set

$$\sin\phi = \frac{\sin\psi \left(ac + \left(\frac{a-c}{2}\right)^2 \sin^2\psi\right)}{cc + \frac{a-c}{2} \cdot \frac{a+3c}{2} \sin^2\psi}, \tag{34.61}$$

we obtain

$$\frac{d\phi}{\sqrt{a^3c - \frac{a-c}{2}\left(\frac{a+3c}{2}\right)^3 \sin^2\phi}} = \frac{d\psi}{\sqrt{c^3a - \left(\frac{a-c}{2}\right)^3 \frac{a+3c}{2} \sin^2\psi}}. \tag{34.62}$$

If, in addition,

$$\sin\psi = \frac{\sin\theta \left(-3ac + \left(\frac{a+3c}{2}\right)^2 \sin^2\theta\right)}{aa - 3\frac{a-c}{2} \cdot \frac{a+3c}{2} \sin^2\theta} \tag{34.63}$$

and

$$\chi = \frac{a-c}{2c}\left(\frac{a+3c}{2a}\right)^3, \tag{34.64}$$

then we have

$$\frac{d\phi}{\sqrt{1 - \chi\sin^2\phi}} = \frac{3\,d\theta}{\sqrt{1 - \chi\sin^2\theta}}. \tag{34.65}$$

---

[26] Jacobi (1969) vol. 1, pp. 31–33.

Note that (34.61) and (34.63) are Jacobi's two cubic transformations, and when applied in succession, they produce the triplication (34.65) for the modulus $k^2 = \chi$, given by (34.64).

Jacobi's second result stated: If we set $a^3 = 2b(1 + a + b)$ and

$$\sin \phi = \frac{\sin \psi \, (1 + 2a + (aa + 2ab + 2b) \sin^2 \psi + bb \sin^4 \psi)}{1 + (aa + 2a + 2b) \sin^2 \psi + b(b + 2a) \sin^4 \psi}, \tag{34.66}$$

we get

$$\int \frac{d\phi}{\sqrt{(a - 2b)(1 + 2a)^2 - (2 - a)(b + 2a)^2 \sin^2 \phi}} = \int \frac{d\psi}{\sqrt{a - 2b - bb(2 - a) \sin^2 \psi}}.$$

Also, if

$$\alpha = \frac{2 - a}{1 + 2a},$$

$$\beta = -\frac{b + 2a}{1 + 2a} \cdot \frac{2 - a}{a - 2b},$$

$$\chi = \frac{2 - a}{a - 2b} \cdot \left(\frac{b + 2a}{1 + 2a} \right)^2,$$

$$\sin \psi = \frac{\sin \theta \, (1 + 2\alpha + (\alpha\alpha + 2\alpha\beta + 2\beta) \sin^2 \theta + \beta\beta \sin^4 \theta)}{1 + (\alpha\alpha + 2\alpha + 2\beta) \sin^2 \theta + \beta(\beta + 2\alpha) \sin^4 \theta}, \tag{34.67}$$

then we have

$$\int \frac{d\phi}{\sqrt{1 - \chi \sin^2 \phi}} = 5 \int \frac{d\theta}{\sqrt{1 - \chi \sin^2 \theta}}. \tag{34.68}$$

Here (34.66) and (34.67) are Jacobi's quintic transformations, and they together produce the quinsection given by (34.68).

In his letter to Legendre of April 12, 1828,[27] Jacobi wrote that he found (34.63) and (34.67) by trial and error. But he explained that he had found the cubic and quintic transformations (34.61) and (34.66) on the basis of the general algebraic theory of transformations he had developed in March 1827. For this theory, he considered the transformation $y = \frac{U}{V}$, where U and V were polynomials in x differing in degree by at most one, and such that

$$\frac{dy}{\sqrt{Y}} = \frac{1}{M} \frac{dx}{\sqrt{X}}, \tag{34.69}$$

where X and Y were quartics in x and y, respectively, and M was a constant depending on the constants in X and Y. In particular, he took $X = (1 - x^2)(1 - k^2 x^2)$ and $Y = (1 - y^2)(1 - \lambda^2 y^2)$. By substituting $y = \frac{U}{V}$ in (34.69), he obtained the relation

[27] ibid. pp. 409–416.

$$\frac{dy}{\sqrt{(1-y^2)(1-\lambda^2 y^2)}} = \frac{\left(V\frac{dU}{dx} - U\frac{dV}{dx}\right)dx}{\sqrt{(V^2-U^2)(V_-^2-\lambda^2 U^2)}}.$$

He noted that if U and V were of degree p, the numerator of the expression on the right was a polynomial of degree $2p-2$, while the expression inside the radical was of degree $4p$. Moreover, since for any number α,

$$(V-\alpha U)\frac{dU}{dx} - \frac{d(V-\alpha U)}{dx}U = V\frac{dU}{dx} - U\frac{dV}{dx},$$

it followed that if any of the factors $V \pm U$, $V \pm \lambda U$ in the denominator had a square factor $(1-\beta x)^2$, then $1-\beta x$ was a factor of the numerator polynomial. Thus, if the denominator was of form $T^2 X$, where X was a quadratic and T was of degree $2p-2$, then

$$M = \frac{T}{V\frac{dU}{dx} - U\frac{dV}{dx}}$$

was a constant depending only on the constants in X and Y. Jacobi noted that the problem of finding $y = \frac{U}{V}$ was determinate because $\frac{U}{V}$ had $2p+1$ constants of which $2p-2$ could be determined by requiring that $(V^2-U^2)(V^2-\lambda^2 U^2) = T^2 X$. This left three undetermined constants, and that number could not be reduced because x could be replaced by $\frac{a+bx}{1+dx}$, resulting in a similar relation. Thus, he looked for polynomials U and V such that $V+U = (1+x)AA$, $V-U = (1-x)BB$, $V+\lambda U = (1+kx)CC$, $V-\lambda U = (1-kx)DD$. He also noted that y was an odd function of x, and hence $U = xF(x^2)$ and $V = \phi(x^2)$. Moreover, the equation (34.69) remained invariant when y was replaced by $\frac{1}{\lambda y}$ and x by $\frac{1}{kx}$. This observation allowed him to determine explicit algebraic relations between k, λ, and the coefficients of U and V. In particular, it was possible to obtain for small values of p (the degree of U) the explicit algebraic relations satisfied by k and λ. These relations are called modular equations. So if either k or λ is given, the other can be found as one of the roots of this equation. The value of M can also be determined. It can be proven that if p is an odd prime, then the modular equation is irreducible and of order $p+1$. Thus, for a given k, there are $p+1$ different values of λ and each one leads to a distinct transformation of order p. We note that Legendre and Jacobi took k^2 to be between 0 and 1. The modular equation gave $p+1$ values of λ of which two were real, one greater than k and the other less than k. Jacobi denoted the smaller value by λ and the larger by λ_1; he called the transformation with the smaller λ the first transformation and the other the second transformation. He noted that when the two transformations were applied one after the other, the result was a multiplication by p of the differential. So if $y = \frac{U}{V}$ was the first transformation and $z = \frac{U_1}{V_1}$ the second, then

$$\frac{dz}{\sqrt{(1-z^2)(1-k^2 z^2)}} = \frac{p\,dx}{\sqrt{(1-x^2)(1-k^2 x^2)}}.$$

Jacobi worked out the algebraic theory of transformations only for the cubic and quintic cases; he needed the theory of elliptic functions to develop the higher-order transformations. He called this the transcendental theory of transformations. He may have obtained from Abel the idea of the elliptic function as the inverse of an elliptic integral, though he unfortunately never discussed this question. Jacobi did not go deeply into modular equations; the algebraic theory of modular equations was developed, starting in the 1850s, by Betti, Brioschi, Hermite, and Kronecker.

In the *Fundamenta Nova*, Jacobi gave details of how he found the first cubic transformation. First set $\frac{a}{c} = 2\alpha + 1$ in (34.61). Then the transformation would take the form: If

$$y = \frac{x(2\alpha + 1 + \alpha^2 x^2)}{(1 + \alpha(\alpha + 2)x^2)},$$ (34.70)

then

$$\frac{dy}{\sqrt{(1 - y^2)(1 - \lambda^2 y^2)}} = \frac{(2\alpha + 1)\, dx}{\sqrt{(1 - x^2)(1 - k^2 x^2)}},$$ (34.71)

where $k^2 = \frac{\alpha^3(2+\alpha)}{2\alpha+1}$ and $\lambda^2 = \frac{\alpha(2+\alpha)^3}{(2\alpha+1)^3}$.

Recall that since $U = xF(x^2)$ and $V = \phi(x^2)$, to derive this cubic transformation, Jacobi could take[28]

$$V = 1 + bx^2 \quad \text{and} \quad U = x(a + a_1 x^2).$$

He then assumed that A was of the form $1 + \alpha x$ so that

$$V + U = (1 + x)AA = 1 + (1 + 2\alpha)x + \alpha(2 + \alpha)x^2 + \alpha\alpha x^3.$$

By equating the powers of x, he had

$$b = \alpha(2 + \alpha), \quad a = 1 + 2\alpha, \quad \text{and} \quad a_1 = \alpha^2.$$

Note that this gives the preceding cubic transformation (34.70). To find the algebraic relation satisfied by k and λ, he changed x into $\frac{1}{kx}$ and y into $\frac{1}{\lambda y}$ in (34.70) to get

$$\frac{\lambda x((2\alpha + 1)\alpha^2 + \alpha^4 x^2)}{\alpha^2 + \alpha^3(\alpha + 2)x^2} = \frac{kx(\alpha(\alpha + 2) + k^2 x^2)}{\alpha^2 + (2\alpha + 1)k^2 x^2}.$$

By equating coefficients of various powers of x, Jacobi found

$$k^2 = \frac{\alpha^3(2 + \alpha)}{2\alpha + 1}, \quad \lambda^2 = \frac{k^6}{\alpha^8} = \alpha\left(\frac{2 + \alpha}{2\alpha + 1}\right)^3.$$

28 ibid. pp. 74–75.

The complementary moduli were then given by

$$k'^2 = 1 - k^2 = \frac{(1-\alpha)(1+\alpha)^3}{2\alpha+1}, \quad \lambda'^2 = \frac{(1+\alpha)(1-\alpha)^3}{(2\alpha+1)^3}.$$

Observe that this immediately gives the modular equation $\sqrt{k\lambda} + \sqrt{k'\lambda'} = 1$. Moreover, he noted that with $D = 1 + \alpha(\alpha+2)x^2$,

$$1 - y = \frac{(1-x)(1-\alpha x)^2}{D}, \quad 1 + y = \frac{(1+x)(1+\alpha x)^2}{D},$$

$$1 - \lambda y = \frac{(1-kx)(1-\frac{kx}{\alpha})^2}{D}, \quad 1 + \lambda y = \frac{(1+kx)(1+\frac{kx}{\alpha})^2}{D},$$

and hence he arrived at the transformation (34.71):

$$\frac{dy}{\sqrt{(1-y^2)(1-\lambda^2 y^2)}} = \frac{(2\alpha+1)\,dx}{\sqrt{(1-x^2)(1-k^2 x^2)}}.$$

Jacobi wrote the modular equation in a slightly different form, by setting $k^{\frac{1}{4}} = u$ and $\lambda^{\frac{1}{4}} = v$, to get

$$u^4 - v^4 + 2uv(1 - u^2 v^2) = 0. \tag{34.72}$$

He showed how to obtain the second transformation from this modular equation. He first wrote (34.70) in terms of u and v by observing that $\frac{k^3}{\lambda} = \alpha^4$ or $\alpha = \frac{u^3}{v}$. Then (34.70) and (34.71) could be rewritten as

$$y = \frac{v(v + 2u^3)x + u^6 x^3}{v^2 + v^3 u^2 (v + 2u^3)x^2}; \tag{34.73}$$

$$\frac{dy}{\sqrt{(1-y^2)(1-v^8 y^2)}} = \frac{v + 2u^3}{v} \frac{dx}{\sqrt{(1-x^2)(1-u^8 x^2)}}. \tag{34.74}$$

Jacobi then observed that the modular equation remained unchanged when u and v were changed to $-v$ and u, respectively. This gave him the second transformation

$$z = \frac{u(u - 2v^3)y + v^6 y^3}{u^2 + u^3 v^2 (u - 2v^3)y^2}; \tag{34.75}$$

$$\frac{dz}{\sqrt{(1-z^2)(1-u^8 z^2)}} = \frac{u - 2v^3}{u} \frac{dy}{\sqrt{(1-y^2)(1-v^8 y^2)}}. \tag{34.76}$$

By the modular equation

$$\left(\frac{v + 2u^3}{v}\right)\left(\frac{u - 2v^3}{u}\right) = \frac{2(u^4 - v^4) + uv(1 - 4u^2 v^2)}{uv} = -3,$$

he obtained triplication formula

$$\frac{dz}{\sqrt{(1-z^2)(1-u^8z^2)}} = \frac{-3\,dx}{\sqrt{(1-x^2)(1-u^8x^2)}}.$$

To get $+3$ instead of -3, it was sufficient to change z to $-z$.
In the case of the quintic transformation, Jacobi set[29]

$$V = 1 + b_1x^2 + b_2x^4, \quad U = x(a_1 + a_2x^2 + a_3x^4), \quad A = 1 + \alpha x + \beta x^2.$$

From the equation $V + U = (1+x)AA$, he found

$$b_1 = 2\alpha + 2\beta + \alpha\alpha, \quad b_2 = \beta(2\alpha + \beta),$$

$$a_1 = 1 + 2\alpha, \quad a_2 = 2\beta + \alpha\alpha + 2\alpha\beta, \quad a_3 = \beta\beta.$$

He gave the details in section 15 of his *Fundamenta*. He presented the modular equation in the form

$$u^6 - v^6 + 5u^2v^2(u^2 - v^2) + 4uv(1 - u^4v^4) = 0.$$

In 1858, Hermite used this relation to solve a quintic equation, just as Viète solved a cubic by means of trigonometric functions,

34.8 Jacobi's Transcendental Theory of Transformations

Euler, Legendre, and others were aware of the fact that the addition formula for elliptic integrals solved the problem of the multiplication or division of an elliptic integral by an integer. In transformation theory, the multiplication was accomplished in two steps. The first step was to apply a transformation that gave a new elliptic integral with a modulus λ^2 smaller than the original modulus k^2. This was followed by a second transformation serving to increase the modulus. Jacobi discovered these facts about transformation theory by the summer of 1827, at least in the cases of the cubic and quintic transformations. To develop the theory in general, he had to invert the elliptic integral and work with elliptic functions. In his December 1827 paper, however, he gave only the first transformation because he did not define elliptic functions of a complex variable.[30] It was after he introduced complex periods in the spring of 1828 that he was able to develop the complete transformation theory as presented in his *Fundamenta Nova*. He explained how the two transformations arose and also the manner in which they were related to the complementary transformations. To obtain a glimpse of the general theory, we consider the cubic transformation in some detail from the transcendental viewpoint. For the most part, we follow the

[29] ibid. pp. 77–79.
[30] ibid. pp. 39–48.

exposition from Cayley's *Elliptic Functions*.[31] in which he also presented Jacobi's work in streamlined form.

It can be shown by means of the addition formula for the elliptic function $x = \text{sn}(u,k)$ that if $z = \text{sn}(3u,k)$, then

$$z = \frac{3x\left(1 - \frac{x^2}{a_1^2}\right)\left(1 - \frac{x^2}{a_2^2}\right)\left(1 - \frac{x^2}{a_3^2}\right)\left(1 - \frac{x^2}{a_4^2}\right)}{(1 - k^2 a_1^2 x^2)(1 - k^2 a_2^2 x^2)(1 - k^2 a_3^2 x^2)(1 - k^2 a_4^2 x^2)}, \tag{34.77}$$

where $\quad a_1 = \text{sn}\dfrac{4K}{3}, \ a_2 = \text{sn}\dfrac{4iK'}{3}, \ a_3 = \text{sn}\dfrac{4K + i4K'}{3}, \ a_4 = \text{sn}\dfrac{-4K + i4K'}{3}.$

Also, it follows from a formula of Legendre that a_1, a_2, a_3, a_4 are the roots of

$$3 - 4(1 + k^2)x^2 + 6k^2 x^4 - k^4 x^8 = 0.$$

Note that Legendre knew that (34.77) was an integral of the differential equation

$$\frac{dz}{\sqrt{(1 - z^2)(1 - k^2 z^2)}} = \frac{3\,dx}{\sqrt{(1 - x^2)(1 - k^2 x^2)}}. \tag{34.78}$$

Now from Jacobi's algebraic theory presented in the Section 34.7, it follows that the first transformation has the form

$$y = \frac{\frac{x}{M}\left(1 - \frac{x^2}{a_1^2}\right)}{1 - k^2 a_1^2 x^2}, \tag{34.79}$$

where M is to be determined. Recall that Jacobi required the existence of a polynomial A such that $V - U = (1 - x)A^2$ where $y = \frac{U}{V}$. This means that the value $x = 1$ can be required to correspond to $y = 1$. Taking these values for x and y in (34.79), we see that

$$M = -\frac{1 - a_1^2}{a_1^2(1 - k^2 a_1^2)},$$

$$1 - y = \frac{(1 - k^2 a_1^2 x^2) - \frac{x}{M}\left(1 - \frac{x^2}{a_1^2}\right)}{D},$$

where $D = 1 - k^2 a_1^2 x^2$. We can rewrite the numerator of $1 - y$ as

$$(1 - x)\left(1 - \left(\frac{1}{M} - 1\right)x - \frac{x^2}{Ma_1^2}\right).$$

---

[31] Cayley (1895) pp. 206–210.

Now let $A = \frac{1-x}{f}$ so that, for consistency, we require that

$$1 - \left(\frac{1}{M} - 1\right)x - \frac{x^2}{Ma_1^2} = 1 - \frac{2}{f}x + \frac{x^2}{f^2}.$$

Equating coefficients, we get

$$\frac{2}{f} = -\frac{1 - k^2 a_1^4}{1 - a_1^2} \quad \text{and} \quad \frac{1}{f^2} = \frac{1}{Ma_1^2} = \frac{1 - k^2 a_1^2}{1 - a_1^2}. \tag{34.80}$$

These relations are consistent because, by the addition formula and periodicity of sn u,

$$\text{sn}\,\frac{8K}{3} = -\text{sn}\,\frac{4K}{3} = \frac{2\text{sn}\left(\frac{4K}{3}\right)\text{cn}\left(\frac{4K}{3}\right)\text{dn}\left(\frac{4K}{3}\right)}{1 - k^2\text{sn}^4\left(\frac{4K}{3}\right)},$$

$$\text{or} \quad 2\sqrt{1 - a_1^2}\sqrt{1 - k^2 a_1^2} = -(1 - k^2 a_1^4).$$

Hence,

$$1 + y = \frac{(1 + x)(1 + \frac{x}{f})^2}{D}.$$

The next step is to determine λ by using the invariance of the transformation (34.79) under the change x to $\frac{1}{kx}$ and y to $\frac{1}{\lambda y}$. This gives

$$\lambda = M^2 k^3 a_1^4 = \frac{k^3(1 - a_1^2)^2}{(1 - k^2 a_1^2)^2}.$$

Note that since a_1 is real, we have $1 - a_1^2 < 1 - k^2 a_1^2$ and λ is smaller than k. It is also easy to check that

$$1 - \lambda y = \frac{(1 - kx)(1 - kfx)^2}{D}; \quad 1 + \lambda y = \frac{(1 + kx)(1 + kfx)^2}{D}. \tag{34.81}$$

It follows that

$$\frac{M\,dy}{\sqrt{(1 - y^2)(1 - \lambda^2 y^2)}} = \frac{dx}{\sqrt{(1 - x^2)(1 - k^2 x^2)}}. \tag{34.82}$$

Then, by means of an algebraic calculation, obtain $\sqrt{\lambda k} + \sqrt{\lambda' k'} = 1$.

Now for the second transformation, we require that if it is applied after the first, we get triplication. Note that (34.79) implies (34.82). Therefore, we want a transformation

$$z = \frac{3My\left(1 - \frac{y^2}{\theta^2}\right)}{1 - \lambda^2\theta^2 y^2} \tag{34.83}$$

such that

$$\frac{dz}{\sqrt{(1-z^2)(1-k^2z^2)}} = \frac{3M\,dy}{\sqrt{(1-y^2)(1-\lambda^2y^2)}}. \tag{34.84}$$

Thus, (34.78) must hold in this case. Next note that if the value of y given by (34.79) when substituted in (34.83) were to produce (34.77), then (34.78) would hold true. Moreover, it can be shown that if we take $\theta = -\frac{a_2a_3a_4}{Ma_1^2}$, then

$$1 - \frac{y}{\theta} = \frac{\left(1-\frac{x}{a_2}\right)\left(1-\frac{x}{a_3}\right)\left(1-\frac{x}{a_4}\right)}{D},$$

$$1 - \lambda\theta y = \frac{(1-ka_2x)(1-ka_3x)(1-ka_4x)}{D},$$

and there are similar formulas for $1 + \frac{y}{\theta}, 1 + \lambda\theta y$ where the sign of x is changed. So this value of θ in (34.83) indeed produces the desired result (34.84). Moreover θ is related to λ as a_1 to k, that is, θ is a solution of

$$3 - 4(1+\lambda^2)\theta^2 + 6\lambda^2\theta^4 - \lambda^4\theta^8 = 0.$$

In fact, it can be shown that θ may be taken to be the purely imaginary value

$$a_2 = \operatorname{sn}\left(\frac{4iK'}{3}\right).$$

This implies that θ^2 is real and negative and that (34.83) is a real transformation. Transformations similar to (34.79), wherein a_1 is replaced by a_3 or a_4, contain complex numbers.

In general, for an odd integer n, Cayley gave Jacobi's transformation formulas in the form[32]

$$y = \frac{\frac{x}{M}\prod_{s=1}^{n}\left(1-\frac{x^2}{\operatorname{sn}^2 2s\omega}\right)}{\prod_{s=1}^{n}\left(1-k^2(\operatorname{sn}^2 2s\omega)x^2\right)},$$

where m_1 and m_2 were integers and

$$\omega = \frac{m_1K+m_2iK'}{n}.$$

Denoting the denominator on the right-hand side by D, he showed that under the conditions

$$\lambda = k^n \prod_{s=1}^{n} \operatorname{sn}^4(K-2s\omega), \tag{34.85}$$

---

[32] ibid. pp. 251–255.

$$\lambda' = \frac{k'^n}{\prod_{s=1}^{n} dn^4(2s\omega)},$$

$$M = (-1)^{\frac{n-1}{2}} \prod_{s=1}^{n} \frac{sn^2(K - 2s\omega)}{sn^2(2s\omega)}, \tag{34.86}$$

the expressions for $1 - y$, $1 + y$, $1 - \lambda y$, $1 + \lambda y$ were consistent with each other and with the expression for y:

$$(1 - y)D = (1 - x) \prod_{s=1}^{n} \left(1 - \frac{x}{sn(K - 2s\omega)}\right)^2,$$

$$(1 + y)D = (1 + x) \prod_{s=1}^{n} \left(1 + \frac{x}{sn(K - 2s\omega)}\right)^2,$$

$$(1 - \lambda y)D = (1 - kx) \prod_{s=1}^{n} \left(1 - kx\, sn(K - 2s\omega)\right)^2,$$

$$(1 + \lambda y)D = (1 + kx) \prod_{s=1}^{n} \left(1 + kx\, sn(K - 2s\omega)\right)^2.$$

These equations implied the differential equation (34.84). He also rewrote the transformation formulas in the form

$$sn\left(\frac{u}{M}, \lambda\right) = \frac{sn\,u}{M} \left(\prod_{s=1}^{n} \left(1 - \frac{sn^2 u}{sn^2(2s\omega)}\right)\right) \div D, \tag{34.87}$$

$$cn\left(\frac{u}{M}, \lambda\right) = cn\,u \left(\prod_{s=1}^{n} \left(1 - \frac{sn^2 u}{sn^2(K - 2s\omega)}\right)\right) \div D,$$

$$dn\left(\frac{u}{M}, \lambda\right) = sn\,u \left(\prod_{s=1}^{n} (1 - k^2 sn^2(K - 2s\omega) sn^2 u)\right) \div D,$$

where

$$D = \prod_{s=1}^{n} (1 - k^2 sn^2(2s\omega) sn^2 u), \quad \text{and} \quad sn\,u = sn(u, k).$$

The real transformations corresponded to the cases $\omega = \frac{K}{n}$ and $\omega' = \frac{iK'}{n}$. Then, by applying the imaginary transformation, Jacobi obtained the transformations for the moduli ω' and λ'. This meant that the transformation for $\frac{K}{n}$, that is the first transformation, was converted to the form of the second transformation, arising from $\frac{iK'}{n}$.

Cayley presented Jacobi's relation between the complete integrals K and Λ by observing[33] that for $\omega = \frac{K}{n}$ the least positive value for which $\operatorname{sn}\left(\frac{u}{M}, \lambda\right)$ vanished in (34.87) was given by $\frac{u}{M} = 2\Lambda$, while on the right-hand side it was given by $u = \frac{2K}{n}$. Hence,

$$2M\Lambda = \frac{2K}{n} \quad \text{or} \quad \frac{K}{nM} = \Lambda. \tag{34.88}$$

Note that this relation came from the first transformation, since ω was taken to be $\frac{K}{n}$. Jacobi denoted the value of M by M_1 in the second transformation, where ω was taken to be $\frac{iK'}{n}$. Since $\operatorname{sn}^2(2s\omega)$ was negative in this case, Jacobi noted that the smallest value of u for which the right-hand side of (34.87) vanished was given by $u = 2K$. Hence, he obtained

$$2M_1\Lambda_1 = 2K \quad \text{or} \quad \frac{K}{M_1} = \Lambda_1. \tag{34.89}$$

On the other hand, the transformations for the complementary moduli gave Jacobi the relations

$$\Lambda' = \frac{K'}{M} \quad \text{and} \quad \Lambda_1 = \frac{K'}{nM_1}. \tag{34.90}$$

The first relation combined with (34.88) produced the modular equation

$$\frac{\Lambda'}{\Lambda} = n\frac{K'}{K},$$

while the second together with (34.89) gave

$$\frac{K'}{K} = n\frac{\Lambda'_1}{\Lambda_1}.$$

Jacobi also found transformations easily derivable from the first and second transformations; he named these supplementary transformations and used them to obtain product expansions for elliptic functions. For example, he started with the second transformation

$$\operatorname{sn}\left(\frac{u}{M_1}, \lambda_1\right) = \frac{\operatorname{sn} u}{M_1}\left(\prod_{s=1}^{n}\left(1 - \frac{\operatorname{sn}^2 u}{\operatorname{sn}^2\left(\frac{2siK'}{n}\right)}\right)\right) \div \prod_{s=1}^{n}\left(1 - \frac{\operatorname{sn}^2 u}{\operatorname{sn}^2\left(\frac{(2s-1)iK'}{n}\right)}\right), \tag{34.91}$$

where $\operatorname{sn} u = \operatorname{sn}(u, k)$. He changed k into λ so that λ_1 then changed to k. Denoting the new value of M_1 by M', he had the relations

$$M_1 = \frac{K}{\Lambda_1}, \quad M' = \frac{\Lambda}{K} = \frac{1}{nM}, \quad \text{or} \quad n = \frac{1}{MM'}.$$

---

[33] ibid. pp. 261–281.

Then, replacing u by $\frac{u}{M}$ in the transformation obtained after changing k to λ, he reached his first supplementary transformation

$$\operatorname{sn}(nu,k) = nM\operatorname{sn}\left(\frac{u}{M},\lambda\right)\frac{N}{D}, \tag{34.92}$$

where

$$N = \prod_{s=1}^{n}\left(1 - \frac{\operatorname{sn}^2(\frac{u}{M},\lambda)}{\operatorname{sn}^2(\frac{2si\Lambda'}{n},\lambda)}\right)$$

and

$$D = \prod_{s=1}^{n}\left(1 - \frac{\operatorname{sn}^2(\frac{u}{M},\lambda)}{\operatorname{sn}^2(\frac{(2s-1)i\Lambda'}{n},\lambda)}\right).$$

Similarly, Jacobi had formulas for the functions sn and dn.

34.9 Jacobi: Infinite Products for Elliptic Functions

In 1828–1829, Jacobi obtained his initial infinite products for the elliptic functions sn, cn, and dn.[34] To do this, he took his order n supplementary transformations for these functions, such as (34.92) for sn, and let the integer n tend to infinity. He noted that since k^2 was less than 1, it followed that k^n tended to zero and hence by equation (34.85) $\lambda = 0$, $\operatorname{am}(u,\lambda) = u$, and $\operatorname{sn}(\theta,\lambda) = \sin\theta$. This then implied that the corresponding complete integral Λ was equal to $\frac{\pi}{2}$. Moreover, since by (34.88) and (34.90) $\Lambda = \frac{K}{nM}$ and $\Lambda' = \frac{K'}{M}$, it followed that

$$nM = \frac{2K}{\pi}, \quad \frac{\Lambda'}{n} = \frac{K'}{nM} = \frac{\pi K'}{2K}. \tag{34.93}$$

Jacobi also had

$$\operatorname{am}\left(\frac{u}{nM},\lambda\right) = \operatorname{am}\left(\frac{u}{nM},0\right) = \frac{u}{nM} = \frac{\pi u}{2K}$$

and he set

$$\operatorname{sn}\left(\frac{u}{M},\lambda\right) = \sin\left(\frac{\pi u}{2K}\right) = y.$$

Replacing nu by u in (34.92), he let $n \to \infty$ to obtain the product formula:

$$\operatorname{sn} u = \frac{2Ky}{\pi}\frac{\left(1 - \frac{y^2}{\sin^2\frac{i\pi K'}{K}}\right)\left(1 - \frac{y^2}{\sin^2\frac{2i\pi K'}{K}}\right)\left(1 - \frac{y^2}{\sin^2\frac{3i\pi K'}{K}}\right)\cdots}{\left(1 - \frac{y^2}{\sin^2\frac{i\pi K'}{2K}}\right)\left(1 - \frac{y^2}{\sin^2\frac{3i\pi K'}{2K}}\right)\left(1 - \frac{y^2}{\sin^2\frac{5i\pi K'}{2K}}\right)\cdots}.$$

[34] Jacobi (1969) vol. 1, pp. 141–146.

In a similar way, he got

$$\operatorname{cn} u = \sqrt{1 - y^2} \, \frac{\left(1 - \dfrac{y^2}{\cos^2 \frac{\pi i K'}{K}}\right)\left(1 - \dfrac{y^2}{\cos^2 \frac{2\pi i K'}{K}}\right)\left(1 - \dfrac{y^2}{\cos^2 \frac{3\pi i K'}{K}}\right)\cdots}{\left(1 - \dfrac{y^2}{\sin^2 \frac{\pi i K'}{2K}}\right)\left(1 - \dfrac{y^2}{\sin^2 \frac{3\pi i K'}{2K}}\right)\left(1 - \dfrac{y^2}{\sin^2 \frac{5\pi i K'}{2K}}\right)\cdots}, \qquad (34.94)$$

and

$$\operatorname{dn} u = \frac{\left(1 - \dfrac{y^2}{\cos^2 \frac{\pi i K'}{2K}}\right)\left(1 - \dfrac{y^2}{\cos^2 \frac{3\pi i K'}{2K}}\right)\left(1 - \dfrac{y^2}{\cos^2 \frac{5\pi i K'}{2K}}\right)\cdots}{\left(1 - \dfrac{y^2}{\sin^2 \frac{\pi i K'}{2K}}\right)\left(1 - \dfrac{y^2}{\sin^2 \frac{3\pi i K'}{2K}}\right)\left(1 - \dfrac{y^2}{\sin^2 \frac{5\pi i K'}{2K}}\right)\cdots}. \qquad (34.95)$$

Recall that Abel obtained his similar product formula (34.27) for $\phi\alpha$ using a different method. Jacobi then set $e^{-\frac{\pi K'}{K}} = q$, $u = \frac{2Kx}{\pi}$, and $y = \sin x$ to obtain

$$\sin \frac{m\pi i K'}{K} = \frac{q^m - q^{-m}}{2i} = \frac{i(1 - q^{2m})}{2q^m}, \qquad (34.96)$$

$$\cos \frac{m\pi i K'}{K} = \frac{q^m + q^{-m}}{2} = \frac{1 + q^{2m}}{2q^m}, \qquad (34.97)$$

$$1 - \frac{y^2}{\sin^2 \frac{m\pi i K'}{K}} = 1 + \frac{4q^{2m}\sin^2 x}{(1 - q^{2m})^2} = \frac{1 - 2q^{2m}\cos 2x + q^{4m}}{(1 - q^{2m})^2}, \qquad (34.98)$$

$$1 - \frac{y^2}{\cos^2 \frac{m\pi i K'}{K}} = 1 - \frac{4q^{2m}\sin^2 x}{(1 + q^{2m})^2} = \frac{1 + 2q^{2m}\cos 2x + q^{4m}}{(1 + q^{2m})^2}. \qquad (34.99)$$

He was then able to rewrite the products as

$$\operatorname{sn} \frac{2Kx}{\pi} = \frac{2AK}{\pi}\sin x$$
$$\cdot \frac{(1 - 2q^2\cos 2x + q^4)(1 - 2q^4\cos 2x + q^8)(1 - 2q^6\cos 2x + q^{12})\cdots}{(1 - 2q\cos 2x + q^2)(1 - 2q^3\cos 2x + q^6)(1 - 2q^5\cos 2x + q^{10})\cdots},$$

$$\operatorname{cn} \frac{2Kx}{\pi} = B\cos x$$
$$\cdot \frac{(1 + 2q^2\cos 2x + q^4)(1 + 2q^4\cos 2x + q^8)(1 + 2q^6\cos 2x + q^{12})\cdots}{(1 - 2q\cos 2x + q^2)(1 - 2q^3\cos 2x + q^6)(1 - 2q^5\cos 2x + q^{10})\cdots},$$

$$\operatorname{dn} \frac{2Kx}{\pi} = C$$
$$\cdot \frac{(1 + 2q\cos 2x + q^2)(1 + 2q^3\cos 2x + q^6)(1 + 2q^5\cos 2x + q^{10})\cdots}{(1 - 2q\cos 2x + q^2)(1 - 2q^3\cos 2x + q^6)(1 - 2q^5\cos 2x + q^{10})\cdots}.$$

Here

$$A = \left\{ \frac{(1-q)(1-q^3)(1-q^5)\cdots}{(1-q^2)(1-q^4)(1-q^6)\cdots} \right\}^2,$$

$$B = \left\{ \frac{(1-q)(1-q^3)(1-q^5)\cdots}{(1+q^2)(1+q^4)(1+q^6)\cdots} \right\}^2,$$

$$C = \left\{ \frac{(1-q)(1-q^3)(1-q^5)\cdots}{(1+q)(1+q^3)(1+q^5)\cdots} \right\}^2.$$

Jacobi set $x = \frac{\pi}{2}$ and observed that since sn $K = 1$ and dn $K = \sqrt{1-k^2 \text{ sn}^2 K} = \sqrt{1-k^2} = k'$,

$$k' = C \cdot C = C^2 \quad \text{or} \quad C = \sqrt{k'}.$$

To rewrite these formulas in a more useful form, he changed x to $x + \frac{i\pi K'}{2K}$ in the first equation, so that by the addition formula

$$\text{sn}\left(\frac{2Kx}{\pi} + iK' \right) = \frac{1}{k \, \text{sn} \frac{2Kx}{\pi}};$$

$$\cos 2\left(x + i\frac{\pi K'}{2K} \right) = \frac{e^{2ix - \frac{\pi K'}{K}} + e^{-2ix + \frac{\pi K'}{K}}}{2} = \frac{1}{2}\left(q e^{2ix} + \frac{1}{q} e^{-2ix} \right);$$

$$e^{i\left(x + \frac{i\pi K'}{(2K)} \right)} = \sqrt{q}\, e^{ix}.$$

Note that the first product formula could be written as

$$\text{sn} \frac{2Kx}{\pi} = \frac{2AK}{\pi}\left(\frac{e^{ix} - e^{-ix}}{2i} \right)$$

$$\cdot \frac{(1-q^2 e^{2ix})(1-q^2 e^{-2ix})(1-q^4 e^{2ix})(1-q^4 e^{-2ix})\cdots}{(1-q e^{2ix})(1-q e^{-2ix})(1-q^3 e^{2ix})(1-q^3 e^{-2ix})\cdots}. \quad (34.100)$$

Observe that after applying $x \to x + i\frac{\pi K'}{2K}$, the formula would become

$$\frac{1}{k \, \text{sn} \frac{2Kx}{\pi}} = \frac{2AK}{\pi}\left(\frac{\sqrt{q}\, e^{ix} - \frac{1}{\sqrt{q}} e^{-ix}}{2i} \right)$$

$$\cdot \frac{(1-q^3 e^{2ix})(1-q e^{-2ix})(1-q^5 e^{2ix})(1-q^3 e^{-2ix})\cdots}{(1-q^2 e^{2ix})(1-e^{-2ix})(1-q^4 e^{2ix})(1-q^2 e^{-2ix})\cdots}. \quad (34.101)$$

Multiply equations (34.100) and (34.101) to obtain Jacobi's result

$$\frac{1}{k} = \frac{1}{\sqrt{q}} \left(\frac{AK}{\pi}\right)^2 \quad \text{or} \quad A = \frac{\pi \sqrt[4]{q}}{\sqrt{k} K}.$$

Jacobi then set $x = \frac{\pi}{2}$ in (34.100) and applied $C = \sqrt{k'}$ to get

$$1 = \frac{2AK}{\pi} \left\{ \frac{(1+q^2)(1+q^4)(1+q^6)\cdots}{(1+q)(1+q^3)(1+q^5)\cdots} \right\}^2 = \frac{2\sqrt{k'}AK}{\pi B}.$$

Jacobi was then in a position to rewrite the products:

$$\text{sn}\frac{2Kx}{\pi} = \frac{2q^{\frac{1}{4}}}{\sqrt{k}} \sin x \frac{(1-2q^2\cos 2x + q^4)(1-2q^4\cos 2x + q^8)\cdots}{(1-2q\cos 2x + q^2)(1-2q^3\cos 2x + q^6)\cdots}, \quad (34.102)$$

$$\text{cn}\frac{2Kx}{\pi} = \sqrt{\frac{k'}{k}} \cdot 2q^{\frac{1}{4}} \frac{\cos x(1+2q^2\cos 2x + q^4)(1+2q^4\cos 2x + q^8)\cdots}{(1-2q\cos 2x + q^2)(1-2q^3\cos 2x + q^6)\cdots},$$

$$\qquad\qquad (34.103)$$

$$\text{dn}\frac{2Kx}{\pi} = \sqrt{k'} \cdot \frac{(1+2q\cos 2x + q^2)(1+2q^3\cos 2x + q^6)\cdots}{(1-2q\cos 2x + q^2)(1-2q^3\cos 2x + q^6)\cdots}. \quad (34.104)$$

Thus, from the products for A, B, and C, Jacobi had infinite products for $\frac{2K}{\pi}$, k' and k:

$$\frac{2K}{\pi} = \left\{ \frac{(1-q^2)(1-q^4)(1-q^6)\cdots}{(1-q)(1-q^3)(1-q^5)\cdots} \right\}^2 \cdot \left\{ \frac{(1+q)(1+q^3)(1+q^5)\cdots}{(1+q^2)(1+q^4)(1+q^6)\cdots} \right\}^2,$$

$$k' = \left\{ \frac{(1-q)(1-q^3)(1-q^5)\cdots}{(1+q)(1+q^3)(1+q^5)\cdots} \right\}^4,$$

$$k = 4\sqrt{q} \left\{ \frac{(1+q^2)(1+q^4)(1+q^6)\cdots}{(1+q)(1+q^3)(1+q^5)\cdots} \right\}^4.$$

After obtaining $\frac{2K}{\pi}$ as an infinite product, Jacobi applied the triple product identity to express this product as a theta series:[35]

$$\sqrt{\frac{2K}{\pi}} = 1 + 2q + 2q^4 + 2q^9 + 2q^{16} + \cdots. \quad (34.105)$$

This formula laid the basis for Jacobi's results on the sums of squares, two of which reproved theorems of Fermat.

[35] ibid. p. 235.

34.10   Jacobi: Sums of Squares

In 1750, Euler suggested that problems on sums of squares could most naturally be studied through the series whose powers were squares. In his 1828 paper on elliptic functions,[36] Jacobi followed Euler's suggestion, with great success. Though primarily an analyst, Jacobi had a strong interest in number theory, leading him to perceive that his famous formula (34.105) could be employed to obtain Fermat's theorems on sums of two and four squares. In fact, Jacobi also found analytic formulas implying results for sums of six and eight squares. He arrived at all these results, including (34.105), through his product expansions of the doubly-periodic elliptic functions. Recall that in an analogous manner, Euler evaluated the zeta values at the even integers by means of the infinite product expansions of the singly periodic trigonometric functions. Also note that the period K of the elliptic function was obtained as a value of a theta function, while, as in the Madhava–Leibniz formula, the period π of a trigonometric function was expressed as a value of an L-series.

To derive the formulas necessary to work with sums of squares, Jacobi first took the logarithmic derivatives of the product expansions for the elliptic functions sn, cn, and dn.[37] First note

$$\log(1 - 2q^m \cos 2x + q^{2m}) = \log(1 - q^m e^{2ix}) + \log(1 - q^m e^{-2ix})$$

$$= -\sum_{l=1}^{\infty} \frac{q^{lm} \cos 2lx}{l}.$$

Combining this relation with the geometric series $1 - q^l + q^{2l} - \cdots = \frac{1}{1+q^l}$ gives us Jacobi's formulas; he simply wrote them down without details:

$$\log \operatorname{sn} \frac{2Kx}{\pi} = \log \left\{ \frac{2\sqrt[4]{q}}{\sqrt{k}} \sin x \right\} + \frac{2q \cos 2x}{1+q} + \frac{2q^2 \cos 4x}{2(1+q^2)} + \frac{2q^3 \cos 6x}{3(1+q^3)} + \cdots,$$

$$\log \operatorname{cn} \frac{2Kx}{\pi} = \log \left\{ 2\sqrt[4]{q} \sqrt{\frac{k'}{k}} \cos x \right\} + \frac{2q \cos 2x}{1-q} + \frac{2q^2 \cos 4x}{2(1+q^2)} + \frac{2q^3 \cos 6x}{3(1-q^3)} + \cdots,$$

$$\log \operatorname{dn} \frac{2Kx}{\pi} = \log \sqrt{k'} + \frac{4q \cos 2x}{1-q^2} + \frac{4q^3 \cos 6x}{3(1-q^6)} + \frac{4q^5 \cos 10x}{5(1-q^{10})} + \cdots.$$

To obtain the derivatives of these formulas, Jacobi observed that

$$\frac{d}{dx} \log \operatorname{sn} \left(\frac{2Kx}{\pi} \right) = \frac{2k'K}{\pi} \frac{\operatorname{cn}\left(\frac{2Kx}{\pi} \right)}{\operatorname{cn}\left(K - \frac{2Kx}{\pi} \right)}, \tag{34.106}$$

[36]   ibid. pp. 255–263, especially p. 262.
[37]   ibid. pp. 155–170.

$$-\frac{d}{dx}\log \operatorname{cn}\left(\frac{2Kx}{\pi}\right) = \frac{2K}{\pi} \cdot \frac{\operatorname{sn}\left(\frac{2Kx}{\pi}\right)}{\operatorname{sn}\left(K - \frac{2Kx}{\pi}\right)}, \tag{34.107}$$

$$-\frac{d}{dx}\log \operatorname{dn}\left(\frac{2Kx}{\pi}\right) = \frac{2k^2K}{\pi}\operatorname{sn}\left(\frac{2Kx}{\pi}\right)\operatorname{sn}\left(K - \frac{2Kx}{\pi}\right). \tag{34.108}$$

Thus, he obtained

$$\frac{2k'K}{\pi} \cdot \frac{\operatorname{cn}\frac{2Kx}{\pi}}{\operatorname{cn}(K - \frac{2Kx}{\pi})} = \cot x - \frac{4q\sin 2x}{1+q} - \frac{4q^2\sin 4x}{1+q^2} - \frac{4q^3\sin 6x}{1+q^3} - \cdots,$$

$$\frac{2K}{\pi} \cdot \frac{\operatorname{sn}\frac{2Kx}{\pi}}{\operatorname{sn}(K - \frac{2Kx}{\pi})} = \tan x + \frac{4q\sin 2x}{1-q} + \frac{4q^2\sin 4x}{1+q^2} + \frac{4q^3\sin 6x}{1-q^3} + \cdots,$$

$$\frac{2k^2K}{\pi}\operatorname{sn}\frac{2Kx}{\pi}\operatorname{sn}\left(K - \frac{2Kx}{\pi}\right) = \frac{8q\sin 2x}{1-q^2} + \frac{8q^3\sin 6x}{1-q^6} + \frac{8q^5\sin 10x}{1-q^{10}} + \cdots.$$

Note that when $x = \frac{\pi}{4}$ in the second equation, we get the Lambert series for $\frac{2K}{\pi}$:

$$\frac{2K}{\pi} = 1 + \frac{4q}{1-q} - \frac{4q^3}{1-q^3} + \frac{4q^5}{1-q^5} - \cdots.$$

Also, the derivative of the second equation at $x = 0$ gives us the Lambert series for the square of $\frac{2K}{\pi}$:

$$\left(\frac{2K}{\pi}\right)^2 = 1 + \frac{8q}{1-q} + \frac{16q^2}{1+q^2} + \frac{24q^3}{1-q^3} + \cdots.$$

By further manipulation of the products, using differentiation and series expansions, Jacobi obtained formulas for the cubes and fourth powers, as given in sections 40–42 of his *Fundamenta Nova*:

$$\left(\frac{2K}{\pi}\right)^3 = 1 + 16\sum_{n=1}^{\infty}\frac{n^2q^n}{1+q^{2n}} - 4\sum_{n=1}^{\infty}(-1)^{n-1}\frac{(2n-1)^2q^{2n-1}}{1-q^{2n-1}}, \tag{34.109}$$

$$\left(\frac{2K}{\pi}\right)^4 = 1 + 16\sum_{n=1}^{\infty}\frac{n^3q^n}{1+(-1)^{n-1}q^n}. \tag{34.110}$$

The reader may observe that, by expressing the Lambert series in the last four equations as power series in q, we obtain the number of representations of an integer as the sum of two, four, six, and eight squares.

In the final paragraph of his *Fundamenta*, Jacobi gave a number theoretic interpretation of his analytic formula for the sums of four squares, but he did not write down interpretations for the other formulas. In 1865, Henry Smith gave these explicitly, in sections 95 and 127 of his report on number theory:[38]

The number of representations of any uneven (or unevenly even) number by the form $x^2 + y^2$ is the quadruple of the excess of the number of its divisors of the form $4n + 1$, above the number of its divisors of the form $4n + 3$.

The number of representations of any number N as a sum of four squares is eight times the sum of its divisors if N is uneven, twenty-four times the sum of its uneven divisors if N is even.

The number of representations of any number N as a sum of six squares is $4 \sum (-1)^{\frac{\delta-1}{2}} (4\delta'^2 - \delta^2)$, δ denoting any uneven divisor of N, δ' its conjugate divisor. In particular if $N \equiv 1, \mod 4$, the number of representations is $12 \sum (-1)^{\frac{\delta-1}{2}}$; if $N \equiv -1, \mod 4$, it is $-20 \sum (-1)^{\frac{\delta-1}{2}} \delta^2$.

The number of representations of any uneven number as a sum of eight squares is sixteen times the sum of the cubes of its divisors; for an even number it is sixteen times the excess of the cubes of the even divisors above the cubes of the uneven divisors.

In his July 1828 paper in *Crelle's Journal*, Jacobi gave a beautiful application of (34.105) to derive a very efficient proof of the transformation formula for a theta function:[39]

$$\sqrt{\frac{1}{x}} = \frac{1 + 2e^{-\pi x} + 2e^{-4\pi x} + 2e^{-9\pi x} + 2e^{-16\pi x} + \cdots}{1 + 2e^{-\frac{\pi}{x}} + 2e^{-\frac{4\pi}{x}} + 2e^{-\frac{9\pi}{x}} + 2e^{-\frac{16\pi}{x}} + \cdots}. \tag{34.111}$$

Cauchy found this in 1817, and Poisson did so in 1823, though Jacobi referred only to Poisson. Jacobi observed that if the moduli k and k' were interchanged, then K and K' would also be interchanged. Thus, with $x = \frac{K'}{K}$, (34.105) implied

$$\sqrt{\frac{2K'}{\pi}} = 1 + 2e^{-\frac{\pi}{x}} + 2e^{-\frac{4\pi}{x}} + 2e^{-\frac{9\pi}{x}} + 2e^{-\frac{16\pi}{x}} + 2e^{-\frac{25\pi}{x}} + \cdots.$$

Dividing (34.105) by this equation gave him the required transformation. As we shall see in the next section, in 1836 Cauchy applied (34.111) to evaluate a Gauss sum, and in 1840 he provided a more succinct argument.

It is interesting to note that Euler foresaw, albeit vaguely, Jacobi's manner of proof for the four squares theorem and the importance of the transformation of the theta function. In a letter to Goldbach of August 17, 1750,[40] Euler discussed the series

$$1 - x + x^4 - x^9 + x^{16} - x^{25} + \cdots.$$

He wrote that he had approximately evaluated to several decimal places this series for values of x close to 1, a remarkable calculation since the series is very slowly convergent. He commented that it would be very useful if a method could be found for efficiently summing the series for such values. And the transformation of theta functions accomplishes just this task. Moreover, in the same letter Euler mentioned Fermat's remarkable theorem that every number could be expressed as a sum of three

---

[38] See Smith (1965b).
[39] Jacobi (1969) vol. 1, p. 260.
[40] Fuss (1968) vol. 1, pp. 530–532.

triangular numbers, four squares, five pentagonal numbers, and so on. He remarked that the most natural way to prove this proposition might be to show that the coefficient of every power of x must be positive in the series:

$$(1 + x + x^3 + x^6 + \cdots)^3, \quad (1 + x + x^4 + x^9 + \cdots)^4, \quad \text{and so on.}$$

34.11 Cauchy: Theta Transformations and Gauss Sums

Cauchy's 1817 derivation of his transformation of the theta function depended upon the theorem now known as the Poisson summation formula.[41] Cauchy was the first to discover this result, and he did so in the course of his work on the theory of waves. For the Poisson summation formula, consult Section 20.4. Independent of Fourier's earlier work, Cauchy also discovered the reciprocity of the Fourier cosine transform, given by

$$f(x) = \sqrt{\frac{2}{\pi}} \int_0^\infty \phi(t) \cos tx \, dt; \qquad \phi(x) = \sqrt{\frac{2}{\pi}} \int_0^\infty f(t) \cos tx \, dt.$$

He gave his summation formula in the form of the relation

$$\sqrt{\alpha} \sum f(n\alpha) = \sqrt{\beta} \sum \phi(n\beta),$$

where $\alpha\beta = 2\pi$, and the summation was taken over all integers. Cauchy obtained his transformation formula by setting $f(x)$ equal to the function he called the reciprocal function, $e^{-\frac{x^2}{2}}$, and then setting $\phi(x) = e^{-\frac{x^2}{2}}$ in the summation formula. He then took $\alpha = \sqrt{2}a$ and $\beta = \sqrt{2}b$ and stated the transformation as

$$a^{\frac{1}{2}} \left(\frac{1}{2} + e^{-a^2} + e^{-4a^2} + e^{-9a^2} + \cdots \right) = b^{\frac{1}{2}} \left(\frac{1}{2} + e^{-b^2} + e^{-4b^2} + \cdots \right) \quad (34.112)$$

when

$$ab = \pi. \quad (34.113)$$

Note that (34.112) describes the transformation of the theta function (or theta constant)

$$\sum_{n=-\infty}^{\infty} e^{\pi i n^2 \tau}$$

under the mapping $\tau \to -\frac{1}{\tau}$.

Cauchy applied a very interesting idea to evaluate Gauss sums from (34.112). Taking n to be an integer, he set $\tau = \frac{2}{n} + i\frac{\alpha^2}{\pi}$ in the transformation formula, and let $\alpha \to 0$. The asymptotic behavior of the two sides of the formula then yielded the result. Note that the theta function is analytic in the upper half plane and every point of the real line is a singular point.

---

[41] Cauchy (1882–1974) Ser. 1, vol. 1, pp. 5–318, especially pp. 300–303.

In 1840, Cauchy published his quadratic Gauss sum evaluation in *Liouville's Journal*. He noted that (34.112) could be rewritten as

$$a\left(\frac{1}{2}+e^{-a^2}+e^{-4a^2}+\cdots\right)=\sqrt{\pi}\left(\frac{1}{2}+e^{-\frac{\pi^2}{a^2}}+e^{-\frac{4\pi^2}{a^2}}+\cdots\right).$$

For $a=\alpha$, an infinitely small number, this reduced to

$$\alpha\left(\frac{1}{2}+e^{-\alpha^2}+e^{-4\alpha^2}+\cdots\right)=\frac{\sqrt{\pi}}{2}.$$

Cauchy remarked that this step could be verified by the fact that the limit as $\alpha\to 0$ of the product

$$\alpha\left(1+e^{-\alpha^2}+e^{-4\alpha^2}+\cdots\right)$$

was the integral

$$\int_0^\infty e^{-x^2}\,dx=\frac{\sqrt{\pi}}{2}.$$

With n a positive integer and $a^2=-\frac{2\pi}{n}\sqrt{-1}$, $b^2=\frac{n\pi}{2}\sqrt{-1}$, Cauchy could obtain

$$e^{-(n+k)^2a^2}=e^{-k^2a^2};\qquad\qquad(34.114)$$

$$e^{-(2m)^2b^2}+e^{-(2m+1)^2b^2}=1+e^{-\frac{n\pi}{2}\sqrt{-1}}.\qquad\qquad(34.115)$$

He then set $a^2=\alpha^2-\frac{2\pi}{n}\sqrt{-1}$ and $b^2=\beta^2+\frac{n\pi}{2}\sqrt{-1}$ where α and β were infinitely small numbers and where $2\beta=n\alpha$. The last condition was needed to satisfy the requirement $ab=\pi$. After substituting these values of a^2 and b^2 in (34.112), he multiplied the equation by $n\alpha=2\beta$ and remarked that the result was

$$a^{\frac{1}{2}}\Delta=b^{\frac{1}{2}}\left(1+e^{-\frac{n\pi}{2}\sqrt{-1}}\right),\qquad\qquad(34.116)$$

where Δ was the Gauss sum

$$\Delta=1+e^{\frac{2\pi\sqrt{-1}}{n}}+e^{4\cdot\frac{2\pi\sqrt{-1}}{n}}+\cdots+e^{(n-1)^2\cdot\frac{2\pi\sqrt{-1}}{n}}.\qquad\qquad(34.117)$$

From (34.116) and (34.113) Cauchy completed his evaluation:

$$\Delta=\frac{\pi^{\frac{1}{2}}}{a}\left(1+e^{-\frac{n\pi}{2}\sqrt{-1}}\right)$$

$$=\frac{n^{\frac{1}{2}}}{2}(1+\sqrt{-1})\left(1+e^{-\frac{n\pi}{2}\sqrt{-1}}\right).\qquad\qquad(34.118)$$

To see why (34.116) holds true, we note that by (34.114),

$$n\alpha(e^{-a^2} + e^{-4a^2} + e^{-9a^2} + \cdots) = n\alpha \sum_{k=1}^{n} e^{\frac{2\pi i k^2}{n}} \sum_{s=0}^{\infty} e^{-(k+sn)^2 \alpha^2}. \qquad (34.119)$$

Moreover,

$$n\alpha \sum_{s=0}^{\infty} e^{-(k+sn)^2 \alpha^2} = \int_0^{\infty} e^{-x^2}\, dx = \frac{\sqrt{\pi}}{2},$$

and hence the expression (34.119) equals $\frac{\Delta\sqrt{\pi}}{2}$. And, by using (34.115), we can show that

$$2\beta(1 + e^{-b^2} + e^{-4b^2} + e^{-9b^2} + \cdots) = \left(1 + e^{-\frac{n\pi}{2}\sqrt{-1}}\right)\frac{\sqrt{\pi}}{2}.$$

Thus, Cauchy's equation (34.116) is verified.

In the first part of his 1859 *Report on the Theory of Numbers*, Smith noted that Cauchy's method could be applied to derive the more general reciprocity relation for Gauss sums.[42] He set

$$\psi(k,n) = \sum_{s=0}^{n-1} e^{\frac{2\pi i k s^2}{n}}$$

and took

$$a^2 = \alpha^2 - \frac{2m\pi i}{n}, \qquad b^2 = \beta^2 + \frac{ni\pi}{2m}, \qquad i = \sqrt{-1}$$

in (34.112) to find the reciprocity relation

$$\psi(m,n) = \frac{1}{4}\sqrt{\frac{n}{m}}(1+i)\psi(-n,4m). \qquad (34.120)$$

He also observed that from $\psi(-4v, 4m) = 4\psi(-v, m)$ and (34.120), it followed that

$$\psi(m, 4v) = 2\sqrt{\frac{v}{m}}(1+i)\psi(-v, m),$$

so that the case with even n would depend upon the case with odd n. Note that it was essentially this expression for the reciprocity of Gauss sums that Schaar obtained in 1850 by using Fourier series. See Section 19.8.

Henry John Stephen Smith, son of an Irish lawyer, studied at Oxford, where mathematics was not then popular. He independently read in detail the number theoretic work of Gauss, Dirichlet, Eisenstein, Jacobi, Kummer, and others; he became

---

[42] See Smith (1965b) p. 54, footnote. This page also contains several references to Cauchy's papers on this topic.

the most outstanding British number theorist of the nineteenth century. An active member of the British Association for the Advancement of Science, he wrote his well-known report on number theory for the association. The report covered developments in number theory up to the 1850s. In spite of his important researches in number theory and elliptic functions, he worked alone without a following and was mostly ignored in his lifetime. Smith is most noted for his 1867 work was on the representation of numbers as sums of squares. Although it established Eisenstein's unproved theorems on sums of five and seven squares, this important paper remained unnoticed. In fact, as late as 1883, the Paris Academy offered a prize for the proof of Eisenstein's results. Fortunately, this brought Smith's work to the notice of mathematicians and also succeeded in gaining some prominence for the 18-year-old Hermann Minkowski (1864–1909), who offered his own highly original paper on the topic.

34.12 Eisenstein: Reciprocity Laws

Even before his great 1847 paper laying the foundations for a new theory of elliptic functions, Eisenstein used Abel's formulas to make some original applications of elliptic functions to number theory. In 1845, Eisenstein published "Application de l'algèbre à l'arithmétique transcendante," in which he used circular and elliptic functions to prove the quadratic and biquadratic reciprocity laws.[43] We review some of the then-known number theoretic results results upon which Eisenstein based his work: Let p be an odd prime. Following Gauss, divide the residues modulo p, namely $1, 2, \ldots, p - 1$, into two classes: $r_1, r_2, \ldots, r_{\frac{p-1}{2}}$ and $-r_1, -r_2, \ldots, -r_{\frac{p-1}{2}}$, so that every residue falls into exactly one class. Eisenstein took one class to be $1, 2, \ldots, \frac{p-1}{2}$. Note that then $-1, -2, \ldots, -\frac{p-1}{2}$ are identical (mod p) to $p - 1, p - 2, \ldots, \frac{p+1}{2}$. A number a, prime to p, is called a quadratic residue modulo p if the equation

$$x^2 \equiv a \;(\text{mod } p) \tag{34.121}$$

has a solution; otherwise, a is a quadratic nonresidue. Eisenstein used a result of Euler now known as Euler's criterion, proved by Euler in a paper read to the Berlin Academy in 1747.[44] The result stated that a number a, prime to p, is a quadratic residue if and only if

$$a^{\frac{p-1}{2}} \equiv 1 \;(\text{mod } p). \tag{34.122}$$

From Fermat's theorem, $a^{p-1} \equiv 1 \;(\text{mod } p)$, and hence $a^{\frac{p-1}{2}} \equiv \pm 1 \;(\text{mod } p)$. With this, Euler's criterion can be proved: If a satisfies (34.121), then (34.122) follows by Fermat's theorem. From the fact that there are exactly $\frac{p-1}{2}$ quadratic residues (mod p), it follows that $x^{\frac{p-1}{2}} \equiv 1 \;(\text{mod } p)$ has at least $\frac{p-1}{2}$ solutions. However, the equation of degree $\frac{p-1}{2}$ has at most $\frac{p-1}{2}$ solutions. Hence these comprise all the solutions.

[43] Eisenstein (1975) vol. 1, pp. 291–298.
[44] Eu. I-2 pp. 62–85. E 134.

To state the law of quadratic reciprocity, we define the Legendre symbol $\left(\frac{a}{p}\right)$ by the equation

$$a^{\frac{p-1}{2}} \equiv \left(\frac{a}{p}\right) \pmod{p}. \tag{34.123}$$

Note that if a is a multiple of p, then we set $\left(\frac{a}{p}\right) = 0$. The law of quadratic reciprocity states that if p and q are odd primes, then

$$\left(\frac{p}{q}\right)\left(\frac{q}{p}\right) = (-1)^{\frac{p-1}{2} \cdot \frac{q-1}{2}}. \tag{34.124}$$

Note that (34.124) is equivalent to the statement that if q is a quadratic residue (mod p), then p is a quadratic residue (mod q) except when both p and q are of the form $4n + 3$. In the latter case, q is a quadratic residue (mod p), if and only if p is not a quadratic residue (mod q).

To begin his proof of the law of quadratic reciprocity,[45] Eisenstein let r denote a number in $1, 2, \ldots, \frac{p-1}{2}$. Then

$$qr \equiv \pm r' \pmod{p}, \tag{34.125}$$

where r' was also contained in $1, 2, \ldots, \frac{p-1}{2}$. Eisenstein observed that since sine was an odd periodic function,

$$\sin\frac{2\pi qr}{p} = \pm \sin\frac{2\pi r'}{p}. \tag{34.126}$$

Therefore (34.125) could be rewritten as

$$qr \equiv r' \frac{\sin\frac{2\pi qr}{p}}{\sin\frac{2\pi r'}{p}} \pmod{p}. \tag{34.127}$$

Substituting the $\frac{p-1}{2}$ different values of r in (34.127) and multiplying, he obtained

$$q^{\frac{p-1}{2}} \Pi r \equiv \Pi r' \prod_{k=1}^{\frac{p-1}{2}} \frac{\sin\frac{2\pi qk}{p}}{\sin\frac{2\pi k}{p}} \pmod{p}. \tag{34.128}$$

Eisenstein saw that Πr and $\Pi r'$ were identical and concluded that

$$q^{\frac{p-1}{2}} \equiv \prod_{k=1}^{\frac{p-1}{2}} \frac{\sin\frac{2\pi qk}{p}}{\sin\frac{2\pi k}{p}} \pmod{p}. \tag{34.129}$$

---

[45] ibid. p. 292.

Note that by Euler's criterion, Eisenstein had found a trigonometric expression for the Legendre symbol $\left(\frac{q}{p}\right)$. By reversing the roles of p and q, he obtained

$$p^{\frac{q-1}{2}} \equiv \prod_{l=1}^{\frac{q-1}{2}} \frac{\sin \frac{2\pi pl}{q}}{\sin \frac{2\pi l}{q}} \pmod{q}. \tag{34.130}$$

At this juncture, Eisenstein employed Euler's factorization, given in Section 15.5,

$$\frac{\sin px}{\sin x} = \frac{(-1)^{\frac{p-1}{2}}}{2^{p-1}} \prod_{k=1}^{\frac{p-1}{2}} \left(\sin^2 x - \sin^2 \frac{2\pi k}{p} \right),$$

to conclude that the product on the right side of (34.130) equaled

$$C \prod_{k=1}^{\frac{p-1}{2}} \prod_{l=1}^{\frac{q-1}{2}} \left(\sin^2 \frac{2\pi l}{q} - \sin^2 \frac{2\pi k}{p} \right), \tag{34.131}$$

where

$$C = \frac{(-1)^{\frac{p-1}{2}\cdot\frac{q-1}{2}}}{2^{\frac{(p-1)(q-1)}{2}}}.$$

For Euler's factorization, see Sections 15.4 and 15.5. Next, by symmetry, Eisenstein had a similar product for (34.129) with the same constant C, but with factors of the form

$$\sin^2 \frac{2\pi k}{p} - \sin^2 \frac{2\pi l}{q}.$$

Thus, each factor in (34.131) was the negative of the corresponding factor in the product for the expression in (34.129) and the number of such factors was $\frac{(p-1)}{2} \cdot \frac{q-1}{2}$. So Eisenstein could obtain the product in (34.130) by multiplying the product in (34.129) by $(-1)^{\frac{p-1}{2}\cdot\frac{q-1}{2}}$. Therefore, employing Euler's criterion, Eisenstein had the reciprocity law

$$\left(\frac{p}{q}\right) \cdot \left(\frac{q}{p}\right) = (-1)^{\frac{p-1}{2}\cdot\frac{q-1}{2}}.$$

Eisenstein gave a similar proof of the biquadratic (quartic) reciprocity law, but used the lemniscatic function instead of the sine function. Again, we consider the backdrop to his work. Even while he was working on the *Disquisitiones Arithmeticae*, Gauss started thinking about extending quadratic reciprocity to cubic and quartic residues. It appears that he very quickly realized that to state these reciprocity laws he had to extend the field of rational numbers by cube roots and fourth roots of unity. It is not clear when Gauss found the law of biquadratic or quartic reciprocity. On October

23, 1813, his mathematical diary noted,[46] "The foundation of the general theory of biquadratic residues which we have sought for with utmost effort for almost seven years but always unsuccessfully at last happily discovered the same day on which our son is born." Strangely, in a letter of April 30, 1807, to Sophie Germain (1776–1831), Gauss had made a similar claim, challenging her to determine the cubic and quartic residue character of 2.[47] Perhaps he discovered the theorem in 1807 and proved it in 1813. In any case, Germain obtained some good results on this problem; she found the quartic character of −4. And Gauss wrote that, especially given the obstacles to women working in mathematics, he was very impressed with her accomplishments. However, Germain's main contribution to number theory was in connection with Fermat's last theorem; she discovered and applied the Germain primes p such that $2p + 1$ was also prime.

Gauss published two papers on biquadratic reciprocity, in 1828 and 1832.[48] The first paper contained a thorough treatment of the biquadratic character of 2 with respect to a prime $p = 4s + 1$. Note that by Euler's criterion, −1 is then a quadratic residue (mod p); further more, p can be expressed as $a^2 + b^2$. Gauss denoted the two solutions of $x^2 \equiv -1 \pmod{p}$ by f and $-f$. He also took a to be odd and b to be even, and he took their signs such that $a \equiv 1 \pmod 4$ and $b \equiv af \pmod{p}$. His theorem stated that 2 satisfied $2^{\frac{p-1}{4}} \equiv 1,\ f - 1,\ -f \pmod{p}$ where $\frac{b}{2}$ was of the form 0, 1, 2, 3, (mod 4), respectively.

In his second paper, Gauss proved that where m and n were integers, the ring $\mathbb{Z}[\sqrt{-1}]$ consisting of $m + ni$, was a unique factorization domain. In 1859, Smith commented on this result:[49] "By thus introducing the conception of imaginary quantity into arithmetic, its domain, as Gauss observes, is indefinitely extended; nor is this extension an arbitrary addition to the science, but is essential to the comprehension of many phenomena presented by real integral numbers themselves." It is clear from Gauss's second paper that since primes of the form $4s + 1$, where s is a positive integer, can be expressed as a sum of two squares, $a^2 + b^2 = (a + ib)(a - ib)$, they are not prime in the ring $\mathbb{Z}[i]$. However, primes of the form $4s + 3$ cannot be factored in $\mathbb{Z}[i]$. We see, therefore, that there are three classes of primes in $\mathbb{Z}[i]$: (a) primes of the form $i^k(4s + 3)$; (b) primes of the form $a + ib$ such that their norm, $N(a + ib) = a^2 + b^2$, is a prime of the form $4s + 1$ in $\mathbb{Z}$; (c) the primes $i^k(1 + i)$, whose norm is 2. Let $m = a + ib$ be a prime such that $a + b$ is an odd integer and $N(m) = p$. In this case, any number n, not a multiple of m in $\mathbb{Z}[i]$, leaves $p - 1$ possible residues when divided by m. For such m and n, the quartic symbol $\left(\frac{n}{m}\right)_4$ takes the values ± 1 and $\pm i$ and is today defined by

$$\left(\frac{n}{m}\right)_4 = n^{\frac{p-1}{4}} \pmod m.$$

[46] See Dunnington (2004) p. 484.
[47] Gauss (1863–1927) vol. 10, part 1, pp. 70–74.
[48] Gauss (1981) pp. 511–586.
[49] Smith (1965b) p. 71.

To prove quartic reciprocity, Eisenstein divided the $p-1$ residues into four classes, with $\frac{p-1}{4}$ residues in each class, such that when r was in one class, ir, $-r$, $-ir$ each fell into a different one of the other classes.[50] He noted that for any n in $\mathbb{Z}[i]$

$$nr \equiv r', ir', -r', -ir' \pmod{m}, \tag{34.132}$$

where r' was in the same class as r.

He set

$$\omega = 4 \int_0^1 \frac{dx}{\sqrt{(1-x^4)}}.$$

Then by the periodicity of the lemniscatic function sl z, for m prime, Eisenstein had

$$\frac{\text{sl}(\frac{nr\omega}{m})}{\text{sl}(\frac{r'\omega}{m})} = 1, i, -1, \text{ or } -i,$$

corresponding to the four cases in (34.132). Hence in all cases,

$$nr \equiv r' \frac{\text{sl}(\frac{nr\omega}{m})}{\text{sl}(\frac{r'\omega}{m})} \pmod{m}.$$

From this he got the formula analogous to (34.129),

$$n^{\frac{p-1}{4}} \equiv \prod_r \frac{\text{sl}(\frac{nr\omega}{m})}{\text{sl}(\frac{r\omega}{m})} \pmod{m}. \tag{34.133}$$

To obtain a formula with m and n interchanged, Eisenstein chose n to be another complex prime $c+id$ with $c+d$ odd and norm q. He divided the residues of nonmultiples of n into four classes represented by ρ, $i\rho$, $-\rho$, $-i\rho$ and concluded that

$$m^{\frac{q-1}{4}} \equiv \prod_\rho \frac{\text{sl}(\frac{m\rho\omega}{n})}{\text{sl}(\frac{\rho\omega}{n})} \pmod{n}. \tag{34.134}$$

Gauss defined the concept of a primary number so that he could express his results in unambiguous form. A number $c+id$, where $c+d$ was odd, was called primary if d was even and $c+d-1$ was evenly even (that is, divisible by 4). This definition was adopted by Eisenstein. We remark by the way that Gauss also suggested a slightly different definition of a primary number, useful in some circumstances; this definition was employed by Dirichlet. It is easy to show and Gauss of course knew that $c+id$, with $c+d$ odd, was primary if and only if $c+id \equiv l \pmod{2+2i}$. It follows that the product of primary numbers is primary and that the conjugate of a primary number is primary. In his work on the division of the lemniscate, Abel showed that for a primary number m

50 Eisenstein (1975) vol. 1, pp. 294–297.

$$\text{sl}\left(\frac{m\omega}{4}\right) = 1.$$

He also proved that

$$\frac{\text{sl}(mv)}{\text{sl}\,v} = \frac{\phi(x^4)}{\psi(x^4)}, \tag{34.135}$$

where $x = \text{sl}\,v$ and where $\phi(x)$ and $\psi(x)$ were polynomials of degree $\frac{p-1}{4}$. See equation (34.45). Eisenstein improved on this by proving that $\psi(x)$ in (34.135) satisfied

$$\psi(x) = i^v x^{\frac{p-1}{4}} \phi\left(\frac{1}{x}\right) \tag{34.136}$$

for some integer v; he also showed that when m was primary, $v = 0$. To prove (34.136), he noted that $y = \frac{x\phi(x^4)}{\psi(x^4)}$ satisfied the differential equation

$$\frac{dy}{\sqrt{1 - y^4}} = \frac{m\,dx}{\sqrt{1 - x^4}}. \tag{34.137}$$

He set $y = \frac{1}{\eta}$, $x = \frac{1}{i^\mu \xi}$ where μ was an integer yet to be determined. This change of variables converted (34.137) to

$$\frac{i^\mu\,d\eta}{\sqrt{(\eta^4 - 1)}} = \frac{m\,d\xi}{\sqrt{(\xi^4 - 1)}}. \tag{34.138}$$

Eisenstein took μ such that (34.138) would be equivalent to

$$\frac{d\eta}{\sqrt{(1 - \eta^4)}} = \frac{m\,d\xi}{\sqrt{(1 - \xi^4)}}$$

and concluded

$$\eta = i^\mu \frac{\xi\psi\left(\frac{1}{\xi^4}\right)}{\phi\left(\frac{1}{\xi^4}\right)};$$

this immediately implied (34.136). Thus, by (34.135) and (34.136) he obtained

$$\text{sl}(mv) = x\frac{\phi(x^4)}{i^v x^{p-1}\phi\left(\frac{1}{x^4}\right)}. \tag{34.139}$$

Eisenstein next set $v = \frac{\omega}{4}$, so that $\text{sl}\,v = 1$ and for primary m, $\text{sl}(mv) = 1$. Thus, for m primary in (34.136), he had $1 = i^{-v}$. He then assumed that n was also a primary prime so that

$$\frac{\text{sl}(nv)}{\text{sl}\,v} = \frac{f(x^4)}{x^{q-1}f\left(\frac{1}{x^4}\right)},$$

where $f(x)$ was a polynomial of degree $\frac{q-1}{4}$. He set

$$\alpha = \text{sl}\left(\frac{r\omega}{m}\right), \quad \beta = \text{sl}\left(\frac{\rho\omega}{n}\right),$$

so that the solutions of $\phi(x^4) = 0$ were of the form $\pm\alpha$, $\pm i\alpha$ and those of $f(x^4) = 0$ were of the form $\pm\beta$, $\pm i\beta$. Thus, he arrived at

$$\frac{\text{sl}(mv)}{\text{sl}\, v} = \frac{\Pi(x^4 - \alpha^4)}{\Pi(1 - \alpha^4 x^4)}, \quad \frac{\text{sl}(nv)}{\text{sl}\,(v)} = \frac{\Pi(x^4 - \beta^4)}{\Pi(1 - \beta^4 x^4)}.$$

When he combined these formulas with (34.133) and (34.134), he obtained

$$n^{\frac{p-1}{4}} \equiv \frac{\Pi(\alpha^4 - \beta^4)}{\Pi(1 - \beta^4 \alpha^4)} \pmod{m},$$

$$m^{\frac{p-1}{4}} \equiv \frac{\Pi(\beta^4 - \alpha^4)}{\Pi(1 - \alpha^4 \beta^4)} \pmod{n}.$$

Eisenstein observed that since there were $\frac{p-1}{4} \cdot \frac{q-1}{4}$ factors in the products, the fundamental theorem on biquadratic residues, or quartic reciprocity, followed immediately.

Eisenstein studied the polynomial $\phi(x)$ in even greater detail later in his 1845 paper[51] "Beiträge zur Theorie der elliptischen Functionen, I." For primary m, he proved that

$$\phi(x) = x^{p-1} + A_1 x^{p-5} + \cdots + m,$$

and showed that all the coefficients $A_1, A_2, \ldots, m$ were divisible by m. Then, in 1850, he published a paper using a generalization of what we now call Eisenstein's criterion to prove the irreducibility of $\phi(x)$. Suppose $f(x) = a_0 x^n + a_1 x^{n-1} + \cdots + a_n$, where $a_j \in \mathbb{Z}[i]$. Also suppose m is a prime in $\mathbb{Z}[i]$ such that m divides $a_1, \ldots, a_n$, but does not divide a_0, and m^2 does not divide a_n. Then $f(x)$ is irreducible over $\mathbb{Z}[i]$. Eisenstein included a statement and proof of this theorem in an 1847 letter to Gauss. But in 1846, Theodor Schönemann, a student of Jacobi and of the Swiss geometer Jakob Steiner, published this theorem for the case where $\mathbb{Z}[i]$ was replaced by $\mathbb{Z}$; this particular case is now known as Eisenstein's criterion or the Schönemann-Eisenstein criterion. Eisenstein acknowledged this work in his 1850 paper.[52]

34.13 Liouville's Theory of Elliptic Functions

Liouville's contributions to this topic are mainly contained in his lectures, published by Borchardt in 1880.[53] However, as we mentioned earlier, Liouville began to grapple with elliptic functions as early as the 1840s. We briefly discuss his early thoughts,

[51] Eisenstein (1975) vol. 1, pp. 299–324.
[52] See Lemmermeyer (2000) p. 254.
[53] Liouville (1880).

contained in his numerous notebooks.[54] When Hermite remarked to Liouville that
one could use Fourier series to prove Jacobi's theorem on the ratio of two independent
periods of a function, Liouville was apparently motivated to prove this and wrote it up
in his notebook on August 1, 1844. A few pages later he included a more direct proof,
supposing that f had real periods α and α', independent over the rationals. Then, using
the fact that α was a period, he noted that f had a Fourier expansion

$$f(x) = \sum A_j \cos\left(\frac{2j\pi x}{\alpha} + \epsilon_j\right).$$

Since α' was also a period, he had

$$A_j \cos\left(\frac{2j\pi x}{\alpha} + \epsilon_j\right) = A_j \cos\left(\frac{2j\pi x}{\alpha} + \epsilon_j + \frac{2j\pi\alpha'}{\alpha}\right).$$

Thus, Liouville concluded that either $A_j = 0$ or $\frac{2j\pi\alpha'}{\alpha} = 2m\pi$, where m was an
integer. The last equation implied that α and α' were commeasurable, or dependent
over the rationals.

These ideas soon led to the statement and proof of the theorem now famous as
Liouville's theorem, that a bounded entire function is a constant. He first proved this
for doubly-periodic functions, using Fourier series. He then extended it to functions
bounded on the Riemann sphere. Assuming the result for periodic functions, he proved
the extension by taking an analytic function $f(z)$ and assuming $|f(z)| \leq M$ for all z.
Then the function $f(\operatorname{sn} z)$, f composed with the Jacobi elliptic function, would be
a doubly-periodic bounded function and hence a constant. Liouville noted that an
application of this theorem was that every algebraic equation had to have a root.
He argued that if $p(x)$ was a polynomial and $\frac{1}{p(x)}$ did not become infinite for any
complex x, then the same would be true for $\frac{1}{p(\operatorname{sn} x)}$, and this was a contradiction. It is
interesting that though Liouville never published this application, it is usually the first
one to be given in textbooks.

Liouville also proved that an elliptic function could not have only one simple pole.
He noted that, on the other hand, if there were two simple poles, then the function
would reduce to the usual elliptic function. In this connection, Liouville showed in his
notebooks that if ϕ had two simple poles, α and β, then there would be a constant D
such that

$$u = \big(\phi(\alpha + x) - D\big) + \big(\phi(\alpha - x) - D\big)$$

was a solution to

$$\left(\frac{du}{dx}\right)^2 = a + bu + cu^2 + du^3 + eu^4.$$

---

[54] See Lützen (1990) chapter 13.

This meant that u was the inverse of the elliptic integral:

$$x = \int \frac{du}{\sqrt{a + bu + cu^2 + du^3 + eu^4}}.$$

The following theorems of Liouville on elliptic functions commonly appear in modern treatments of the topic:

- The number of poles equals the number of zeros, counting multiplicity.
- The sum of the zeros minus the sum of the poles (in a period parallelogram) is a period of the function.
- The sum of the residues is equal to zero. Liouville proved this for functions with only two poles.
- A doubly-periodic function with only one simple pole does not exist.

Liouville wrote that within his new approach, "integrals which have given rise to the elliptic functions and even moduli disappear in a way, leaving only the periods and the points for which the functions become zero or infinite." This important new principle, that a function may be largely defined by its singularities, was greatly extended by Riemann in his remarkable works on functions of a complex variable.

We present Liouville's proofs based on Borchardt's notes. Liouville considered a doubly-periodic function $\phi(z)$ with periods 2ω and $2\omega'$ so that its values were completely defined by its values in the region

$$z = z_0 + u\omega + u'\omega', \quad -1 \le u \le 1, \quad -1 \le u' \le 1.$$

We would now refer to this region as the period parallelogram P_{z_0}. Liouville than assumed that $z = \alpha, z = \alpha_1, z = \alpha_2, \ldots, z = \alpha_{n-1}$ were the n roots of the equation $\phi(z) = \pm\infty$ in this region. Then there would exist constants $G, G_1, \ldots, G_{n-1}$ so that

$$\phi(z) - \left\{ \frac{G}{z - \alpha} + \frac{G_1}{z - \alpha_1} + \frac{G_2}{z - \alpha_2} + \cdots + \frac{G_{n-1}}{z - \alpha_{n-1}} \right\} \tag{34.140}$$

was finite at $\alpha_1, \alpha_2, \ldots, \alpha_n$. In the case where there were multiple roots, so that the (say i) values

$\alpha_p, \alpha_q, \ldots, \alpha_s$ coincided, the sum of simple fractions

$$\frac{G_p}{z - \alpha_p} + \frac{G_q}{z - \alpha_q} + \cdots + \frac{G_s}{z - \alpha_s}$$

had to be replaced by

$$\frac{G_p}{z - \alpha_p} + \frac{G_q}{(z - \alpha_p)^2} + \cdots + \frac{G_s}{(z - \alpha_p)^i}. \tag{34.141}$$

Liouville designated the sum of the fractions as the fractional part of $\phi(z)$ and denoted it by $[\phi(z)]$. He noted that this fractional part played an important role in the calculus of residues, and he showed that a doubly-periodic function without a fractional part was a constant. Liouville did not refer to poles, but we would now say that a doubly-periodic function must have poles. Note that in (34.140) all the poles are simple and in (34.141), α_p is a pole of order i.

Liouville next proved that there could be no doubly-periodic function with a fractional part, $[\phi(z)] = \frac{G}{z-\alpha}$. In other words, there did not exist a doubly-periodic function with just one simple pole in the period parallelogram. To prove this, Liouville set $z - \alpha = t$ so that the fractional part at α would be given by

$$[\phi(z)]^{\alpha} = [\phi(\alpha+t)]^0 = \frac{G}{t}.$$

Similarly,

$$[\phi(\alpha-t)]^0 = -\frac{G}{t},$$

so that

$$[\phi(\alpha+t)+\phi(\alpha-t)]^0 = 0,$$

and therefore $\phi(\alpha+t)+\phi(\alpha-t) = 2c$, where c was a constant. Liouville then set $f(t) = \phi(\alpha+t)-c$, to get $f(t) = -f(-t)$. Since 2ω and $2\omega'$ were periods of f, he obtained

$$f(\omega) = -f(-\omega) = -f(-\omega+2\omega) = -f(\omega) = 0, \tag{34.142}$$

$$f(\omega') = 0, \quad f(\omega+\omega') = 0. \tag{34.143}$$

He then defined a new function $F(t) = f(t) f(t+\omega)$, noting that this function had no singularities; the zeros cancelled with the poles, based on (34.142) and (34.143). This implied that there were constants k, k', and k'' such that

$$f(t) f(t+\omega) = k, \quad f(t) f(t+\omega') = k', \quad f(t) f(t+\omega+\omega') = k''. \tag{34.144}$$

Liouville changed t to $t+w$ in the third equation, obtaining

$$f(t+\omega) f(t+\omega') = k''.$$

Finally, multiplying the first two equations and dividing by the fourth he arrived at

$$\left(f(t)\right)^2 = \frac{k\,k'}{k''}.$$

This implied that $\phi(z)$ was a constant and the result was proved.

Liouville then gave a simple construction of a doubly-periodic function with periods 2ω and $2\omega'$ and poles at α and β. He set

$$\phi(z) = \sum_{i=-\infty}^{\infty} f(z+2i\omega'),$$

where

$$f(z) = \frac{1}{\cos \frac{\pi}{\omega}(z - h) - \cos \frac{\pi}{\omega}h'} \quad \text{and} \quad \alpha = h + h', \ \beta = h - h'.$$

Liouville next analyzed the zeros of an elliptic function. He observed that a doubly-periodic function $\phi(z)$, with poles at α and β, could not have only one simple zero because its reciprocal would have one simple pole, an impossibility. He then showed that $\phi(z)$ could not have three zeros. Supposing a and b to be two of the zeros, he took another function $\psi(z)$ with periods 2ω and $2\omega'$ and poles at a and b. He also set $\psi_1(z) = \psi(z) - \psi(\alpha)$. Clearly, $\psi_1(z)\phi(z)$ had only one pole at β, implying that $\psi_1(z)\phi(z) = $ constant. Now if ϕ had another zero at c, then ψ_1 would have a pole at c. This contradiction proved that $\phi(z)$ had zeros only at a and b. Liouville also proved that if two functions $\phi(z)$ and $\phi_1(z)$ had the same periods with simple poles at α and β, then there existed constants c, c', such that $\phi_1(z) = c\phi(z) + c'$. To prove this he set

$$[\phi(z)] = \frac{G}{z - \alpha} + \frac{H}{z - \beta}, \quad [\phi_1(z)] = \frac{G_1}{z - \alpha} + \frac{H_1}{z - \beta}, \tag{34.145}$$

so that

$$[G\phi_1(z) - G_1\phi(z)] = \frac{GH_1 - G_1H}{z - \beta}.$$

Hence, the result:

$$G\phi_1(z) - G_1\phi(z) = \text{constant}. \tag{34.146}$$

Liouville proceeded to prove that for a doubly-periodic function ϕ, the sum of the zeros was equal to the sum of the poles, modulo some period of the function. He assumed that ϕ had poles at α and β, so that $\phi(\alpha + \beta - z)$ also had poles at α and β. Hence, by the previous result,

$$\phi(z) = c\phi(\alpha + \beta - z) + c'.$$

Replacing z by $\alpha + \beta - z$, he got the relation

$$\phi(\alpha + \beta - z) = c\phi(z) + c',$$

and by subtraction

$$(1 + c)(\phi(z) - \phi(\alpha + \beta - z)) = 0.$$

Liouville noted that if $c = -1$, then

$$\phi(z) + \phi(\alpha + \beta - z) = c'.$$

To prove this impossible, he set $z = \frac{\alpha+\beta}{2} + t$, and $\phi\left(\frac{\alpha+\beta}{2} + t\right) - \frac{1}{2}c' = f(t)$, so that $f(t) = -f(-t)$. From this and seeing that 2ω and $2\omega'$ were periods of $f(t)$, it followed that

$$f(0) = f(\omega) = f(\omega') = f(\omega + \omega') = 0.$$

Since f could not have four roots, he obtained the required contradiction and therefore

$$\phi(z) = \phi(\alpha + \beta - z).$$

By taking the reciprocal of ϕ, Liouville saw that $\phi(z) = \phi(a + b - z)$. He thus arrived at the relation

$$\phi(z) = \phi(\alpha + \beta - z) = \phi(\alpha + \beta - a - b + z)$$

and concluded that

$$\alpha + \beta = a + b + 2m\omega + 2m'\omega'.$$

Liouville went on to show that any doubly-periodic function could be written in terms of functions of the form ϕ. He presented the details for functions with simple poles. Suppose ψ is a function with periods 2ω and $2\omega'$ and

$$[\psi(z)] = \frac{A}{z - \alpha} + \frac{A_1}{z - \alpha_1} + \frac{A_2}{z - \alpha_2} + \cdots.$$

Denote by $\phi(z; \alpha, \alpha_1)$ the function with the same periods as ψ and with simple poles at α and α_1 and let

$$[\phi(z; \alpha, \alpha_1)] = \frac{G_1}{z - \alpha} - \frac{G_1}{z - \alpha_1},$$

$$[\phi(z; \alpha_1, \alpha_2)] = \frac{G_2}{z - \alpha_1} - \frac{G_2}{z - \alpha_2},$$

$$[\phi(z; \alpha_2, \alpha_3)] = \frac{G_3}{z - \alpha_2} - \frac{G_3}{z - \alpha_3} \quad \text{and so on.}$$

Then

$$[\psi(z)] = B_1 [\phi(z; \alpha, \alpha_1)] + B_2 [\phi(z; \alpha_1, \alpha_2)] + B_3 [\phi(z; \alpha_2, \alpha_1)] + \cdots$$

$$\text{with} \quad B_1 = \frac{A}{G_1}, \quad B_2 = \frac{A + A_1}{G_2}, \quad B_3 = \frac{A + A_1 + A_2}{G_3}, \text{etc.}$$

Thus,

$$\psi(z) = B + B_1\phi(z; \alpha, \alpha_1) + B_2\phi(z; \alpha, \alpha_2) + B_3\phi(z; \alpha, \alpha_3) + \cdots.$$

This theorem was then employed to prove that any doubly-periodic function had exactly as many zeros as poles. Here let $\phi(z; \alpha, \beta; a, b)$ denote a function with periods $2w$, $2w'$; poles at α, β; zeros at a, b, with $\alpha + \beta = a + b$. Suppose also that $\psi(z)$ is doubly-periodic with poles at $z = \alpha, \alpha_1, \ldots, \alpha_{n-1}$ and zeros at $z = a, a_1, a_2, \ldots, a_{i-1}$. If $i < n$, Liouville arbitrarily chose $n - i - 1$ numbers $a_i, a_{i+1}, \ldots, a_{n-2}$ and determined $b, b_1, \ldots, b_{n-2}$ by the system of equations

$$b = \alpha + \alpha_1 - a,$$
$$b_1 = \alpha_2 + b - a_1,$$
$$b_2 = \alpha_3 + b_1 - a_2,$$
$$\vdots$$
$$b_{n-2} = \alpha_{n-1} + b_{n-3} - a_{n-2}.$$

He next defined $w(z)$ as

$$\phi(z; \alpha, \alpha_1; a, b) \cdot \phi(z; \alpha_2, b; a_1, b_1) \cdot \phi(z; a_3, b_1; a_2, b_2) \cdot \phi(z; a_{n-1}, b_{n-3}; a_{n-2}, b_{n-2})$$

and noted that $w(z)$ had poles at

$$\alpha, \alpha_1, \alpha_2, \ldots, \alpha_{n-1}$$

and zeros at

$$a, a_1, a_2, \ldots, a_{n-2}, b_{n-2}.$$

If $i < n$, then the function $\frac{w(z)}{\psi(z)}$ had no poles but had zeros at $a_i, \ldots, a_{n-2}, b_{n-2}$, an impossibility. Similarly, for $i > n$, he took the function $\frac{\psi(z)}{w(z)}$ to get a similar contradiction. Thus, $i = n$, $\psi(z) = cw(z)$, and $\psi(z)$ had as many zeros as poles. Also, since ψ had zeros at $z = a, a_1, a_2, \ldots, a_{n-1}$ and $w(z)$ had zeros at $z = a, a_1, \ldots, a_{n-2}, b_{n-2}$, he could conclude that $b_{n-2} = a_{n-1}$. Liouville substituted these values in his system of equations to arrive at

$$\alpha + \alpha_1 + \alpha_2 + \cdots + \alpha_{n-1} = a + a_1 + a_2 + \cdots + a_{n-1}.$$

This implied that the sum of the zeros differed from the sum of the poles by $2m\omega + 2m'\omega'$, where m and m' were integers. In the applications, Liouville derived the differential equation and addition formula satisfied by a function ϕ with simple poles at α and β. He also explained how to obtain the Abel and Jacobi elliptic functions from his general results.

34.14 Hermite's Theory of Elliptic Functions

Charles Hermite had a life-long interest in the theory of elliptic functions. In fact, we can credit Hermite with the complex analytic proofs of the basic results that a doubly-periodic function must have poles and that the sum of the residues in a period parallelogram must be zero. However, even before he gave these proofs in a lost paper of 1849, of which a report by Cauchy exists,[55] Hermite published two papers on elliptic functions. In the second of these, given in 1848 in the *Cambridge and Dublin Mathematical Journal*,[56] Hermite took the ratio of functions with the same period and then found sufficient conditions for this ratio to be doubly-periodic.

In this 1848 paper, he wrote that he had been motivated to develop his approach to elliptic functions by Arthur Cayley's 1845 paper[57] in which Cayley had started with a double infinite product to establish the properties of elliptic functions. Hermite, by contrast, began with the ratio of two periodic functions and he explained later[58] why he did this: He observed that Liouville's theorem amounted to the result that if a periodic entire function

$$f(x) = \sum_{m=-\infty}^{\infty} A_m e^{2m \frac{i\pi x}{\omega_1}}$$

with period ω_1 had another fundamental period ω_2, then $f(x)$ was a constant. Thus, he found it natural to research the fraction

$$f(x) \equiv \frac{\sum_{m=-\infty}^{\infty} A_m e^{2m \frac{i\pi x}{\omega_1}}}{\sum_{m=-\infty}^{\infty} B_m e^{2m \frac{i\pi x}{\omega_1}}}. \tag{34.147}$$

Clearly, $f(x + \omega_1) = f(x)$. In order to establish sufficient conditions for double periodicity, Hermite supposed that

$$f(x + \omega_2) = f(x), \tag{34.148}$$

where Im $\frac{\omega_2}{\omega_1} > 0$. Equation (34.148) led him tp propose to determine the conditions on A_m and B_m such that

$$\sum_{m=-\infty}^{\infty} A_m e^{2m \frac{i\pi x}{\omega_1}} \sum_{m=-\infty}^{\infty} B_m e^{2m \frac{i\pi(x+\omega_2)}{\omega_1}} = \sum_{n=-\infty}^{\infty} B_n e^{2n \frac{i\pi x}{\omega_1}} \sum_{n=-\infty}^{\infty} A_n e^{2n \frac{i\pi(x+\omega_2)}{\omega_1}}. \tag{34.149}$$

With μ taken to be an arbitrarily chosen integer, equating coefficients of $e^{2\mu \frac{i\pi x}{\omega_1}}$ on each side of (34.149), and setting $q_1 = e^{i\pi \frac{\omega_2}{\omega_1}}$, he arrived at

$$\sum_{m=-\infty}^{\infty} A_{\mu-m} B_m q_1^{2m} = \sum_{n=-\infty}^{\infty} A_{\mu-n} B_n q_1^{\mu-n}. \tag{34.150}$$

[55] Hermite (1905–1917) vol. 1, pp. 75–83.
[56] ibid. pp. 71–73. See also the treatment in Tannery and Molk (1972) vol. 2, pp. 152–158.
[57] Cayley (1889–1898) vol. 1, pp. 136–155.
[58] Hermite (1905–1917) vol. 2, pp. 143–148.

Hermite observed that one way of satisfying equation (34.150) was to make it an identity, so that each term $A_{\mu-m} B_m q_1^{2m}$ was equal to $A_{\mu-n} B_n q_1^{\mu-n}$. Moreover, for all values of μ he had

$$A_{\mu-m} B_m q_1^{2m} = A_{\mu-n} B_n q_1^{2(\mu-n)}. \qquad (34.151)$$

Taking $n = m + k$, where k was an integer, he wrote equation (34.151) as

$$\frac{B_m}{B_{m+k}} q_1^{2m} = \frac{A_{\mu-(m+k)}}{A_{\mu-m}} q_1^{2(\mu-(m+k))}. \qquad (34.152)$$

He next noted that since μ was arbitrary, he could set $\mu - (m + k) = m'$, where m' was independent of m, so that (34.152) became

$$\frac{A_{m'}}{A_{m'+k}} q_1^{2m'} = \frac{B_m}{B_{m+k}} q_1^{2m}, \qquad (34.153)$$

an equation each of whose sides must be a constant. Thus

$$\frac{A_m}{A_{m+k}} q_1^{2m} = \text{const.} \quad \text{and} \quad \frac{B_m}{B_{m+k}} q_1^{2m} = \text{const.},$$

and A_m and B_m were solutions of the same difference equation

$$z_{m+k} = q_1^{2m+\alpha} z_m,$$

with α taking account of the constant; the general solution of this difference equation was, as Hermite noted,

$$z_m = q_1^{\frac{m^2}{k} + \alpha \frac{m}{k} - m} u_m,$$

where u_m satisfied the relation $u_{m+k} = u_m$. Therefore

$$A_m = q_1^{\frac{m^2}{k} + \alpha \frac{m}{k} - m} a_m \qquad (34.154)$$

and

$$B_m = q_1^{\frac{m^2}{k} + \alpha \frac{m}{k} - m} b_m \qquad (34.155)$$

could be substituted in (34.147). However, by taking x to be $x + \beta$ for a suitable constant, Hermite could write (34.154) and (34.155) as

$$A_m = q_1^{\frac{m^2}{k}} a_m, \quad \text{and} \quad B_m = q_1^{\frac{m^2}{k}} b_m$$

with

$$a_{m+k} = a_m, \qquad \text{and} \quad b_{m+k} = b_m.$$

Denoting the numerator and denominator of (34.147) by $\Phi(x)$ and $\Pi(x)$ respectively, he had

$$\Phi(x) = \sum_{m=-\infty}^{\infty} a_m q_1^{\frac{m^2}{k}} e^{2m \frac{i\pi x}{\omega_1}} \tag{34.156}$$

and

$$\Pi(x) = \sum_{m=-\infty}^{\infty} b_m q_1^{\frac{m^2}{k}} e^{2m \frac{i\pi x}{\omega_1}}, \tag{34.157}$$

functions having period ω_1, but with respect to ω_2, producing a factor $e^{-k\frac{ik}{\omega_1}(2x+\omega_2)}$. For example,

$$\Phi(x + \omega_2) = \Phi(x) e^{-k\frac{i\pi}{\omega_1}(2x+\omega_2)} \tag{34.158}$$

and similarly for $\Pi(x + \omega_2)$. To prove (34.158), Hermite noted that changing m to $m - k$ in the sum (34.156) left the sum unchanged. Thus

$$\Phi(x + \omega_2) = \sum_{m=-\infty}^{\infty} a_m q_1^{\frac{m^2}{k}+2m} e^{2m\frac{i\pi x}{\omega_1}}$$

$$= \sum_{m=-\infty}^{\infty} a_m q_1^{\frac{(m-k)^2}{k}+2(m-k)} e^{2(m-k)\frac{i\pi x}{\omega_1}}$$

$$= q_1^{-k} e^{-2k\frac{i\pi x}{\omega_1}} \sum_{m=-\infty}^{\infty} a_m q^{\frac{m^2}{k}} e^{2m\frac{i\pi x}{\omega_1}}$$

$$= \Phi(x) e^{-k\frac{i\pi}{\omega_1}(2x+\omega_2)}.$$

Note that since $a_{m+k} = a_m$, the series for $\Phi(x)$ and $\Pi(x)$ can each be broken up into k separate series. Thus

$$\Phi(x) = \sum_{j=0}^{k} a_j \sum_{m=-\infty}^{\infty} q_1^{\frac{(mk+j)^2}{k}} e^{2(mk+j)\frac{i\pi x}{\omega_1}}. \tag{34.159}$$

Hermite considered the particular case $k = 2$ with

$$\omega_1 = 4K, \quad \omega_2 = 2iK', \quad q_1 = e^{-\frac{\pi K'}{2K}}, \quad q = e^{-\pi \frac{K'}{K}}.$$

In this case, (34.159) simplified to

$$\Phi(x) = a_0 \sum_{m=-\infty}^{\infty} q^{m^2} e^{m\frac{i\pi x}{K}} + a_1 \sum_{m=-\infty}^{\infty} q^{\frac{(2m+1)^2}{4}} e^{(2m+1)\frac{i\pi x}{2K}}$$

$$= a_0 \left(1 + 2q \cos 2\left(\frac{\pi x}{2K}\right) + 2q^4 \cos 4\left(\frac{\pi x}{2K}\right) + 2q^9 \cos 9\left(\frac{\pi x}{4K}\right) + \cdots \right)$$

$$+ a_1 \left(2q^{\frac{1}{4}} \cos \left(\frac{\pi x}{2K}\right) + 2q^{\frac{9}{4}} \cos 3\left(\frac{\pi x}{2K}\right) + 2q^{\frac{25}{4}} \cos 5\left(\frac{\pi x}{4K}\right) + \cdots \right).$$

He denoted the coefficient of a_0 by $\Theta_1(x)$ and the coefficient of a_1 by $H_1(x)$ so that he had

$$\Theta_1 \left(\frac{2Kx}{\pi}\right) = 1 + 2q \cos 2x + 2q^4 \cos 4x + 2q^9 \cos 6x + \cdots, \tag{34.160}$$

$$H_1 \left(\frac{2Kx}{\pi}\right) = 1 + 2q^{\frac{1}{4}} \cos x + 2q^{\frac{9}{4}} \cos 3x + 2q^{\frac{25}{4}} \cos 5x + \cdots. \tag{34.161}$$

Hermite's functions Θ_1 and H_1 are very closely related to Jacobi's two basic theta functions as defined in his paper of 1828, "Suite des notices sur les fonctions elliptiques:"[59]

$$\Theta(x) = 1 - 2q \cos 2x + 2q^4 \cos 4x - 2q^9 \cos 6x + \cdots, \tag{34.162}$$

$$H(x) = 2q^{\frac{1}{4}} \sin x - 2q^{\frac{9}{4}} \sin 3x + 2q^{\frac{25}{4}} \sin 5x - \cdots. \tag{34.163}$$

Jacobi noted some properties of his functions and also observed that his elliptic functions sn, cn, and dn could be expressed in terms of Θ and H:

$$\text{sn}\frac{2Kx}{\pi} = \frac{1}{\sqrt{k}} \frac{H(x)}{\Theta(x)},$$

$$\text{cn}\frac{2Kx}{\pi} = \sqrt{\frac{k'}{k}} \frac{H(x + \frac{\pi}{2})}{\Theta(x)},$$

$$\text{dn}\frac{2Kx}{\pi} = \sqrt{k'} \frac{H(x)}{\Theta(x + \frac{\pi}{2})}.$$

In this way, Hermite's theory of doubly-periodic functions succeeded in reproducing Jacobi's theory of elliptic functions as the ratio of theta functions. See also Exercise 7 of this chapter.

Hermite's 1849 paper, communicated to the Acadèmie des Sciences (Paris), showed how Cauchy's calculus of residues could be applied to derive Liouville's basic results on doubly-periodic functions. All that remains of this paper is Cauchy's account,[60] according to which Hermite proved that a doubly-periodic meromorphic function f

[59] Jacobi (1969) vol. I, p. 256.
[60] Hermite (1905–1917) vol. I, pp. 75–83.

must have more than one simple pole in a period parallelogram. His method was to integrate the function along the boundary of the parallelogram. A period parallelogram is defined as the set of points inside the parallelogram $ABCD$, where A, B, C, D can be represented by the complex numbers $\zeta, \zeta + \omega_1, \zeta + \omega_1 + \omega_2, \zeta + \omega_2$ respectively; ω_1 and ω_2 are the fundamental periods of f. Thus, if we denote the boundary of the parallelogram $ABCD$ by L, then Hermite showed that

$$\int_L f(z)dz = 0. \tag{34.164}$$

Note that L can be chosen so that it does not contain any singularities of f. Hermite's argument is identical to the one found in modern text books. He observed that

$$\int_L f(z)dz = \int_A^B + \int_B^C + \int_C^D + \int_D^A f(z)dz. \tag{34.165}$$

Since ω_2 was a period of $f(z)$, he had

$$\int_C^D f(z)dz = \int_B^A f(z + \omega_2)dz = -\int_A^B f(z)dz; \tag{34.166}$$

because ω_1 was also a period, he could write

$$\int_D^A f(z)dz = \int_C^B f(z + \omega_1)dz = -\int_B^C f(z)dz. \tag{34.167}$$

On subsituting (34.166) and (34.167) in (34.165), Hermite had a proof of (34.164). In their 1859 treatise on elliptic functions, Briot and Bouquet applied Hermite's idea of complex integration to prove basic theorems in elliptic functions theory.[61]

34.15 Exercises

(1) If ϕ is the cosine transform of f, then

$$\sqrt{\alpha}\big(f(\alpha) - f(3\alpha) - f(5\alpha) + f(7\alpha) + f(9\alpha) - \cdots\big)$$
$$= \sqrt{\beta}\big(\phi(\beta) - \phi(3\beta) - \phi(5\beta) + \cdots\big),$$

where $\alpha\beta = \frac{\pi}{4}$; and for $\alpha\beta = \frac{\pi}{6}$

$$\sqrt{\alpha}\big(f(\alpha) - f(5\alpha) - f(7\alpha) + f(11\alpha) + f(13\alpha) - \cdots\big)$$
$$= \sqrt{\beta}\big(\phi(\alpha) - \phi(5\alpha) - \phi(7\alpha) + \cdots\big),$$

---

[61] Briot and Bouquet (1859), especially pp. 79–81.

where the integers $1, 5, 7, 11, 13, \ldots$ are prime to 6. See Ramanujan (2000) p. 63.

(2) Let $\frac{\omega}{2}$ denote the complete lemniscatic integral $\int_0^1 \frac{dx}{\sqrt{1-x^4}}$. Show that then

$$\frac{\omega}{4\pi} = \frac{e^{\frac{\pi}{2}}}{e^\pi - 1} - \frac{e^{\frac{3\pi}{2}}}{e^{3\pi} - 1} + \frac{e^{\frac{5\pi}{2}}}{e^{5\pi} - 1} - \cdots,$$

$$\frac{\omega^2}{4\pi^2} = \frac{e^{\frac{\pi}{2}}}{e^\pi + 1} - \frac{3e^{\frac{3\pi}{2}}}{e^{3\pi} + 1} + \frac{5e^{\frac{5\pi}{2}}}{e^{5\pi} + 1} - \cdots.$$

See Abel (1965) vol. 1, p. 351.

(3) For $f(\alpha)$ and $F(\alpha)$ defined by (34.11) and (34.12), and ω by (34.7), show that

$$f\left(\frac{\alpha\omega}{2}\right) = \frac{4\pi}{\omega}\left(\frac{\cos(\frac{\alpha\pi}{2})}{e^{\frac{\pi}{2}} - e^{-\frac{\pi}{2}}} - \frac{\cos(\frac{3\alpha\pi}{2})}{e^{\frac{3\pi}{2}} - e^{-\frac{3\pi}{2}}} + \frac{\cos(\frac{5\alpha\pi}{2})}{e^{\frac{5\pi}{2}} - e^{-\frac{5\pi}{2}}} - \cdots\right),$$

$$F\left(\frac{\alpha\omega}{2}\right) = \frac{4\pi}{\omega}\left(\frac{\cosh(\frac{\alpha\pi}{2})}{e^{\frac{\pi}{2}} - e^{-\frac{\pi}{2}}} - \frac{\cosh(\frac{3\alpha\pi}{2})}{e^{\frac{3\pi}{2}} - e^{-\frac{3\pi}{2}}} + \frac{\cosh(\frac{5\alpha\pi}{2})}{e^{\frac{5\pi}{2}} - e^{-\frac{5\pi}{2}}} - \cdots\right).$$

(4) Consider the equation

$$\frac{dy}{\sqrt{(1 - y^2)(1 - e^2 y^2)}} = \frac{\sqrt{-n}\, dx}{\sqrt{(1 - x^2)(1 - e^2 x^2)}}.$$

Show that if $n = 3$, then e satisfies the equation $e^2 - 2\sqrt{3}e = 1$; and if $n = 5$, then e satisfies $e^3 - 1 - (5 + 2\sqrt{5})e(e - 1) = 0$. Kronecker called these values of e singular moduli. See Abel (1965) vol. 1, pp. 379–384.

(5) Show that if the equation

$$\frac{dy}{\sqrt{(1 - c^2 y^2)(1 - e^2 y^2)}} = \frac{a\, dx}{\sqrt{(1 - c^2 x^2)(1 - e^2 x^2)}}$$

admits of an algebraic solution in x and y, then a is necessarily of the form $\mu' + \sqrt{-\mu}$ when μ and μ' are rational and μ is positive. For such values of a, the moduli e and c can be expressed in radicals. See Abel (1965) vol. 1, pp. 425–428. Kronecker made a deep study, related to algebraic number theory, of complex multiplication. For a historical discussion of this topic, see Vlăduţ (1991). Also see Takase (1994).

(6) Show that the functions θ and θ_1 below satisfy $\frac{\partial^2 f}{\partial x^2} = \frac{4\partial f}{\partial \omega}$:

$$\theta(x) = 1 - 2e^{-\omega}\cos 2x + 2e^{-4\omega}\cos 4x - 2e^{-9\omega}\cos 6x + \cdots,$$

$$\theta_1(x) = 2e^{-\frac{\omega}{4}}\sin x - 2e^{-\frac{9\omega}{4}}\sin 3x + 2e^{-\frac{25\omega}{4}}\sin 5x - \cdots.$$

See Jacobi (1969) vol. 1, p. 259.

(7) Jacobi defined the theta functions $\left(\text{with } q = e^{-\frac{\pi K'}{K}}\right)$:

$$\Theta\left(\frac{2Kx}{\pi}\right) = 1 - 2q\cos 2x + 2q^4\cos 4x - 2q^9\cos 6x + \cdots,$$

$$H\left(\frac{2Kx}{\pi}\right) = 2q^{\frac{1}{4}}\sin x - 2q^{\frac{9}{4}}\sin 3x + 2q^{\frac{25}{4}}\sin 5x - \cdots.$$

Show that for $u = \frac{2Kx}{\pi}$, we have

$$\Theta(u+2K) = \Theta(-u) = \Theta(u), \quad H(u+2K) = H(-u) = -H(u);$$

$$\Theta(u+2iK') = -e^{\frac{\pi(K'-iu)}{K}}\Theta(u), \quad H(u+2iK') = -e^{\frac{\pi(K'-iu)}{K}}H(u);$$

$$\operatorname{sn} u = \frac{H(u)}{(\sqrt{k}\,\Theta(u))}, \quad \operatorname{cn} u = \frac{\sqrt{k'}\,H(u+K)}{(\sqrt{k}\,\Theta(u))}.$$

See Jacobi (1969) vol. 1, pp. 224–231.

(8) Let $w = 2m\omega + 2n\omega'$, where m and n are integers and let $\tau = \frac{\omega'}{\omega}$, with $\operatorname{Im}\tau \neq 0$. Note that σ' denotes the derivative of σ. Define the Weierstrass sigma function by

$$\sigma(u) = u\prod_{m,n}' \left(1 - \frac{u}{w}\right)e^{\frac{u}{w} + \frac{u^2}{2w^2}},$$

where the product is taken over all m and n except $m = n = 0$. Show that

$$\sigma(u) = e^{2\eta\omega v^2}\frac{2\omega}{\pi}\sin v\pi\prod_{n}\left(\frac{1 - 2h^{2n}\cos 2v\pi + h^{4n}}{(1 - h^{2n})^2}\right),$$

where $v = \frac{u}{2\omega}$, $h = e^{\tau\pi i}$, $\eta = \frac{\pi^2}{12\omega} + \sum\csc^2 n\tau\pi$. Show also that

$$\sigma(u\pm 2\omega) = -e^{\pm 2\eta(u\pm\omega)}\sigma(u), \quad \eta = \frac{\sigma'(\omega)}{\sigma(\omega)},$$

$$\sigma(u\pm 2\omega') = -e^{\pm 2\eta'(u\pm\omega')}\sigma(u), \quad \eta' = \frac{\sigma'(\omega')}{\sigma(\omega')}.$$

Prove Legendre's relation $\eta\omega' - \omega\eta' = \frac{\pi i}{2}$, when $\operatorname{Im}\tau > 0$. See Schwarz (1893) pp. 5–9; note that these are Schwarz's notes of Weierstrass's lectures.

(9) Set $\wp(u) = -\frac{d^2}{du^2}\log\sigma(u)$. Show that $\wp'(\omega) = \wp'(\omega') = \wp'(\omega+\omega') = 0$. Set $\wp(\omega) = e_1$, $\wp(\omega+\omega') = e_2$, $\wp(\omega') = e_3$ and show that

$$(\wp'(u))^2 = 4(\wp(u) - e_1)(\wp(u) - e_2)(\wp(u) - e_3).$$

Also prove that

$$\wp(u) - \wp(v) = -\frac{\sigma(u+v)\sigma(u-v)}{\sigma^2(u)\sigma^2(v)},$$

$$\wp(u \pm v) = -\wp(u) - \wp(v) - \left(\frac{\wp'(u) \mp \wp'(v)}{\wp(u) - \wp(v)}\right)^2.$$

The last result is the addition formula for Weierstrass's $\wp$-function. See Schwarz (1893) pp. 10–14.

(10) Let $k^2 = \frac{e_2 - e_3}{e_1 - e_3}$. Set

$$\sigma_1(u) = e^{-\eta u}\frac{\sigma(\omega + u)}{\sigma(\omega)}, \sigma_3(u) = e^{-\eta' u}\frac{\sigma(\omega' + u)}{\sigma(\omega')}.$$

Prove that

$$\frac{\sigma(u)}{\sigma_3(u)} = \frac{\operatorname{sn}(\sqrt{e_1 - e_3} \cdot u, k)}{\sqrt{e_1 - e_3}}$$

and

$$\frac{\sigma_1(u)}{\sigma_3(u)} = \frac{\operatorname{cn}(\sqrt{e_1 - e_3} \cdot u, k)}{\sqrt{e_1 - e_3}}.$$

See Schwarz (1893) pp. 30–35.

34.16 Notes on the Literature

Abel (2007) contains an English translation of Abel's papers on analysis. A 150-page summary of Abel's mathematical work is given in C. Houzel's article in Laudal and Piene (2002). Consult Prasad (1933) for an interesting account of the work of Abel and Jacobi in elliptic functions. Ramanujan was also a master of the theory of elliptic functions. For his work, see Venkatachaliengar and Cooper (2011). For Ramanujan's prolific work on modular equations, see Berndt's helpful summary in Andrews, Askey, Berndt et al. (1988).

As pointed out in the text, the Eisenstein criterion given in algebra textbooks is due to Schönemann. Refer to Lemmermeyer (2000) for its history. A historical account of this criterion is also available in Cox (2004). The entertaining book by Dörrie (1965), page 19, attributes the criterion to Schönemann alone.

35

Irrational and Transcendental Numbers

35.1 Preliminary Remarks

The ancient Greek mathematicians were aware of the existence of irrational numbers; Eudoxus gave his theory of proportions to deal with that awkward situation. The Greeks also considered the problem of constructing a square with area equal to that of a circle. Later generations of mathematicians probably began to suspect that this was not possible; they were possibly almost certain that π was not rational. The sixteenth-century Indian mathematician and astronomer, Nilakantha, wrote in his *Aryabhatyabhasya*, "If the diameter, measured using some unit of measure, were commensurable with that unit, then the circumference would not allow itself to be measured by means of the same unit so likewise in the case where the circumference is measurable by some unit, then the diameter cannot be measured using the same unit."[1] He gave no indication of a proof in any of his works. It appears that the first proof of the irrationality of π was presented to the Berlin Academy by the Swiss mathematician J. H. Lambert (1728–1777) in a 1768 paper.[2] He demonstrated that if $x \neq 0$ was a rational number, then $\tan x$ was irrational. He deduced this from the continued fraction expansion

$$\tan v = \cfrac{1}{v - \cfrac{1}{\dfrac{3}{v} + \cfrac{1}{\dfrac{5}{v} - \cdots}}}.$$

Then, since $\tan \frac{\pi}{4} = 1$, it followed that π was irrational. Lambert's work was based on some results of Euler, who was a colleague of Lambert for about two years at the Berlin Academy. Later, in his 1794 book on geometry, Legendre gave a completely rigorous and concise presentation of Lambert's proof.[3] In particular, he showed that the continued fraction

---

[1] Yushkevich (1964) p. 169.
[2] Lambert (1768).
[3] Legendre (1894) Note VI, pp. 296–304.

$$\cfrac{m}{n + \cfrac{m_1}{n_1 + \cfrac{m_2}{n_2 + \cfrac{m_3}{n_3 + \cdots}}}},$$

where m_i, n_i were nonzero integers, converged to an irrational number when $\frac{m_i}{n_i} < 1$ for all i beyond some i_0. Legendre went a little further than Lambert by observing that the continued fraction for $\tan x$ also implied that π^2 was irrational.

A century before Lambert, James Gregory tried to prove that π was transcendental. Since he was starting from scratch, it is not surprising that Gregory failed. C. Goldbach and D. Bernoulli carried on a correspondence in the 1720s, in which they mentioned that the series they had discovered could not represent rational numbers or even roots of rational numbers. Thus, in a letter to Goldbach of April 28, 1729,[4] Bernoulli commented, concerning the series

$$\log \frac{m+n}{n} = \frac{m}{n} - \frac{n^2}{2m^2} + \frac{n^3}{2m^3} - \frac{n^4}{4m^4} + \cdots, \quad (m,n \text{ positive integers, } m < n),$$

"Moreover, when summed, these are expressible neither in rational numbers, nor even in radicals or irrational numbers." Unfortunately, he had to admit to Goldbach that he had no proof; a proof of the transcendence of this number follows from a theorem proved by Ferdinand Lindemann in 1882.[5] In reply to Bernoulli, Goldbach remarked[6] that it was not known whether, in general, with f rational, the number

$$\sum_{n=1}^{\infty} \frac{1}{n^2 + fn}$$

could be expressed as a root of a rational number. Note, for example, that when $f = 2$, the sum is $\frac{3}{4}$, while with $f = \frac{1}{2}$, the sum is $4\log(\frac{e}{2})$. Goldbach was probably aware of the first result; though he may not have noticed it, he could have derived the second result from Brouncker's series for $\log 2$. Observe that the second result can also be derived from the Euler–Maclaurin summation formula or from Mengoli's inequalities, given in Exercise 4 of Chapter 20. In a later letter of October 20, Goldbach wrote,[7] "Here follows a series of fractions, such as you requested whose sum is neither rational nor the root of any rational number:

$$\frac{1}{10} + \frac{1}{100} + \frac{1}{10000} + \frac{1}{100000000} + \text{etc.}$$

[4] Fuss (1968) vol. 2, pp. 298–304, especially p. 301.
[5] Lindemann (1882).
[6] Fuss (1968) vol. 2, pp. 312–315, especially p. 313.
[7] ibid. pp. 326–327. Translation from Lützen (1990) p. 514.

The general term is

$$\frac{1}{10^{2^{x-1}}}.\text{''}$$

Neither Goldbach nor Bernoulli could suggest any method for attacking these problems. Kurt Mahler (1903–1988), FRS, was a largely self-taught German mathematician who had to leave Germany and had a long career in England, Australia, and the United States. He learned Chinese and encouraged other mathematicians to read Chinese mathematics. In 1926, he proved a theorem more general than required but as a consequence of which Goldbach's number was necessarily transcendental.[8] In 1938, Rodion Kuzmin (1891–1949) also gave a proof of the transcendence of Goldbach's number.[9]

In his 1748 book, *Introductio in Analysin Infinitorum*, Euler made some insightful remarks on the values of the logarithm function,[10] "Since the logarithms of numbers which are not powers of the base are neither rational nor irrational, it is with justice that they are called transcendental quantities." He did not clearly define his meaning of irrational, but from his examples we gather that he meant numbers expressible by radicals. A clear definition of a transcendental number was given by Legendre in a note in his 1794 book: "It is probable that the number π is not even comprised among algebraic irrationals, that is, it cannot be the root of an algebraic equation of a finite number of terms whose coefficients are rational, but it seems very difficult to prove this proposition rigorously."[11]

The first mathematician to rigorously prove the existence of transcendental numbers was Liouville. In 1840, he published two notes showing that e and e^2 could not be solutions of a quadratic equation. The 1843 publication by P. H. Fuss of the Euler, Goldbach, and Bernoulli correspondence further aroused Liouville's interest in transcendental numbers. He read a note on continued fractions to the French Academy in 1844. Given that a continued fraction was the root of an algebraic equation with integral coefficients (in modern terminology, an algebraic number), he stated the condition that the terms of such a continued fraction had to satisfy. In a subsequent paper presented to The Academy and published in the *Comptes Rendus*,[12] he presented his famous criterion for a number to be algebraic of degree n: If x was such a number, then there existed an $A > 0$ such that for all rational $\frac{p}{q} \neq x$,

$$\left| x - \frac{p}{q} \right| > \frac{A}{q^n}.$$

[8] See Mahler (1982) p. 182. Also see the Kurt Mahler online archive for more information on his outstanding career.

[9] Kuzmin (1938).

[10] Euler (1988) p. 80.

[11] Legendre (1894) Note VI, pp. 303–304. Translation from Lützen (1990) p. 516.

[12] Liouville (1844).

He noted an almost immediate consequence of this:

$$\frac{1}{l} + \frac{1}{l^{(2!)}} + \frac{1}{l^{(3!)}} + \cdots + \frac{1}{l^{(n!)}} + \cdots$$

was a transcendental number, for $l > 1$ any integer.

It is clear that Liouville was attempting to prove the transcendence of e and it must have pleased him that his younger friend Hermite did so in 1873. Hermite used the basic identity

$$\int e^{-z} F(z)\, dz = -e^{-z} \gamma(z),$$

where $F(z)$ and $\gamma(z)$ were polynomials. Note that his can be proved by integration by parts and depends on the fact that $\frac{d}{dz} e^{-z} = -e^{-z}$. By means of this formula, Hermite defined certain polynomials with integer coefficients and employed them to obtain simultaneous rational approximations of e^x, for certain integer values of x. These in turn were sufficient to show that, except for the trivial case, there could be no equation of the form

$$e^{z_0} N_0 + e^{z_1} N_1 + \cdots + e^{z_n} N_n = 0, \tag{35.1}$$

when $z_0, z_1, \ldots, z_n$ and $N_0, N_1, \ldots, N_n$ were all integers. We note that in the last portion of his paper, Hermite used his method to obtain the rational approximations

$$e = \frac{58019}{21344}, \qquad e^2 = \frac{157712}{21344}.$$

He left the problem of proving the transcendence of π to others. And soon afterward, in 1883, Ferdinand Lindemann used Hermite's methods to prove this.[13] Lindemann's theorem was a generalization of Hermite's: If $z_0, z_1, \ldots, z_n$ were distinct algebraic numbers and $N_0, N_1, \ldots, N_n$ were algebraic and not all zero, then equation (35.1) could not hold. The equation $1 + e^{i\pi} = 0$ implied the transcendence of π. Lindemann argued that if π were algebraic, then by the preceding theorem, $1 + e^{i\pi}$ could not equal zero. Lindemann's theorem also implied that when $x \neq 0$ was algebraic, then all the numbers e^x, $\arcsin x$, $\tan x$, $\sin^{-1} x$, and $\tan^{-1} x$ were transcendental. Moreover, if x was not equal to one, then $\log x$ was transcendental. Lindemann's proof was somewhat sketchy, but in 1885 Weierstrass gave a completely rigorous proof.[14] In particular, he noted that Lindemann's theorem followed readily from the particular case in which $N_0, N_1, \ldots, N_n$ were integers. A number of mathematicians, including Hilbert, Hurwitz, Markov, Mertens, Sylvester, and Stieltjes improved and streamlined the proofs of Hermite and Lindemann without introducing any essentially new methods or results.

[13] Lindemann (1992).
[14] Weierstrass (1885).

In his famous 1900 lecture at Paris, David Hilbert (1862–1943) gave a list of twenty-three problems for future mathematicians; the seventh of these was to prove the transcendence of certain numbers:[15]

> I should like, therefore, to sketch a class of problems which, in my opinion, should be attacked as here next in order. That certain special transcendental functions, important in analysis, take algebraic values for certain algebraic arguments, seems to us particularly remarkable and worthy of thorough investigation. Indeed, we expect transcendental functions to assume, in general, transcendental values for even algebraic arguments; and, although it is well known that there exist integral transcendental functions, which even have rational values for all algebraic arguments, we shall still consider it highly probable that the exponential function $e^{i\pi z}$, for example, which evidently has algebraic values for all rational arguments z, will on the other hand always take transcendental values for irrational algebraic values of the argument z. We can also give this statement a geometrical form, as follows: *If, in an isosceles triangle, the ratio of the base angle to the angle at the vertex be algebraic but not rational, the ratio between base and side is always transcendental.* In spite of the simplicity of this statement and of its similarity to the problems solved by Hermite and Lindemann, I consider the proof of this theorem very difficult; as also the proof that *The expression α^β, for an algebraic base α and an irrational algebraic exponent β, e.g., the number $2^{\sqrt{2}}$, or $e^\pi = i^{-2i}$, always represents a transcendental or at least an irrational number.* It is certain that the solution of these and similar problems must lead us to entirely new methods and to a new insight into the nature of special irrational and transcendental numbers.

Hilbert's last comment has certainly turned out to be true. The resolution of Hilbert's seventh problem in the 1930s by the efforts of A. O. Gelfond and T. Schneider and the work of C. L. Siegel, the latter more directly inspired by Hermite and Lindemann, have initiated an era of tremendous growth and development in the theory of transcendental numbers. Hilbert himself was not very hopeful of a proof of his theorem within his lifetime, a theorem, as we have seen, also stated by Euler. Hilbert thought, in fact, that the Riemann hypothesis would be proved first.

The Russian mathematician Aleksandr O. Gelfond (1906–1968) took the first important step toward a proof of the Hilbert-Euler conjecture. He was a colleague at Moscow University of I. I. Privalov, whose influence in complex analysis is evident in Gelfond's work. Gelfond was a student of Aleksandr Khinchin who in 1922–23 studied the metrical properties of continued fractions, in which he obtained important results. Khinchin attracted several researchers to a whole range of problems in analytic number theory through his 1925–1926 seminar on this subject at Moscow. Gelfond's early work was influenced by a result in analytic functions due to Pólya:[16] If an entire function assumes integral values for positive rational integral values of its argument and its growth is restricted by the inequality

$$|f(z)| < C2^{\alpha|z|}, \quad \alpha < 1,$$

then it must be a polynomial. Roughly speaking, this means that a transcendental entire function taking integral values at integers must grow at least as fast as 2^z. Concerning

[15] Yandell (2002) p. 404.
[16] Pólya (1974) vol. 1, pp. 1–16.

the connection of this result with transcendental numbers, in his *Transcendental and Algebraic Numbers*, first published in Russian in 1952, Gelfond wrote,[17]

> There is a very essential relationship between the growth of an entire analytic function and the arithmetic nature of its values for an argument which assumes values in a given algebraic field. If we assume in this connection that the values of the function also belong to some definite algebraic field, where all the conjugates of every value do not grow too rapidly in this field, then this at once places restriction on the growth of the function from below, in other words, it cannot be too small. This situation and its analogs for meromorphic functions can be used with success to solve transcendence problems. The first theorem concerning the relationship between the growth and the arithmetic value of a function was the Pólya theorem.

Hardy, Landau, and Okada successively managed to produce a streamlined proof of this result. We briefly sketch the argument, showing that very old ideas on interpolation have continued to play a role in function theory and transcendence theory. First, prove that if an entire function $f(z)$ satisfies

$$\varlimsup_{r \to \infty} \frac{\log M(r, f)}{r} < \log 2,$$

then

$$f(0) + z\Delta f(0) + \frac{z(z-1)}{2!}\Delta^2 f(0) + \frac{z(z-1)(z-2)}{3!}\Delta^3 f(0) + \cdots$$

converges uniformly to $f(z)$ in any finite region of the plane. Note that this is the Briggs–Harriot–Gregory–Newton interpolation series. Thus, if $f(z)$ is of exponential type less than $\log 2$, then $f(z)$ is represented by the interpolation series and can be evaluated at $z = -1$:

$$f(-1) = f(0) - \Delta f(0) + \Delta^2 f(0) - \cdots.$$

The convergence of the series implies that $|\Delta^n f(0)| < 1$ for $n > N$. Moreover, since $\Delta^n f(0)$ is an integer when $f(0), f(1), f(2), \ldots$ are all integers, we may conclude that $\Delta^n f(0) = 0$ for $n > N$. Thus, $f(z)$ is a polynomial and Pólya's theorem is proved.

In 1929, Gelfond took a step closer to solving Hilbert's problem when he used this type of interpolation series to obtain a key transcendence theorem:[18] For $\alpha \neq 0, 1$ and algebraic, $\alpha^{\sqrt{-p}}$ is transcendental when p is a nonsquare positive integer. In particular, $2^{\sqrt{-2}}$ and $(-1)^{-i} = e^\pi$ are transcendental numbers. Gelfond gave details of only the particular case that e^π is transcendental. We present an outline of Gelfond's proof. First enumerate the Gaussian integers $m + in$ as a sequence $z_0, z_1, z_2, \ldots$, where one term precedes another if its absolute value is smaller; if the absolute values are the same, then the term with the smaller argument comes first. Then expand $e^{\pi z}$ as an interpolation series

[17] Gelfond (1960) p. 97.
[18] For a translation, see Gelfond (1960) chapter 3.

$$\sum_{n=0}^{\infty} A_n(z - z_0) \cdots (z - z_{n-1}),$$

where, by Cauchy's theorem, A_n can be expressed as

$$\sum_{k=0}^{n} \frac{e^{\pi z_k}}{B_k}, \quad \text{where} \quad B_k = \prod_{\substack{j=0 \\ j \neq k}}^{n} (z_k - z_j).$$

This interpolation series converges to $e^{\pi z}$ because of the relatively slow growth of the function and the relatively high density of the interpolation points. Now if Ω_n is the least common multiple of $B_0, B_1, \ldots, B_n$, then, by the distribution of the primes of the form $4n + 1$ and $4n + 3$, it can be established that

$$|\Omega_n| = e^{\frac{n \log n}{2} + O(n)} \quad \text{and} \quad |\frac{\Omega_n}{B_k}| = e^{O(n)}.$$

If one assumes that e^{π} is algebraic, then these estimates can be used to show that either $A_n = 0$ or that

$$|\Omega_n A_n| > e^{-O(n)}.$$

However, from the Cauchy integral for A_n, it follows that

$$|\Omega_n A_n| < e^{-\frac{n \log n}{2} + O(n)}.$$

The two inequalities contradict one another for large enough n, unless $A_n = 0$ for all n larger than some value. Thus, one may argue that the interpolation series is finite and hence is a polynomial. This is a contradiction, so that Gelfond could conclude that his assumption that e^{π} was algebraic was false, proving his result.

In 1930, R. Kuzmin showed that,[19] with some modifications in Gelfond's proof, one could prove the transcendence of $\alpha^{\sqrt{p}}$, with α and p as before. One implication of this was that $2^{\sqrt{2}}$ was transcendent, as Hilbert and Euler had conjectured. Since for general algebraic numbers β (in α^{β}), it was no longer possible to find useful upper bounds for Ω_n, a generalization along these lines was difficult. However, in 1933 K. Boehle was able to prove by this method that if $\alpha \neq 0, 1$ and β was an irrational algebraic number of degree $n \geq 2$, then at least one of the numbers $\alpha, \alpha^{\beta}, \alpha^{\beta^2}, \ldots, \alpha^{\beta^{n-1}}$ had to be transcendental.[20] Carl Ludwig Siegel (1896–1981), who had been a student of Landau at Göttingen, also succeeded in proving Kuzmin's result after seeing Gelfond's proof of the transcendence of $\alpha^{\sqrt{-p}}$. But Siegel did not publish his proof in spite of Hilbert's suggestion that he do so. Siegel also made important and very original contributions to the theory of quadratic forms and to modular forms in several variables. His interest in the history of mathematics led him to study Riemann's cryptic unpublished notes on the zeta function and to discover the Riemann–Siegel formula.

[19] Kuzmin (1930).
[20] Boehle (1933).

Though the method of Gelfond did not generalize to α^β, it suggested new lines of research. Gelfond himself applied it to a new proof of Lindemann's theorem; in 1943, his student A. V. Lototskii used it to show that certain infinite products represented irrational numbers;[21] and in 1932 Siegel showed that if g_2 and g_3 were algebraic numbers, then at least one period of the Weierstrass $\wp$ function, satisfying the equation,

$$\wp'(z)^2 = 4\wp(z)^3 - g_2\wp(z) - g_3,$$

was transcendental.[22] In particular, if $\wp(z)$ allowed complex multiplication, then both periods were transcendental. Siegel's student, Theodor Schneider (1911–1988), developed improved methods, allowing him to prove in 1934 that both periods were transcendental and even their ratio was transcendental, except when $\wp(z)$ permitted complex multiplication.[23]

In 1934, Gelfond published a new method[24] by which he obtained the complete proof of Hilbert's seventh problem. This proof made use of complex analysis, but some years later Gelfond and Linnik gave an interesting elementary proof of a special case, without recourse to analysis, except for Rolle's theorem.[25] In his proof of the seventh problem, Gelfond assumed the result false. Thus, he posited that there existed algebraic numbers α, β, where $\alpha \neq 0, 1$ and β was not rational but $\alpha^\beta = \lambda$ was algebraic. On this assumption, there existed algebraic numbers α and β such that $\beta = \frac{\log \lambda}{\log \alpha}$ was an algebraic irrational number. He then constructed a function

$$f(z) = \sum_{k=0}^{N} \sum_{m=0}^{N} C_{k,m} \alpha^{kz} \lambda^{mz} = \sum_{k,m} C_{k,m} e^{(ak+bm)z} \quad (a = \log \alpha, \ b = \log \lambda),$$

where N was a suitably chosen large integer. Also, the $C_{k,m}$ were such that their absolute values and the absolute values of their conjugates were less than e^{2N^2}. Note that $f(z)$ could not be identically zero because $\frac{b}{a}$ had to be irrational; moreover, the derivative of order s could be expressed as

$$f^{(s)}(z) = a^s \sum_{k=0}^{N} \sum_{m=0}^{N} C_{k,m}(k + \beta m)^s \alpha^{kz} \lambda^{mz}.$$

Gelfond proved that if α^β was an algebraic number, then it was possible to choose the $(N+1)^2$ nonzero algebraic numbers $C_{k,m}$ such that $f^{(s)}(z) = 0$ at $z = 0, 1, \ldots, r_2$ for $0 \leq s \leq r_1$, where r_1 was the greatest integer in $\frac{N^2}{\log N}$ and r_2 was the greatest integer in $\log \log N$. All this then implied that $f(z)$ had zeros of sufficiently high order at $0, 1, \ldots, r_2$. By an ingenious argument using Cauchy's integral formula, Gelfond

[21] Lototskii (1943).
[22] Siegel (1932).
[23] Schneider (1934) II.
[24] Gelfond (1934).
[25] Gelfond and Linnik (1966).

then showed that $f(z)$ had a zero of even higher order at $z = 0$—in fact, of order at least $(N + 1)^2 + 1$. Thus, he could conclude that the nonzero algebraic numbers $C_{k,m}$ satisfied the equations

$$a^{-s} f^{(s)}(0) = \sum_{k,m=0}^{N} C_{k,m}(k + \beta m)^s = 0, \quad 0 \le s \le (N + 1)^2.$$

Taking the first $(N + 1)^2$ equations, he obtained a system of equations with a Vandermonde determinant that had to be zero; this could happen if and only if there were integers

$$m_1, k_1; \quad m_2, k_2 \quad \text{such that} \quad \beta m_1 + k_1 = \beta m_2 + k_2.$$

This relation yielded the conclusion that β was rational, a contradiction to Gelfond's assumption, so that the theorem was proved.

In 1934, Schneider obtained an independent solution of Hilbert's seventh problem.[26] His interest in transcendental numbers was aroused by a lecture of Siegel, who subsequently gave him a list from which to choose a dissertation topic. Schneider selected a problem on transcendental numbers; he reported,[27] "After a few months, I gave him a work of six pages and then was told by Siegel that the work contained the solution of Hilbert's seventh problem." Schneider's proof was different in details from Gelfond's, but it too depended on the construction of an auxiliary function with a large number of zeros at specific points. In fact, both these mathematicians had adopted this technique from a previous work of Siegel on transcendence questions related to the values of Bessel functions. One may go further back and observe that in his proof of the transcendence of e, Hermite had also constructed a function of this kind!

Siegel's work of 1929 introduced another important method in the theory of transcendental numbers. Recall that Hermite's work depended on the fact that $\frac{d}{dx} e^x = e^x$. It was not until 1929, when Siegel published his paper on E-functions, that this idea was generalized to prove the transcendence of values of functions satisfying linear differential equations. E-functions are entire functions

$$\sum_{n=0}^{\infty} \frac{a_n}{n!} z^n,$$

where a_n are algebraic numbers satisfying certain arithmetic conditions. First, for any $\epsilon > 0$, a_n and all its conjugates are $O(n^{\epsilon n})$ as $n \to \infty$ and second, the least common denominator of $a_0, a_1, \ldots, a_n$ is also $O(n^{\epsilon n})$. Siegel considered a system of homogeneous linear differential equations of the first order

$$y_k' = \sum_{l=1}^{m} Q_{kl}(x) y_l, \quad \text{for} \quad (k = 1, \ldots, m),$$

[26] Schneider (1934) I.
[27] Yandell (2002) p. 199.

where the $Q_{kl}(x)$ were rational functions with coefficients in a number field K. To obtain transcendence results, Siegel required that some products of powers of the E-function $E_1, E_2, \ldots, E_m$, in fact, solutions of this system, satisfy a normality condition. Siegel formalized the concept of normality in his 1949 book;[28] its meaning was only implicit in his 1929 paper. In spite of the fact that this condition was difficult to verify, thereby limiting the scope of its application, Siegel was able to employ it to rederive the classical theorem of Lindemann (and Weierstrass). He also proved a new theorem on the transcendence of a class of numbers related to the Bessel function: Observing that

$$K_\lambda(x) = \sum_{n=0}^{\infty} \frac{(-1)^n}{n!\,(\lambda+1)\cdots(\lambda+n)} \left(\frac{x}{2}\right)^n \quad (\lambda \neq -1, -2, \ldots),$$

one may verify that K_λ is an E-function and satisfies the differential equation

$$y'' + \frac{2\lambda+1}{x} y' + y = 0.$$

Siegel's theorem states that if λ is a rational number, $\lambda \neq \pm\frac{1}{2}, -1, \pm\frac{3}{2}, -2, \ldots$, and $\alpha \neq 0$ an algebraic number, then $K_\lambda(\alpha)$ and $K_\lambda'(\alpha)$ are algebraically independent. Note that complex numbers $\zeta_1, \zeta_2, \ldots, \zeta_n$ are called algebraically independent if for every nonzero polynomial $P(x_1, \ldots, x_n)$, in n variables with rational coefficients, we have

$$P(\zeta_1, \ldots, \zeta_n) \neq 0.$$

Otherwise, the ζ_j are algebraically dependent. Thus, if several numbers are algebraically independent, then each of them is transcendental. Therefore, $K_\lambda(\alpha)$ and $K_\lambda'(\alpha)$ are transcendental. Also, since the Bessel function $J_\lambda(x)$ may be expressed as

$$\frac{1}{\Gamma(\lambda+1)} \left(\frac{x}{2}\right)^\lambda K_\lambda(x),$$

it follows that except for $x = 0$, all the zeros of $J_\lambda(x)$ and $J_\lambda'(x)$ are transcendental numbers.

From his theorem, Siegel obtained the transcendence of certain continued fractions by noting that

$$i\sqrt{x}\, \frac{K_\lambda(2i\sqrt{x})}{K_\lambda'(2i\sqrt{x})} = \lambda + 1 + \cfrac{x}{\lambda + 2 + \cfrac{x}{\lambda + 3 + \ldots}}$$

Thus, Siegel's theorem implied that when 2λ was not an odd integer, the continued fraction was transcendental for every nonzero algebraic x. But when 2λ was an odd

---

[28] Siegel (1949).

integer, Lindemann's theorem entailed the transcendence of the continued fraction. Siegel took the special case, when $\lambda = 0$ to obtain a nice result: the transcendence of

$$1 + \cfrac{1}{2 + \cfrac{1}{3 + \dots}}$$

Siegel obtained the Lindemann–Weierstrass theorem using his method: Take algebraic numbers $a_1, \dots, a_m$ linearly independent over the rational number field. The E-functions are $E_k(x) = e^{a_k x}$, $(k = 1, \dots, m)$, and the μ power products take the form $e^{\rho_k x}$ $(k = 1, \dots, \mu)$ with μ different algebraic numbers ρ_k. The system of equations takes the form $y_k' = \rho_k y_k$ $(k = 1, \dots, \mu)$; verifying the normality condition in this case reduces to proving that any equation

$$P_1(x)e^{\rho_1 x} + \dots + P_\mu(x)e^{\rho_\mu x} = 0,$$

where $P_i(x)$ are polynomials, implies that $P_1 = 0, \dots, P_\mu = 0$. This is easy to show, and the Lindemann theorem follows from Siegel's theorem.

Siegel described the historical background to his work:[29]

Lambert's work was generalized by Legendre who considered the power series

$$y = f_\alpha(x) = \sum_{n=0}^{\infty} \frac{x^n}{n!\,\alpha(\alpha+1)\cdots(\alpha+n-1)} \qquad (\alpha \neq 0, -1, -2, \dots)$$

satisfying the linear differential equation of second order $y'' + \alpha y' = y$. He obtained the continued fraction expansion

$$\frac{y}{y'} = \alpha + \cfrac{x}{\alpha + 1 + \cfrac{x}{\alpha + 2 + \dots}}$$

and proved the irrationality of $\frac{y}{y'}$ for all rational $x \neq 0$ and all rational $\alpha \neq 0, -1, -2, \dots$. In the special case $\alpha = \frac{1}{2}$ we have $y = \cosh(2\sqrt{x})$, $y' = \frac{\sinh 2\sqrt{x}}{\sqrt{x}}$, so that Legendre's theorem contains the irrationality of $\frac{\text{tg}\,a}{a}$ for rational $a^2 \neq 0$. In more recent times, Stridsberg proved the irrationality of y and of y', separately, for rational $x \neq 0$ and rational $\alpha \neq 0, -1, \dots$, and Maier showed that neither y nor y' is a quadratic irrationality. Maier's work suggested the idea of introducing more general approximation forms which enabled me to prove that the numbers y and y' are not connected by any algebraic equation with algebraic coefficients, for any algebraic $x \neq 0$ and any rational $\alpha \neq 0, \pm\frac{1}{2}, -1, \pm\frac{3}{2}, \dots$. The excluded case of an integer $\alpha + \frac{1}{2}$ really is an exception, since then the function $f_\alpha(x)$ satisfies an algebraic differential equation of first order whose coefficients are polynomials in x with rational numerical coefficients; this follows from the explicit formulas

$$f_{k+\frac{1}{2}} = \frac{1}{2} \cdot \frac{3}{2} \cdots \left(k - \frac{1}{2}\right) D^k \cosh(2\sqrt{x}),$$

$$f_{\frac{1}{2}-k} = \frac{(-1)^k x^{k+\frac{1}{2}}}{\frac{1}{2} \cdot \frac{3}{2} \cdots (k - \frac{1}{2})} D^{k+1} \sinh(2\sqrt{x}) \qquad (k = 0, 1, 2, \dots).$$

---

[29] ibid. pp. 31–32.

For instance, in case $\alpha = \frac{1}{2}$, the differential equation is $y^2 - xy'^2 = 1$. In the excluded case, however, Lindemann's theorem shows that y and y' are both transcendental for any algebraic $x \neq 0$.

Due to the difficulty in verifying the normality condition, only these examples involving the exponential function and the Bessel function were obtained by this method between the publication of Siegel's paper in 1929 and his book in 1949. Finally, in 1988, F. Beukers, W. D. Brownwell, and G. Heckmann applied differential Galois theory to obtain a more tractable equivalent of the normality condition.[30] They were able to verify the normality condition for a large class of hypergeometric functions. In their theory, the algebraic relations between the solutions of differential equations could be studied by means of the classification of linear algebraic groups. We note that the work of E. Vessiot, G. Fano, and E. Picard on linear differential equations during the late nineteenth century provided the foundation for differential Galois theory. Starting in 1948, E. Kolchin's work, itself based on the earlier 1932 book of J. F. Ritt, brought differential Galois theory to maturity.

In the period 1953–1959, Andrei Shidlovskii (1915–2007), student of Gelfond and teacher of V. A. Oleinikov, made major advances in the theory of E-functions. In 1954, he was able to replace Siegel's normality condition with a certain irreducibility condition, enabling him to work with some E-functions satisfying third- or fourth-order linear differential equations. A year later, he obtained stronger results; we give definitions before stating one of his theorems. Functions $f_1(z), f_2(z), \ldots, f_m(z)$ are homogeneously algebraically independent over $\mathbb{C}(z)$ if $P(f_1(z), \ldots, f_m(z)) \neq 0$ for every nonzero homogeneous polynomial in m variables with coefficients in $\mathbb{C}(z)$. Similarly, complex numbers $w_1, \ldots, w_m$ are said to be homogeneously algebraically independent over the field of algebraic numbers if $P(w_1, \ldots, w_n) \neq 0$ for every nonzero homogeneous polynomial P with algebraic numbers as coefficients. Now suppose

$$y'_k = \sum_{i=1}^{m} Q_{k,i} y_i \quad (k = 1, \ldots, m), \quad Q_{k,i} \in \mathbb{C}(z), \tag{35.2}$$

and suppose $T(z)$ is the least common denominator of all the m^2 rational functions $Q_{k,i}$. Shidlovskii's theorem may then be stated as: Let $f_1(z), f_2(z), \ldots, f_m(z)$ be a set of E-functions that satisfy the system of equations (35.2) and are homogeneously algebraically independent over $\mathbb{C}(z)$, and let ζ be an algebraic number such that $\zeta T(\zeta) \neq 0$. Then the numbers $f_1(\zeta), \ldots, f_m(\zeta)$ are homogeneously algebraically independent.[31]

We may get an idea of the mathematical tradition within which Gelfond, Shidlovskii, and their students did their work by reading Mikhail Gromov's comments on his experience as a student in Russia:[32] "There was a very strong romantic attitude toward science and mathematics: the idea that the subject is remarkable and that it is worth dedicating your life to. ... that is an attitude that I and many

---

[30] Beukers, Brownawell, and Heckman (1988).
[31] See Shidlovskii (1989) chapter 3.
[32] Raussen and Skau (2010) p. 392.

other mathematicians coming from Russia have inherited." The accounts of the Gelfand seminars in Moscow, by Gromov, Landis and others, describe this attitude. The seminars extended to many hours of enthusiastic, colorful, and passionate discussion.[33]

During the 1960s, Alan Baker, a student of Harold Davenport, effected another important and very productive development in transcendental number theory. He proved a substantial generalization of the Gelfond–Schneider theorem of 1934. We may state the latter in the form: If α and β are nonzero algebraic numbers and $\log \alpha$ and $\log \beta$ are independent over the rationals, then for any nonzero algebraic numbers α_1 and β_1,

$$\alpha_1 \log \alpha + \beta_1 \log \beta \neq 0.$$

In 1939, Gelfond obtained an explicit lower bound for $|\alpha_1 \log \alpha + \beta_1 \log \beta|$ in terms of the degrees and height of the four algebraic numbers.[34] In a paper of 1948, Yuri Linnik and Gelfond pointed out that if a lower bound could be obtained for a similar three-term sum, then it would follow that the number of imaginary quadratic fields of class number one was finite; note that this result was one case within Gauss's class number problem. In 1966, Baker began to study this question by means of linear forms in logarithms.[35] In that year, he established that if $\alpha_1, \alpha_2, \ldots, \alpha_n$ were nonzero algebraic numbers such that $\log \alpha_1$, $\log \alpha_2$, $\ldots$, $\log \alpha_n$ were independent over the rationals, then $1, \log \alpha_1, \ldots, \log \alpha_n$ would be independent over the field of algebraic numbers. As a corollary, Baker obtained the generalization of the Gelfond–Schneider theorem: If $\alpha_1, \alpha_2, \ldots, \alpha_n$ are algebraic but not 0 or 1; and if $\beta_1, \beta_2, \ldots, \beta_n$ are algebraic numbers such that $1, \beta_1, \beta_2, \ldots, \beta_n$ are linearly independent over the rationals, then $\alpha_1^{\beta_1} \alpha_2^{\beta_2} \cdots \alpha_n^{\beta_n}$ is transcendental. Baker also found an effectively computable lower bound for the absolute values of a nonvanishing linear form

$$|\beta_0 + \beta_1 \log \alpha_1 + \cdots + \beta_n \log \alpha_n|.$$

This result was applicable to a number of outstanding number theory problems, including Gauss's class number problem.[36] But some of these number theoretic problems were also solved by other methods. In 1967, H. M. Stark solved the class number one problem by means of the theory of modular functions. Two years later, he published another paper, explaining that Kurt Heegner's 1952 solution of this problem was essentially sound.[37] In constructing his proof,[38] Heegner, a secondary school teacher, had made use of his deep understanding of Heinrich Weber's work on modular functions. Perhaps Heegner took it for granted that his readers would be equally familiar with Weber; this may have rendered his proof opaque. Indeed, Serre remarked that he found the paper very difficult to understand.

[33] See, for example, Zdravkovska and Duren (1993) p. 69.
[34] Gelfond (1939).
[35] Baker (1979) chapter 2.
[36] ibid. Chapter 5.
[37] Stark (1967) and (1969).
[38] Heegner (1952).

It is remarkable that mathematicians have managed to learn so much about transcendental numbers, for they have very strange properties. For example, transcendental numbers do not behave well under the usual algebraic operations. Moreover, even though it is true that if a number can be approximated sufficiently well by rational numbers, it must be transcendental, there nevertheless exist transcendental numbers that are not able to be approximated even as well as some quadratic irrational numbers. Indeed, Weil has stated that a preliminary version of Siegel's 1929 paper ended with the remark: "Ein Bourgeois, wer noch Algebra treibt! Es lebe die unbeschkränte Individualität der transzendenten Zahlen!" [It's a bourgeois, who still does algebra! Long live the unrestricted individuality of transcendental numbers!]".[39]

35.2 Liouville Numbers

In a paper of 1851,[40] based on earlier work, Liouville constructed his transcendental numbers by proving that if x was the root of an irreducible polynomial of degree $n > 1$ with integer coefficients

$$f(x) = ax^n + bx^{n-1} + \cdots + gx + h,$$

then there existed a constant $A > 0$ such that $|x - \frac{p}{q}| > \frac{1}{Aq^n}$ for all rational numbers $\frac{p}{q}$. Although the absolute value sign was not in use at that time, Liouville made his meaning clear. To prove this theorem, Liouville supposed that $x, x_1, x_2, \ldots, x_{n-1}$ comprised all the roots of $f(x) = 0$ so that

$$f\left(\frac{p}{q}\right) = a\left(\frac{p}{q} - x\right)\left(\frac{p}{q} - x_1\right)\cdots\left(\frac{p}{q} - x_{n-1}\right).$$

He then set

$$f(p,q) = q^n f\left(\frac{p}{q}\right) = ap^n + bp^{n-1}q + \cdots + hq^n$$

so that he could write

$$\left|\frac{p}{q} - x\right| = \frac{|f(p,q)|}{q^n \left|a\left(\frac{p}{q} - x_1\right)\left(\frac{p}{q} - x_2\right)\cdots\left(\frac{p}{q} - x_n\right)\right|}.$$

Next, since $n > 1$, $f(p,q)$ was a nonzero integer, so that $|f(p,q)| \geq 1$. Moreover,

$$\left|a\left(x_1 - \frac{p}{q}\right)\left(x_2 - \frac{p}{q}\right)\cdots\left(x_n - \frac{p}{q}\right)\right|$$

[39] Weil (1992) p. 53.
[40] Liouville (1851).

was bounded by a maximum value for values of $\frac{p}{q}$ in a neighborhood of x (of, say, radius 1). Liouville denoted that maximum by A. It then became clear that

$$\left|\frac{p}{q} - x\right| > \frac{1}{Aq^n}$$

so that the proof was complete. Note that for points $\frac{p}{q}$ outside the radius 1 around x, one has $|\frac{p}{q} - x| > 1$, so that the result holds. Liouville went on to show that the result was valid even when $n = 1$. In that case, $f(x) = ax + b = 0$ and so

$$\frac{p}{q} - x = \frac{ap + bq}{aq}.$$

If $x \neq \frac{p}{q}$, then $ap + bq \neq 0$, and

$$\left|\frac{p}{q} - x\right| \geq \frac{1}{aq} = \frac{1}{Aq}.$$

Liouville used his theorem to produce examples of transcendental numbers. He argued that a given number x could not be algebraic unless there was a constant A such that $|\frac{p}{q} - x| > \frac{1}{Aq^n}$. He took x to be defined by the series

$$x = \frac{1}{l} + \frac{1}{l^{2!}} + \frac{1}{l^{3!}} + \cdots + \frac{1}{l^{m!}} + \cdots,$$

where l was an integer ≥ 2. He let the partial sum up to the term whose denominator was $l^{m!}$ be $\frac{p}{q}$, so that $q = l^{m!}$. Liouville then observed that

$$x - \frac{p}{q} = \frac{1}{l^{(m+1)!}} + \frac{1}{l^{(m+2)!}} + \cdots \leq \frac{2}{l^{(m+1)!}} = \frac{2}{q^{m+1}}.$$

This inequality followed from the series

$$\frac{1}{l^{(m+1)!}} \left(1 + \frac{1}{l^{m+1}} + \frac{1}{l^{(m+1)(m+2)}} + \cdots\right)$$

$$< \frac{1}{l^{(m+1)!}} \left(1 + \frac{1}{2} + \frac{1}{2^2} + \cdots\right) = \frac{2}{l^{(m+1)!}}.$$

By increasing m, he saw that for any fixed A and n he could not obtain $x - \frac{p}{q} > \frac{1}{Aq^n}$, thus proving that x was transcendental. Liouville also noted the more general case; if he took

$$x = \frac{k_1}{l} + \frac{k_2}{l^{2!}} + \frac{k_3}{l^{3!}} + \cdots + \frac{k_m}{l^{m!}} + \cdots,$$

where $k_1, k_2, \ldots, k_m, \ldots$ were nonzero integers bounded by a constant, then x would be transcendental. He gave the example in which l could take the value 10 and the k_m

could then take values between 1 and 9 inclusive. As an example of a slightly different kind, he considered

$$x = \frac{1}{l} + \frac{1}{l^4} + \frac{1}{l^9} + \cdots + \frac{1}{l^{m^2}} + \cdots.$$

For $q = l^{m^2}$,

$$x - \frac{p}{q} = \frac{1}{l^{(m+1)^2}} + \cdots < \frac{2}{l^{2m+1}q};$$

therefore, x could not be a root of a first-degree equation with rational coefficients and was hence irrational.

35.3 Hermite's Proof of the Transcendence of *e*

In his 1873 paper "Sur la fonction exponentielle," Hermite gave two proofs that e was transcendental.[41] We give his second and more rigorous proof, following his notation for the most part, except that in some places we employ the matrix notation, not explicitly used by Hermite. About fifteen years before Hermite gave his proof, Cayley introduced the matrix notation and some of the elementary algebraic properties of matrices. It was some time, however, before the usefulness of matrices was generally recognized. We first sketch the structure of Hermite's argument. Take a relation of the form

$$e^{z_0} N_0 + e^{z_1} N_1 + \cdots + e^{z_n} N_n = 0, \tag{35.3}$$

where $z_0, z_1, \ldots, z_n$ are distinct nonnegative integers and $N_0, N_1, \ldots, N_n$ are any integers. It is clear that unless all the N are zero, e is an algebraic number. Hermite defined a set of $n(n + 1)$ numbers η_j^i, $i = 0, 1, \ldots, n$ and $j = 1, 2, \ldots, n$, by the equation

$$\eta_j^i = \frac{1}{(m-1)!} \int_{z_0}^{z_j} \frac{e^{-z} f^m(z)}{z - z_i} \, dz, \tag{35.4}$$

where m was some positive integer and

$$f(z) = (z - z_0)(z - z_1)(z - z_2) \cdots (z - z_n). \tag{35.5}$$

He showed that the numbers η_j^i got arbitrarily small as m became large. To demonstrate this fact, Hermite's reasoning was that since e^{-z} was always positive, for any continuous functions $F(z)$, he had

$$\int_{z_0}^{Z} e^{-z} F(z) \, dz = F(\xi) \int_{z_0}^{Z} e^{-z} \, dz = F(\xi)(e^{-z_0} - e^{-Z}),$$

---

[41] Hermite (1900–1917) vol. 3, pp. 150–181.

where ξ lay between z_0 and Z, the limits of integration. By choosing $Z = z_j$ and

$$F(z) = \frac{f^m(z)}{(m-1)!\,(z-z_i)},$$

he obtained

$$\eta_j^i = \frac{f^{m-1}(\xi)}{(m-1)!}\frac{f(\xi)}{\xi-z_i}\left(e^{-z_0} - e^{-z_j}\right).$$

This proved that $\eta_j^i \to 0$ as $m \to \infty$. Thus, the $(n+1) \times n$ matrix $\eta = \left(\eta_j^i\right)$, with η_j^i the element in the ith row and the jth column, depends on m. Denote this dependence by $\eta(m)$. Hermite determined a relation between $\eta(m)$ and $\eta(m-1)$ giving, by iteration, a relation between $\eta(m)$ and $\eta(1)$. Let us write the first relation as

$$\eta(m) = \Theta(m)\eta(m-1), \tag{35.6}$$

where $\Theta(m)$ is an $n+1 \times n+1$ matrix depending on $z_0, z_1, \ldots, z_n$. Thus,

$$\eta(m) = \Theta(m)\Theta(m-1)\cdots\Theta(2)\eta(1),$$

and we write (following Hermite) the element in the ith row and jth column of the matrix Θ as $\theta(j,i)$ where i and j run from 0 to n. Hermite showed that the θs were integers and that

$$\det \Theta(k) = \prod_{0 \le i < j \le n} (z_i - z_j)^2, \quad \text{for} \quad k = 2, \ldots, m. \tag{35.7}$$

He then obtained an explicit expression for the elements of $\eta(1)$ in a suitable form: Let ζ denote any one of $z_0, z_1, \ldots, z_n$. He set

$$F(z) = \frac{f(z)}{z - \zeta}$$

and

$$\int e^{-z}F(z)\,dz = -e^{-z}\gamma(z), \tag{35.8}$$

where $\gamma(z) = F(z) + F'(z) + F''(z) + \cdots + F^{(n)}(z)$. Hermite noted that if

$$f(z) = z^{n+1} + p_1 z^n + p_2 z^{n-1} + \cdots + p_{n+1},$$

then

$$F(z) = z^n + (\zeta + p_1)z^{n-1} + (\zeta^2 + p_1\zeta + p_2)z^{n-2} + \cdots, \tag{35.9}$$

and the coefficients of the two polynomials were integers. From this he could conclude that

$$\gamma(z) = \Phi(z,\zeta) = z^n + \phi_1(\zeta)z^{n-1} + \phi_2(\zeta)z^{n-2} + \cdots + \phi_n(\zeta), \qquad (35.10)$$

with $\phi_i(\zeta)$ a monic polynomial in ζ of degree i and with integer coefficients. We may let Φ denote the matrix with entry $\Phi(z_j, z_i)$ in the ith row and jth column where i and j run from 0 to n. Again, these entries must be integers and $\det \Phi = \det \Theta$. From (35.8) and (35.10), he then had

$$\int_{z_0}^{Z} \frac{e^{-z}f(z)}{z-\zeta}\,dz = e^{-z_0}\Phi(z_0,\zeta) - e^{-Z}\Phi(Z,\zeta), \qquad (35.11)$$

giving Hermite the required values of the entries of $\eta(1)$. For the final step, let $X = \Theta(m)\ldots\Theta(2)\Phi$ so that elements η_j^i of $\eta(m)$ are given by

$$\eta_j^i = e^{-z_0}X_{i0} - e^{-z_j}X_{ij}, \qquad (35.12)$$

where the integers X_{ij} are the entries of X. Note that (35.12) gives rational approximations of $e^{z_j - z_0}$ for j running from 1 to n. Now, by (35.3) and (35.12),

$$\begin{aligned}
e^{z_1}\eta_1^i N_1 + e^{z_2}&\eta_2^i N_2 + \cdots + e^{z_n}\eta_n^i N_n \\
&= e^{-z_0}\left(e^{z_1}N_1 + e^{z_2}N_2 + \cdots + e^{z_n}N_n\right)X_{i0} \\
&\quad - (X_{i1}N_1 + X_{i2}N_2 + \cdots + X_{in}N_n) \\
&= -(X_{i0}N_0 + X_{i1}N_1 + X_{i2}N_2 + \cdots + X_{in}N_n).
\end{aligned}$$

Hermite argued that, since the X_{ij} and the N_i were integers, the term on the right-hand side was an integer, but the term on the left-hand side could be made arbitrarily small because of the η_j^i. Therefore, he concluded that

$$X_{i0}N_0 + X_{i1}N_1 + \cdots + X_{in}N_n = 0, \quad i = 0, 1, \ldots, n.$$

We can write this system of equations as $XN = 0$, where the components of the vector N are $N_0, N_1, \ldots, N_n$. Since $\det X = (\det \Theta)^{m-1}\det \Phi = (\det \Theta)^m \neq 0$, we must have $N = 0$. This completes our outline of Hermite's proof.

Now let us see how Hermite obtained the basic formulas (35.6) and (35.7). To prove (35.6), Hermite showed that

$$\begin{aligned}
\int_{z_0}^{Z} \frac{e^{-z}f^{m+1}(z)}{z-\zeta}\,dz = m\theta(z_0,\zeta)\int_{z_0}^{Z}\frac{e^{-z}f^m(z)}{z-z_0}\,dz &+ m\theta(z_1,\zeta)\int_{z_0}^{Z}\frac{e^{-z}f^m(z)}{z-z_1}\,dz \\
+\cdots + m\theta(z_n,\zeta)\int_{z_0}^{Z}&\frac{e^{-z}f^m(z)}{z-z_n}\,dz, \qquad (35.13)
\end{aligned}$$

where $\theta(z, \zeta)$ was of the form

$$\theta(z, \zeta) = z^n + \alpha_1(\zeta)z^{n-1} + \alpha_2(\zeta)z^{n-2} + \cdots + \alpha_n(\zeta), \qquad (35.14)$$

with $\alpha_1(\zeta), \alpha_2(\zeta), \ldots, \alpha_n(\zeta)$ monic polynomials in ζ, with integer coefficients, and where Z and ζ took values in $z_0, z_1, \ldots, z_n$. For this, he needed the auxiliary formula that there existed polynomials $\theta(z)$ and $\theta_1(z)$ of degree n, such that

$$\int \frac{e^{-z}G(z)f(z)}{z-\zeta} \, dz = \int \frac{e^{-z}G(z)\theta_1(z)}{f(z)} \, dz - e^{-z}G(z)\theta(z), \qquad (35.15)$$

where $G(z) = \left(f(z)\right)^m$. After taking the derivative of (35.15) and multiplying across by $\frac{f(z)}{G(z)}$, he had only to determine $\theta_1(z)$ and $\theta(z)$ so that

$$\frac{f(z)}{z-\zeta} f(z) = \theta_1(z) + \left[1 - \frac{G'(z)}{G(z)}\right] f(z)\theta(z) - f(z)\theta'(z). \qquad (35.16)$$

He set $z = z_i$ in this equation, and got $0 = \theta_1(z_i) - mf'(z_i)\theta(z_i)$ or

$$\theta_1(z_i) = mf'(z_i)\theta(z_i), \quad i = 0, 1, \ldots, n. \qquad (35.17)$$

Once the values $\theta(z_i)$ were found, the $n + 1$ values determined the polynomials $\theta_1(z)$ and $\theta(z)$. To this end, he divided equation (35.16) by $f(z)$ to get

$$\frac{f(z)}{z-\zeta} = \frac{\theta_1(z)}{f(z)} + \left[1 - \frac{G'(z)}{G(z)}\right]\theta(z) - \theta'(z). \qquad (35.18)$$

Next, by (35.17), the fractional part of $\left[1 - \frac{G'(z)}{G(z)}\right]\theta(z)$ cancelled with $\frac{\theta_1(z)}{f(z)}$; and hence to determine $\theta(z)$ Hermite had to consider only the polynomial part of $\left[1 - \frac{G'(z)}{G(z)}\right]\theta(z)$. So he supposed

$$\theta(z) = \alpha_0 z^n + \alpha_1 z^{n-1} + \alpha_2 z^{n-2} + \cdots + \alpha_n.$$

By taking the logarithmic derivative of $G(z)$, he obtained

$$\frac{G'(z)}{G(z)} = \frac{m}{z - z_0} + \frac{m}{z - z_1} + \cdots + \frac{m}{z - z_n} = \frac{s_0}{z} + \frac{s_1}{z^2} + \frac{s_3}{z^3} + \cdots, \qquad (35.19)$$

where $s_i = m(z_0^i + z_1^i + \cdots + z_n^i)$. Thus, comparing the coefficients of the polynomials on the two sides of (35.18) and using (35.9), he got the relations

$$1 = \alpha_0,$$
$$\zeta + p_1 = \alpha_1 - \alpha_0(s_0 + n),$$
$$\zeta^2 + p_1\zeta + p_2 = \alpha_2 + \alpha_1(s_0 + n - 1) - \alpha_0 s_1,$$

$$\vdots$$

These equations yielded the required coefficients of $\theta(z)$:

$$\alpha_0 = 1,$$
$$\alpha_1 = \zeta + p_1 + s_0 + n,$$
$$\alpha_2 = \zeta_2 + (s_0 + n - 1)\zeta_1 + (s_0 + n)(s_0 + n - 1) + s_1,$$
$$\vdots$$

where $\zeta_2 = \zeta^2 + p_1\zeta + p_2$, $\zeta_1 = \zeta + p_1$. Thus, α_i was shown to be a monic polynomial of degree i in ζ, and Hermite could write

$$\theta(z) = \theta(z,\zeta) = z^n + \alpha_1(\zeta)z^{n-1} + \alpha_2(\zeta)z^{n-2} + \cdots + \alpha_n(\zeta). \tag{35.20}$$

But in order to derive (35.13), Hermite set the limits of integration in (35.15) from z_0 to Z, where Z was one of the values $z_0, z_1, \ldots, z_n$; he arrived at

$$\int_{z_0}^{Z} \frac{e^{-z}G(z)f(z)}{z - \zeta}\, dz = \int_{z_0}^{Z} e^{-z}G(z)\frac{\theta_1(z)}{f(z)}\, dz. \tag{35.21}$$

By (35.17),

$$\frac{\theta_1(z)}{f(z)} = \frac{m\theta(z_0)}{z - z_0} + \frac{m\theta(z_1)}{z - z_1} + \cdots + \frac{m\theta(z_n)}{z - z_n}, \tag{35.22}$$

and then, in order to recognize the dependence on ζ, he wrote, as in (35.20), $\theta(z_i) = \theta(z_i, \zeta)$. And when (35.22) was substituted in (35.21), he got (35.13).

Now, in order to prove (35.7), observe that from (35.20), $\det \Theta$ can obtained by multiplying the determinants

$$\begin{vmatrix} z_0^n & z_0^{n-1} & \cdots & 1 \\ z_1^n & z_1^{n-1} & \cdots & 1 \\ \vdots & \vdots & \cdots & \vdots \\ z_n^n & z_n^{n-1} & \cdots & 1 \end{vmatrix} \times \begin{vmatrix} 1 & 1 & \cdots & 1 \\ \alpha_1(z_0) & \alpha_1(z_1) & \cdots & \alpha_1(z_n) \\ \vdots & \vdots & \cdots & \vdots \\ \alpha_n(z_0) & \alpha_n(z_1) & \cdots & \alpha_n(z_n) \end{vmatrix},$$

completing Hermite's proof of (35.7).

35.4 Hilbert's Proof of the Transcendence of *e*

In 1893, Hilbert's presented a very efficient proof[42] of the transcendence of e. His elegant proof was based on ideas of Lindemann, Weierstrass, and Paul Gordan. To begin the proof, take ρ to be a positive integer, and set

$$I = z^\rho \left[(z - 1)(z - 2) \cdots (z - n)\right]^{\rho+1} e^{-z}.$$

[42] Hilbert (1893).

Suppose e is not transcendental. Then we may set $a, a_1, a_2, \ldots, a_n$ to be integers such that

$$a + a_1 e + a_2 e^2 + \cdots + a_n e^n = 0.$$

Then

$$\frac{a}{\rho!} \int_0^\infty I \, dz + \frac{a_1 e}{\rho!} \int_1^\infty I \, dz + \frac{a_2 e^2}{\rho!} \int_2^\infty I \, dz + \cdots + \frac{a_n e^n}{\rho!} \int_n^\infty I \, dz$$

$$+ \left(\frac{a_1 e}{\rho!} \int_0^1 I \, dz + \frac{a_2 e^2}{\rho!} \int_0^2 I \, dz + \cdots + \frac{a_n e^n}{\rho!} \int_0^n I \, dz \right) = 0, \quad (35.23)$$

or $P_1 + P_2 = 0$, where P_2 is the sum inside the parentheses and P_1 is the part outside. In the term

$$\frac{a_k e^k}{\rho!} \int_k^\infty I \, dz,$$

with $k \geq 1$, contained in P_1, change z to $z + k$. We then have

$$\frac{a_k e^k}{\rho!} \int_0^\infty e^{-(z+k)} z^{\rho+1} (z+k)^\rho (z+k-1)^{\rho+1} \cdots (z+1)^{\rho+1} (z-1)^{\rho+1} \cdots (z+k-n)^{\rho+1} \, dz$$

$$= \frac{a_k}{\rho!} \int_0^\infty e^{-z} z^{\rho+1} \sum t_m z^m \, dz,$$

where $\sum t_m z^m$ is a polynomial in z with integer coefficients. Take one term in the sum, and evaluate as a gamma integral to get

$$\frac{a_k t_m}{\rho!} \int_0^\infty e^{-z} z^{\rho+m+1} \, dz = \frac{a_k t_m (\rho + m + 1)!}{\rho!}.$$

Therefore,

$$\frac{a_k e^k}{\rho!} \int_k^\infty I \, dz$$

is an integer divisible by $\rho + 1$. The first term $\frac{a}{\rho!} \int_0^\infty I \, dz$ in P_1 is easily seen to be

$$\pm a \, (n!)^{\rho+1} \pmod{\rho + 1},$$

and hence

$$P_1 = \pm a \, (n!)^{\rho+1} \pmod{\rho + 1}.$$

Take $\rho + 1$ to be a large prime so that $a \, (n!)^{\rho+1}$ is not divisible by $\rho + 1$. Notice that we can obviously choose $a \neq 0$, if e is algebraic. As for P_2, we can make it as small as we like. But $a \, (n!)^{\rho+1}$ is a nonzero integer, contradicting (35.23), and hence e cannot be algebraic. Note that Hermite's original proof still has the advantage that it obtains rational approximations of e raised to integer powers.

35.5 Exercises

(1) Show that the sum of Goldbach's series is given exactly by

$$\sum_{n=1}^{\infty} \frac{1}{n^2 + \frac{p}{q}n} = \frac{q}{p} \left(\psi \left(\frac{p}{q} \right) + \gamma + \frac{q}{p} \right),$$

where $\psi(x)$ is Gauss's digamma function. Observe that the value of $\psi(\frac{p}{q})$ may be explicitly calculated for integer values of p and q, as Goldbach used them. See the results of Gauss in Chapter 17, Exercise 11.

(2) Suppose λ is a rational number not equal to a negative integer. Let

$$\phi_\lambda(z) = \sum_{n=0}^{\infty} \frac{z^n}{(\lambda+1)_n}$$

and let ξ be a nonzero algebraic number. Show that $\phi_\lambda(\xi)$ is transcendental. Siegel stated this result without proof in his 1929 paper. The first published proof is due to Shidlovskii, dating from 1954. See Shidlovskii (1989) p. 185.

(3) Suppose that the E-functions $f_1(z), \ldots, f_m(z)$ are algebraically independent over $\mathbb{C}(z)$ and form a solution of the system of linear differential equations

$$y_k' = Q_{k,0} + \sum_{i=0}^{m} Q_{k,i} y_i, \quad k = 1, \ldots, m; \quad Q_{k,i} \in \mathbb{C}(z).$$

Let ξ be an algebraic number such that $\xi T(\xi) \neq 0$, with $T(\xi)$ as defined earlier. Show that under these conditions, the numbers $f_1(\xi), \ldots, f_m(\xi)$ are algebraically independent. See Shidlovskii (1989) p. 123.

(4) Let

$$f(z) = \sum_{n=0}^{\infty} z^{2^n}.$$

Show that if α is algebraic and $0 < |\alpha| < 1$, then $f(\alpha)$ is transcendental. This result is due to Kurt Mahler (1903–1985); though mostly self-taught, he regarded himself as a student of Siegel in his research. For this and other results, see the paper by J. H. Loxton and A. J. van der Poorten in Baker and Masser (1977) pp. 211–226.

(5) Show that if

$$f_1(z) = \sum_{n=0}^{\infty} z^{2^n} \quad \text{and} \quad f_2(z) = \sum_{n=0}^{\infty} z^{3^n},$$

then for any two algebraic numbers α_1, α_2 in $0 < |z| < 1$, $f_1(\alpha_1)$ and $f_2(\alpha_2)$ are algebraically independent. See Kubota (1977).

(6) Show that for an algebraic number $\beta \neq 0, 1$, the two numbers defined by the hypergeometric series $F(\frac{1}{2}, \frac{1}{2}, 1, \beta)$ and $F(-\frac{1}{2}, \frac{1}{2}, 1, \beta)$ are algebraically independent over the rationals. See Chudnovsky and Chudnovsky (1988).

(7) Show that if α is algebraic and $0 < |\alpha| < 1$, the theta series $\sum_{n \geq 0} \alpha^{n^2}$ is transcendental. Recall that Liouville had shown this number to be irrational for $\alpha = \frac{1}{l}$, where l was an integer > 1. See Nesterenko (2006) in Bolibruch, Osipov, and Sinai (2006).

35.6   Notes on the Literature

See Lützen (1990) pp. 511–526 for a very interesting history of Liouville's work on transcendental numbers and some early history of such numbers. For the development of the theory of transcendental numbers in the 1970s and 1980s, see the articles in Baker (1988) and Baker and Masser (1977). Gelfond (1960) also contains some historical remarks on transcendental numbers.

An English translation of a portion of Hermite's 1873 paper may be found in Smith (1959) vol. I, pp. 97–106. Hilbert (1970) vol. I, pp. 1–4, is a reprint of Hilbert's short proofs of the transcendence of e and π. These proofs were also presented by Felix Klein (1911), as lecture seven of his 1893 Evanston lectures. For the proof of the transcendence of e, see Klein (1911) pp. 53–55.

Yandell (2002) gives an entertaining popular account of Hilbert's problems and those who made contributions to the solutions. See also Browder (1976) to read articles by experts on the mathematical developments connected with Hilbert's problems (up to 1975).

36

Value Distribution Theory

36.1 Preliminary Remarks

Value distribution theory addresses the problem of measuring the solution set of the equation $f(z) = b$, where f is some analytic function in some domain D and b is any complex number. For example, when f is a polynomial of degree n, the fundamental theorem of algebra, proved by Gauss and others, states that $f(z) = b$ has n solutions for a given b. The converse of this is an easier proposition. Algebraists since Descartes and Harriot recognized the important property of polynomials, that if $a_1, a_2, \ldots, a_n$ were a finite sequence of numbers, then there would be a polynomial of degree n, $(x - a_1)(x - a_2) \cdots (x - a_n)$, with zeros at exactly these numbers. After Euler found the infinite product factorization of the trigonometric and other functions, mathematicians could raise the more general question of the existence of a function with an infinite sequence $a_1, a_2, a_3, \ldots$ as its set of zeros. Of course, it was almost immediately understood that the product $\prod_{n=1}^{\infty}(x - a_n)$ might not converge; in special instances such as the gamma function, the proper modification was also determined, in order to ensure convergence. Gauss and Abel treated infinite products with some care in their work on the gamma and elliptic functions. But the answer to the general question had to wait for the development of the foundations of the theory of functions of a complex variable. In fact, Weierstrass, one of the founders of this theory, published an important 1876 paper, "Zur Theorie der eindeutigen analytischen Funktionen," dealing with the problem.[1]

Karl Weierstrass (1815–1897) studied law at Bonn University, but after four years he failed to get a degree. With Christoph Gudermann as his mathematics teacher, Weierstrass became a Gymnasium teacher in 1841. Gudermann was a researcher in the area of power series representation of elliptic functions, and Weierstrass in turn made power series the basic technique in his work in complex analysis. His great accomplishment was the construction of a theory of Abelian functions; the 1854 publication of the first installment of his theory, secured him a professorship at Berlin. Weierstrass was a great teacher and had many great students, including H. A. Schwarz,

---

[1] Weierstrass (1894–1927) vol. 2, pp. 76–101.

G. Cantor, Leo Königsberger, and Sonya Kovalevskaya, whom he held in high regard. The Mathematics Genealogy Project counts his mathematical descendants as 16,585; the author would be included in that number. In order to lay a firm foundation for his theories, Weierstrass carefully developed the basic concepts of infinite series and infinite products.

Weierstrass took a sequence $\{a_n\}$ such that $\lim_{n=\infty} |a_n| = \infty$ was the only limit point. Note that if c is a finite limit point of the sequence, then a nonconstant function with zeros at $\{a_n\}$ cannot be analytic at c. And since a polynomial $f(x)$ is analytic at every finite x, it is reasonable to require that our function be analytic in the complex plane; Weierstrass called such a function an entire function and specified that $\lim_{n\to\infty} |a_n| = \infty$. He observed Cauchy's result that in the case of a finite number of zeros, an entire function f with zeros at $a_1, a_2, a_3, \ldots, a_n$ would take the form $f(x) = e^{g(x)}$ $\prod_{k=1}^{n}(x - a_k)$, with g an entire function. Note that it is now standard practice to denote a complex variable by z or w but Weierstrass used x. Weierstrass noted that for infinite sequences, one might make the product conditionally convergent by arranging the factors in a particular order, but this was not possible in general. As an example, he gave the product always divergent for $x \neq 0$:

$$(1 + x)\left(1 + \frac{x}{2}\right)\left(1 + \frac{x}{3}\right)\cdots.$$

Now, as discussed in Chapter 17, the reciprocal of the gamma function has zeros at the negative integers and by Euler's definition, attributed by Weierstrass to Gauss,

$$\frac{1}{\Gamma(x)} = \prod_{n=1}^{\infty}\left\{\left(1 + \frac{x}{n}\right)\left(\frac{n+1}{n}\right)^{-x}\right\}$$

or

$$\frac{1}{\Gamma(x)} = \prod_{n=1}^{\infty}\left\{\left(1 + \frac{x}{n}\right)e^{-x\log(\frac{n+1}{n})}\right\}.$$

In this context, instead of $\ln z$, we use the notation $\log z$, the logarithm of a complex number z; this is a multivalued function whose principal value is such that for $x > 0$, $\log x = \ln x$. Next, as we mentioned in Section 17.5, F. W. Newman had explicitly observed that convergence required the exponential factor $e^{-\frac{x}{n}}$. Although Weierstrass may not have been familiar with Newman's paper, he wrote that the product for $\frac{1}{\Gamma(x)}$ directed him toward a method for achieving convergence. He realized that with each factor $(1 + \frac{x}{a_n})$ it was necessary to include an exponential factor

$$e^{\frac{x}{a_n} + \frac{1}{2}\left(\frac{x}{a_n}\right)^2 + \cdots + \frac{1}{m_n}\left(\frac{x}{a_n}\right)^{m_n}},$$

where m_n was chosen in such a way that the product converged. For this purpose, Weierstrass defined the primary factors

$$E(x, 0) = (1 - x) \quad \text{and} \quad E(x, m) = (1 - x)e^{\sum_{r=1}^{m}\left(\frac{x^r}{r}\right)}, \quad m = 1, 2, 3, \ldots.$$

Since

$$1 - x = e^{\log(1-x)} = e^{-\sum_{r=1}^{\infty} \left(\frac{x^r}{r} \right)} \quad \text{for } |x| < 1,$$

he had

$$E(x,m) = e^{-\sum_{r=1}^{\infty} \frac{x^{m+r}}{m+r}} \quad \text{for } |x| < 1.$$

Thus, m_n could be chosen so that for any fixed x, the series $\sum_{n=1}^{\infty} \left| \frac{x}{a_n} \right|^{m_n+1}$ converged. In fact, note that $m_n = n$ would work, since $\left| \frac{x}{a_n} \right|^n < \left(\frac{1}{2} \right)^n$ as long as $|a_n| > 2|x|$. Also, because $\lim_{n \to \infty} |a_n| = \infty$, this inequality would be valid for all but a finite number of a_n. Weierstrass proved that

$$\sum_{n=1}^{\infty} \log E \left(\frac{x}{a_n}, m_n \right)$$

converged absolutely and uniformly in any disk $|x| \le R$, and so did the product

$$\prod_{n=1}^{\infty} E \left(\frac{x}{a_n}, m_n \right).$$

This product was an entire function with zeros at exactly $a_1, a_2, a_3, \ldots$.

The French mathematician Edmond Laguerre (1834–1886) used Weierstrass's product to classify transcendental entire functions according to their genus, just as polynomials may be classified by their degree. He defined a product to be of genus m if the integer m_n in each primary factor was a fixed integer m. Thus, a product of genus 0 is of the form $\prod_{n=1}^{\infty} (1 - \frac{x}{a_n})$ while a product of genus 1 takes the form $\prod (1 - \frac{x}{a_n}) e^{-\frac{x}{a_n}}$. As we have seen in a different context, Laguerre was motivated by a desire to extend to transcendental functions the classical results on polynomials of Descartes, Newton, and others. Recall that in the course of discovering his extension of the Descartes rule of signs, Newton showed that if a polynomial with real coefficients $c_0 + c_1 x + c_2 x^2 + \cdots + c_n x^n$ had all roots real, then

$$(r + 1)(n - r + 1)c_{r+1}c_{r-1} \le r(n - r)c_r^2, \quad r = 1, 2, \ldots, n - 1;$$

the inequality would be strict when all roots were not equal. Note that by taking n infinite, we have the inequalities

$$(r + 1)c_{r+1}c_{r-1} < rc_r^2, \quad r = 1, 2, 3, \ldots.$$

Laguerre raised the question: Given a transcendental entire function $f(x) = \sum_{n=0}^{\infty} c_n x^n$ with all real roots and real coefficients, will the coefficients satisfy these inequalities? In a paper of 1882, he stated that the result was true for $f(x)$ of genus 1. In the same year, he proved that, if $\frac{f'(x)}{x^n f(x)}$ went to zero as $|x| \to \infty$, then $f(x)$ was of genus n.[2] Laguerre also investigated the relationship between the zeros of a function of genus 0 or 1, with real zeros, and the zeros of its derivative.

In an 1883 paper[3] on entire functions, Poincaré looked for a connection between the growth of a function and its genus p. He proved that for every $\epsilon > 0$,

$$\lim_{x \to \infty} \frac{|f(x)|}{e^{\epsilon |x|^{p+1}}} = 0,$$

and if

$$f(x) = \sum_{n=0}^{\infty} c_n x^n,$$

then

$$\lim_{n \to \infty} c_n \Gamma \left(\frac{n + \epsilon + 1}{p + 1} \right) = 0.$$

These results suggested that in order to measure growth of a function, one required a concept more refined than its genus. Consider the case of a monic polynomial $g(x)$ of degree n. For large x, $|g(x)|$ behaves like $|x|^n$. So if $M(r, g)$ is the maximum value of $|g(x)|$ on $|x| = r$, then

$$\lim_{r \to \infty} \frac{\log M(r, g)}{\log r} = n.$$

In 1896,[4] Émile Borel defined the order ρ of a transcendental entire function f:

$$\rho = \varlimsup_{r \to \infty} \frac{\log \log M(r, f)}{\log r},$$

a concept implicitly contained in Hadamard's 1893 work on the Riemann zeta function.[5] Now Riemann had introduced the entire function

$$\xi(s) = \Gamma(1 + \frac{s}{2})(s - 1)\pi^{-\frac{s}{2}}\zeta(s) \tag{36.1}$$

and by a brilliantly intuitive argument obtained its product formula. Hadamard's work on entire functions was motivated by the desire to provide justification for some of Riemann's results.

Jacques Hadamard (1865–1963) studied at the École Normal, where his teachers included the outstanding mathematicians J. Tannery, Hermite, Picard, P. Appell, and G. Goursat. Hadamard wrote his doctoral thesis on the Taylor series of complex analytic functions, deriving results on the relation of the coefficients with the location of the singularities and with the radius of convergence. In his report on the thesis,

---

[2] Laguerre (1972) vol. 1, pp. 167–170, pp. 171–173.
[3] Poincaré (1883).
[4] Borel (1896).
[5] Hadamard (1893).

Picard wrote that the abstract results appeared to lack practical value; Hermite suggested that Hadamard look for applications. Fifty years later, Hadamard recalled,[6] "At that time, I had none [no applications] available. Now, between the time my manuscript was handed in and the day when the thesis was defended, I became aware of an important question which had been proposed by the Académie des Sciences as a prize subject; and precisely the results in my thesis gave the solution of that question. I had been led solely by my feeling of the interest of the problem and it led me the right way." The problem posed by the Académie was to prove Riemann's unproved assertions. Hadamard used his result on the relation between the coefficients and the growth of the function to prove that the exponent of convergence of the zeros the function in (36.1) was at most 1. This effectively established Riemann's product formula for $\xi(s)$.

Hadamard's 1893 work implicitly contained the factorization theorem for functions of finite order: If $f(z)$ is of order ρ, then

$$f(z) = z^k P(z) e^{Q(z)},$$

where $Q(z)$ is a polynomial of degree $q \leq \rho$ and $P(z)$ is a product of genus $p \leq \rho$. Moreover, the order of $P(z)$ is equal to the exponent of convergence ρ_1 of the zeros z_n of P and $\rho_1 \leq \rho$. Note that the exponent of convergence is the infimum of the positive numbers α such that $\sum_{n=1}^{\infty} |z_n|^{-\alpha}$ converges. The work of Hadamard and Borel also implied a formula connecting the coefficients c_n of the Taylor series expansion of an entire function with the order of the function. In 1902, this was explicitly stated by the Finnish mathematician Ernst Lindelöf, son of Lorenz Lindelöf (1870–1946), a student of Mellin, as the relation

$$\rho = \varlimsup_{n \to \infty} \frac{-n \log n}{\log |c_n|}.$$

Lindelöf's interest in entire functions was aroused by his contact with Hadamard and others when he stayed in Paris in 1893–94 and then in 1898–99. When he returned to the University of Helsingfors in Finland, Lindelöf communicated this interest to his students, including Frithiof and Rolf Nevanlinna, who made fundamental contributions to the value distribution theory of meromorphic functions. Rolf Nevanlinna (1895–1980) founded value distribution theory as a quantitative generalization of Picard's theorem.

Picard (1856–1941) proved that for an entire function $f(x)$, the equation $f(x) = a$ had a solution for every complex number a with at most one exception. The value of e^x is never 0, illustrating that exceptions might exist. Picard proved this in 1879 by an ingenious application of the multivalued inverse of the elliptic modular function $k^2(\tau)$; note that $k^2(\tau)$ had earlier been studied by Abel, Jacobi, Hermite, Schwarz, and others. Speaking at his Jubilee celebration of 1936, Hadamard praised Picard's teaching as masterly. Referring to Picard's theorem, Hadamard addressed his teacher:[7] "All

---

[6] Maz'ya and Shaposhnikova (1998) p. 56.
[7] ibid. p. 36.

mathematicians know, on the other hand, what a marvelous stimulus for research your mysterious and disconcerting theorem on entire functions was, and still is, because the subject has lost nothing of its topicality. I can say that I owe to it a great part of the inspiration of my first years of work." Indeed, Picard went on to extend his result to functions with an essential singularity at infinity. This theorem is a vast generalization of the Sokhotskii–Casorati–Weierstrass theorem that every complex number is a limit of the values assumed by a function in any neighborhood of an essential singularity.

Now in the case of a polynomial $f(x)$ of degree n, for every a, the equation $f(x) = a$ has n roots, counting multiplicity. Picard's theorem predicts that for a transcendental entire function $f(x)$, the equation $f(x) = a$ has an infinite number of solutions with at most one exceptional number a. It was then natural to seek a more precise measure of the number of solutions, or to inquire about their density. In 1896, Borel proved[8] that, for entire functions of finite order, if f was of nonintegral order ρ, then the exponent of convergence of the zeros of $f - a$ equaled ρ for all complex numbers a. If f was of integral order ρ, then the same result would hold with at most one exceptional value of a, in which case the exponent of convergence was less than ρ.

With the development of complex function theory, attempts were made to prove Picard's theorem without using elliptic modular functions. Borel, Landau, Bloch and R. Nevanlinna found such proofs, opening up new paths in function theory and making the topic among the most popular in the mathematics of the early twentieth century. In working with meromorphic functions, Nevanlinna's difficulty in extending the results for entire functions to meromorphic functions was the lack of a concept corresponding to the maximum modulus of a function. Interestingly, in 1899, Jensen derived the basic formula for obtaining this expanded concept.[9] He proved that if f was meromorphic in $|z| \le r$ with zeros at a_j and poles at b_k inside $|z| < r$, then

$$\log|f(0)| = \frac{1}{2\pi} \int_0^{2\pi} \log|f(r\,e^{i\phi})|\,d\phi - \sum \log \frac{r}{|a_j|} + \sum \log \frac{r}{|b_k|},$$

where sums were taken over all the zeros and all the poles respectively. Jensen thought that the formula might be important in studying the zeros of the Riemann zeta function and in particular in the proof of the Riemann hypothesis. In fact, Jensen's formula was useful in simplifying proofs of results in both prime number theory and entire functions.

In 1925, R. Nevanlinna defined the analog of the maximum modulus, the characteristic function of a meromorphic function f, as the sum of two functions:[10] the mean proximity function, measuring the average closeness of f to a given complex number a; and the counting function, measuring the frequency with which f assumed the value a. In the same year, he went on to prove two fundamental theorems on the characteristic function, and his brother F. Nevanlinna recast them in a

[8] Borel (1896).
[9] Jensen (1899).
[10] Nevanlinna (1925).

geometric context. The latter approach was further developed and extended in 1929 by T. Shimizu; by Lars Ahlfors in papers of the 1930s; and in 1960 by S. S. Chern. In the 1980s, Charles Osgood and Paul Vojta observed a close analogy between Nevanlinna theory and Diophantine approximation. The clarification and precise delineation of this analogy has had important consequences for both topics.

It is somewhat surprising that Jacobi had already found Jensen's result in 1827. Jacobi stated it only for polynomials, but as Landau pointed out, his argument can be extended to the general case. Jacobi used Fourier series to obtain his formula and was inspired by the work of Marc-Antoine Parseval (1755–1836) in Lagrange series. Parseval derived formulas for roots of equations in terms of definite integrals. Jacobi carried this program further by finding integral expressions for sums of powers of any number of roots of an equation, given in increasing order. Incidentally, Jacobi mentioned in his paper that as early as 1777, Euler had discovered the "Fourier coefficients;" Riemann was apparently not aware of this fact when he wrote his 1853 thesis on trigonometric series.

36.2 Jacobi on Jensen's Formula

Jacobi took a polynomial $f(x) = a + bx + cx^2 + \cdots + x^p$ with real coefficients and[11]

$$\log(U^2 + V^2) = \phi\left(r\,e^{+x\sqrt{-1}}\right) + \phi\left(r\,e^{-x\sqrt{-1}}\right),$$

where $\phi(x) = \log f(x)$. He denoted the zeros of f by $\alpha', \alpha'', \alpha''', \ldots, \alpha^{(p)}$. These were taken in increasing order of absolute values, and Jacobi considered three separate cases: (i) r greater than all the roots, (ii) r less than all the roots, and (iii) r between $\alpha^{(k)}$ and $\alpha^{(k+1)}$. His formula is stated as

$$\frac{1}{2\pi}\int_{-\pi}^{\pi}\log(U^2+V^2)dx = \begin{cases} p\log r^2 \text{ in the first case,} \\ \log a^2 \text{ in the second case,} \\ k\log r^2 + \log(\alpha^{(k+1)})^2 + \log(\alpha^{(k+2)})^2 + \cdots + \log(\alpha^{(p)})^2 \\ \quad \text{in the third case.} \end{cases}$$

$$(36.2)$$

Summarizing Jacobi's argument, suppose $\phi(x) = \log(a + bx + cx^2 + \cdots + x^p)$, where the coefficients $a, b, c, \ldots$ of the pth degree polynomial are real. Let $(x - \alpha')(x - \alpha'')(x - \alpha''') \cdots (x - \alpha^{(p)})$ represent the factorization of the polynomial. Since the polynomial has real coefficients, the complex roots appear in conjugate pairs. It is clear that

$$\phi(r\,e^{+x\sqrt{-1}}) + \phi(r\,e^{-x\sqrt{-1}})$$
$$= \log\left\{(a + br\,\cos x + cr^2\,\cos 2x + \cdots + r^p\,\cos px)^2\right.$$
$$\left. + (br\,\sin x + cr^2\,\sin 2x + \cdots + r^p\,\sin px)^2\right\}.$$

---

[11] Jacobi (1969) vol. 6, pp. 12–20.

Denote the expression inside the chain brackets by $U^2 + V^2$ so that

$$U^2 + V^2 = \begin{cases} a^2 + b^2r^2 + c^2r^4 + \cdots + r^{2p}, \\ + \quad 2r\cos x(ab + bcr^2 + cdr^4 + c\ldots), \\ + \quad 2r^2\cos 2x(ac + bdr^2 + cer^4 + \cdots), \\ + \quad 2r^2\cos 3x(ad + ber^2 + dfr^4 + \cdots), \\ + \quad \cdots. \end{cases} \tag{36.3}$$

Now if $f(x) = a + bx + cx^2 + dx^3 + \cdots$, then the values of the Fourier integrals for $f(re^{ix})$ are given by

$$a^2 + b^2r^2 + c^2r^4 + d^2r^6 + \cdots = \frac{1}{2\pi}\int_{-\pi}^{\pi} f\left(re^{+x\sqrt{-1}}\right) f\left(re^{-x\sqrt{-1}}\right) dx,$$

$$ab + bcr^2 + cdr^4 + der^6 + \cdots = \frac{1}{2\pi r}\int_{-\pi}^{\pi} f\left(re^{+x\sqrt{-1}}\right) f\left(re^{-x\sqrt{-1}}\right) \cos x \, dx,$$

$$ac + bdr^2 + cer^4 + dfr^6 + \cdots = \frac{1}{2\pi r^2}\int_{-\pi}^{\pi} f\left(re^{+x\sqrt{-1}}\right) f\left(re^{-x\sqrt{-1}}\right) \cos 2x \, dx,$$

$$\cdots.$$

$$(36.4)$$

Note that $U^2 + V^2 = f(re^{+x\sqrt{-1}}) f(re^{-x\sqrt{-1}})$ and (36.3) gives the Fourier expansion of this function so that the Fourier coefficients can be computed by (36.4). We also have

$$\log(r e^{\sqrt{-1}x} - \alpha') = \log r e^{\sqrt{-1}x} + \log\left(1 - \frac{\alpha'}{r}e^{-\sqrt{-1}x}\right) \quad \text{when } r > |\alpha'|,$$

$$= \log(-\alpha') + \log\left(1 - \frac{r}{\alpha'}e^{\sqrt{-1}x}\right) \quad \text{when } |\alpha'| > r.$$

The logarithms on the right-hand sides can be expanded as a series by

$$-\log(1 - t) = t + \frac{1}{2}t^2 + \frac{1}{3}t^3 + \frac{1}{4}t^4 + \cdots, \qquad |t| < 1.$$

Jacobi used the above facts to give the series expansions for $\log(U^2 + V^2)$:

1. With r greater than the absolute values of all the roots:

$$p\log r^2 - 2\sum_{1}^{p}\left(\frac{\alpha}{r}\cos x + \frac{\alpha^2}{2r^2}\cos 2x + \frac{\alpha^3}{3r^3}\cos 3x + \frac{\alpha^4}{4r^4}\cos 4x + \cdots\right).$$

2. With r smaller than the absolute values of all the roots:

$$\log a^2 - 2\sum_{1}^{p}\left(\frac{r}{\alpha}\cos x + \frac{r^2}{2\alpha^2}\cos 2x + \frac{r^3}{3\alpha^3}\cos 3x + \frac{r^4}{4\alpha^4}\cos 4x + \cdots\right),$$

where a is the constant term in the polynomial $f(x)$.

3. When $|\alpha^{(k)}| < r < |\alpha^{(k+1)}|$:

$$k \log r^2 - 2 \sum_{1}^{k} \left(\frac{\alpha}{r} \cos x + \frac{\alpha^2}{2r^2} \cos 2x + \frac{\alpha^3}{3r^3} \cos 3x + \frac{\alpha^4}{4r^4} \cos 4x + \cdots \right)$$

$$+ 2 \sum_{k+1}^{p} \left(\frac{1}{2} \log \alpha^2 - \frac{r}{\alpha} \cos x - \frac{r^2}{2\alpha^2} \cos 2x - \frac{r^3}{2\alpha^3} \cos 3x - \frac{r^4}{4\alpha^4} \cos 4x - \cdots \right),$$

where, as Jacobi explained, $\sum_{m}^{n} \psi(\alpha)$ denoted the sum of the quantities

$$\psi(\alpha^{(m)}), \psi(\alpha^{(m+1)}), \ldots, \psi(\alpha^{(n)}).$$

Jacobi integrated the series for $\log(U^2 + V^2)$ over the interval $(-\pi, \pi)$ to obtain

$$\int_{-\pi}^{\pi} \log(U^2 + V^2)\, dx$$

in the three cases. All the cosine terms vanished, and he obtained the formula (36.2). Jacobi also found formulas for $\sum_{1}^{k} \alpha^n$ and $\sum_{k+1}^{p} \frac{1}{\alpha^n}$ in terms of the nth Fourier coefficients of $\log(U^2 + V^2)$ and $\arctan \frac{V}{U}$.

36.3 Jensen's Proof

Jensen's 1899 rediscovery of Jacobi's formula succeeded in connecting the modulus of an analytic function with its zeros, and this occurred at just the right moment to fill a need in the theory of analytic functions.[12] Jensen himself mentioned the possibility of applying his result to a proof of the Riemann hypothesis. Though he apparently did not pursue this topic further, and though his work has not as yet made a significant dent in the Riemann hypothesis, his formula is fundamental for the theory of entire functions. His proof, similar in some respects to Jacobi's, started with the formula

$$\log \left| 1 - \frac{z}{a} \right| = - \sum_{\nu=1}^{\infty} \frac{r^\nu}{2\nu} \left(\frac{e^{\nu\theta i}}{a^\nu} + \frac{e^{-\nu\theta i}}{\overline{a}^\nu} \right), \quad r = |z| < |a|,$$

$$= \log \frac{r}{|a|} - \sum_{\nu=1}^{\infty} \frac{1}{2\nu r^\nu} \left(a^\nu e^{-\nu\theta i} + \overline{a}^\nu e^{\nu\theta i} \right), \quad r > |a|.$$

By integrating, he obtained

$$\frac{1}{2\pi} \int_{0}^{2\pi} \log \left| 1 - \frac{z}{a} \right| d\theta = \begin{cases} \log \frac{r}{|a|}, & \text{for} \quad r > |a|, \\ 0, & \text{for} \quad r < |a|. \end{cases} \tag{36.5}$$

Next he supposed $f(z)$ to be meromorphic in $|z| \leq r$ with zeros at $a_1, a_2, a_3, \ldots, a_n$ and poles at $b_1, b_2, \ldots, b_m$ in $|z| < r$ and with no singularities on $|z| = r$. Then he could express $f(z)$ in the form

---

[12] Jensen (1899).

$$f(z) = f(0) \frac{\prod_{k=1}^{n} \left(1 - \frac{z}{a_k}\right)}{\prod_{k=1}^{m} \left(1 - \frac{z}{b_k}\right)} e^{f_1(z)}, \tag{36.6}$$

where

$$f_1(z) = \sum_{\nu=1}^{\infty} B_\nu z^\nu \qquad \text{for} \quad |z| \le r. \tag{36.7}$$

He took the real part of the logarithm of each side of (36.6), integrated over $(0, 2\pi)$, and applied (36.5) to get

$$\frac{1}{2\pi} \int_0^{2\pi} \log |f(r\,e^{i\theta})|\,d\theta = \log |f(0)| + \log r^{n-m} \frac{|b_1||b_2|\cdots|b_m|}{|a_1||a_2|\cdots|a_n|}.$$

Note that the constant term in f_1 is zero, and hence there is no contribution from the integral of f_1.

36.4 Bäcklund Proof of Jensen's Formula

The Finnish mathematician R. J. Bäcklund, a student of Ernst Lindelöf, is credited with using a conformal mapping to prove Jensen's theorem in 1916 or 1918.[13] This proof first assumed $g(z)$ to be analytic without zeros in $|z| \le R$ and used Cauchy's integral formula to compute $\log g(0)$ as an integral. Then for a function f with zeros at $a_1, a_2, \ldots, a_n$ in $|z| \le R$, consider a new function with no zeros in $|z| \le R$:

$$g(z) = f(z) \frac{R^2 - \overline{a_1} z}{R(z - a_1)} \cdot \frac{R^2 - \overline{a_2} z}{R(z - a_2)} \cdots \frac{R^2 - \overline{a_n} z}{R(z - a_n)}. \tag{36.8}$$

Since

$$\left| \frac{R^2 - \overline{a_k} z}{R(z - a_k)} \right| = 1 \quad \text{for} \quad |z| = R,$$

we have

$$|g(R\,e^{i\theta})| = |f(R\,e^{i\theta})|.$$

To fill out the details, start with Cauchy's integral formula

$$\log g(0) = \frac{1}{2\pi i} \int_{|w|=R} \log g(w) \frac{dw}{w} = \frac{1}{2\pi} \int_0^{2\pi} \log g(R\,e^{i\theta})\,d\theta. \tag{36.9}$$

---

[13] Bäcklund (1918).

Now apply the expression for g in (36.8) and take the real part to find that

$$\log \frac{|f(0)| R^n}{|a_1||a_2| \cdots |a_n|} = \frac{1}{2\pi} \int_0^{2\pi} \log |f(R e^{i\theta})| \, d\theta.$$

This proof considered only analytic functions, but if one takes the case where $f(z)$ has poles at $b_1, b_2, \ldots, b_m$, one need merely multiply the right-hand side of (36.9) by

$$\frac{R(z - b_1)}{R^2 - \overline{b_1} z} \cdot \frac{R(z - b_2)}{R^2 - \overline{b_2} z} \cdots \frac{R(z - b_m)}{R^2 - \overline{b_m} z}$$

to obtain the result. Although it is difficult to ascertain exactly where Bäcklund gave this proof, we note that the use of the conformal mapping

$$\frac{R^2 - \overline{a} z}{R(z - a)}$$

is a beautiful and efficient innovation because it vanishes at $z = a$ and its value on $|z| = R$ is 1.

36.5 R. Nevanlinna's Proof of the Poisson–Jensen Formula

Rolf Nevanlinna gave an important extension of Jensen's formula; this result became the foundation of his theory of meromorphic functions.[14] Suppose $f(x)$ is a meromorphic function in $|x| \leq \rho$ $(0 < \rho < \infty)$ with zeros and poles at a_h $(h = 1, \ldots, \mu)$ and b_k $(k = 1, \ldots, \nu)$, respectively. Let $x = r e^{i\phi}$, $f(x) \neq 0, \infty$, and $r < \rho$. Then

$$\log |f(r e^{i\phi})| = \frac{1}{2\pi} \int_0^{2\pi} \log |f(\rho e^{i\theta})| \frac{\rho^2 - r^2}{\rho^2 + r^2 - 2\rho r \cos(\theta - \phi)} \, d\theta$$

$$- \sum_{h=1}^{\mu} \log \left| \frac{\rho^2 - \overline{a_n} x}{\rho(x - a_n)} \right| + \sum_{k=1}^{\nu} \log \left| \frac{\rho^2 - \overline{b_k} x}{\rho(x - b_k)} \right|. \qquad (36.10)$$

Nevanlinna called this the Poisson–Jensen formula, and his proof employed Green's formula

$$\int_\Gamma \left(U \frac{\partial V}{\partial n} - V \frac{\partial U}{\partial n} \right) ds = - \int_\Gamma (U \, \Delta V - V \, \Delta U) \, \partial\sigma. \qquad (36.11)$$

Here U and V are twice continuously differentiable functions in a connected domain G with boundary Γ formed by a finite number of analytic arcs. The symbol Δ denotes the Laplacian and $\frac{\partial}{\partial n}$ represents the derivative in the direction normal to the boundary but pointing to the interior of G.

---

[14] Nevanlinna (1974).

Take U to be a real-valued function $u(z)$ harmonic in $G \cup \Gamma$ except for logarithmic singularities at $z = c_1, c_2, \ldots, c_p$, so that

$$u(z) = \lambda_k \log |z - c_k| + u_k(z),$$

where λ_k is real, and u_k is continuous at $z = c_k$. Now let $V = g(z, x)$, where g denotes Green's function for the domain G with the singularity at an interior point $z = x$. This function is completely defined by the two conditions: The sum $g(z, x) + \log |z - x|$ is harmonic at all points interior to the domain G; also, $g(z, x)$ vanishes on the boundary Γ. Green's formula (36.11), can be applied to U and V as chosen earlier if the points c_k and x are excluded by means of small circles around these points. Then, when the radii of these circles are allowed to tend to zero after the application of Green's formula, we get

$$u(x) = \frac{1}{2\pi} \int_\Gamma u(z) \frac{\partial g(z, x)}{\partial n} \, ds - \sum_{k=1}^{p} \lambda_k \, g(c_k, x). \tag{36.12}$$

Nevanlinna then took a meromorphic function f in the domain G with zeros at a_h $(h = 1, \ldots, \mu)$ and poles at b_k $(k = 1, \ldots, \nu)$. Then the function $u(z) = \log |f(z)|$ satisfied the conditions for (36.12) to hold so that he had

$$\log |f(x)| = \frac{1}{2\pi} \int_\Gamma \log |f(z)| \frac{\partial g(z, x)}{\partial n} \, ds - \sum_{h=1}^{\mu} g(a_n, x) + \sum_{k=1}^{\nu} g(b_k, x). \tag{36.13}$$

Nevanlinna noted that this important formula permitted him to compute the modulus, $|f|$, at any point inside G by using its values on the boundary of G and the location of the poles and zeros of f inside G. He took G to be a circle of radius ρ about the origin so that Green's function would be given by

$$g(z, x) = \log \left| \frac{\rho^2 - \bar{x} z}{\rho(x - z)} \right|.$$

By substituting this g, with $z = \rho e^{i\theta}$ and $x = r e^{i\theta}$, in (36.13), Nevanlinna obtained his Poisson–Jensen formula (36.10).

Furthermore, to obtain Jensen's formula, he took $x = 0$ in (36.10), assuming that $f(0) \neq 0$ or ∞. Note that a slight modification was necessary in the cases $f(0) = 0$ or ∞. Thus, if

$$f(x) = c_\lambda x^\lambda + c_{\lambda+1} x^{\lambda+1} + \cdots,$$

and if $c_\lambda \neq 0$, then f had to be replaced by $x^{-\lambda} f$.

He therefore had

$$\log |f(0)| = \frac{1}{2\pi} \int_0^{2\pi} \log |f(\rho e^{i\theta})| \, d\theta - \sum_{h=1}^{\mu} \frac{\rho}{|a_n|} + \sum_{k=1}^{\nu} \frac{\rho}{|b_k|}. \tag{36.14}$$

In his 1964 book on meromorphic functions, W. K. Hayman noted that the idea in the Bäcklund proof could be extended to yield a simple derivation of the Poisson–Jensen formula.[15] Let g be an analytic function without zeros or poles in $|z| < R$. Now note that the mapping

$$w = \frac{R(\zeta - z)}{R^2 - \bar{z}\,\zeta}$$

maps the disk $|\zeta| \le R$ conformally onto the unit disk and takes the point $\zeta = z$ to $w = 0$. This gave Hayman

$$\frac{dw}{w} = \frac{d\zeta}{\zeta - z} + \frac{\bar{z}\,d\zeta}{R^2 - \bar{z}\,\zeta} = \frac{(R^2 - |z|^2)\,d\zeta}{(R^2 - \bar{z}\,\zeta)(\zeta - z)},$$

so that the result of Cauchy's theorem,

$$\log g(0) = \frac{1}{2\pi i} \int_{|w|=R} \log g(w)\,\frac{dw}{w},$$

could be replaced by

$$\log g(z) = \frac{1}{2\pi i} \int_{|\zeta|=R} \log g(\zeta)\,\frac{(R^2 - |z|^2)\,d\zeta}{(R^2 - \bar{z}\,\zeta)(\zeta - z)}.$$

Taking real parts of this formula and setting $z = r\,e^{i\theta}$ and $\zeta = R\,e^{i\phi}$,

$$\log\left|g(r\,e^{i\theta})\right| = \frac{1}{2\pi} \int_0^{2\pi} \log\left|g(R\,e^{i\phi})\right| \frac{(R^2 - r^2)\,d\phi}{R^2 - 2Rr\,\cos(\theta - \phi) + r^2}.$$

Following the Bäcklund approach, Hayman took

$$g(\zeta) = f(\zeta) \prod_{k=1}^{\mu} \left(\frac{R(\zeta - b_k)}{R^2 - \bar{b}_k\,\zeta}\right) \prod_{k=1}^{\nu} \left(\frac{R^2 - \bar{a}_k\,\zeta}{R(\zeta - a_k)}\right),$$

and the Poisson–Jensen formula followed.

36.6 Nevanlinna's First Fundamental Theorem

In his 1913 thesis,[16] Georges Valiron (1884–1955), student of Borel and teacher of Laurent Schwartz, expressed the sums on the right-hand side of (36.14) as integrals by means of the counting functions $n(r,0)$ and $n(r,\infty)$. These functions denote the number of zeros and poles, counting multiplicity, of $f(x)$ in $|x| \le r$. To efficiently implement Valiron's idea,[17] Nevanlinna applied the Stieltjes integral to get

[15] Hayman (1964) chapter 1.
[16] Valiron (1913).
[17] Nevanlinna (1925).

$$\sum \log \frac{\rho}{|a_h|} = \int_0^\rho \log \frac{\rho}{r} d\, n(r, 0)$$

$$= \int_0^\rho \frac{n(r, 0)}{r} dr \quad \text{and}$$

$$\sum \log \frac{\rho}{|b_k|} = \int_0^\rho \frac{n(r, \infty)}{r} dr.$$

He could then express (36.14) in the form

$$\log |f(0)| = \frac{1}{2\pi} \int_0^{2\pi} \log |f(\rho e^{i\theta})| \, d\theta - \int_0^\rho \frac{n(r, 0)}{r} dr + \int_0^\rho \frac{n(r, \infty)}{r} dr.$$

$$(36.15)$$

Nevanlinna went on to write this in symmetric form, where he had the large values of the function on one side and the small values on the other. For that purpose, he set

$$\log^+ \alpha = \log \alpha, \quad \alpha \geq 1,$$
$$= 0, \quad 0 \leq \alpha < 1,$$

so that for $x > 0$, $\log x = \log^+ x - \log^+ \frac{1}{x}$. He then defined the mean proximity function

$$m(\rho, f) = \frac{1}{2\pi} \int_0^{2\pi} \log^+ |f(\rho e^{i\theta})| \, d\theta$$

and the function

$$N(\rho, f) = \int_0^\rho \frac{n(r, \infty)}{r} dr.$$

Thus, he was able to rewrite (36.15) as

$$\log |f(0)| = m(\rho, f) - m\left(\rho, \frac{1}{f}\right) + N(\rho, f) - N\left(\rho, \frac{1}{f}\right),$$

or

$$T(\rho, f) = T\left(\rho, \frac{1}{f}\right) + \log |f(0)|, \qquad (36.16)$$

where

$$T(\rho, f) = m(\rho, f) + N(\rho, f).$$

Note that the term $m(\rho, f)$ is an average of $\log |f|$ on $|z| = \rho$ for large values of $|f|$, while the term $N(\rho, f)$ deals with the poles. So $T(\rho, f)$ acts as a measure of the large values of $|f|$ in $|z| \leq \rho$, while $T(\rho, \frac{1}{f})$ does the same for the small values of $|f|$. The function $T(\rho, f)$ has been named the Nevanlinna characteristic function, and it plays a fundamental role in the theory of meromorphic functions.

In the preceding formulation, we considered small and large values of f. More generally, Nevanlinna considered values of f close and/or equal to any fixed number a by defining

$$N(r,a) = N\left(r, \frac{1}{f-a}\right) = \int_0^r \frac{n(t,a)}{t}\, dt, \qquad (36.17)$$

$$m(r,a) = m\left(r, \frac{1}{f-a}\right) = \frac{1}{2\pi} \int_0^{2\pi} \log^+ \left| \frac{1}{f(r\, e^{i\theta}) - a} \right| d\theta, \qquad (36.18)$$

$$T\left(r, \frac{1}{f-a}\right) = m(r,a) + N(r,a). \qquad (36.19)$$

By a simple argument, he showed that

$$|T(r,f) - T(r, f-a)| \le \log^+ |a| + \log 2.$$

Combining this inequality with Jensen's formula (36.16), Nevanlinna arrived at his first fundamental theorem:

$$T\left(r, \frac{1}{f-a}\right) = T(r,f) - \log|f(0) - a| + \epsilon(r,a), \qquad (36.20)$$

$$|\epsilon(r,a)| \le \log^+ |a| + \log 2.$$

Additionally, he proved that $N(r,a)$ and $T\left(r, \frac{1}{f-a}\right)$ were increasing convex functions of $\log r$. The result for $N(r,a)$ followed from (36.17), since

$$\frac{d\, N(r,a)}{d\, \log r} = n(r,a). \qquad (36.21)$$

Nevanlinna's proof for the convexity of $T\left(r, \frac{1}{(f-a)}\right)$ was rather lengthy, but in 1930 Henri Cartan obtained a simpler proof by first showing that

$$T(r,f) = \frac{1}{2\pi} \int_0^{2\pi} N(r, e^{i\theta})\, d\theta + \log^+ |f(0)|. \qquad (36.22)$$

Since this immediately implied that

$$\frac{d\, T(r,f)}{d\, \log r} = \frac{1}{2\pi} \int_0^{2\pi} n(r, e^{i\theta})\, d\theta,$$

the theorem was proved.

Note that the characteristic function $T(r,f)$ was Nevanlinna's analog of the maximum modulus $\log M(r,f)$, long sought after by complex function theorists. Recall that the logarithm of the maximum modulus, $\log M(r,f)$, was one of the essential objects in the study of entire functions; it was investigated by Hadamard, Borel, E. Lindelöf, and others. The efforts to extend the theory of entire functions to meromorphic functions required a suitable analog of $\log M(r, f)$ and Nevanlinna provided just that. Incidentally, in 1896 Hadamard proved that $\log M(r, f)$ was a convex function of $\log r$; this is usually known as the Hadamard three circles theorem.

36.7　Nevanlinna's Factorization of a Meromorphic Function

By use of the Poisson–Jensen formula, Nevanlinna was able to present a simple form of a canonical factorization of meromorphic functions of finite order. He gave this proof in the third chapter of his book, *Le théorème de Picard–Borel*,[18] and we outline his argument. Note that a function meromorphic in the complex plane is said to be of order ρ if $\rho = \varlimsup\limits_{r \to \infty} \frac{\log T(r, f)}{\log r}$.

Nevanlinna stated the theorem: Suppose $f(x)$ is a meromorphic function of finite order with zeros and poles at $a_1, a_2, \ldots$ and $b_1, b_2, \ldots$, respectively. Let q be a integer such that

$$\lim_{r=\infty} \frac{T(r)}{r^{q+1}} = 0.$$

Then

$$f(x) = x^\alpha \, e^{\sum_0^q c_\nu x^\nu} \lim_{\rho=\infty} \frac{\prod_{|a_\nu| < \rho} \left(1 - \frac{x}{a_\nu}\right) e^{\frac{x}{a_\nu} + \cdots + \frac{1}{q}\left(\frac{x}{a_\nu}\right)^q}}{\prod_{|b_\nu| < \rho} \left(1 - \frac{x}{b_\nu}\right) e^{\frac{x}{b_\nu} + \cdots + \frac{1}{q}\left(\frac{x}{b_\nu}\right)^q}},$$

where α is an integer.

To prove this, assuming $f(0) \neq 0$, Nevanlinna differentiated the Poisson–Jensen formula $q + 1$ times to get

$$D^{q+1} \log f(x) = \sum_{|a_\mu| < \rho} \frac{(-1)^q q!}{(x - a_\mu)^{q+1}} - \sum_{|b_\nu| < \rho} \frac{(-1)^q q!}{(x - b_\nu)^{q+1}} + S_\rho(x) + T_\rho(x),$$

where

$$S_\rho(x) = q! \sum_{|a_\mu| < \rho} \left(\frac{\overline{a_\mu}}{\rho^2 - a_\mu x}\right)^{q+1} - q! \sum_{|b_\nu| < \rho} \left(\frac{\overline{b_\nu}}{\rho^2 - b_\nu x}\right)^{q+1},$$

$$I_\rho(x) = \frac{(q+1)!}{2\pi} \int_0^{2\pi} \log \left|f(\rho \, e^{i\theta})\right| \frac{2\rho \, e^{i\theta} \, d\theta}{(\rho \, e^{i\theta} - x)^{q+1}}.$$

He then showed that $S_\rho(x)$ and $I_\rho(x)$ uniformly converged to zero for $|x| \leq r$ as $\rho \to \infty$. Taking this for granted, we have

---

[18] Nevanlinna (1974).

$$D^{q+1} \log f(x) = (-1)^{q+1} q! \lim_{\rho \to \infty} \left\{ \sum_{|b_\nu| < \rho} \left(\frac{1}{x - b_\nu} \right)^{q+1} - \sum_{|a_\mu| < \rho} \left(\frac{1}{x - a_\mu} \right)^{q+1} \right\}.$$

Because of the uniform convergence, he could integrate $q + 1$ times to get

$$\log f(x) = \sum_0^q c_\nu x^\nu + \lim_{\rho = \infty} \sum_{|a_\mu| < \rho} \left[\log \left(1 - \frac{x}{a_\mu} \right) + \frac{x}{a_\mu} + \cdots + \frac{1}{q} \left(\frac{x}{a_\mu} \right)^q \right]$$

$$- \sum_{|b_\nu| < \rho} \left[\log \left(1 - \frac{x}{b_\nu} \right) + \frac{x}{b_\nu} + \cdots + \frac{1}{q} \left(\frac{x}{b_\nu} \right)^q \right].$$

The result follows after exponentiation, since in case $f(0) = 0$, $f(0)$ can be replaced by $\frac{f(x)}{x^\alpha}$ for a suitable positive integer α.

36.8 Picard's Theorem

In his 1953 *A Mathematician's Miscellany*, later published with additional material as *Littlewood's Miscellany*, J. E. Littlewood raised and answered the question:[19] whether a dissertation of 2 lines could deserve and get a Fellowship." He answered in the affirmative, giving examples, including Picard's theorem, for which there was a one-line statement and a one-line proof:

(Theorem.) An integral [entire] function $f(z)$ never 0 or 1 is a constant.
(Proof.) $\exp\{i\Omega(f(z))\}$ is a bounded integral function.

Littlewood explained that $\tau = \Omega(w)$ was the inverse of the modular function $w = k^2(\tau)$, arising in the theory of elliptic functions. The function $k^2(\tau)$ gave an analytic map from the half-plane $\{\tau \in \mathbb{C} : \text{Im}\,\tau > 0\}$ onto $\mathbb{C} \setminus \{0, 1\}$. Although the inverse Ω was many-valued, for any branch of it, $\Omega(f(z))$ extended analytically to give an entire function from $\mathbb{C}$ into $\{\tau \in \mathbb{C} : \text{Im}\,\tau > 0\}$. Note further that this argument implies that $\exp\{i\Omega(f(z))\}$ is a bounded analytic function; hence, by Liouville's theorem, it is a constant. Therefore, f is a constant. This was Picard's proof, but recall that in 1879, the study of the inverse of the modular function Ω was not well established. So Picard used some care to prove that it was possible to define a single-valued branch of $\Omega(f(z))$ on the complex plane.[20] Littlewood continued by imagining what a referee's report could have been:

Exceedingly striking and a most original idea. But, brilliant as it undoubtedly is, it seems more odd than important; an isolated result, unrelated to anything else, and not likely to lead anywhere.

It was clearly difficult to foresee the large number of interesting developments of complex function theory that would arise from Picard's theorem.

[19] Littlewood (1986) p. 40.
[20] Picard (1879).

36.9 Borel's Theorem

Recall the Hadamard–Borel factorization theorem: An entire function of finite order $f(z)$ can be written in the form $z^k P(z) e^{Q(z)}$, where $Q(z)$ is a polynomial and $P(z)$ is the canonical product constructed from the zeros of f. Note here that the order of the entire function $P(z)$ is equal to the exponent of convergence of the zeros of f. We may deduce from this theorem, since $e^{Q(z)}$ is of integral order, that if f is of nonintegral order ρ, then the order of $P(z)$ must be ρ. This in turn implies that for an entire function $f(z)$ of nonintegral order ρ and any complex number x, the exponent of convergence of the zeros of $f(z) - x$ is also ρ. In keeping with the notation of Weierstrass, Valiron, and Nevanlinna, we sometimes employ x to represent a complex number or variable. In 1900, Borel showed[21] that for entire functions of integral order ρ, the exponent of convergence of the zeros of $f(z) - x$ was equal to ρ except for at most one value of x. These exceptions became known as the Borel exceptional values.

Outlining Borel's proof,[22] first suppose a and b are two exceptional values of x. Then by Hadamard's theorem

$$f(z) - a = z^{\alpha_1} P_1(z) e^{Q_1(z)} \quad \text{and} \quad f(z) - b = z^{\alpha_2} P_2(z) e^{Q_2(z)}, \tag{36.23}$$

where Q_1 and Q_2 are polynomials of degree ρ and P_1 and P_2 are canonical products of order less than ρ. By subtracting the equations and multiplying by e^{-Q_2}, we have

$$z^{\alpha_1} P_1 e^{Q_1 - Q_2} = z^{\alpha_2} P_2 + (b - a) e^{-Q_2}.$$

The term on the right-hand side has order equal to ρ and hence the polynomial $Q_1 - Q_2$ must be of degree ρ. Now differentiate the equation

$$z^{\alpha_1} P_1 e^{Q_1} - z^{\alpha_2} P_2 e^{Q_2} = b - a$$

to get

$$\left(z^{\alpha_1} P_1 Q_1' + (z^{\alpha_1} P_1)' \right) e^{Q_1} - \left(z^{\alpha_2} P_2 Q_2' + (z^{\alpha_2} P_2)' \right) e^{Q_2} = 0.$$

The coefficients of e^{Q_1} and e^{Q_2} are entire functions of order less than ρ, since the order of the derivative does not exceed the order of the function. So we can factorize these coefficients by the Hadamard–Borel theorem to obtain

$$z^{\alpha_3} P_3 e^{Q_3} e^{Q_1} - z^{\alpha_4} P_4 e^{Q_4} e^{Q_2} = 0,$$

where Q_3 and Q_4 are polynomials of degree at most $\rho - 1$, with P_3 and P_4 canonical products of orders less than ρ. Now rewrite the last equation as

$$e^{Q_1 - Q_2 + Q_3 - Q_4} \equiv z^{\alpha_4 - \alpha_3} \frac{P_4}{P_3}.$$

[21] See Valiron (1949).
[22] Borel (1900).

The degree of the polynomial $Q_1 - Q_2 + Q_3 - Q_4$ is ρ, and hence the left-hand side is an entire function of order ρ. On the other hand, the order of the function on the right-hand side is less than ρ. This contradiction proves the theorem.

36.10 Nevanlinna's Second Fundamental Theorem

In a paper of 1925, in what is now called his second fundamental theorem, R. Nevanlinna gave a far-reaching generalization of Picard's theorem.[23] Nevanlinna's result showed that the term $N(r, a)$ was the dominant part of the characteristic function and that most of the roots of the equation $f(z) = a$ were simple. In his influential 1929 book on the Picard-Borel theorem, he discussed his theorem. He supposed $f(x)$ to be a meromorphic function and $z_1, z_2, \ldots, z_q$ $(q \geq 3)$ distinct complex numbers, finite or not. Then

$$(q - 2)T(r, f) < \sum_{v=1}^{q} N(r, z_v) - N_1(r) + S(r), \tag{36.24}$$

where

$$N_1(r) = N\left(r, \frac{1}{f'}\right) + (2N(r, f) - N(r, f'))$$

and where the expression S satisfied:

1. For any positive number λ,

$$\int_{r_0}^{r} \frac{S(t)}{t^{\lambda+1}}\, dt = O\left(\int_{r_0}^{r} \frac{\log T(t, f)}{t^{\lambda+1}}\, dt\right). \tag{36.25}$$

2. Moreover,

$$S(r) < O\left(\log T(r, f) + \log r\right) \tag{36.26}$$

except for a set of finite linear measure. And if $f(x)$ was of finite order, that is,

$$\varlimsup_{r \to \infty} \frac{\log T(r, f)}{\log r} < \infty, \quad \text{then}$$

$$S(r) = O(\log r) \tag{36.27}$$

without restriction.

The proof of this theorem is lengthy and requires the computation of several estimates, the most important of which shows that $m\left(r, \frac{f'}{f}\right)$ is in general negligible in comparison with $T(r, f)$. For this quantity Nevanlinna proved that

[23] Nevanlinna (1925).

$$m\left(r, \frac{f'}{f}\right) = O\left(\log\left(r\,T(r,f)\right)\right),$$

except on a set of finite linear measure when f is of infinite order, and

$$m\left(r, \frac{f'}{f}\right) \doteq O(\log r),$$

without restriction, when f is of finite order.

To derive Picard's theorem, suppose f is an entire function that does not assume the values a and b. Take $q = 3$ and $z_1 = a$, $z_2 = b$, and $z_3 = \infty$ in (36.24) and since $N_1(r)$ is positive, we have $T(r, f) < S(r)$, contradicting (36.26). Thus, Picard's theorem is proved.

Nevanlinna combined the two fundamental theorems to derive an elegant extension of Picard's theorem. By the first fundamental theorem

$$\lim_{r=\infty} \frac{m(r,a) + N(r,a)}{T(r,f)} = 1.$$

He set

$$\delta(a) = \lim_{r=\infty} \frac{m(r,a)}{T(r,f)} = 1 - \overline{\lim_{r=\infty}} \frac{N(r,a)}{T(r,f)},$$

and by the second fundamental theorem

$$\sum_{\nu=1}^{q} \delta(a_\nu) \le 2.$$

Observe that from this, Borel's theorem can be deduced. If a is a Borel exceptional value of an entire function, the reader may easily verify that $\delta(a) = 1$ and that $\delta(\infty) = 0$. By the preceding inequality, we know that there cannot be more than one exceptional value, completing the derivation.

36.11   Exercises

(1) Suppose that all the roots of

$$f(x) = a_0 + a_1 x + a_2 x^2 + \cdots + a_n x^n$$

are real and that $\theta(x)$ is an entire function of genus 0 or 1. Suppose also that $\theta(x)$ is real for real x and all its zeros are real and negative. Prove that all the roots of

$$g(x) = a_0\theta(0) + a_1\theta(1)x + \cdots + a_n\theta(n)x^n$$

are real; in fact, that $f(x)$ and $g(x)$ have the same number of positive zeros and the same number of negative zeros. See Laguerre (1972) vol. 1, p. 201.

(2) Show that if $f(z) = \sum a_n z^n$ is an entire function of finite order ρ, then

$$\rho = \limsup_{n\to\infty} \frac{n \log n}{\log(\frac{1}{|a_n|})}.$$

See Lindelöf (1902).

(3) Let $f(z) = \sum a_n z^n$ be an entire function of finite order and let

$$m(r) = \max(|a_n| r^n), \quad n = 0, 1, 2, \ldots.$$

Prove that

$$\lim_{r\to\infty} \frac{\log M(r, f)}{\log m(r, f)} = 1.$$

See Valiron (1949), p. 32.

(4) Let $f(z)$ be of finite order ρ and of finite type $\tau = \lim_{r\to\infty} \frac{\log M(r)}{r^\rho}$. If

$$L = \limsup_{r\to\infty} r^{-\rho} n(r, f) \quad \text{and} \quad l = \liminf_{r\to\infty} r^{-\rho} n(r, f),$$

then $Le^{\frac{l}{L}} \le e\rho\tau$. See Shah (1948) and Boas (1954) p. 16. Swarupchand Mohanlal Shah (1905–1996) received his appointment at Aligarh Moslem University (India) from André Weil, who served there as department head from 1931 to 1933. In 1942, Shah received his Ph.D. from the University of London under Hardy's student Titchmarsh. Returning to Aligarh, Shah served as head of the department from 1953 to 1958 when he reached the mandatory retirement age in India; he then took up a second mathematics career in the United States. He taught for more than twenty years in the United States, at Kansas and at Kentucky. Shah published hundreds of papers in complex analysis and gave a boost to a number of young mathematicians by encouraging them and collaborating with them.

(5) Show that if $a \ne 0$ and $f(x) = a + a_1 x + \cdots$ is analytic at the origin, then there is a number L, depending only on a and a_1, such that if $f(x)$ is analytic in the disk $|x| < L$, then $f(x)$ must take the value 0 and/or 1 somewhere in the disk. See Landau (1904). In 1905, Constantin Carathéodory found an expression for L in terms of the fundamental branch of the inverse of the elliptic modular function. Carathéodory used what is now called Schwarz's lemma, a result he extracted from Schwarz's work; he showed its importance, thereby elevating it to the status of an important lemma. Georg Pick then generalized this lemma.

(6) Suppose $f(z)$ is meromorphic and has only a finite number of poles, and that $f(z)$, $f^{(l)}(z)$ have only a finite number of zeros for some $l \ge 2$. Show that then

$$f(z) = \frac{P_1(z) e^{P_2(z)}}{P_3(z)},$$

with P_1, P_2, P_3 polynomials. Furthermore, if $f(z)$ and $f^{(l)}(z)$ have no zeros, then either $f(z) = e^{Az+B}$ or else $f(z) = (Az + B)^{-n}$. This result is due to J. Clunie. See Hayman (1964) p. 67.

(7) Let $f(z)$ be a meromorphic function of order ρ, where $0 \leq \rho \leq \frac{1}{2}$; $\delta(a, f) > 0$ when $\rho = 0$ and $\delta(a, f) \geq 1 - \cos \pi\rho$ when $\rho > 0$. Show that then a is the only deficient value of $f(z)$; in particular, a meromorphic function of order zero can have at most one deficient value. This result is due to the German mathematician Oswald Teichmüller (1913–1943) for functions with positive poles and negative zeros; to the Russian mathematician A. A. Goldberg for the general case. See Hayman (1964) p. 114.

36.12   Notes on the Literature

For references to works on entire functions, the reader may consult Borel (1900), Valiron (1949), and Boas (1954). For Nevanlinna theory, see Nevanlinna (1974), a reprint of his 1929 book, and Hayman (1964). Neuenschwander (1978) gives a history of the Casorati–Weierstrass theorem; Picard's theorem is a far-reaching refinement of this theorem. Littlewood's witty comments on Picard's theorem first appeared in *A Mathematician's Miscellany*. Littlewood (1986) was put out by his friend Béla Bollobás who also wrote a twenty-two page foreword; it contains a reprint of Littlewood's 1953 book along with photographs and some additional material by Littlewood.

See Cherry and Ye (2001), M. Ru (2001), and Bombieri and Gubler (2006) for treatments of the remarkable analogy between the Diophantine equations and value distribution or Nevanlinna theory. This parallel has been worked out in some detail and has led to significant advances in both areas.

37

Univalent Functions

37.1 Preliminary Remarks

Weierstrass constructed a theory of functions using power series as the basic object, in contrast with Riemann, who studied analytic functions as mappings, specifically conformal mappings. The Bieberbach conjecture was rooted in these dual aspects of analytic function theory; it simultaneously viewed a function as a mapping and as a series. Thus, Bieberbach considered conformal mappings, such as those studied by Riemann, and then speculated on the magnitude of the coefficients, assuming the first two to be zero and one, respectively. A function f analytic in a domain D, an open and connected subset of the complex plane, is called univalent in D if it does not assume any value more than once. A univalent function f maps D conformally onto its image domain $f(D)$. Riemann was the first to study conformal mappings in the context of complex function theory. In his 1851 doctoral dissertation, he stated his famous theorem,[1] now called the Riemann mapping theorem, that any simply connected proper subdomain D of the complex plane could be conformally mapped onto the unit disk $|z| < 1$. Note here that the mapping must be one-to-one and analytic. This mapping f is unique if we require that for a given point z_0 in the domain D, $f(z_0) = 0$ and $f'(z_0) > 0$. Observe that since the inverse of a univalent function is also univalent, it is of interest to consider functions univalent on the unit disk. We denote by S the set of normalized univalent functions on the unit disk, that is, univalent functions for which $f(0) = 0$ and $f'(0) = 1$. The Taylor expansion of f would take the form

$$f(z) = z + a_2 z^2 + a_3 z^3 + \cdots + a_n z^n + \cdots . \tag{37.1}$$

In a paper of 1916,[2] Ludwig Bieberbach (1886–1982) proved that $|a_2| \leq 2$ and then, in a footnote, conjectured that $|a_n| \leq n$. Attempts to prove this conjecture led to valuable developments in the theory of analytic functions of one variable, lending it additional significance. Louis de Branges's 1984 proof of this conjecture concluded

[1] Riemann (1990) pp. 35–75 or Riemann (2004) pp. 1–39.
[2] Bieberbach (1916).

363

an era in the theory of functions, comparable, albeit on a smaller scale, to the 350-year era in number theory brought to an end by Andrew Wiles's 1994 resolution of Fermat's problem.

Riemann gave a sketch of a proof of his mapping theorem, but in 1871 his student, F. Emil Prym, found a flaw in his line of reasoning, even apart from Riemann's use of an unproved variational principle rigorously established by Hilbert only a half century later. Note that George Green had made use of this principle in his famous work of 1828 on electricity and magnetism.[3] In spite of its shaky foundations, however, the significance of Riemann's mapping theorem was immediately recognized. In 1867, Dirichlet's student Elwin B. Christoffel (1829–1900) showed that the upper half plane could be conformally mapped onto polygonal regions by means of functions defined by integrals.[4] Note that the upper half plane may be mapped onto the unit disk by a fractional linear transformation. About two years later, Christoffel's result was independently rediscovered by H. A. Schwarz. In the 1870s, Carl Neumann and Schwarz used potential theoretic methods to prove the mapping theorem for regions bounded by analytic arcs. In the years around 1900, Hilbert brought renewed attention to the Riemann mapping problem and its generalization, the uniformization theorem, with the statement of his twenty-second problem[5] and with his proofs of the Dirichlet principle.

In 1907, Paul Koebe (1882–1945) and Henri Poincaré proved the uniformization theorem that every simply connected Riemann surface was conformal to one of the three: the unit disk, the complex plane, or the extended complex plane.[6] Poincaré's work was a continuation of methods and ideas he had developed in the early 1880s, when he established the theory of Fuchsian and Kleinian groups and the related theory of automorphic functions. Felix Klein played an equally important role in this development. In fact, Klein and Poincaré corresponded regularly in 1881–82 while creating these theories by differing approaches and techniques. In one of his proofs of the uniformization theorem, Koebe showed that the set S of normalized univalent functions was a normal family. Now a family F of analytic functions defined on a domain D is called normal if every sequence of functions f_n in F has a subsequence converging uniformly on each compact subset of D. The concept of a normal family is due to Paul Montel (1876–1975), a student of Borel and Lebesgue. In a June 1935 letter to Zermelo, Carathéodory discussed the history of this concept:[7]

> The word and the notion "normal family" comes from Montel, who had shaped it around 1904. This notion has emerged from a further development of the Weierstrass double-series theorem stemming from Stieltjes (around 1895). If one notes that for all *analytic* functions $f(z)$, which are regular for $|z| < 1$ and satisfy the condition $|f(z)| < 1$ there, all coefficients of the power series $a_0 + a_1 z + \cdots = f(z)$ are uniformly limited, it follows that from every set $\{f(z)\}$ of such functions one can choose a uniformly convergent sequence on every circle $|z| \leq r < 1$. This led Montel to give the name "normal families" to all sets of functions which possess an analogous

---

[3] Green (1970) pp. 23–41.
[4] Christoffel (1867) p. 97.
[5] Yandell (2002) p. 417.
[6] Koebe (1907–1908) and Poincaré (1907).
[7] Georgiadou (2004) p. 82.

property. So, one was able to show that all functions which are regular in a domain G and are $\neq 0, 1$ constitute a normal family; the Picard theorem follows on from here easily. The notion of the limiting oscillation which allows us to speak of families that are normal in a *point* comes from me.

Constantin Carathéodory (1873–1950) was a German mathematician of Greek descent. He initially studied engineering at the Military School of Belgium and was involved with the construction of the Assiut dam in Egypt. Abandoning his engineering career due to an increasing attraction to mathematics, Carathéodory attended H. A. Schwarz's Berlin colloquia; he received his doctoral degree from Göttingen in 1904 under Minkowski for a thesis on the calculus of variations; his peripatetic career was then spent at a number of institutions in Germany, Greece, and the United States. Hans Rademacher was his 1916 student at Göttingen. Carathéodory's interest in function theory was aroused when Pierre Boutroux, nephew of H. Poincaré, visited Göttingen in 1905. Boutroux was then trying to simplify E. Borel's recent proof of Picard's theorem on entire functions and discussed this problem with Carathéodory. In his autobiographical notes, Carathéodory recalled this encounter:[8]

> Boutroux had noticed that his proof was successful only because in the case of conformal mappings there was a remarkable rigidity, which, by the way, he was not able to put into formulae. Boutroux's discovery did not let me rest and six weeks later I was able to prove Landau's sharpening of the Picard theorem in a few lines by using the theorem which is today called the lemma of Schwarz. I produced this theorem with the help of Poisson's integral; only through Erhard Schmidt, whom I had informed of my findings, did I learn not only that the theorem already exists in the work of Schwarz, but also that it can be gained by absolutely elementary means. Indeed, the proof, which Schmidt informed me about, cannot be improved. Thus, I gained a further field of activity apart from the calculus of variations.

Schmidt's proof of Schwarz's lemma, a form of which was used by Schwarz in 1869 for his proof of the Riemann mapping theorem, is the one usually found in complex analysis textbooks. It was Carathéodory, however, who revealed the importance of the lemma by giving several significant applications of it. It is due to his efforts that Schwarz's lemma and its generalizations became so useful in complex function theory.

In his important 1912 paper, Carathéodory applied Schwarz's lemma to prove a result on kernel convergence, a key concept within geometric function theory.[9] Suppose $G_1, G_2, \ldots, G_n, \ldots$ is an infinite sequence of simply connected domains in the complex plane, containing the origin but not coinciding with the whole complex plane. Suppose also that $f_n(z)$ is a conformal mapping of the unit disk onto the domain G_n with $f_n(0) = 0$ and $f_n'(0) > 0$. Carathéodory's theorem related the geometric behavior of the domains G_n with the analytic behavior of the functions f_n; this result was later employed by Löwner (Loewner) to develop his parametric method for the study of univalent functions. Carathéodory applied it to determine the boundary behavior of conformal mappings. As he wrote in his letter to Hilbert in connection with this theorem, "A first application of this theorem is, for instance,

---

[8] ibid. p. 63.
[9] Carathéodory (1912).

the proof of continuity of the conformal mapping as a function of its boundary, even if the boundary is a non-analytic curve and the Cauchy theorem cannot be applied." To state Carathéodory's convergence theorem, first suppose the origin is an interior point of $\bigcap G_n$; then the kernel of the sequence $\{G_n\}$ is defined as the largest domain G containing the origin such that each compact subset of G is contained in every G_n, with the possible exception of a finite number of G_n. Note that it is easy prove that G exists. Next, if the origin is not an interior point of $\bigcap G_n$, then the kernel is defined by $G = \{0\}$. The sequence $\{G_n\}$ is said to converge to the kernel G if every subsequence of $\{G_n\}$ has G as kernel. When convergence occurs, either $G = \{0\}$ or G is simply connected. Also, let $\{f_n\}$ be a sequence of univalent functions on the unit disk with $f_n(0) = 0$ and $f_n' > 0$; moreover, let f_n map the unit disk to G_n. On this basis, the theorem states that a sequence of functions $\{f_n\}$ converges uniformly on compact subsets of the unit disk to a function f if and only if $\{G_n\}$ converges to the kernel $G \neq \mathbb{C}$. If convergence occurs, then either $G = \{0\}$, in which case $f = 0$, or $G \neq \{0\}$, in which case f is a conformal mapping from the unit disk to G. To prove this theorem, Carathéodory used Schwarz's lemma combined with Koebe's one-quarter theorem. The latter theorem was not fully proved until Bieberbach did so in 1916; for Carathéodory's convergence theorem, Koebe's weaker result, for some positive constant not necessarily $\frac{1}{4}$, was sufficient.

Carathéodory's results were used a decade later by Löwner to construct his parametric theory of univalent functions. The Czech mathematician Karel Löwner (1893–1968) was a student of Georg Pick, and his name was later spelled Karl Löwner and then, after emigration to America, Charles Loewner. He studied in the German section of the University of Prague, writing his thesis in 1917 on convex conformal mappings under the direction of Pick, who himself did notable work in complex analysis. Pick's invariant form of the Schwarz lemma appears in several books on geometric function theory. Note that Pick was a student of Weierstrass's student Königsberger. Löwner's thesis contained interesting results on the growth of convex univalent functions and their derivatives. He also proved that the Bieberbach conjecture would hold for the subclass of convex univalent functions, and in fact, $|a_n| \leq 1$. In an important paper of 1923,[10] Löwner developed a powerful method for dealing with the class of univalent functions. Bieberbach was very impressed by this method and inserted "I" (i.e., part I) in the title of Löwner's paper, implying that Löwner should work further in this area; unfortunately, Löwner did not return to the coefficient problem. In his paper, he defined a subset S_1 of S consisting of single slit mappings, univalent functions mapping the unit disk onto the complex plane minus one analytic Jordan arc extending to infinity. Using Carathéodory's theorem, he showed that S_1 was a dense subset of S in the topology defined by uniform convergence on compact subsets. Next, he proved that any function in S_1 could be obtained from the identity mapping by a series of successive infinitesimal transformations. He gave a fairly simple differential equation to effect this transformation; in fact, he gave two forms of this equation, one of which he himself used to prove that $|a_i| \leq i$ for $i = 2, 3$; the other form was used by de Branges to derive the complete result.

[10] Loewner (1988) pp. 45–64.

An alternative approach to the coefficient problem for univalent functions, using area inequalities, was initiated in a 1914 paper[11] by the Swedish-American mathematician Thomas Hakon Gronwall (1877–1932); Bieberbach's work was an independent discovery of the same idea. This method would play a part in the development of univalent functions and in the proof of the Bieberbach conjecture. Gronwall received his doctor's degree in 1898 under Mittag-Leffler, but also learned from mathematicians such as H. von Koch, I. Fredholm, and E. Phragmén. He received an engineering degree from Berlin in 1902 and then worked at various steel works in the United States. In 1912 he returned to his first love and in the next two years published almost two dozen papers ranging over the topics of Fourier series, analytic functions, conformal mappings, and special functions. Consequently, he was invited to Princeton as an instructor in 1913, and was promptly promoted. Gronwall soon left Princeton to take up a number of other pursuits, but not before J. W. Alexander (1888–1971), the famous topologist, had completed his thesis on univalent functions under him. In fact, Alexander had been a protégé of O. Veblen (1880–1960) and had already published couple of papers in topology when Veblen suggested that he do his thesis in analysis under Gronwall. Apparently, Veblen feared that topology might be a passing fad!

In 1916, Bieberbach rediscovered one of Gronwall's area inequalities and employed it to prove his theorem on the second coefficient in the Taylor expansion of a normalized univalent function.[12] In this paper, he also obtained a result on the growth of a univalent function; he later used this to prove that $|a_n| = O(n^2)$. Then in 1923, Littlewood improved on this, showing that the order of the nth coefficient had to be n. Seven years later, Littlewood made another significant contribution to this topic in collaboration with his pupil Paley. R. E. A. C. Paley (1907–1933), graduated from Trinity College in 1929. He wrote his dissertation under Littlewood on nondifferentiable functions and was elected to a Trinity Fellowship in 1930. He was quickly blossoming into one of the leading British mathematicians of his generation when his life was cut short by a skiing accident in the Rocky Mountains. In his very brief career, he published almost thirty papers in several aspects of analysis and collaborated with such outstanding mathematicians as Littlewood, N. Wiener, and A. Zygmund. Littlewood and Paley proved that the coefficients of any odd univalent function in S are bounded by a constant independent of the function.[13] More precisely, for all $F \in S$, and

$$F(z) = z + c_3 z^3 + c_5 z^5 + \cdots, \tag{37.2}$$

there exists an absolute constant A independent of F such that $|c_{2n+1}| \leq A$. In a footnote they observed, "No doubt the true bound is by $A = 1$." This conjecture makes sense in light of an earlier result of I. I. Privalov, that $A = 1$ for odd starlike functions. A set $E \subset \mathbb{C}$ is called starlike with respect to a point $w_0 \in E$ if the line segment joining w_0 to every point $w \in E$ lies entirely in E. A starlike function is a conformal mapping of the unit disk onto a domain starlike with respect to the origin.

---

[11] Gronwall (1914).
[12] Bieberbach (1916).
[13] Littlewood and Paley (1932).

This conjecture implied the Bieberbach conjecture, but it was proved false in 1933 by the Hungarian mathematicians M. Fekete and G. Szegő.[14] They used Löwner's theory to establish that

$$|c_5| \leq \frac{1}{2} + e^{-\frac{2}{3}} = 1.013\ldots, \tag{37.3}$$

and that the inequality was sharp. A modification of the Paley-Littlewood conjecture was suggested by a result of the French mathematician Jean Dieudonné (1906–1992). Dieudonné, a founding member of the Bourbaki group, proved in 1931[15] that if an odd univalent function is real on the real axis, then

$$|c_{2n-1}| + |c_{2n+1}| \leq 2, \quad \text{and} \quad |c_3| \leq 1. \tag{37.4}$$

Then in 1936, M. S. Robertson applied the method of Fekete to prove that

$$|c_3| + |c_5| \leq 2, \tag{37.5}$$

even when $F(z)$ was not real on the real axis.[16] Combining this with Dieudonné's result, Robertson conjectured that the Littlewood-Paley conjecture was true on the average for an odd univalent function:

$$\sum_{k=1}^{n} |c_{2k-1}|^2 \leq n. \tag{37.6}$$

Observe that this implies the Bieberbach conjecture: If $f \in S$ is given by (37.1), and the odd function $F(z) = (f(z^2))^{\frac{1}{2}}$ by (37.2), then the relation between the coefficients of these two functions is given by

$$a_n = c_1 c_{2n-1} + c_3 c_{2n-3} + \cdots + c_{2n-1} c_1, \quad n \geq 1. \tag{37.7}$$

De Branges's proof of the Bieberbach conjecture in actuality demonstrated a more general result, Milin's conjecture; this concerned the logarithmic coefficients of the univalent function. Thus, de Branges's method did not directly yield Bieberbach's conjecture. We note that logarithmic coefficients in connection with univalent functions were first considered by Helmut Grunsky (1904–1986). Grunsky was an excellent analyst with a long and varied career; Ahlfors remarked that his thesis on extremal problems in conformal mappings was "a truly remarkable piece of work."[17] In 1939, while at Berlin, Grunsky showed[18] that an analytic function

$$g(z) = z + b_0 + \frac{b_1}{z} + \cdots$$

[14] Fekete and Szegő (1933).
[15] Dieudonné (1931).
[16] Robertson (1936).
[17] Ahlfors (1982) vol. 1, p. 498.
[18] Grunsky (1939).

in a neighborhood of ∞ would extend to an injective and analytic function in the disk $|z| > 1$ (i.e. $g \in \Sigma$), if and only if its Grunsky coefficients, defined by

$$\log \frac{g(z) - g(\zeta)}{z - \zeta} = -\sum_{k=1}^{\infty} \sum_{l=1}^{\infty} b_{kl} z^{-k} \zeta^{-l}, \tag{37.8}$$

satisfied the Grunsky inequalities

$$\left| \sum_{k=1}^{\infty} \sum_{l=1}^{\infty} b_{kl} x_k x_l \right| \leq \sum_{k=1}^{\infty} \frac{|x_k|^2}{k}, \tag{37.9}$$

where $\{x_k\}$ was a sequence of complex numbers. Clearly, the Grunsky coefficients provided a characterization of the property of univalence. Grunsky's proof of the theorem employed contour integration and was not difficult, although the expressions of the Grunsky coefficients b_{kl} in terms of the coefficients b_k of g were very complicated. Perhaps this is one reason that the effectiveness of Grunsky's inequality was not noticed until around 1960 when it was used by Z. Charzynski and M. Schiffer to reprove the result[19] that $|a_4| \leq 4$. In 1955, Schiffer and P. R. Garabedian had already proved[20] $|a_4| \leq 4$ by means of a powerful variational technique developed by Schiffer in the 1930s. Soon after the work of Charzynski and Schiffer, a generalization of Gronwall's area theorem in terms of the Grunsky coefficients was noted by a number of mathematicians, including J. A. Jenkins, Milin, and C. Pommerenke. Schiffer had already made this observation in 1948. For Pommerenke's formulation, let $g \in \Sigma$, and let $x_1, x_2, \ldots, x_m$ be complex numbers not all zero. Then

$$\sum_{k=1}^{\infty} \left| \sum_{l=1}^{m} b_{kl} x_l \right|^2 \leq \sum_{k=1}^{m} \frac{1}{k} |x_k|^2, \tag{37.10}$$

where equality holds if and only if the area of $\mathbb{C} \backslash g(|z| > 1)$, that is, the complement of the image of $|z| > 1$, is zero. Note that when $m = \infty$, (37.10) and (37.9) are equivalent.

In a 1964 paper,[21] I. M. Milin applied the area method to study the properties of $\{A_n(\zeta)\}$, defined by

$$\log \frac{z - \zeta}{F(z) - F(\zeta)} = \sum_{n=1}^{\infty} A_n(\zeta) z^{-n}, \tag{37.11}$$

where $F \in \Sigma$. Soon after this, I. E. Bazilevich worked directly with $\log \left(\frac{f(z)}{z} \right)$ and proved an interesting inequality about its coefficients. In his account of the motivations behind his conjecture, Milin wrote,[22] "In this way I developed the conviction that the property of univalence reveals itself rather simply through area theorems or other methods in the form of restrictions on the coefficients of the logarithmic

---

[19] Charzynski and Schiffer (1960).
[20] Garabedian and Schiffer (1955).
[21] Milin (1964).
[22] Milin (1986).

function (37.11) and $\log \frac{f(z)}{z} = 2 \sum_{n=1}^{\infty} \gamma_n z^n$, and that it is necessary to construct an 'apparatus of exponentiation' to transfer the restrictions from logarithmic coefficients to coefficients of the functions themselves."

It was with this in mind that in 1965, N. A. Lebedev and Milin worked out the exponential inequality:[23] If $\sum_{k=1}^{\infty} A_k z^k$ is an arbitrary power series with positive radius of convergence and

$$\exp\left(\sum_{k=1}^{\infty} A_k z^k\right) = \sum_{k=0}^{\infty} D_k z^k,$$

then

$$\sum_{k=0}^{n-1} |D_k|^2 \le n \exp\left\{\frac{1}{n} \sum_{v=1}^{n-1} \sum_{k=1}^{v} \left(k|A_k|^2 - \frac{1}{k}\right)\right\}. \qquad (37.12)$$

Now note that if we write $\left(\frac{f(z)}{z}\right)^{\frac{1}{2}} = \sum c_{2n+1} z^{2n+1}$, then

$$\left(\frac{f(z)}{z}\right)^{\frac{1}{2}} = \exp\left(\frac{1}{2} \log \frac{f(z)}{z}\right)$$

implies that for γ_n, as defined in Milin's quotation,

$$\sum_{n=0}^{\infty} c_{2n+1} z^n = \exp\left\{\sum_{n=1}^{\infty} \gamma_n z^n\right\}. \qquad (37.13)$$

Applying the Lebedev–Milin inequality, we obtain

$$\sum_{k=0}^{n-1} |c_{2k+1}|^2 \le n \exp\left\{\frac{1}{n} \sum_{v=1}^{n} \sum_{k=1}^{v} \left(k|\gamma_k|^2 - \frac{1}{k}\right)\right\}. \qquad (37.14)$$

Milin observed this inequality in 1970 in the course of writing his book on univalent functions.[24] He perceived that if the inequalities

$$\sum_{v=1}^{n} \sum_{k=1}^{v} \left(k|\gamma_k|^2 - \frac{1}{k}\right) \le 0, \quad n = 1, 2, 3, \dots, \qquad (37.15)$$

were true, then Robertson's conjecture (and hence Bieberbach's conjecture) followed from (37.14). For Koebe's function $f_\theta(z)$, given by (37.25), $|\gamma_k| = \frac{1}{k}$ so that equality holds in (37.15). Milin did not state (37.15) as a conjecture in his book, although he

[23] Lebedev and Milin (1965).
[24] See Milin (1986) p. 111.

had evidence to support it. For instance, inspired by a result of Pommerenke, Milin obtained the equality

$$\sum_{k=1}^{n} |\gamma_k|^2 \le \sum_{k=1}^{n} \frac{1}{k} + \delta, \quad \delta < 0.312. \tag{37.16}$$

It was in a 1972 paper[25] that A. Z. Grinshpan, with the approval of Lebedev and Milin, referred to inequality (37.15) as Milin's conjecture.

Lebedev, Milin, E. G. Emelyanov, and Grinshpan were members of the Leningrad or St. Petersburg school of geometric function theory. In 1984, these mathematicians and other members of the Leningrad Seminar joined together and exerted considerable effort to reformulate in classical form de Branges's proof of the Milin conjecture, making it more accessible to the community of geometric function theorists. The Leningrad school was founded by G. M. Goluzin (1906–1952). Goluzin entered Leningrad University in 1924 and remained there in various capacities until his death. He was appointed professor of mathematics in 1938, and from then on he led the seminar and built up a school of function theorists. Goluzin made major contributions to the theory of univalent functions and developed a variation on Schiffer's technique of interior variations, applying it to several problems and deriving a number of deep results. In an early paper, he applied Löwner's parametric method to obtain a sharp bound on $|\arg f'(z)|$ for $f \in S$. Another easily stated theorem of Goluzin is that $|a_n| < \frac{3}{4} en$, an improvement on Littlewood; of course, he derived the Goluzin inequality: If

$g \in \Sigma$, z_ν lie in the set $|z| > 1$ and $\gamma_\nu \in \mathbb{C}$, $\nu = 1, 2, \dots, n$, then

$$\left| \sum_{\mu=1}^{n} \sum_{\nu=1}^{n} \gamma_\mu \gamma_\nu \log \frac{g(z_\mu) - g(z_\nu)}{z_\mu - z_\nu} \right| \le \sum_{\mu=1}^{n} \sum_{\nu=1}^{n} \gamma_\mu \overline{\gamma_\nu} \log \frac{1}{1 - (z_\mu \overline{z_\nu})^{-1}}. \tag{37.17}$$

We observe that this inequality can be derived from Grunsky's; conversely, this implies Grunsky. In 1972, the American mathematician C. H. FitzGerald exponentiated Goluzin's inequality to obtain what is now called FitzGerald's inequality, from which he derived several coefficient inequalities.[26] For example, he showed that $|a_n| < \sqrt{\frac{7}{6}} n < 1.081\, n$, and in 1978, D. Horowitz, using the same method, made an improvement,[27] obtaining

$$|a_n| < \left(\frac{1,659,164,137}{681,080,400} \right)^{\frac{1}{14}} n \approx 1.0657\, n.$$

Until de Branges, this was the best result on the Bieberbach conjecture for all n.

[25] Grinshpan (1972).
[26] FitzGerald (1972).
[27] Horowitz (1978).

37.2   Gronwall: Area Inequalities

In his 1914 paper, Gronwall derived results on the growth of a univalent function and its derivative.[28] These depended on the measure of the area of the image of a disk under the conformal transformation given by the univalent function. Gronwall gave two main applications of this idea. In the first, he assumed $f(x) = \sum_{n=1}^{\infty} a_n x^n$ to be univalent and the area of the image of the unit disk under f to be at most A. Then for $|x| \le r < 1$, he showed that

$$|f(x)| \le \sqrt{\frac{A}{\pi}} \sqrt{\log \frac{1}{1 - r^2}}.$$

He used the change of variables formula for an integral to conclude that the area $A(r)$ of the image of the disk $|x| \le r < 1$ was

$$A(r) = \int_0^r d\rho \int_0^{2\pi} |f'(\rho e^{i\theta})|^2 \rho \, d\theta = \pi \sum_{n=1}^{\infty} n |a_n|^2 r^{2n}. \qquad (37.18)$$

Note that

$$f'(\rho e^{i\theta}) = \sum_{n=1}^{\infty} n a_n \rho^{n-1}$$

and term by term integration is possible because of absolute convergence. From (37.18) Gronwall concluded, after letting r tend to one, that

$$\pi \sum_{n=1}^{\infty} n |a_n|^2 \le A. \qquad (37.19)$$

Using (37.19),

$$|f(x)| \le \sum_{n=1}^{\infty} |a_n| r^n,$$

and the Cauchy–Schwarz inequality, he found the required result:

$$|f(x)|^2 \le \sum_{n=1}^{\infty} n |a_n|^2 \sum_{n=1}^{\infty} \frac{r^{2n}}{n} = \frac{A}{\pi} \log \frac{1}{1 - r^2}.$$

For the second application, Gronwall considered a function

$$f(x) = \frac{1}{x} + \sum_{n=1}^{\infty} a_n x^n,$$

[28] Gronwall (1914).

where, without the term $\frac{1}{x}$, the series converged for $|x| < 1$. Such series had been discussed earlier, but Gronwall derived an important inequality for them, called the area theorem. For this purpose, it is convenient to let $z = \frac{1}{x}$ and consider the class Σ of functions

$$g(z) = z + b_0 + \frac{b_1}{z} + \frac{b_2}{z^2} + \cdots, \tag{37.20}$$

analytic and one-to-one in $|z| > 1$ except for a simple pole at ∞ with residue 1. For these functions, Gronwall proved that

$$\sum_{n=1}^{\infty} n|b_n|^2 \le 1. \tag{37.21}$$

To verify this, he again applied the area method. He did not give all the details, but one may apply Green's theorem to see that if the closed curve C_r is the image of $|z| = r > 1$ under $g(z)$, then it encloses a positive area

$$0 < \frac{1}{2} \int_0^{2\pi} \overline{g(re^{i\theta})}\, g'(re^{i\theta}) re^{i\theta}\, d\theta$$
$$= \pi \left\{ r^2 - \sum_{n=1}^{\infty} n|b_n|^2 r^{-2n} \right\}. \tag{37.22}$$

The necessary inequality follows by letting $r \to 1^+$.

37.3 Bieberbach's Conjecture

In 1916, apparently unaware of Gronwall's earlier work, Bieberbach reproved the area theorem and deduced his inequality for the second coefficient of functions in the set S of normalized univalent functions.[29] To achieve this, he used an idea he called Faber's trick: Supposing f is a function in S, then $F(z) = (f(z))^{\frac{1}{2}}$ is an odd univalent function. To prove this, observe that $f(z)$ vanishes only at $z = 0$ and hence a single valued branch of the square root can be chosen in

$$F(z) = z(1 + a_2 z^2 + a_3 z^4 + \cdots)^{\frac{1}{2}}.$$

Clearly, $F(z)$ is odd. It is univalent because if $F(z_1) = F(z_2)$, then $f(z_1^2) = f(z_2^2)$; moreover, the univalence of $f(z)$ implies $z_1 = \pm z_2$. If $z_1 = -z_2$, then $F(z_1) = F(z_2) = -F(z_1)$. This implies $F(z_1) = 0$ or $z_1 = 0$, proving the result. To apply the area theorem, Bieberbach noted that

$$F(z) = z + \frac{1}{2} a_2 z^3 + \cdots,$$

---

[29] Bieberbach (1916).

and used $F(z)$ to construct a function $g(z)$ in the class Σ:

$$g(z) = \frac{1}{F(\frac{1}{z})} = z - \frac{1}{2}a_2\frac{1}{z} + \cdots = z + \sum_{n=1}^{\infty} b_n z^{-n}.$$

Hence, by (37.21), he obtained $|b_1| \leq 1$ or $|a_2| \leq 2$. In a footnote, Bieberbach went on to conjecture that $|a_n| \leq n$ for the coefficients of a normalized univalent function on the unit disk. He was able to verify Koebe's conjecture that the image of the open unit disk under any $f \in S$ would always contain a circle of radius $\frac{1}{4}$ with the origin as center. More precisely, he showed that if $f(z) \neq w$ in $|z| < 1$, then $|w| \geq \frac{1}{4}$ (an improvement on Koebe). This was an immediate corollary of the bound for a_2. For if $f(z) \neq w$, then

$$\frac{w f(z)}{w - f(z)} = z + \left(a_2 + \frac{1}{w}\right)z^2 + \cdots \quad \in S$$

and hence

$$\left|a_2 + \frac{1}{w}\right| \leq 2, \quad \left|\frac{1}{w}\right| \leq 2 + |a_2| \leq 4 \text{ or } |w| \geq \frac{1}{4}. \tag{37.23}$$

The example (often called Koebe's function)

$$f(z) = \frac{z}{(1-z)^2} = z + 2z^2 + 3z^3 + \cdots + nz^n + \cdots, \tag{37.24}$$

or more generally

$$f_\theta(z) = \frac{z}{(1 - e^{i\theta}z)^2} = z + 2e^{i\theta}z^2 + 3e^{2i\theta}z^3 + \cdots + ne^{(n-1)\theta}z^n + \cdots, \tag{37.25}$$

shows that $|a_2| = 2$ actually occurs for functions in S. We can also write (37.24) as

$$w = f(z) = \frac{1}{4}\left(\frac{1+z}{1-z}\right)^2 - \frac{1}{4}.$$

From this representation, it is easy to see that $f(z)$ maps $|z| < 1$ conformally onto the w-plane cut from $-\frac{1}{4}$ to $-\infty$ along the negative real axis. Note that $f(z) \neq -\frac{1}{4}$ in $|z| < 1$. Moreover, because $f_\theta(z) = e^{-i\theta} f(ze^{i\theta})$, this function maps $|z| < 1$ conformally onto the w-plane cut radially from $-\frac{1}{4}e^{-i\theta}$ to $-\infty e^{-i\theta}$.

In this paper, Bieberbach obtained another important result on the growth of a normalized univalent function $f(z)$:

$$\frac{r}{(1+r)^2} \leq |f(z)| \leq \frac{r}{(1-r)^2}, \quad |z| = r, \quad (0 < r < 1). \tag{37.26}$$

37.4 Littlewood: $|a_n| \le en$

In 1923, Littlewood proved that Bieberbach's conjecture was correct up to the order of magnitude. His paper with the result given in the title of this section appeared in 1925.[30] Littlewood derived his inequality for the coefficients a_n from the inequality,

$$\frac{1}{2\pi} \int_0^{2\pi} |f(re^{i\theta})| \, d\theta < \frac{r}{1-r}, \quad 0 < r < 1, \tag{37.27}$$

where $f \in S$. He considered the univalent function

$$\phi(z) = (f(z^2))^{\frac{1}{2}} = z + b_3 z^3 + \cdots$$

and by (37.26) concluded that

$$|\phi(te^{i\theta})| \le \frac{t}{1-t^2}.$$

This result in turn implied that ϕ transformed $|z| \le t < 1$ to a region whose area $A(t)$ was less that $\frac{\pi t^2}{(1-t^2)^2}$. He combined this with the equation

$$\pi \sum_{n=1}^{\infty} n|b_n|^2 t^{2n} = \int_0^t r \, dr \int_0^{2\pi} |\phi'(re^{i\theta})|^2 \, d\theta = A(t)$$

to derive the inequality

$$\sum_{n=1}^{\infty} n|b_n|^2 t^{2n-1} \le \frac{t}{(1-t^2)^2}.$$

Integrating from 0 to $r < 1$, he obtained

$$\sum_{n=1}^{\infty} |b_n|^2 r^{2n} \le \frac{r^2}{1-r^2}.$$

He next observed that the series on the left-hand side of this inequality was given by

$$\frac{1}{2\pi} \int_0^{2\pi} |\phi(re^{i\theta})| \, d\theta = \frac{1}{2\pi} \int_0^{2\pi} |f(r^2 e^{2i\theta})| \, d\theta = \frac{1}{2\pi} \int_0^{2\pi} |f(r^2 e^{i\psi})| \, d\psi.$$

At this point, to derive the necessary result for a_n, Littlewood could apply Cauchy's formula, with $r = 1 - \frac{1}{n}$, to obtain

[30] Littlewood (1925).

$$|a_n| = \frac{1}{2\pi}\left|\int_{|z|=r}\frac{f(z)}{z^{n+1}}\,dz\right| \leq \frac{1}{2\pi r^n}\int_0^{2\pi}|f(re^{i\theta})|\,d\theta \leq \frac{1}{r^{n-1}(1-r)},$$

(37.28)

$$\leq \left(1+\frac{1}{n-1}\right)^{n-1} \quad n < en.$$

In the same paper, Littlewood also showed that if $M(r, f)$ denoted the maximum of $|f(z)|$ on the circle $|z| = r$, and f was univalent, then for $\lambda > \frac{1}{2}$,

$$\frac{1}{2\pi}\int_{-\pi}^{\pi}|f(re^{i\theta})|^\lambda d\theta \leq A_\lambda \rho^\lambda(1-\rho)^{-2\lambda+1}.$$

(37.29)

37.5   Littlewood and Paley on Odd Univalent Functions

Littlewood and Paley stated their main theorem of 1932:[31] If

$$f(z) = z + a_3 z^3 + a_5 z^5 + \cdots$$

is an odd univalent function, then there is an absolute constant A, such that $|a_n| \leq A$.

To prove this result, they used Bieberbach's growth theorem for univalent functions, an inequality from Littlewood's 1925 paper, and a new inequality given by

$$\frac{1}{2\pi}\int_{-\pi}^{\pi}|\sigma'(\rho e^{i\theta})|^2 d\theta \leq C\rho^{-1}(1-\rho)^{-1}M^2(\rho^{\frac{1}{2}},\sigma),$$

(37.30)

where C denoted a constant and σ was in S, the set of normalized univalent functions. In their proof of this, they assumed that $\sigma(z) = z + c_2 z^2 + c_3 z^3 + \cdots$ and applied Gronwall's formula to conclude that the area of the image of $|z| < \rho$ under σ was given by

$$\pi \sum n|c_n|^2\rho^{2n} \leq \pi M^2(\rho,\sigma).$$

Using this they arrived at the required result:

$$2\pi \sum n^2|c_n|^2\rho^{2n} \leq 2\pi \, \text{Max}(n\rho^n) \sum n|c_n|^2\rho^n$$

$$\leq \frac{A\rho}{1-\rho}AM^2(\rho^{\frac{1}{2}},\sigma),$$

for some absolute constant A. We note that in Littlewood and Paley's paper, every absolute constant was denoted by the same symbol, A. We shall follow their convention. They constructed two other univalent functions related to $f(z)$, defined by the relations

---

[31] Littlewood and Paley (1932).

$$\phi(z) = \left(f(\sqrt{z})\right)^2 = z + 2a_3 z^2 + \cdots,$$

$$\psi(z) = \left(f(z^3)\right)^{\frac{1}{3}} = z + \frac{1}{3}a_3 z^7 + \cdots.$$

They proved the univalence of $\phi(z)$ by noting that $\phi(z) = w$ implied $f(\sqrt{z}) = \pm\sqrt{w}$. They reasoned that since f was odd, only a pair of equal and opposite values were possible for $\sqrt{z}$, and hence only one value was possible for z. They also used a simple argument to demonstrate $\psi(z)$ to be univalent.

To prove their theorem, Littlewood and Paley applied Cauchy's theorem to the coefficients of $f'(z)$:

$$|na_n| \leq \frac{\rho^{-n+1}}{2\pi} \int_{-\pi}^{\pi} |f'(\rho e^{i\theta})| d\theta.$$

Thus, it was sufficient for them to show that

$$\int_{-\pi}^{\pi} |f'(\rho e^{i\theta})| \, d\theta < \frac{A}{1-\rho}. \tag{37.31}$$

They noted that by combining (37.31) with the inequality for na_n, $\rho = 1 - \frac{1}{n}$, they obtained the required inequality for a_n. To prove (37.31), they observed that since $f(z) = \psi^3(z^{\frac{1}{3}})$, it followed that for $z = \rho e^{i\theta}$,

$$\frac{1}{2\pi} \int_{-\pi}^{\pi} |f'(z)| d\theta = \frac{1}{6\pi} \int_{-3\pi}^{3\pi} |f'(z)| d\theta = \frac{\rho^{-\frac{2}{3}}}{6\pi} \int_{-3\pi}^{3\pi} |\psi^2(z^{\frac{1}{3}}\psi'(z^{\frac{1}{3}})| d\theta.$$

On applying the Cauchy–Schwarz inequality to the last integral, they found

$$\frac{1}{2\pi} \int_{-\pi}^{\pi} |f'(z)| d\theta \leq \rho^{-\frac{2}{3}} \left(\frac{1}{6\pi} \int_{-3\pi}^{3\pi} |\psi(z^{\frac{1}{3}})|^4 d\theta\right)^{\frac{1}{2}} \left(\frac{1}{6\pi} \int_{-3\pi}^{3\pi} |\psi'(z^{\frac{1}{3}})|^2 d\theta\right)^{\frac{1}{2}}. \tag{37.32}$$

Denoting the two integrals on the right by P and Q, respectively, and applying $|\psi(z)|^4 = |\phi(z^2)|^{\frac{2}{3}}$ combined with the change of variables $t = 2\theta$, they estimated

$$P = \frac{1}{12\pi} \int_{-6\pi}^{6\pi} |\phi(\rho^2 e^{it})|^{\frac{2}{3}} dt = \frac{1}{2\pi} \int_{-\pi}^{\pi} < A\rho^{\frac{4}{3}}(1-\rho^2)^{-\frac{1}{3}}. \tag{37.33}$$

The last inequality followed from Littlewood's inequality (37.29). To estimate Q, they first used (37.30) to arrive at

$$Q = \frac{1}{2\pi} \int_{-\pi}^{\pi} |\psi'(\rho^{\frac{1}{3}} e^{it})|^2 dt \leq A\rho^{-\frac{1}{3}}(1-\rho^{\frac{1}{3}})^{-1} M^2(\rho^{\frac{1}{3}}, \psi). \tag{37.34}$$

An application of the growth estimate to $M^2(\rho^{\frac{1}{3}}, \psi)$ would not produce the necessary result, so they used $\psi^3(z^{\frac{1}{3}}) = \phi^{\frac{1}{2}}(z^2)$ to get

$$M^2(\rho^{\frac{1}{3}}, \psi) = M^2(\rho^2, \phi^{\frac{1}{6}}) < \rho^{\frac{2}{3}}(1 - \rho^2)^{-\frac{2}{3}}.$$

Combining this with (37.34), they obtained

$$Q < A\rho^{\frac{1}{3}}(1 - \rho^{\frac{1}{3}})^{-1}(1 - \rho^2)^{-\frac{2}{3}} < A(1 - \rho)^{-\frac{5}{3}}.$$

Taking this inequality with (37.33) and (37.32) gave the required result:

$$\frac{1}{2\pi} \int_{-\pi}^{\pi} |f'(\rho e^{i\theta})| d\theta < A\rho^{-\frac{2}{3}} \rho^{\frac{2}{3}}(1 - \rho^2)^{-\frac{1}{6}}(1 - \rho)^{-\frac{5}{6}} < A(1 - \rho)^{-1}.$$

This completed their ingenious proof that the coefficients of odd univalent functions were bounded.

37.6   Karl Löwner and the Parametric Method

Carathéodory's theorem was used a decade later by Löwner to construct his parametric theory of univalent functions. To describe Löwner's method, we must first define slit mappings. A single-slit mapping is a function mapping a domain conformally onto the complex plane minus a single Jordan arc. Löwner showed[32] that such mappings were dense in S, the set of all conformal mappings of the unit disk with $f(0) = 0$ and $f'(0) = 1$. More exactly, for each $f \in S$, there exists a sequence of single-slit mappings $f_n \in S$ such that $f_n \to f$ uniformly on compact subsets of the unit disk.

We follow Duren's presentation to summarize the argument from Löwner's 1923 paper.[33] It is sufficient to consider functions f mapping the unit disk onto a domain bounded by a (closed) analytic Jordan curve because, for any $f \in S$, the function $f(rz), 0 < r < 1$, is also univalent with the required image and $f_r(z) = \frac{1}{r} f(rz) \in S$. By letting $f \to 1^-$, we get functions $f_r \in S$ such that $f_r \to f$ uniformly on compact subsets of the unit disk. So assume that $f \in S$ maps the unit disk onto a domain G bounded by an analytic Jordan curve C. Choose a point $w_0 \in C$ and let Γ be any Jordan curve from ∞ to w_0. Denote by Γ_n the Jordan curve consisting of Γ followed by a part of C joining w_0 to a point $w_n \in C$. Let G_n represent the complement of Γ_n in the complex plane, and let g_n map the unit disk onto G_n with $g_n(0) = 0$ and $g_n'(0) > 1$. We note that such a function g_n exists, by the Riemann mapping theorem. Now choose a sequence of points $w_n \in C$ such that $w_n \to w_0$ and $\Gamma_n \subset \Gamma_{n+1}$. Then G is the kernel of the sequence $\{G_n\}$. By Carathéodory's kernel convergence theorem, we must have $g_n \to f$ uniformly on compact subsets of the unit disk and, by Cauchy's theorem, $g_n'(0) \to f'(0) = 1$. Hence, $f_n = \frac{g_n}{g_n'(0)}$ is a sequence of single-slit

[32] Loewner (1988) pp. 45–63.
[33] Duren (1983).

mappings converging to f uniformly on compact subsets of the unit disk. Thus, we conclude with Löwner that the single-slit mappings are dense in S.

Now suppose $f \in S$ is a single-slit mapping taking the unit disk onto a domain G, the complement of a Jordan arc Γ extending from a point w_0 in the complex plane to ∞. Also suppose that $w = \psi(t)$, $0 \leq t < T$ is a continuous, one-to-one parametrization of Γ with $\psi(0) = w_0$. Let Γ_t denote that part of Γ from $\psi(t)$ to ∞, and let G_t represent the complement of Γ_t. Let $g_t(z) = g(z, t)$ be the conformal mapping of the unit disc onto G_t, with $g(0, t) = 0$ and $g'(0, t) = \gamma(t) > 0$, so that $g(z, t)$ has the series expansion

$$g_t(z) = g(z, t) = \gamma(t) \left\{ z + c_1(t) z^2 + c_2(t) z^3 + \cdots \right\}, \tag{37.35}$$

where $g(z, 0) = f(z)$. By an application of the Schwarz lemma, $\gamma(t)$ may be seen to be a monotonically increasing function of t. Thus, by reparametrization, we can take $\gamma(t) = e^t$. Moreover, T will then be ∞. So we can write

$$g_t(z) = g(z, t) = e^t \left\{ z + \sum_{n=2}^{\infty} b_n(t) z^n \right\}, \quad 0 \leq t < \infty, \tag{37.36}$$

in what is called the standard parametrization. Löwner then considered the family of mappings

$$f_t(z) = g_t^{-1}(f(z)) = e^{-t} \left\{ z + \sum_{n=2}^{\infty} a_n(t) z^n \right\}, \quad 0 \leq t < \infty. \tag{37.37}$$

It is easy to see that the functions f_t map the unit disk onto the unit disk minus an arc extending inward from the boundary, and that $e^t f_t \in S$. By using the growth estimates of Bieberbach and Gronwall, he was able to conclude that

$$\lim_{t \to \infty} e^t f_t(z) = f(z). \tag{37.38}$$

And it is obvious that $f_0(z) = z$, the identity function. So the function $e^t f_t(z)$ starts at the identity function $f(z) = z$ and ends at $f(z) \in S$ as $t \to \infty$. Löwner determined the differential equation satisfied by this one parameter family of functions,

$$\frac{\partial f_t}{\partial t} = -f_t \frac{1 + \chi(t) f_t}{1 - \chi(t) f_t}, \tag{37.39}$$

where $\chi(t)$ was a continuous complex valued function with $|\chi(t)| = 1$, $0 \leq t < \infty$. He also gave the equation satisfied by the family of functions $g_t(z)$. By (37.37), $g_t(f_t(z)) = f(z)$. Setting $\zeta = f_t(z)$, we have $g_t(\zeta) = f(z)$; take the derivative with respect to t to get

$$\frac{\partial g_t}{\partial \zeta} \frac{\partial \zeta}{\partial t} + \frac{\partial g_t}{\partial t} = 0.$$

When this is substituted in (37.39), the result is the differential equation for $g_t(z)$:

$$\frac{\partial g_t}{\partial t} = \frac{\partial g_t}{\partial z} z \frac{1 + \chi(t)z}{1 - \chi(t)z}, \qquad 0 \le t < \infty, \tag{37.40}$$

where $g_0(z) = f(z)$ and $\lim_{t \to \infty} g_t(z) = z$.

Löwner applied his parametric method to the Bieberbach conjecture. In his paper he deduced only that $|a_2| \le 2$ and $|a_3| \le 3$.[34] Bieberbach suggested that Löwner call his paper part I of a work in progress, since it was clear that he had a general method applicable to the coefficient problem. To understand Löwner's derivation of the inequalities for the second and third coefficients, note that since the class of univalent functions S is invariant under rotation, it is sufficient to prove that $\mathrm{Re}(a_3) \le 3$. Now substitute the series (37.37) for f_t into the differential equation (37.39) and equate the coefficients of z^2 and z^3 on both sides to get the two relations

$$a_2'(t) = -2e^{-t}\chi(t) \tag{37.41}$$

and

$$a_3'(t) = -2e^{-2t}[\chi(t)]^2 - 4e^{-t}\chi(t) a_2(t). \tag{37.42}$$

Since $a_2(0) = 0$, and $\lim_{t \to \infty} a_n(t) = a_n$, where a_n is the nth Taylor coefficient of the univalent function $f \in S$, we may integrate equation (37.41) to get

$$a_2 = \int_0^\infty a_2'(t)\, dt = -2 \int_0^\infty e^{-t}\chi(t)\, dt; \tag{37.43}$$

hence

$$|a_2| \le 2 \int_0^\infty e^{-t}\, dt = 2 \quad \text{because} \quad |\chi(t)| = 1.$$

Substituting (37.41) into (37.42) then produces

$$a_3'(t) = 2a_2(t) a_2'(t) - 2e^{-2t}[\chi(t)]^2\,; \tag{37.44}$$

integrate to obtain

$$a_3 = 4\left(\int_0^\infty \chi(t) e^{-t}\, dt \right)^2 - 2 \int_0^\infty \chi^2(t) e^{-2t}\, dt.$$

We next set $\chi(t) = e^{i\theta(t)}$ to get

$$\mathrm{Re}(a_3) = 4\left\{ \left(\int_0^\infty \cos\theta(t) e^{-t}\, dt \right)^2 - \left(\int_0^\infty \sin\theta(t) e^{-t}\, dt \right)^2 \right.$$
$$\left. - \int_0^\infty \cos^2\theta(t) e^{-2t}\, dt \right\} + 1.$$

---

[34] Loewner (1988) pp. 62–63.

By the Cauchy–Schwarz inequality

$$\left(\int_0^\infty \cos\theta(t)\,e^{-t}\,dt\right)^2 \le \int_0^\infty \cos^2\theta(t)\,e^{-t}\,dt \int_0^\infty e^{-t}\,dt \le \int_0^\infty \cos^2\theta(t)\,e^{-t}\,dt,$$

so that

$$\mathrm{Re}(a_3) < 4\int_0^\infty \cos^2\theta(t)\left(e^{-t}-e^{-2t}\right)dt + 1$$

$$< 4\int_0^\infty \left(e^{-t}-e^{-2t}\right)dt + 1 = 3.$$

Löwner also wrote down expressions for $a_n'(t)$ and then, after integration, the expressions for $a_n(t)$. However, these are generally too complex to be conveniently utilized.

37.7 De Branges: Proof of Bieberbach

In proving the Bieberbach conjecture, by first proving the Milin conjecture, de Branges applied Löwner's theory of 1923.[35] Though it may appear that Löwner's theory could have been applied at any time, it was not until Milin's contribution that there was a route connecting Löwner to Bieberbach; recall that Milin stated his conjecture for logarithmic coefficients only in 1971. De Branges's great insight was to use special functions to prove Milin's conjecture. And it took the boldness of an independent thinker such as de Branges to make such an attempt. As we have mentioned, de Branges was a functional analyst. He had developed the theory of square summable power series within that context and wished to apply it to various problems, including the Bieberbach conjecture. His extensive functional analytic machinery, so useful to his insights and manner of thought, proved to be a roadblock for others attempting to understand his proof. In the spring of 1984, de Branges presented his proof to the members of the Leningrad (St. Petersburg) geometric function theory seminar. The members of the seminar generously expended a good deal of effort to help him simplify it and express it in classical form. This version of the proof was soon written up by Milin and published as a preprint by the Steklov Institute in Leningrad. FitzGerald and Pommerenke used this preprint to obtain further technical simplifications, also independently found by de Branges. It is the simplified form of the proof that we shall discuss here.

Consider the logarithmic coefficients of the function $g_t(z)$ defined by equation (37.36). Thus,

$$\log\left(\frac{g(z,t)}{e^t z}\right) = \sum_{k=1}^\infty c_k(t)z^k, \quad |z| < 1, \tag{37.45}$$

with $0 \le t < \infty$, and $c_k(0) = 2\gamma_k$, where $2\gamma_k$ are the logarithmic coefficients of the function $f \in S$. Here recall that $g(z,t) = g_t(z)$. If equation (37.45) is differentiated

---

[35] de Branges (1985).

with respect to t and then z and the results are substituted in (37.40) and simplified, we get

$$
\begin{aligned}
1 + \sum_{n=1}^{\infty} c_n'(t) z^n &= \frac{1 + \chi(t)z}{1 - \chi(t)z} \left(1 + \sum_{n=1}^{\infty} n c_n(t) z^n \right) \\
&= \left(1 + 2\chi(t)z + 2\chi(t)^2 z^2 + \cdots \right) \left(1 + \sum_{n=1}^{\infty} n c_n(t) z^n \right).
\end{aligned}
\tag{37.46}
$$

Equate the coefficients of z^n on both sides to get the differential equations satisfied by the coefficients $c_n(t)$:

$$
c_n'(t) = 2\chi(t)^n + n c_n(t) + 2 \sum_{m=1}^{n-1} \chi(t)^{n-m} m c_m(t), \quad n = 1, 2, \ldots .
\tag{37.47}
$$

Note that this is a differential equation for logarithmic coefficients; recall that Löwner had obtained similar differential equations for the coefficients of $g_t(z)$, although they quickly became unwieldy when one tried to solve for them inductively. To prove Milin's conjecture, de Branges made effective use of these differential equations, by introducing some special functions. Recall the Milin conjecture:

$$
\sum_{m=1}^{n} \left(m |c_m(0)|^2 - \frac{4}{m} \right) (n - m + 1) \le 0, \quad n = 1, 2, 3, \ldots .
$$

De Branges defined a function

$$
\phi(t) = \sum_{m=1}^{n} \left(m |c_m(t)|^2 - \frac{4}{m} \right) \tau_{n,m}(t),
\tag{37.48}
$$

where certain properties were required of $\tau_{n,m}(t)$, including

$$
\tau_{n,m}(0) = n - m + 1, \quad m = 1, 2, \ldots, n.
\tag{37.49}
$$

Now, if we can choose $\tau_{n,m}(t)$, such that $\phi'(t) \ge 0$ and $\phi(\infty) = 0$, then we would automatically have $\phi(0) \le 0$, Milin's conjecture. To compute $\phi'(t)$, first set

$$
b_0(t) = 0, \quad b_m(t) = \sum_{m=1}^{n} m c_m(t) \chi(t)^{-m}, \quad m = 1, 2, \ldots
\tag{37.50}
$$

and set

$$
\tau_{n,n+1}(t) \equiv 0, \quad \text{for } 0 \le t < \infty,
\tag{37.51}
$$

so that, by a straightforward calculation,

$$\phi'(t) = \sum_{m=1}^{n} \left(|b_m - b_{m-1}|^2 - 4 \right) \frac{\tau'_{n,m}}{m} + \sum_{m=1}^{n} \left(2|b_m|^2 + 4\text{Re}(b_m) \right) \left(\tau_{n,m} - \tau_{n,m+1} \right).$$

(37.52)

This expression for $\phi'(t)$ takes a very simple form if the functions $\tau_{n,m}(t)$ satisfy the difference differential equation

$$\tau_{n,m} - \tau_{n,m+1} = -\frac{\tau'_{n,m}}{m} - \frac{\tau'_{n,m+1}}{m+1}.$$

(37.53)

In that case,

$$\phi'(t) = -\sum_{m=1}^{n} |b_m + b_{m-1} + 2|^2 \frac{\tau'_{n,m}}{m},$$

(37.54)

and Milin's conjecture is proved, provided that de Branges's system of functions $\tau_{n,m}$ satisfying (37.49), (37.51), and (37.53) also satisfy

$$\tau'_{n,m}(t) \leq 0, \quad 0 \leq t < \infty, \quad m = 1, 2, \ldots, n.$$

(37.55)

Thus, from these equations we must determine the form of the functions $\tau_{n,m}(t)$. We may solve successively for $\tau_{n,n}$, $\tau_{n,n-1}$, and so on. So by (37.49), (37.51), and (37.53), we may write

$$\tau_{n,n} = -\frac{\tau'_{n,n}}{n} \quad \text{or}$$
$$\tau_{n,n}(t) = Ae^{-nt} = e^{-nt},$$

because $\tau_{n,n}(0) = 1$. Next we solve

$$\frac{\tau'_{n,n-1}}{n-1} + \tau_{n,n-1} = 2e^{-nt}$$

to get

$$\tau_{n,n-1}(t) = -2(n-1)e^{-nt} + 2ne^{-(n-1)t}.$$

Note that in general we obtain

$$\tau_{n,m}(t) = m \sum_{k=0}^{n-m} \frac{(-1)^k (2m+k+1)_k (2m+2k+2)_{n-m-k}}{(m+k)k!\,(n-m-k)!} e^{-(m+k)t},$$

(37.56)

when $m = 1, 2, \ldots, n$, since $\tau_{n,n+1} \equiv 0$.

Now recall that the truth of the nth Milin inequality implies the truth of the Bieberbach conjecture for $(n + 1)$th coefficient. Thus, to show that $|a_2| \leq 1$, it is enough to check that $\tau'_{1,1}(t) = -e^{-t}$ is negative, and this fact is obvious. For the third coefficient, we need to check the derivatives of two polynomials in e^{-t}:

$$\tau_{2,1}(t) = -2e^{-2t} + 4e^{-t} \quad \text{and} \quad \tau_{2,2}(t) = e^{-2t}.$$

Observe that their derivatives are

$$4e^{-t}(e^{-t} - 1) \leq 0 \quad \text{and} \quad -2e^{-2t} < 0 \quad (\text{for } 0 \leq t < \infty),$$

respectively. Hence, $|a_3| \leq 3$. In this manner, de Branges verified the Bieberbach conjecture up to the sixth coefficient, although the computations for the last two cases were complicated.

At this point in early February 1984, the stage was set for de Branges to request his colleague at Purdue, Walter Gautschi, a numerical analyst with an interest in special functions, to check the calculations by computer. Gautschi was swamped with work at the time, but he was unable to resist the challenge; he attended de Branges's seminar and reported,[36] "I was immediately struck by the clarity, freshness, and elegance of Louis's talk and began to appreciate how those inequalities came about. To my delight, they could be written in terms of orthogonal polynomials – currently a subject very much on my mind." Gautschi developed the necessary algorithms and managed to verify the Milin conjecture up to $n = 30$. Wondering if the inequalities could be proved analytically, he consulted Richard Askey. It turned out Askey and George Gasper had proved a slightly more general inequality less than a decade earlier.[37]

As it turned out, $\tau'_{n,m}(t)$ could be expressed as a sum of Jacobi polynomials:

$$\tau'_{n,m}(t) = -me^{-mt} \sum_{k=0}^{n-m} P_k^{(2m,0)}(1 - 2e^{-t}). \tag{37.57}$$

In a 1976 paper, Askey and Gasper had proved[38] that for any real $\alpha > -1$,

$$\sum_{k=0}^{n} P_k^{(\alpha,0)}(x) \geq 0, \quad -1 \leq x \leq 1. \tag{37.58}$$

This immediately implied de Branges's inequalities: $\tau'_{n,m}(t) \leq 0$ for $0 \leq t < \infty$. Askey and Gasper's investigation of these sums of Jacobi polynomials arose out of their study of several classical inequalities for trigonometric functions. Their insight was that the correct generalization for these classical inequalities was in the context of Jacobi polynomials, within which the powerful machinery of hypergeometric functions could be applied.

---

[36] Gautschi (1986).
[37] Askey and Gasper (1976).
[38] ibid.

One proof of the Askey–Gasper inequality employed a theorem of Clausen on the square of a $_2F_1$ hypergeometric function and also a connection coefficient result of Gegenbauer. Note that we have stated these results in the exercises of Chapters 23 and 24. This second result is also known as the Gegenbauer–Hua formula. Askey has pointed out[39] that Gegenbauer's formula having been forgotten, Hua rediscovered it[40] in the course of his work in harmonic analysis, carried out in the 1940s and 1950s. Note that Askey also rediscovered this formula in the 1960s. Hua Loo-Keng (1910–1985) taught himself mathematics, and by the age of 19, he was writing papers; these came to the notice of a professor at Qinghua University in Beijing. Hua was consequently appointed to a position at that university, and his career was launched. In 1936, he traveled to Cambridge to work with Hardy, Littlewood, and Davenport on problems in additive number theory. Later, he did research in several complex variables, automorphic functions, and group theory. His wide interests helped him lead the development of modern mathematics in China; in fact, in the 1960s, he turned his attention to mathematical problems with immediate practical applicability. Hua's student, Chen Jing-Run (1933–1996) made important contributions to the Goldbach conjecture.

37.8 Exercises

(1) Show that if $0 \leq t \leq 1$ and $\alpha > -2$, then

$$_3F_2 \left(\begin{matrix} -n, n+\alpha+2, \frac{\alpha+1}{2} \\ \frac{\alpha+3}{2}, \alpha+1 \end{matrix} ; t \right) > 0.$$

See the article by Askey and Gasper in Baernstein (1986).

(2) Show that if

$$|a_1| \geq \sum_{i=2}^{\infty} |a_i|, \text{ then the function } \sum_{n=1}^{\infty} a_n z^n / n$$

maps the interior of the unit circle upon a star-shaped region with center at the origin.

(3) Show that, with the same condition on the coefficients as in Exercise 2, the function $\sum_{n=1}^{\infty} \frac{a_n}{n^2} z^n$ maps the interior of the unit circle upon a convex region. See Alexander (1915) for this and for Exercise 2.

(4) Prove that if f is an analytic mapping of the unit disk into itself and if z, z_1, z_2 are in the unit disk, then

$$\frac{|f(z_1) - f(z_2)|}{|1 - \overline{f(z_1)} f(z_2)|} \leq \frac{|z_1 - z_2|}{|1 - \overline{z_1} z_2|};$$

$$\frac{|f'(z)|}{1 - |f(z)|^2} \leq \frac{1}{1 - |z|^2}.$$

[39] Schoenberg (1988) vol. 1, p. 192.
[40] Hua (1981) pp. 38–39.

Since the Poincaré metric $ds = \frac{2|dz|}{1-|z|^2}$ defines a noneuclidean length element (infinitesimal) in the unit disk, these inequalities may be interpreted to mean that the analytic mapping f decreases the noneuclidean distance between two points and the noneuclidean length of an arc. This invariant form of Schwarz's lemma is due to Georg Pick; see Pick (1915). Georg Pick (1859–1942), professor at Prague for forty-five years, was Königsberger's student and Löwner's teacher; as mentioned earlier, he died in the Theresienstadt concentration camp.

(5) This exercise and the next mention results in geometric function theory, taking a direction different from that discussed in the text. If Γ is a finitely generated Kleinian group with region of discontinuity Ω, then Ω/Γ is of finite type. Read the proof of this theorem in Ahlfors (1982) vol. 2, pp. 273–290. Note that this proof has a gap, later filled by Lipman Bers and Ahlfors himself. The theory of Kleinian groups was initiated by Poincaré. A nice history of this topic is given by Gray (1986). Poincaré (1985) translated into English by Stillwell, offers many important papers of Poincaré in this area and also provides a useful introduction by putting the papers into historical perspective and relating their results to some modern work. For Poincaré's pioneering work on topology, the reader may enjoy the article by Karanbir S. Sarkaria in James (1999) pp. 123–167.

(6) If a Kleinian group is generated by N elements, then Area $(\Omega/\Gamma) \leq 4\pi(N-1)$ and Ω/Γ has at most $84(N-1)$ components. Bers proved this theorem in 1967; read a proof in Bers (1998) vol. 1, pp. 459–477.

37.9   Notes on the Literature

Gray (1994) presents a history of the Riemann mapping theorem. Duren (1953) is an extremely readable and clear exposition and explanation of results on univalent functions; it was written just before the Bieberbach conjecture was proved. The books of Hayman (1994) and Gong (1999) give proofs of the conjecture. Several articles in Baernstein et al. (1986) and the paper by Fomenko and Kuzmina (1986) deal with the historical aspects of the Bieberbach conjecture and proof. FitzGerald (1985) and Pommerenke (1985) contain the reactions to the proof by two experts in univalent function theory.

38

Finite Fields

38.1 Preliminary Remarks

Finite fields are of fundamental importance in pure as well as applied mathematics. Applications to coding, combinatorial design, and switching circuits have been made since the mid-1900s. Gauss himself first conceived of the theory of finite fields between 1796 and 1800, although he published only some of his work in this area, so that it did not exert the influence it might have. Gauss's work arose in the context of divisibility problems in number theory. The origins of these questions may, in turn, be traced to the work of Fermat who pursued the topic in the course of tackling a problem on perfect numbers posed to him by the amateur mathematician Frénicle de Bessy, through Mersenne.[1] This question boiled down to showing that $2^{37} - 1$ was not prime. In a letter to Frénicle dated October 18, 1640, Fermat wrote that a prime p would divide $a^n - 1$ for some n dividing $p - 1$; this is now called Fermat's little theorem. Moreover, if $N = nm$, where n was the smallest such number, then p would also divide $a^N - 1$. The second part of the result is easy to understand by observing that

$$a^N - 1 = a^{mn} - 1 = (a^n - 1)\left(a^{(m-1)n} + a^{(m-2)n} + \cdots + a^n + 1\right).$$

Fermat intended to write a treatise on his number theoretic work, but never did so.

Because Fermat failed to publish his proofs, Euler had to rediscover them; this effort, like many of his projects, stretched over decades. He investigated the structure of the set of integers modulo a prime p, among many other questions. He conjectured but did not completely prove that there existed an integer a such that $a, a^2, \ldots, a^{p-1}$ modulo p produced all the integers $1, 2, \ldots, p - 1$, though not in that order. Such an integer a is called a primitive root of the equation $x^{p-1} = 1 \pmod{p}$, or in Gauss's notation $x^{p-1} \equiv 1 \pmod{p}$.

In modern terminology, letting $\mathbb{Z}_p$ denote the integers modulo p, and setting $\mathbb{Z}_p^{\times} = \mathbb{Z}_p - \{0\}$, then Euler's conjecture was that $\mathbb{Z}_p^{\times}$ would be cyclic. In fact,

---

[1] For a discussion of Fermat's work on number theory and references for the little theorem, see Weil (1983); for perfect numbers, see pp. 53–56.

Euler thought that he had proved this proposition; he presented his efforts to the St. Petersburg Academy in 1772.[2] However, Gauss pointed out in Article 56 of his *Disquisitiones Arithmeticae* that Euler's proof was incomplete; Gauss proved the theorem in full in the 1790s and published his proof in his *Disquisitiones*. Euler worked with $\mathbb{Z}[x]$, the ring of polynomials with coefficients in $\mathbb{Z}$, as did Lagrange. In 1768, Lagrange proved the basic theorem that any such polynomial of degree m would have at most m roots modulo p.[3]

Gauss was able to delve deeply into the theory of the ring of polynomials over finite fields when he perceived that the number theoretic properties of this ring were analogous to those of the ring of integers; the irreducible polynomials here played the role of the prime numbers. Gauss proved the fundamental theorem that every irreducible polynomial $P(x) \neq x$ of degree m in $\mathbb{Z}_p[z]$ must divide $X^{p^m-1} - 1$ in $\mathbb{Z}_p[x]$. Gauss also gave a formula for the number of irreducible polynomials of degree m. In his derivation, he applied Möbius inversion without an explicit statement of the general formula. In 1832, August Möbius (1790–1868), a student of Pfaff and Gauss, published this formula,[4] although it was not much noted. In fact, in 1857, Dedekind[5] and Liouville[6] published proofs of the inversion formula without reference to Möbius.

Also during the period 1796–1800, Gauss studied the Galois theory of cyclotomic extensions of the field $\mathbb{Z}_p$ by explicitly constructing the subfields of the splitting field of the polynomial $x^\nu - 1$ over $\mathbb{Z}_p$, where ν was a positive integer not divisible by p. He saw this theory as analogous to his cyclotomic theory over the field of rational numbers. Gauss applied his results to obtain a new proof of the law of quadratic reciprocity. He intended to include his work on the extensions of finite fields as the eighth section of his 1801 *Disquisitiones*, but omitted it to make room for his theory of binary quadratic forms, completed after 1798. In fact, binary quadratic forms occupied more than half of the published book, so that Gauss could include in the text only references to his unpublished work on finite fields.

In 1830, Galois published his theory of algebraic extensions of finite fields[7] by using numbers analogous to complex numbers, known in the nineteenth century as Galois imaginaries. These numbers were required in order to extend the field $\mathbb{Z}_p$. For example, if a polynomial $F(x)$ of degree ν was irreducible over $\mathbb{Z}_p$, then Galois assumed i to be the imaginary solution of $F(x) = 0$; he then showed that the set consisting of the p^ν expressions $a_0 + a_1 i + \cdots + a_{\nu-1} i^{\nu-1}$ could be given the structure of a field and that $F(x)$ could be completely factored in this field. It is interesting to note that Gauss preferred to avoid imaginary roots. For example, in the unpublished eighth section of the *Disquisitiones*, he wrote,[8]

[2] Eu. I-3 pp. 240–281. E 449 § 37.
[3] Lagrange (1867–1892) vol. 2, pp. 667–669. Gauss refers to this work in Gauss (1965) p. 27.
[4] Möbius (1832).
[5] Dedekind (1857).
[6] Liouville (1857).
[7] Galois (1830).
[8] Frei (2007) p. 180. See also Gauss (1981) p. 607.

It is clear that the congruence $\xi \equiv 0$ does not have real roots if ξ has no factors of dimension one; but nothing prevents us from decomposing ξ, nevertheless, into factors of two, three or more dimensions, whereupon, in some sense, *imaginary* roots could be attributed to them. Indeed, we could have shortened incomparably all our following investigations, had we wanted to introduce such imaginary quantities by taking the same liberty some more recent mathematicians have taken; but nevertheless, we have preferred to deduce everything from [first] principles. Perhaps, we shall explain our view on this matter in more detail on another occasion.

In 1845, Theodor Schönemann, unaware of Galois's work, published a paper on algebraic extensions of $\mathbb{Z}_p$.[9] By application of his theory, he partially recovered Kummer's result on the factorization of a prime $q \neq p$ in the cyclotomic field generated by a pth root of unity. Schönemann also applied his theory to prove the irreducibility of the cyclotomic polynomial $x^{q-1} + x^{q-2} + \cdots + x + 1$. In 1857, Dedekind began to develop a theory of finite fields, in order to generalize Kummer's theory of ideals in cyclotomic fields and to place it on a firm logical foundation. In 1871, Dedekind published his first version of this generalization as his theory of algebraic numbers. In this work, given a polynomial irreducible over the rational numbers, he delineated the relation between the factorization of this polynomial modulo p and the prime ideal factorization of the ideal generated by p in the number field arising out of the polynomial. We remark that Dedekind was familiar with the work of Schönemann and of Galois. Note that Galois's 1830 paper was republished by Liouville in the 1840s and that J. A. Serret's 1854 algebra book[10] discussed the work of Galois in detail.

Richard Dedekind (1831–1916) was the Ph.D. student of Gauss and mathematical friend of Dirichlet. Dedekind performed the valuable service of editing the works of Riemann and Gauss and the lectures of Dirichlet. In his 1857 paper, Dedekind carefully showed that results in elementary number theory, including Fermat's theorem of Euler's generalization that $a^{\phi(m)} \equiv 1 \bmod m$, with a and m relatively prime, could be carried over to the ring of polynomials over finite fields. Dedekind also stated and proved the corresponding law of quadratic reciprocity. In 1902, Hermann Kühne proved the general reciprocity theorem in $F_q[x]$, a finite field with $q = p^m$ elements.[11] This theorem was rediscovered in 1925 by Friedrich K. Schmidt, and then again in 1932 by Leonard Carlitz. The reciprocity laws are more easily proved for polynomials in $F_q[x]$ than for integers. In 1914, Heinrich Kornblum (1890–1914), student of Landau, further developed this analogy by defining L-functions for $F_p[x]$ and proving an analog of Dirichlet's theorem on primes in arithmetic progressions. Unfortunately, Kornblum was killed in World War I, but Landau published the work in 1919. It may also be of interest to note that Gauss used the zeta function for $\mathbb{Z}_p[x]$ to find a formula for the number of irreducible polynomials of degree n, although he did not express it in such terms. Gauss's formula implies that if (n) denotes the number of irreducible polynomials of degree n, then

[9] Schönemann (1845).
[10] Serret (1854).
[11] For more detail and references on this topic, see Frei (2007). See also Rosen (2002) chapters 2–4.

$$(n) - \frac{p^n}{n} = O\left(\frac{p^{n/2}}{n}\right);$$

set $x = p^n$, then $n = \log_p x$, to obtain

$$(n) = \frac{x}{\log_p x} + O\left(\frac{\sqrt{x}}{\log_p x}\right).$$

Note the similarity in appearance between this equation and the conjectured form of the number of primes less than x, following from the unproven Riemann hypothesis on the nontrivial zeros of the Riemann zeta function. Note that in 1973, Pierre Deligne established the Riemann hypothesis for the zeta function of smooth projective varieties over finite fields.[12] He based his proof on the novel framework for algebraic geometry created by Alexander Grothendieck and his collaborators, including Deligne. Weil conjectured this theorem in 1949, so that it was also known as Weil's conjecture. Even earlier, special cases of the Riemann hypothesis over finite fields had been proved; Emil Artin (1898–1962) presented the earliest example in 1921, when he defined and studied the zeta function of a quadratic extension of the field $\mathbb{Z}_p(x)$ and proved the Riemann hypothesis for that case. Artin's advisor, Gustav Herglotz (1881–1953), proposed the problem after reading Kornblum's posthumously published thesis.

38.2   Euler's Proof of Fermat's Little Theorem

In 1736, in his second paper on number theory, Euler presented an inductive proof of Fermat's theorem.[13] He stated the result that if $a^p - a$ is divisible by p, then $(a + 1)^p - (a + 1)$ is also divisible by p. To see this, consider that by the binomial theorem,

$$(a + 1)^p = a^p + \frac{p}{1}a^{p-1} + \frac{p(p-1)}{1 \cdot 2}a^{p-2} + \cdots + \frac{p}{1}a + 1$$

or

$$(a + 1)^p - a^p - 1 = \frac{p}{1}a^{p-1} + \frac{p(p-1)}{1 \cdot 2}a^{p-2} + \cdots + \frac{p}{1}a.$$

The right-hand side has p as a factor in each term, and therefore p divides the left-hand side

$$(a + 1)^p - a^p - 1 = (a + 1)^p - (a + 1) - a^p + a.$$

This implies the required result, that if p divides $a^p - a$, it must also divide $(a + 1)^p - (a + 1)$. Since the result is true for $a = 1$, it is true for all positive integers a. Moreover, if p does not divide a, then since p divides $a(a^{p-1} - 1)$, we obtain the result that p divides $a^{p-1} - 1$.

---

[12] See Katz (1976) for a sketch of Deligne's proof.
[13] Eu. I-2 pp. 33–37. E 54.

For Euler's multiplicative proof of Fermat's theorem,[14] we follow the concise presentation in Gauss's *Disquisitiones*.[15] Suppose a prime p does not divide a positive integer a. Then there are at most $p - 1$ different remainders when $1, a, a^2, \ldots$ are divided by p. So let a^m and a^n have the same remainder with $m > n$. Then $a^{m-n} - 1$ is divisible by p. Let t be the least integer such that p divides $a^t - 1$. If $t = p - 1$, our proof is complete. If $t \neq p - 1$, then $1, a, a^2, \ldots, a^{t-1}$ have t distinct remainders when divided by p. Thus, we can choose an integer b, not divisible by p and not among $1, a, a^2, \ldots, a^{t-1}$ modulo p, and consider the numbers $b, ab, a^2b, \ldots, a^{t-1}b$. Each of these numbers also leaves a different remainder after division by p and each of these is different from the previous set of remainders. Hence, we have $2t \leq p - 1$ remainders. If $2t = p - 1$, then our proof is complete. If not, then we can continue the process until some multiple of t is $p - 1$. This completes the proof of the theorem.

38.3 Gauss's Proof That $\mathbb{Z}_p^\times$ Is Cyclic

In one of his two proofs that $\mathbf{Z}_p^\times$ is cyclic, Gauss used a theorem of Lagrange: Assuming $A \not\equiv 0$ modulo p, the congruence

$$Ax^m + Bx^{m-1} + Cx^{m-2} + \cdots + Mx + N \equiv 0 \pmod{p}$$

has at most m noncongruent solutions. Gauss presented a proof of this result,[16] similar to that of Lagrange but more succinct. It is easy to see that a congruence of degree 1 has at most one solution. Assume, with Gauss, that m is the lowest degree for which the result is false; then $m \geq 2$. Suppose that the preceding congruence has at least $m + 1$ roots, $\alpha, \beta, \gamma, \ldots$, where $0 \leq \alpha < \beta < \gamma < \cdots \leq p - 1$. Now set $y = x + \alpha$ so that the congruence takes the form

$$A'y^m + B'y^{m-1} + C'y^{m-2} + \cdots + M'y + N' \equiv 0 \pmod{p}.$$

Note that this congruence has at least $m + 1$ solutions $0, \beta - \alpha, \gamma - \alpha, \ldots$. Since $y \equiv 0$ is a solution, we must have $N' \equiv 0$. Thus,

$$y(A'y^{m-1} + B'y^{m-2} + C'y^{m-3} + \cdots + M') \equiv 0 \pmod{p}.$$

If y is replaced by any of the m values $\beta - \alpha, \gamma - \alpha, \ldots$, then the identity is satisfied but y is not zero. This means that the m values $\beta - \alpha, \gamma - \alpha, \ldots$ are solutions of the $m - 1$ degree congruence

$$A'y^{m-1} + B'y^{m-2} + C'y^{m-3} + \cdots + M' \equiv 0, \quad (A' \equiv A \not\equiv 0).$$

This contradicts the statement that m is the least integer for which the result is false, proving the theorem. Gauss thought that this theorem was significant; in his

[14] Eu. I-2 pp. 493–518. E 262.
[15] Gauss (1965) art. 49–50 or Gauss (1863–1927) vol. I, pp. 40–42.
[16] Gauss (1965) art. 42 or Gauss (1863–1927) vol. I, pp. 34–35.

Disquisitiones he discussed its history, pointing out that Euler had found special cases, Legendre had given it in his dissertation, and that in 1768 Lagrange had been the first to state and prove it.

Gauss used this result to prove the proposition that there always exist primitive $(p-1)$th roots of unity modulo p.[17] Suppose that $p - 1 = a^\alpha b^\beta c^\gamma \cdots$, where $a, b, c, \ldots$ are distinct primes. The first step is to show the existence of integers $A, B, C, \ldots$ of orders $a^\alpha, b^\beta, c^\gamma, \ldots$ respectively. Note that by Lagrange's theorem given above, the congruence

$$x^{\frac{p-1}{a}} \equiv 1 \pmod{p}$$

has at most $\frac{p-1}{a}$ solutions. Hence there is an integer g, $1 \leq g \leq p - 1$, not a solution of the congruence. Now let h be an integer, $1 \leq h \leq p - 1$, congruent to $g^{\frac{p-1}{a^\alpha}}$. It is clear that $h^{a^\alpha} \equiv 1 \pmod{p}$, but that no power d less that a^α will give $h^d \equiv 1$. This is because d must take the form a^j with $j < k$, so that $h^d = g^{\frac{p-1}{a^{\alpha-j}}} \not\equiv 1$ by the definition of g. We may take A to be h and similarly find $B, C, \ldots$. We can now show that the order of $y = ABC \cdots$ is $p - 1$. To see this, suppose, without loss of generality, that the order of y divides $\frac{p-1}{a}$. Since $b^\beta, c^\gamma, \ldots$ also divide $\frac{p-1}{a}$, it follows that

$$1 \equiv y^{\frac{p-1}{a}} \equiv A^{\frac{p-1}{a}} B^{\frac{p-1}{a}} C^{\frac{p-1}{a}} \cdots \equiv A^{\frac{p-1}{a}} \pmod{p}.$$

This implies that a^α divides $\frac{p-1}{a}$, an impossibility. Thus, the theorem is proved. This argument of Gauss can be extended to arbitrary finite fields. By a different argument, Gauss showed that the number of primitive roots of unity modulo p was equal to $\phi(p-1)$.

38.4 Gauss on Irreducible Polynomials Modulo a Prime

To count the number of irreducible polynomials of a given degree modulo a prime p, Gauss started with the observation that the number of monic polynomials (mod p)

$$x^n + Ax^{n-1} + Bx^{n-2} + Cx^{n-3} + \cdots$$

was p^n, because each of the n coefficients $A, B, C, \ldots$ would take exactly p values

$$0, 1, 2, \ldots, p - 1.$$

Thus, there were p polynomials of degree 1, all irreducible. Gauss then remarked that it followed from the theory of combinations that the number of (monic) reducible degree-two polynomials was $\frac{p(p+1)}{2}$. So the number of irreducible ones would be given by

$$p^2 - \frac{p(p+1)}{2} = \frac{p^2 - p}{2}.$$

---

[17] Gauss (1965) art. 55 or Gauss (1863–1927) vol. I, pp. 44–46.

To determine the irreducible polynomials of higher degree, Gauss devised a notation and method.[18] He let (a) denote the number of irreducible polynomials of degree a and (a^α) the number of polynomials of degree αa factorizable into α irreducible polynomials of degree a. Gauss represented the number of polynomials of degree $\alpha + 2\beta + 3\gamma + \cdots$ with α factors of degree 1, β factors of degree 2, γ factors of degree 3, etc. by $(1^\alpha 2^\beta 3^\gamma 4^\delta \cdots)$. It followed that

$$(1^\alpha 2^\beta 3^\gamma 4^\delta \cdots) = (1^\alpha)(2^\beta)(3^\gamma)(4^\delta) \cdots .$$

Again, Gauss remarked that the theory of combinations implied that

$$(a^\alpha) = \frac{(a)}{1} \cdot \frac{[(a)+1]}{2} \cdot \frac{[(a)+2]}{3} \cdots \frac{[(a)+\alpha-1]}{\alpha}.$$

Though Gauss did not bother, it is not difficult to prove this: Let $p_1(x), p_2(x), \ldots, p_{(a)}(x)$ be the irreducible polynomials of degree a. In a factorization of a polynomial of degree αa, let y_i factors be $p_i(x)$, $i = 1, 2, \ldots, (a)$. Then (a^α) will be the number of nonnegative solutions of the equation

$$y_1 + y_2 + \cdots + y_{(a)} = \alpha,$$

the solution of which yields the same value given by Gauss. He then noted that

$$p = (1),$$
$$p^2 = (1^2) + (2),$$
$$p^3 = (1^3) + (1 \cdot 2) + (3),$$
$$p^4 = (1^4) + (1^2 \cdot 2) + (1 \cdot 3) + (2^2) + (4),$$

and so on. Using the formula for (a^α), he found the following eight values:

$$(1) = p, \qquad (4) = \frac{p^4 - p^2}{4},$$

$$(7) = \frac{p^7 - p}{7},$$

$$(2) = \frac{p^2 - p}{2}, \qquad (5) = \frac{p^5 - p}{5},$$

$$(8) = \frac{p^8 - p^4}{8}.$$

$$(3) = \frac{p^3 - p}{3}, \qquad (6) = \frac{p^6 - p^3 - p^2 + p}{6},$$

Solving these equations led him to:

$$p = (1), \qquad\qquad p^5 = 5(5) + (1),$$

$$p^2 = 2(2) + (1), \qquad\qquad p^6 = 6(6) + 3(3) + 2(2) + (1),$$

$$p^3 = 3(3) + (1), \qquad\qquad p^7 = 7(7) + (1),$$

$$p^4 = 4(4) + 2(2) + (1), \qquad p^8 = 8(8) + 4(4) + 2(2) + (1).$$

[18] Gauss (1981) pp. 609–611.

These results then suggested to Gauss that

$$p^n = \alpha(\alpha) + \beta(\beta) + \gamma(\gamma) + \delta(\delta) + \cdots,$$

where $\alpha, \beta, \gamma, \delta, \ldots$ were all the divisors of n. He sketched a proof, using generating functions. Although there are a few missing lines in Gauss's manuscript, it is not difficult to fill in the details. He wrote that the product

$$\left(\frac{1}{1-x}\right)^{(1)} \left(\frac{1}{1-x^2}\right)^{(2)} \left(\frac{1}{1-x^3}\right)^{(3)} \cdots \qquad (38.1)$$

could be developed into the series

$$1 + Ax + Bx^2 + \cdots = P, \qquad (38.2)$$

where $A = p$, $B = p^2$, $C = p^3, \ldots$. Then by taking the logarithmic derivative of (38.1), he got

$$\frac{x\,dP}{P\,dx} = \frac{(1)x}{1-x} + \frac{2(2)x^2}{1-x^2} + \frac{3(3)x^3}{1-x^3} + \cdots. \qquad (38.3)$$

The required result followed after expanding the terms as an infinite series and equating coefficients. Gauss gave no details, but to prove (38.2), let f denote a monic irreducible polynomial. Since p^n is the number of monic polynomials of degree n, we see that by unique factorization of polynomials

$$\frac{1}{1-px} = 1 + px + p^2 x^2 + \cdots + p^n x^n + \cdots \qquad (38.4)$$

$$= \sum_{n=0}^{\infty} (\text{number of monic polynomials of degree } n)\, x^n$$

$$= \prod_f \left(1 + x^{\deg f} + x^{\deg f^2} + \cdots\right)$$

$$= \prod_f (1 - x^{\deg f})^{-1}$$

$$= \prod_{d=1}^{\infty} (1 - x^d)^{-(d)},$$

where the notation $\prod_f$ stands for the product over all irreducible polynomials. This proves Gauss's assertion that product (38.1) equals the series (38.2). Now by (38.3)

$$\frac{px}{1-px} = \sum_{d=1}^{n} \frac{d(d)x^d}{1-x^d} = \sum_{n=1}^{\infty} \left(\sum_{d\mid n} d(d)\right) x^n.$$

Gauss then equated the coefficients of x^n on each side to get

$$p^n = \sum_{d|n} d(d).$$ (38.5)

He then inverted this formula to get (n) in terms of p^n, stating that if $n = a^\alpha b^\beta$ $c^\gamma \cdots$ where $a, b, c, \ldots$ were distinct primes, then

$$n(n) = p^n - \sum p^{\frac{n}{a}} + \sum p^{\frac{n}{ab}} - \sum p^{\frac{n}{abc}} + \cdots.$$ (38.6)

Gauss wrote that, for example, when $n = 36$, he had

$$36(36) = p^{36} - p^{18} + p^{12} + p^6.$$

As a corollary of (38.6), Gauss observed that if $n = a^\alpha$, with a prime, then

$$p^n \equiv p^{\frac{n}{a}} \pmod{n}.$$

And for $\alpha = 1$ and a prime to p, the result was

$$p^{a-1} \equiv 1 \pmod{a}.$$

Note that it is also easy to see from (38.6) that $n(n) > 0$. Thus, there are irreducible polynomials of every degree n. Gauss gave no proof of the inversion (38.6) of (38.5). This means that Gauss knew the Möbius inversion formula before 1800 when he wrote up his researches on polynomials over the integers modulo p; Möbius's paper appeared in 1832.[19]

38.5 Galois on Finite Fields

Although the French mathematician Évariste Galois's (1811–1832) research career lasted less than four years, his accomplishments have had lasting value and importance. Galois's premature death came about as a result of a tragic duel, the cause of which is not fully understood, but was possibly related to Galois's political activities. In his number theory report of 1859–66, Smith wrote on Galois:[20]

> His mathematical works are collected in Liouville's Journal, vol. xl. p. 381. Obscure and fragmentary as some of these papers are, they nevertheless evince an extraordinary genius, unparalleled, perhaps, for its early maturity, except by that of Pascal. It is impossible to read without emotion the letter in which, on the day before his death and in anticipation of it, Galois endeavours to rescue from oblivion the unfinished researches which have given him a place for ever in the history of mathematical science.

Galois published his first paper in 1828 on purely periodic continued fractions,[21] a topic studied by Euler and Lagrange in the 1760s. Euler had shown that a periodic

---

[19] Möbius (1832).
[20] Smith (1965b) p. 149.
[21] For an English translation, see Neumann (2011) pp. 36–47.

continued fraction would satisfy a quadratic equation with integer coefficients. Lagrange proved the more difficult converse, that a quadratic irrational number, a number satisfying a quadratic equation with integer coefficients, could be expressed as a periodic continued fraction. Galois explicitly proved a theorem implicitly in Lagrange: For integers $a_0, a_1, \ldots, a_n$, if the continued fraction

$$a_0 + \cfrac{1}{a_1+} \cfrac{1}{a_2+} \cdots \cfrac{1}{a_n+} \cfrac{1}{a_0+} \cdots \cfrac{1}{a_n+} \cfrac{1}{a_0+} \cdots$$

is a solution of a polynomial equation with integer coefficients, then the continued fraction

$$-\cfrac{1}{a_n+} \cfrac{1}{a_{n-1}+} \cdots \cfrac{1}{a_0+} \cfrac{1}{a_n+} \cdots \cfrac{1}{a_0+} \cfrac{1}{a_n+} \cdots$$

is also a solution of the same polynomial equation.

From a very early age, Galois had plans to develop the theory of algebraic extensions of fields. In this context, in 1830 Galois wrote his paper "Sur la théorie des nombres,"[22] rediscovering Gauss's unpublished results in this area and creating the theory of finite fields. Galois started with an equation $F(x) = 0$ modulo p, with $F(x)$ having integer coefficients and irreducible modulo p. Note that by this he meant that there could not exist polynomials $\phi(x)$, $\psi(x)$, and $\chi(x)$ with integer coefficients such that

$$\phi(x) \cdot \psi(x) = F(x) + p\chi(x).$$

After the initial portion of the paper, Galois omitted the modulo p, writing simply $F(x) = 0$. This means that he was assuming that the coefficients of the polynomials were taken from the finite field, integers modulo p. Galois argued that since $F(x)$ was irreducible, the equation $F(x) = 0$ had no solutions in integers (more precisely and in modern terms, no solutions in the finite field). He supposed F to be of degree ν and denoted by i an imaginary solution of $F(x) = 0$. Galois explained this imaginary solution by drawing an analogy with complex numbers. He let α denote any one of the $p^\nu - 1$ expressions

$$a + a_1 i + a_2 i^2 + \cdots + a_{\nu-1} i^{\nu-1}, \tag{38.7}$$

where $a, a_1, a_2, \ldots, a_{\nu-1}$ took values in the finite field, that is, $a_i = 0, 1, \ldots, p-1$, though all the as could not be zero. Then $\alpha, \alpha^2, \alpha^3, \ldots$ would all be expressions of the form (38.7), since if the degree of i were ν or higher, then $F(x) = 0$ could be used to express i^ν in the form (38.7). Next, since there were only $p^\nu - 1$ different such expressions, Galois had $\alpha^k = \alpha^l$ for two different integers k and l, or $\alpha^l(\alpha^{k-l} - 1) = 0$ (for $k > l$). From the irreducibility of F, Galois arrived at $\alpha^{k-l} = 1$. Letting n be the least positive integer such that $\alpha^n = 1$, Galois noted that $1, \alpha, \alpha^2, \ldots, \alpha^{n-1}$ were distinct; moreover, if β were another expression of the form (38.7), then $\beta, \beta\alpha, \ldots, \beta\alpha^{n-1}$ would be n additional elements distinct from each other and from the α^j. Moreover, if $2n < p^\nu - 1$, then there would be yet another element in (38.7) distinct from the known $2n$ elements. Because this process could be continued, Galois concluded that n divided $p^\nu - 1$, meaning that $\alpha^{p^\nu-1} = 1$ for every α of the form (38.7). This is Galois's generalization of Fermat's theorem. Here Galois also observed that by the known methods of number theory (in fact, by Gauss's published argument outlined earlier), there was a primitive root α for which $n = p^\nu - 1$. Moreover, any primitive root had to satisfy a congruence of degree ν irreducible modulo p.

Note that Galois's generalization of Fermat's theorem implied that all members of (38.7), including 0, were roots of the polynomial $x^{p^\nu} - x$. And every irreducible $F(x)$ would divide $x^{p^\nu} - x$ modulo p. We now continue to follow Galois. Because

$$(F(x))^{p^n} = F\left(x^{p^n}\right) \quad (\text{mod } p),$$

the roots of $F(x) = 0$ had to be $i, i^p, i^{p^2}, \ldots, i^{p^\nu-1}$. Thus, he saw that all the roots of $x^{p^\nu} = x$ were polynomials in any root α of an irreducible polynomial of degree ν. To find all the irreducible factors of $x^{p^\nu} - x$, he factored out all polynomials dividing $x^{p^\mu} - x$ for $\mu < \nu$. The remaining product of polynomials was then a product of irreducible polynomials of degree ν. Galois pointed out that, since each of their roots was expressible in terms of a single root, these were obtainable through Gauss's method. Recall that Galois did not write modulo p repeatedly because he saw the coefficients of F to be elements of the finite field $\mathbb{Z}_p$.

Galois then gave an example in which $p = 7$ and $\nu = 3$. He here showed how to find the generator α of the multiplicative group of this field, as well as the irreducible polynomial equation satisfied by α. He noted that $x^3 - 2$ was an irreducible polynomial of degree 3 modulo 7 and hence the roots of $x^{7^3} - x$ would be $a + a_1 i + a_2 i^2$ where a, a_1, a_2 took values $0, 1, \ldots, 6$ and $i^3 = 2$. Galois denoted i by $\sqrt[3]{2}$ and then wrote the roots as

$$a + a_1 \sqrt[3]{2} + a_2 \sqrt[3]{4}.$$

To find a primitive root of $x^{7^3} - x$, Galois noted that $7^3 - 1 = 2 \cdot 3^2 \cdot 19$, so that he needed primitive roots of the three equations:

$$x^2 = 1, \quad x^{3^2} = 1, \quad x^{19} = 1.$$

He observed that $x = -1$ was a primitive root of the first equation. He cleverly noted that

$$x^{3^2} - 1 = (x^3 - 1)(x^3 - 2)(x^3 - 4) \quad (\text{mod } 7),$$

so that where $i^3 = 2$, i was a primitive root of the second equation. For the third equation, Galois took $x = a + a_1 i$ so that

$$(a + a_1 i)^{19} = 1.$$

Expanding by the binomial theorem, referred to by Galois as Newton's formula, and employing

$$a^{m(p-1)} = 1, \quad a_1^{m(p-1)} = 1, \quad i^3 = 2, \quad p = 7,$$

he reduced the expression modulo 7 to

$$3\left[a - a^4 a_1^3 + (a^5 a_1^2 + a^2 a_1^5)i^2\right] = 1, \tag{38.8}$$

so that

$$3a - 3a^4 a_1^3 = 1, \quad a^5 a_1^2 + a^2 a_1^5 = 0. \tag{38.9}$$

Galois saw that equations (38.8) and (38.9) were satisfied (modulo 7) by $a = -1$ and $a_1 = 1$. He therefore concluded that $-1 + i$ was a primitive root of $x^{19} = 1$. Thus, the product $i - i^2$ of the three primitive roots, $-1, i$, and $-1 + i$, was a primitive root of $x^{7^3 - 1} = 1$. By eliminating i from

$$i^3 = 2 \quad \text{and} \quad \alpha = i - i^2,$$

he obtained the irreducible equation for the primitive root α,

$$\alpha^3 - \alpha + 2 = 0.$$

Thus, α would generate all the nonzero elements of a finite field of 7^3 members. Galois ended his paper with the observation that an arbitrary polynomial $F(x)$ of degree n has n real or imaginary roots. The real roots, assuming no multiple roots, could be found from the greatest common divisor of $F(x)$ and $x^{p-1} - 1$. Note that this can be obtained by means of the Euclidean algorithm. And the imaginary roots of degree 2 could be obtained from the greatest common divisor of $F(x)$ and $x^{p^2-1} - 1$; this process could clearly be continued.

38.6 Dedekind's Formula

With his characteristic systematic approach, in his paper of 1857, Dedekind explained how to develop the theory of polynomials over finite fields such that the analogy with the ring of integers was completely clear. We present just one formula from his paper, deriving an elegant expression for the product of all irreducible polynomials of degree d. The arguments given by Galois showed that if $F_d(x)$ was the product of all irreducible polynomials of degree d, then

$$x^{p^n} - x = \prod_{d \mid n} F_d(x).$$

By the multiplicative form of the Möbius inversion formula, a proof of which Dedekind provided, since it was not generally known in 1857, he obtained

$$F_n(x) = \frac{(x^{p^n} - x) \prod \left(x^{p^{\frac{n}{ab}}} - x\right) \cdots}{\prod \left(x^{p^{\frac{n}{a}}} - x\right) \prod \left(x^{p^{\frac{n}{abc}}} - x\right) \cdots}.$$

Here $a, b, c, \ldots$ denoted the distinct prime divisors of n. Thus,

$$\prod \left(x^{p^{\frac{n}{a}}} - x\right) = \left(x^{p^{\frac{n}{a}}} - x\right)\left(x^{p^{\frac{n}{b}}} - x\right)\left(x^{p^{\frac{n}{c}}} - x\right) \cdots.$$

With the use of the Möbius function μ, Dedekind's formula can now be written as

$$F_n(x) = \prod_{d|n} \left(x^{p^{\frac{n}{d}}} - x\right)^{\mu(d)}.$$

Note that Möbius stated his inversion formula for a sum; Dedekind extended it to cover a product. We observe that the symbol μ for the Möbius function was introduced by Mertens in 1874.[23]

38.7 Finite Field Analogs of the Gamma and Beta Integrals

In Chapter 17, we noted that the gamma integral is given by

$$\int_0^\infty t^{x-1} e^{-t} dt$$

and the beta integral by

$$\int_0^1 t^{x-1}(1-t)^{y-1} dt.$$

Next observe that $\psi(t) = e^{ct}$ is the general nonzero and real-valued solution of the functional equation

$$\psi(x + y) = \psi(x)\psi(y). \tag{38.10}$$

To verify this, suppose ψ is nonzero, continuous, and satisfies (38.10). Clearly, $\psi(0) = 1$ and $\psi(x) > 0$. Let $\psi(1) = e^c$, with c some constant. The argument used by Euler in Section 4.3 can be applied here to show that for integers m and n,

$$\psi\left(\frac{m}{n}\right) = e^{c\frac{m}{n}}. \tag{38.11}$$

Since ψ is continuous, (38.11) implies that

$$\psi(t) = e^{ct}.$$

---

[23] Mertens (1874b).

Similarly, $\chi(t) = t^a$ is the general nonzero solution of

$$\chi(xy) = \chi(x)\chi(y); \tag{38.12}$$

to see this, observe that $\psi(t) = \chi(e^t)$ satisfies (38.10).

We mention that Kac proved that to show that $\psi(t) = e^{ct}$, it is sufficient to assume that ψ satisfying (38.10) is integrable.[24] To prove this, choose a real number a such that $\int_0^a \psi(t)dt \neq 0$. Then

$$\psi(x)\int_0^a \psi(t)dt = \int_0^a \psi(x+t)dt = \int_x^{a+x} \psi(t)dt$$

and thus

$$\psi(x) = \frac{\int_x^{a+x} \psi(t)dt}{\int_0^a \psi(t)dt}. \tag{38.13}$$

Equation (38.13) implies that if ψ is integrable and satisfies (38.10), then it must be continuous; hence, $\psi(t) = e^{ct}$.

Now observe that the gamma function is the integral of the product of a function satisfying (38.12) and a function satisfying (38.10). Moreover, the beta integral is the integral of two functions, t^{x-1} and $(1-t)^{y-1}$, both of which satisfy (38.12); also, the sum of t and $1-t$ is one.

We may now look for the finite field analogs of the gamma and beta integrals. Gauss, Jacobi, and Eisenstein first determined these analogs, working, as we shall, in the finite field $\mathbb{Z}_p$, the integers modulo a prime p. A function defined in $\mathbb{Z}_p$ has to be periodic, with period p; it is natural, then, to consider

$$\psi(x) = e^{\frac{2\pi i k x}{p}}, \qquad k = 0, 1, 2, \ldots, p-1$$

as the p basic functions satisfying (38.10). Thus, ψ is a map satisfying (38.10):

$$\psi : \mathbb{Z}_p \to \mathbb{C}.$$

Since it satisfies $\psi(x+y) = \psi(x)\psi(y)$, we denote ψ an additive character. Similarly, a multiplicative character χ is a map

$$\chi : \mathbb{Z}_p \to \mathbb{C}$$

such that $\chi(mn) = \chi(m)\chi(n)$ and $\chi(0) = 0$. We have seen how multiplicative characters can be constructed: Let g generate the cyclic group $\mathbb{Z}_p \setminus \{0\}$ and let $\gamma(n)$ denote the integer (mod $(p-1)$) such that

$$g^{\gamma(n)} \equiv n \pmod{p}. \tag{38.14}$$

---

[24] Kac (1936–1937).

Then

$$g^{\gamma(mn)} \equiv mn \equiv g^{\gamma(m)}g^{\gamma(n)} \equiv g^{\gamma(m)+\gamma(n)} \quad (\mathrm{mod}\ p)$$

and

$$\gamma(mn) = \gamma(m) + \gamma(n) \quad (\mathrm{mod}\ (p-1)).$$

If $\omega = e^{\frac{2\pi i k}{p-1}}$, $k = 1, 2, \ldots, p-1$, then multiplicative characters will be defined by $\chi(n) = \omega^{\gamma(n)}$, $n \in \mathbb{Z}_p \setminus \{0\}$, and $\chi(0) = 0$.

Now because integrals become sums when defined on finite sets, the sums corresponding to the gamma and beta integrals would be given respectively by

$$\sum_{n=0}^{p-1} \chi(n)\psi(n), \tag{38.15}$$

$$\sum_{n=0}^{p-1} \chi_1(n)\chi_2(1-n), \tag{38.16}$$

where χ_1 and χ_2 are two multiplicative characters. The two sums (38.15) and (38.16) are designated, respectively, Gauss and Jacobi sums; Gauss sums have also been called Lagrange resolvents. The reasons for these names can be understood by delving into the history of this topic.

In 1770–1771, Lagrange published, in two parts, his paper, "Réflections sur la résolution algébriques des équations,"[25] in the Berlin Academy Journal. In this paper, Lagrange discussed the various methods that had previously been used to solve, by radicals, equations of degree 2, 3 and 4. He also analyzed these equations, and general equations of the nth degree, by means of permutations of the roots. In article 86 of his paper, he wrote that the simplest expression that solved equations of degrees 2, 3 and 4 could be written as

$$x_1 + \omega x_2 + \omega^2 x_3 + \cdots + \omega^{n-1} x_n, \tag{38.17}$$

where $x_1, x_2, \ldots, x_n$ represented the n roots of the equation of degree n and $\omega \neq 1$ was a root of $\omega^n - 1 = 0$. The expression (38.17) is called the Lagrange resolvent. The idea of this resolvent was also discovered by Alexandre-Théophile Vandermonde (1735–1796) at about the same time. Both Lagrange and Vandermonde applied their ideas to solve the equation

$$\frac{x^p - 1}{x - 1} = 0,$$

where p was a prime. Of course, for $p = 2, 3, 5, 7$, the equation is not difficult to solve; when $p = 3, 5, 7$, the substitution $y = x + \frac{1}{x}$ reduces the equations to those of degree 1, 2, 3 respectively. The same substitution reduces the equation for $p = 11$ to one

[25] Lagrange (1770–1771).

of degree 5, an equation that apparently stumped Lagrange. However, Vandermonde constructed a solution for this fifth-degree equation.[26] His idea was to choose the ten roots of $\frac{x^{11}-1}{x-1} = 0$ in (38.17) in a certain order; he took the kth term x_k, $1 \le k \le 10$, to be $\beta^{\nu(k)}$, where β was a primitive tenth root of unity and $\nu(k)$ was the least positive integer such that

$$2^{\nu(k)} \equiv k \quad \bmod 11.$$

Note that 2 is a primitive root modulo 11. Thus, Vandermonde had used a Gauss sum to solve $x^{11} - 1 = 0$.

From entry 37 in his mathematical journal, it appears that Gauss may have rediscovered the idea of a Lagrange resolvent; it is very possible that Gauss was unaware of Vandermonde's paper on the solution of algebraic equations when he began work on the cyclotomic equation $x^n - 1 = 0$. Gauss's motivation in studying this equation was number theoretic. In the preface to his *Disquisitiones Arithmeticae*, he wrote,[27]

> The theory of the division of the circle or of a regular polygon treated in Section VII *of itself* does not pertain to Arithmetic but the *principles* involved depend uniquely on Higher Arithmetic. This will perhaps prove unexpected to geometers, but I hope they will be equally pleased with the new results that derive from this treatment.

One such result was that the roots of the equation $x^n - 1 = 0$ could be expressed in terms of radicals. In article 359 of his *Disquisitiones*, he wrote,

> Everyone knows that the most eminent geometers have been ineffectual in the search for a general solution of equations higher than the fourth degree, or (to define the search more accurately) for the REDUCTION OF MIXED EQUATIONS TO PURE EQUATIONS. And there is little doubt that this problem does not so much defy modern methods of analysis as that it proposes the impossible (cf. what we said on this subject in *Demonstratio nova*, art. 9). Nevertheless it is certain that there are innumerable mixed equations of every degree which admit a reduction to pure equations, and we trust the geometers will find it gratifying if we show that our equations are always of this kind. But because of the length of this discussion we will present here only the most important principles necessary to show the possibility of our claim and reserve for another time a more complete consideration worthy of this argument.

As a particular case, Gauss proved his famous result on constructible regular polygons; the solutions of the corresponding equations have only square roots as radicals. Gauss used results on products of Gauss sums in his analysis in the *Disquisitiones*; though he did not present complete proofs there, he wrote that he indeed had them. In fact, he gave such proofs in an unpublished paper,[28] "Disquisitionum circa aequations puras ulterior evolutio."

In a February 1827 letter to Gauss,[29] Jacobi discussed at length his discovery of some properties of Gauss sums. Apparently, Gauss did not give a reply to this letter,

---

[26] Vandermonde (1770–1771), especially pp. 415–416.
[27] Gauss (1965) p. xx.
[28] Gauss (1863–1927) vol. 2, pp. 243–265.
[29] Jacobi (1969) vol. 7, pp. 393–402.

perhaps because the proofs of most of the results mentioned by Jacobi were contained in Gauss's paper on pure equations. Jacobi then published a paper on this topic,[30] in which he observed the analogy between the gamma or beta integrals and the Gauss or Jacobi sums; in fact, Jacobi also stated an analog of the multidimensional integral of Dirichlet.

Turning to a discussion of the properties of Gauss and Jacobi sums, we follow Eisenstein's 1844 exposition[31] because of its very systematic approach. Eisenstein was apparently unaware of Jacobi's 1837 paper when he published this work. Note that in 1846, Jacobi republished his paper in *Crelle's Journal*.[32]

Eisenstein started with the sum, with α and β integers

$$\phi(\alpha, \beta) = \sum_{k=1}^{p-1} e^{\frac{2\pi i \alpha\, v(k)}{p-1}} \cdot e^{\frac{2\pi i \beta k}{p}}, \tag{38.18}$$

where $v(k)$ was defined by (38.14). Following Gauss, Eisenstein denoted $v(k)$ by "Ind. k," a notation introduced by Gauss in article 59 of his *Disquisitiones*. Note that the first term in the sum in (38.18) denotes a multiplicative character; we denote it by $\chi(k)$ so that we may write (38.18) as

$$\phi(\alpha, \beta) = g_\beta(\chi). \tag{38.19}$$

Eisenstein noted some simple properties of these Gauss sums:

(a) $\phi(\alpha, \beta) = \phi(\alpha', \beta')$, when $\alpha \equiv \alpha'$ (mod $p-1$) and $\beta \equiv \beta'$ (mod p);
(b) $\phi(\alpha, 0) = 0$, when $\alpha \not\equiv 0$ (mod $p-1$);
(c) $\phi(0, \beta) = -1$, when $\beta \not\equiv 0$ (mod p);
(d) $\phi(0, 0) = p - 1$.

We indicate a proof of (b), where the condition $\alpha \not\equiv 0$ (mod $p-1$) means that there exists a $k \in \mathbb{Z}_p \setminus \{0\}$ such that $\chi(k) \neq 1$. Then

$$\phi(\alpha, 0) = g_0(\chi) = \sum_{s=1}^{p-1} \chi(s) = \sum_{s=1}^{p-1} \chi(sk) = \chi(k) \sum_{s=1}^{p-1} \chi(s).$$

Since $\chi(k) \neq 1$, we have $\sum_{s=1}^{p-1} \chi(s) = 0$. Note that sk and s run over all integers $1, 2, \ldots, p-1$.

We call χ a trivial character if $\chi(k) = 1$ for all $k \in \mathbb{Z}_p \setminus \{0\}$; this corresponds to $\alpha = 0$. Eisenstein proved for nontrivial characters χ that when $\beta \not\equiv 0$ (mod p),

$$g_\beta(\chi) = \chi(\beta^{-1}) g_1(\chi); \tag{38.20}$$

[30] Jacobi (1837).
[31] Eisenstein (1844).
[32] Jacobi (1846).

note that, since $\beta \neq 0 \pmod{p}$, β^{-1} exists. Eisenstein's proof was that

$$g_\beta(\chi) = \sum_{s=1}^{p-1} \chi(s) \, e^{\frac{2\pi i \beta s}{p}} = \sum_{s=1}^{p-1} \chi(\beta^{-1}s) \, e^{\frac{2\pi i s}{p}}$$

$$= \chi(\beta^{-1}) \sum_{s=1}^{p-1} \chi(s) \, e^{\frac{2\pi i s}{p}} = \chi(\beta^{-1}) \, g_1(\chi).$$

We observe that s could be replaced by $\beta^{-1}s$ because $\beta^{-1}s$ runs through all integers $1, 2, \ldots, p-1$, as does s. Also note that we can write $\chi(\beta^{-1}) = \overline{\chi}(\beta)$ because

$$\chi(\beta^{-1})\chi(\beta) = \chi(1) = 1 \quad \text{and} \quad \overline{\chi}(\beta)\chi(\beta) = 1.$$

So by using $g(x)$ for $g_1(x)$, we can rewrite (38.20) as

$$g_\beta(\chi) = \overline{\chi}(\beta) \, g(x), \quad \beta \not\equiv 0 \pmod{p}. \tag{38.21}$$

Eisenstein next proved that for $\alpha \neq 0 \pmod{p-1}$,

$$\phi(\alpha, 1) \, \phi(-\alpha, 1) = (-1)^\alpha p$$

or for nontrivial χ,

$$g(\chi) \, g(\overline{\chi}) = \chi(-1)p. \tag{38.22}$$

To prove (38.22), Eisenstein observed that

$$g(\overline{\chi}) \, g(\chi) = \sum_{m=1}^{p-1} \overline{\chi}(m) \sum_{n=1}^{p-1} \chi(n) \, e^{\frac{2\pi i(m+n)}{p}}.$$

Given an m, he noted, for each n there must exist a σ such that $n = \sigma m$. Thus

$$g(\overline{\chi}) \, g(\chi) = \sum_{m=1}^{p-1} \overline{\chi}(m) \sum_{\sigma=1}^{p-1} \chi(\sigma m) e^{\frac{2\pi i(\sigma+1)m}{p}}$$

$$= \sum_{\sigma=1}^{p-1} \chi(\sigma) \sum_{m=1}^{p-1} \chi(m)\overline{\chi}(m) \, e^{\frac{2\pi i(\sigma+1)m}{p}}$$

$$= \sum_{\sigma=1}^{p-1} \chi(\sigma) \sum_{m=1}^{p-1} e^{\frac{2\pi i(\sigma+1)m}{p}}. \tag{38.23}$$

He went on to point out that when $\sigma = p - 1$, the inner sum in (38.23) was equal to $p - 1$ and for other values of σ, it was -1. Thus

$$g(\overline{\chi}) \, g(\chi) = -\sum_{\sigma=1}^{p-2} \chi(\sigma) + (p-1)\chi(p-1)$$

$$= -\sum_{\sigma=1}^{p-1} \chi(\sigma) + p\chi(p-1)$$

$$= 0 + p\chi(-1).$$

Eisenstein then established the important relationship between Gauss and Jacobi sums, proving that for nontrivial multiplicative characters χ and η, such that $\chi\eta$ would also be nontrivial,

$$\frac{g(\chi) \, g(\eta)}{g(\chi\eta)} = \sum_{\sigma=1}^{p-2} \chi(\sigma)\left(\overline{\chi\eta}\right)(\sigma+1). \tag{38.24}$$

Note that we can rewrite the right-hand side of (38.24) as

$$\sum_{\sigma=1}^{p-2} \chi\left((\sigma+1)^{-1}\right) \eta\left(\sigma(\sigma+1)^{-1}\right)$$

$$= \sum_{\sigma=1}^{p-2} \chi\left((\sigma+1)^{-1}\right) \eta\left(1 - (\sigma+1)^{-1}\right)$$

$$= \sum_{n=2}^{p-1} \chi(n) \, \eta(1-n)$$

$$= \sum_{n=1}^{p-1} \chi(n) \, \eta(1-n). \tag{38.25}$$

We have given (38.25) as the definition of the Jacobi sum in (38.16); denote it by

$$J(\chi, \eta) = \sum_{n=1}^{p-1} \chi(n) \, \eta(1-n). \tag{38.26}$$

Thus

$$J(\chi, \eta) = \frac{g(\chi) \, g(\eta)}{g(\chi\eta)}. \tag{38.27}$$

To prove (38.24), Eisenstein observed that

$$g(\eta)g(\chi) = \sum_{m,n=1}^{p-1} \eta(m) \, \chi(n) e^{\frac{2\pi i(m+n)}{p}}$$

$$= \sum_{m,\sigma=1}^{p-1} \eta(m) \, \chi(m\sigma) e^{\frac{2\pi i(\sigma+1)m}{p}} \tag{38.28}$$

and that, in (38.28) because $\chi \eta$ was nontrivial, the part of the sum corresponding to $\sigma = p - 1$ was

$$\sum_{m=1}^{p-1} (\chi \eta)(m) \chi(-1) = 0. \tag{38.29}$$

He could apply (38.21) to find that the sum in (38.28) for values of σ (other than $p - 1$) produced

$$g(\chi) g(\eta) = \sum_{\sigma=1}^{p-2} \chi(\sigma)(\overline{\chi \eta})(\sigma + 1) \sum_{m=1}^{p-1} (\chi \eta)(m) e^{\frac{2\pi i m}{p}},$$

thus verifying (38.24). Next, from (38.22) and (38.27), Eisenstein perceived that when χ, η, and $\chi \eta$ were nontrivial,

$$J(\chi, \eta) J(\overline{\chi}, \overline{\eta}) = \frac{g(\chi) g(\overline{\chi}) g(\eta) g(\overline{\eta})}{g(\chi \eta) g(\overline{\chi \eta})}$$

$$= \frac{\chi(-1) p \, \eta(-1) p}{(\chi \eta)(-1) p}. \tag{38.30}$$

Eisenstein gave two corollaries of (38.30): for the cases $p = 4f+1$ and $p = 3f+1$. For $p = 4f + 1$, he noted that the character $\chi(n) = e^{\frac{2\pi i \nu(n)}{4f}}$ was of order $4f$, meaning that $4f$ was the smallest integer such that $\chi^{4f}(n) = 1$ for all $n \in \mathbb{Z}_p \setminus \{0\}$. Therefore,

$$\eta(n) = \chi^f(n) = e^{\frac{2\pi i \nu(n)}{4}} = i^{\nu(n)}.$$

Thus he had, with A and B integers,

$$J(\eta, \eta) = \sum_{n=1}^{4f} \eta(n) \eta(1 - n) = \sum_{n=1}^{4f} \eta(n - n^2) = \sum_{n=1}^{4f} i^{\nu(n-n^2)} = A + Bi$$

and hence

$$p = A^2 + B^2. \tag{38.31}$$

Similarly, for $p = 3f + 1$, Eisenstein argued that there was a character η of order 3. So

$$\eta(n) = \left(e^{\frac{2\pi i}{3}}\right)^{\nu(n)}$$

and

$$J(\eta, \eta) = \sum_{n=1}^{3f} \left(e^{\frac{2\pi i}{3}}\right)^{\nu(n-n^2)} = A + B e^{\frac{2\pi i}{3}}$$

and he could conclude that

$$p = A^2 - AB + B^2. \tag{38.32}$$

Note that (38.32) implies that

$$4p = (2A - B)^2 + 3B^2 = (A - 2B)^2 + 3A^2 = (A + B)^2 + 3(A - B)^2.$$

Now since one of B, A, or $A - B$ must be divisible by 3, we have

$$4p = X^2 + 27Y^2, \tag{38.33}$$

with X and Y integers.

In his 1827 letter to Gauss, Jacobi mentioned (38.22), saying that it would prove the assertion left unproved by Gauss, due to a lack of space, in article 360 of his *Disquisitiones*. Jacobi also noted formulas (38.31) and (38.33) in his letter; Gauss had given a lengthy proof of (38.32) in his *Disquisitiones*.

38.8 Weil: Solutions of Equations in Finite Fields

The Gauss and Jacobi sums defined in Section 37.7 were taken on the field $\mathbb{Z}_p$, where p was a prime. In a paper of 1890,[33] L. Stickelberger, a student of Weierstrass and Kummer, defined these sums on general finite fields with $q = p^m$ elements. Now to define Gauss sums on an arbitrary finite field with, say, p^m elements, it is necessary to define an additive character on F and a multiplicative character on $F^\times = F \setminus \{0\}$. We define the additive character as a function ψ in a manner consistent with our definition of an extension of F: Let $\alpha \in F$ and let the trace Tr of α be

$$\mathrm{Tr}(\alpha) = \alpha + \alpha^p + \alpha^{p^2} + \cdots + \alpha^{p^{n-1}}.$$

Note that

$$\big(\mathrm{Tr}(\alpha)\big)^p = \alpha^p + \alpha^{p^2} + \cdots + \alpha^{p^n}$$

$$= \mathrm{Tr}(\alpha),$$

because $\alpha^{p^n} = \alpha$. Now, based on Galois's theory, developed in Section 37.5, of which Gauss was also aware,[34] since $\mathrm{Tr}(\alpha)$ is a solution of $x^p = x$, it follows that $\mathrm{Tr}(\alpha) \in \mathbb{Z}_p$. It is then not difficult to show that

$$\mathrm{Tr} : F \to \mathbb{Z}_p$$

is an onto map and that

$$\mathrm{Tr}(\alpha + \beta) = \mathrm{Tr}(\alpha) + \mathrm{Tr}(\beta). \tag{38.34}$$

[33] Stickelberger (1890).
[34] Gauss (1981) pp. 616–618.

Now set

$$\psi(\alpha) = e^{\frac{2\pi i \operatorname{Tr}(\alpha)}{p}} \tag{38.35}$$

so that

$$\psi(\alpha + \beta) = \psi(\alpha)\psi(\beta),$$

showing ψ to be additive. Weil took the additive character $\psi : F \to \mathbb{C}$ to be a nonzero function satisfying

$$\psi(x + y) = \psi(x)\psi(y).$$

Clearly, $\psi(0) = 1$ and $\psi(1) = e^{\frac{2\pi i k}{p}}$, where k is an integer.

Next, to define a multiplicative character on F, note that $F^\times = F \setminus \{0\}$ is a cyclic group of order $q - 1$. Let us here take g to be the generator of $F \setminus \{0\}$. A mapping $\chi : F^\times \to \mathbb{C}^\times$ that satisfies

$$\chi(\alpha\beta) = \chi(\alpha)\chi(\beta)$$

is called a multiplicative character that is completely determined by its value on g, equal to a $(q - 1)$th root of unity. Denote by χ_0 the character whose value at g is 1; χ_0 represents the trivial character or principal character. The definition of χ_0 may be extended to all F by specifying $\chi_0(0) = 1$; all nontrivial multiplicative characters are extended by taking $\chi(0) = 0$. Observe that any multiplicative character may be written as

$$\chi(g) = e^{\frac{2\pi i s}{q-1}}, \quad s = 0, 1, \ldots, q - 1. \tag{38.36}$$

With χ as given in (38.35), the Gauss sum belonging to a character χ is defined by

$$g(x) = \sum_{x \in F} \chi(x)\psi(x). \tag{38.37}$$

Stickelberger also defined the Jacobi sum, though he called it the Eisenstein sum, for a general finite field:

$$J(\chi, \eta) = \sum_{a \in F} \chi(a)\,\eta(1 - a). \tag{38.38}$$

In fact, Eisenstein had defined a more general sum for multiplicative characters $\chi_1, \chi_2, \ldots, \chi_s$:

$$J(\chi_1, \chi_2, \ldots, \chi_s) = \sum_{a_1 + \cdots + a_s = 1} \chi_1(a_1)\chi_2(a_2) \cdots \chi_s(a_s) \tag{38.39}$$

and applied it to obtain a new proof of the law of biquadratic reciprocity.[35]

---

[35] Eisenstein (1975) vol. 1, pp. 141–163.

At the beginning of his long paper, Stickelberger showed that (38.22) and (38.27) would hold for a general finite field with q elements: For a nontrivial character χ on F,

$$g(\chi)\,g(\overline{\chi}) = \chi(-1)q \tag{38.40}$$

and for nontrivial χ, η, and $\chi\eta$,

$$J(\chi,\eta) = \frac{g(\chi)\,g(\eta)}{g(\chi\eta)}. \tag{38.41}$$

Stickelberger's proofs of (38.40) and (38.41) follow the same lines as the argument given by Eisenstein for (38.22) and (38.27). Note that since

$$\overline{g(\chi)} = \chi(-1)\,g(\overline{\chi}),$$

(38.40) implies that

$$|g(\chi)|^2 = q \quad \text{or} \quad |g(\chi)| = q^{\frac{1}{2}}. \tag{38.42}$$

Weil applied Gauss and Jacobi sums to determine the number of solutions in F of the equation

$$a_0 x_0^{n_0} + a_1 x_1^{n_2} + \cdots + a_r x_r^{n_r} = b, \tag{38.43}$$

where $a_0, a_1, \ldots, a_r \in F^{\times}$ and $b \in F$. In his 1972 lectures on the history of number theory, Weil talked about how he was led to this problem:[36]

> In 1947, in Chicago, I felt bored and depressed, and, not knowing what to do, I started reading Gauss's two memoirs on biquadratic residues, which I had never read before. The Gaussian integers occur in the second paper. The first one deals essentially with the number of solutions of equations $ax^4 - by^4 = 1$ in the prime field modulo p, and with the connection between these and certain Gaussian sums; actually the method is exactly the same that is applied in the last section of the *Disquisitiones* to the Gaussian sums of order 3 and the equations $ax^3 - by^3 = 1$. Then I noticed that similar principles can be applied to all equations of the form $ax^m + bt^n + cz^r + \cdots = 0$, and that this implies the truth of the so-called "Riemann hypothesis" (of which more later) for all curves $ax^n + by^n + cz^n = 0$ over finite fields, and also a "generalized Riemann hypothesis" for varieties in projective space with a "diagonal" equation $\sum a_i x_i^n \equiv 0$. This led me in turn to conjectures about varieties over finite fields, some of which have been proved later by Dwork, Grothendieck, M. Artin, and Lubkin, and some of which are still open.

Concerning the open conjectures, Weil added an epilogue, received in June 1973:[37]

> Reference has been made above to my conjectures of 1948, which included the extension of the "Riemann hypothesis" to algebraic varieties of arbitrary dimension over finite fields.

> Those conjectures have now been proved by Deligne. In the meanwhile, he had also shown, in conjunction with the work of Ihara, that their truth would imply the truth of Ramanujan's conjecture on the τ-function, which has been described above as "very much of an open problem."

> Number theory is not standing still.

---

[36] Weil (1974), especially pp. 106 and 110. Weil (1979) vol. III, p. 298.
[37] Weil (1979) vol. III, p. 302.

André Weil began his 1949 paper, "Numbers of solutions of equations in finite fields"[38] with some very valuable historical perspective and insight:

The equations to be considered here are those of the type (38.43). Such equations have an interesting history. In art. 358 of the *Disquisitiones*,[39] Gauss determines the Gaussian sums (the so-called cyclotomic periods) of order 3, for a prime of the form $p = 3n + 1$, and at the same time obtains the numbers of solutions for all congruences $ax^3 - by^3 \equiv 1 \pmod{p}$. He draws attention himself to the elegance of his method, as well as to its wide scope; it is only much later, however, viz. in his first memoir on biquadratic residues,[40] that he gave in print another application of the same method; there he treats the next higher case, finds the number of solutions of any congruence $ax^4 - by^4 \equiv 1 \pmod{p}$, for a prime of the form $p = 4n + 1$, and derives from this the biquadratic character of 2 mod p, this being the ostensible purpose of the whole highly ingenious and intricate investigation. As an incidental consequence ("*coronidis loco*"), he also gives in substance the number of solutions of any congruence $y + 2 \equiv ax^4 - b \pmod{p}$; this result includes as a special case the theorem stated as a conjecture ("*observatio per inductionem facta gravissima*") in the last entry of his *Tagebuch*;[41] and it implies the truth of what has lately become known as the Riemann hypothesis, for the function-field defined by that equation over the prime field of p elements.

Gauss's procedure is wholly elementary, and makes no use of the Gaussian sums, since it is rather his purpose to apply it to the determination of such sums. If one tries to apply it to more general cases, however, calculations soon become unwieldy, and one realizes the necessity of inverting it by taking Gaussian sums as a starting pint. The means for doing so were supplied, as early as 1827, by Jacobi, in a letter to Gauss.[42] But Lebesgue, who in 1837 devoted two papers[43] to the case $n_0 = \cdots = n_r$ of equation (38.43), did not succeed in bringing out any striking result. The whole problem seems then to have been forgotten until Hardy and Littlewood found it necessary to obtain formulas for the number of solutions of the congruence $\sum_i x_i^n \equiv b \pmod{p}$ in their work on the singular series for Waring's problem;[44] they did so by means of Gaussian sums. More recently, Davenport and Hasse[45] have applied the same method to the case $r = 2, b = 0$ of equation (1) as well as to other similar equations; however, as they were chiefly concerned with other aspects of the problem, and in particular with its relation to the Riemann hypothesis in function-fields, the really elementary character of their treatment does not appear clearly.

As equations of type (38.43) have again recently been the subject of some discussion,[46] it may therefore serve a useful purpose to give here a brief but complete exposition of the topic. This will contain nothing new, except perhaps in the mode of presentation of the final results, which will lead to the statement of some conjectures concerning the numbers of solutions of equations over finite fields, and their relation to the topological properties of the varieties defined by the corresponding equations over the field of complex numbers.

To determine the number of solutions in F of (38.43), following Weil,[47] suppose that (38.43) has only one variable: $x^n = u$, where $u \in F$. Now, letting $N(u)$ be the number of solutions of $x^n = u$, and with $n \mid q - 1$, we must prove

[38] Weil (1949). Weil (1979) vol. 1, p. 399.
[39] Gauss (1863–1927) vol. 1, pp. 445–449.
[40] ibid. vol. II, pp. 67–92.
[41] ibid. vol. X_1, p. 571.
[42] Jacobi (1969) vol. VII, pp. 393–400; also see vol. VI, pp. 254–274.
[43] Lebesegue (1837) and (1838).
[44] Hardy and Littlewood (1922).
[45] Davenport and Hasse (1935).
[46] See, e.g, Hua and Vandiver (1948).
[47] Weil (1949).

$$N(u) = \sum_{\chi^n = \chi_0} \chi(u), \qquad (38.44)$$

where the sum is taken over all characters of order d and $d \mid n$.

Next, to prove (38.44), suppose that $s = \frac{q-1}{n}$ in (38.36) and $\chi(g) = e^{\frac{2\pi i s}{q-1}}$; in that case, $\chi_0, \chi, \chi^2, \ldots, \chi^{n-1}$ are the n characters of orders that divide n. If $u = 0$ in (38.44), then $N(u) = 1$, $x = 0$ being the only solution for that case. But the right-hand side of (38.44) is also equal to 1, because $\chi_0(0) = 1$ is the only nonzero term in the sum. Again, if $u \neq 0$ and $x^n = u$ has a solution, then $u = a^n$ for some $a \in F$. Thus, in this case, $N(u) = n$, since the n solutions are $a, ba, \cdots, b^{n-1}a$, where $b \in F$ and $b^n = 1$. And the right-hand side of (38.44) is also equal to n in this case, because $\chi(u) = \chi(a^n) = \chi^n(a) = 1$, and there must be n such terms in the sum. Taking $u \neq 0$, if $x^n = u$ does not have a solution, then $N(u) = 0$. Moreover, if g denotes the generator of $F \setminus \{0\}$, then $u = g^s$, for some s that does not divide n, and then $\chi(u) \neq 1$. Therefore, the right-hand side of (38.44) is also zero, completing Weil's proof.

We now turn to an equation with two variables:

$$ax^m + by^n + c \equiv 0 \pmod{q}. \qquad (38.45)$$

In a letter of January 1932 to Hasse, Davenport showed how to solve (38.45) using Gauss and Jacobi sums:[48]

My dear Helmut,
I promised to send you my treatment of the congruence

$$(1) \qquad ax^m + by^n + c \equiv 0 \pmod{p}.$$

Let $\chi_1, \ldots, \chi_{m-1}$ be the nonprincipal characters for which $\chi^m = \chi_0$, the principal character. It is easily seen that

$$1 + \chi_1(t) + \cdots + \chi_{m-1}(t)$$

is precisely the number of solutions of $x^m \equiv t$. Hence the number of solutions of (1) is

$$N = \sum_t \{1 + \chi_1(t) + \cdots + \chi_{m-1}(t)\} \left\{ 1 + X_1 \left(-\frac{at+c}{b} \right) + \cdots + X_{n-1} \left(-\frac{at+c}{b} \right) \right\}$$

where $X_1, \ldots X_{n-1}$ are the n.-p. [nonprincipal] characters for which $X^n = \chi_0$. Hence

$$N = p + \sum_{r=1}^{m-1} \sum_{s=1}^{n-1} \sum_t X_r(t) X_s \left(-\frac{at+c}{b} \right).$$

The sums t can be easily expressed in terms of generalized Gaussian sums

$$\tau(\chi) = \sum_v \chi(v) e(v), \quad e(x) = e^{\frac{2\pi i x}{p}}.$$

---

[48] Hasse and Davenport (2014) p. 19.

These have the property

$$\overline{\chi}(u)\tau(\chi) = \sum_v \chi(v)e(uv).$$

Hence

$$\sum_t \chi(t)X(at+c) = \frac{1}{\tau(\overline{X})} \sum_{t,v} \chi(t)e((at+c)v)\overline{X}(v)$$

$$= \frac{\tau(\chi)}{\tau(\overline{X})} \sum_v \overline{\chi}(av)\overline{X}(v)e(cv)$$

$$= \frac{\tau(\chi)\tau(\overline{\chi}\,\overline{X})}{\tau(\overline{X})}\overline{\chi}(a)\chi X(c).$$

Therefore

$$N = p + \sum_{r=1}^{m-1}\sum_{s=1}^{n-1} \frac{\tau(\chi_r)\tau(\overline{\chi}_r\overline{X}_s)}{\tau(\overline{X}_s)} X_r\left(\frac{c}{a}\right) X_s\left(-\frac{c}{b}\right)$$

$$= p + \vartheta\sqrt{p}(m-1)(n-1) \quad \text{since} \quad |\tau| = \sqrt{p}, \ |\vartheta| \le 1$$

$$> 0 \quad \text{if} \quad p > (m-1)^2(n-1)^2.$$

Quite trivial!

Here Davenport assumed that both m and n divide $p-1$. It is not difficult to prove that if $d = (m, p-1)$, then the equation $x^m = t \pmod p$ has the same number of solutions as $x^d = t \pmod p$. He moreover assumed that the product of two characters χ_i and χ_j was not the identity; we shall see that this assumption also does not seriously affect the calculation of the number of solutions of (38.45).

Next, to calculate the number of solutions of equation (38.43) in general, Weil let N denote the number of solutions when $b = 0$. Although Weil did not do so, here assume that $n_i \mid q-1$, $i = 0, 1, \ldots, r$; as noted before, this restriction does not produce a loss of generality. For if $n \nmid p-1$, then we replace n by $d = (n, p-1)$. Thus

$$N = \sum_{a_0u_0} N(x^{n_0} = u_0)N(x^{n_1} = u_1)\cdots N(x^{n_r} = u_r). \tag{38.46}$$

Let η_i, $i = 0, 1, \ldots, r$, denote any character whose order divides n_i, meaning that $\eta_i^{n_i} = \chi_0$ on $F^\times$. Recall that χ_0 denotes the trivial character. Then (38.46) can be rewritten as

$$N = \sum_{\eta_0,\ldots,\eta_r} \sum_{a_0u_0+\cdots+a_ru_r=0} \eta_0(u_0)\eta_1(u_1)\cdots\eta_r(u_r), \tag{38.47}$$

where the first sum is taken over all characters η_i of order n_i, $i = 0, 1, \ldots, r$.

Now if we let $t_i = a_iu_i$, $i = 0, 1, \ldots, r$, then the sum in (38.47) can then be written as

$$\sum_{\eta_0,\ldots,\eta_r} \eta_0(a_0^{-1})\eta_1(a_1^{-1})\cdots\eta_r(a_r^{-1}) \sum_{t_0+\cdots+t_r=0} \eta_0(t_0)\cdots\eta_r(t_r). \tag{38.48}$$

Observe that if all characters in the inner sum (38.48) are trivial, then that part of the sum is q^r, because the first r values in the equation $t_0 + t_1 + \cdots + t_r = 0$ can be chosen arbitrarily so that the value of t_r is fixed. And if some but not all of those characters are trivial, then the sum corresponding to those characters is zero. Thus, for example, if $\eta_0, \ldots, \eta_j$ are trivial characters and $\eta_{j+1}, \ldots, \eta_r$ are all nontrivial, then the part of the sum corresponding to these latter characters would take the form

$$\sum_t \eta_{j+1}(t) \sum_t \eta_{j+2}(t) \cdots \sum_t \eta_r(t)$$

and the value of each of these sums is zero.

If all characters are nontrivial in the inner sum, and so is the product $\eta_0, \eta_1 \cdots \eta_r$, then the value of that part of the sum in (38.48) is zero. To verify this, observe that in the inner sum of (38.48) the condition on t_i can be written as

$$t_1 + t_2 + \cdots + t_r = -t_0.$$

Set

$$t_i = -t_0 s_i, \quad i = 1, 2, \ldots, r,$$

so that

$$\sum_{t_0 + \cdots + t_r = 0} \eta(t_0) \cdots \eta_r(t_r) = \eta_0(-1) \sum_{t_0 \neq 0} (\eta_0 \eta_1 \cdots \eta_r)(-t_0) J(\eta_1, \eta_2, \ldots, \eta_r).$$

$$(38.49)$$

But $\eta_0 \eta_1 \cdots \eta_r$ is a nontrivial character, so the sum on the right-hand side of (38.49) is zero.

The final case to be considered is that in which $\eta_0, \eta_1, \ldots \eta_r$ are nontrivial, but $\eta_0 \eta_1 \cdots \eta_r$ is trivial. In this case, we get

$$\eta_0(-1) \sum_{t_0 \neq 0} (\eta_0 \eta_1 \cdots \eta_r)(-t_0) = \eta_0(-1)(q - 1),$$

so that in this instance, (38.49) takes the form

$$\sum_{t_0 + \cdots + t_r = 0} \eta_0(t_0) \cdots \eta_r(t_r) = \eta_0(-1)(q - 1) J(\eta_1, \eta_2, \ldots, \eta_r). \quad (38.50)$$

We therefore get Weil's equation that

$$N = q^r + \eta_0(-1)(q - 1) \sum_{\eta_1, \ldots, \eta_r} \eta_0(a_0^{-1}) \cdots \eta_r(a_1^{-1}) J(\eta_1, \eta_2, \ldots, \eta_r), \quad (38.51)$$

where $\eta_i, i = 0, 1, \ldots, r$ are all nontrivial and $\eta_0 \eta_1 \cdots \eta_r = \chi_0$, the trivial character.

Observe that $\eta_1, \eta_2 \ldots, \eta_r$ and $\eta_1 \eta_2 \cdots \eta_r$ are all nontrivial. In this situation, it can be shown that

$$J(\eta_1, \eta_2, \ldots, \eta_r) = \frac{g(\eta_1)g(\eta_2)\cdots g(\eta_r)}{g(\eta_1 \eta_2 \cdots \eta_r)}, \tag{38.52}$$

an equation that can be demonstrated in exactly the same way as (38.41). Since $\eta_0 \eta_1 \cdots \eta_r = \chi_0$, the trivial character, it follows that $\eta_1 \eta_2 \cdots \eta_r = \eta_0^{-1}$. Now we multiply the right-hand side of (38.52) by $\frac{g(\eta_0)}{g(\eta_0)}$ and arrive at

$$J(\eta_1, \eta_2, \ldots, \eta_r) = \frac{g(\eta_0)g(\eta_1)\cdots g(\eta_r)}{g(\eta_0)g(\eta_0^{-1})}$$

$$= \frac{\chi_0(-1)}{q} \cdot g(\eta_0) \cdots g(\eta_r), \tag{38.53}$$

verifying the case in which all characters are nontrivial. Weil, however, proved equation (38.53) in a slightly different manner. He used (38.40) to obtain the Fourier expansion of a nontrivial character $\eta(x)$ on F:

$$\eta(x) = \frac{g(\eta)}{q} \sum_t \overline{\eta}(t) \, \overline{\psi}(tx). \tag{38.54}$$

Substituting (38.54) for $\eta_1(t_0), \ldots \eta_r(t_r)$ in the inner sum of (38.48), he simplified the resulting expression to derive (38.53) . Weil next applied (38.42) to (38.51) and (38.53) to arrive at

$$|N - q^r| \le M(q - 1)q^{\frac{r-1}{2}}, \tag{38.55}$$

where M was defined as the number of nontrivial characters η_i of order n_i, $i = 0$, $1, \ldots, r$ such that $\eta_0 \eta_1 \cdots \eta_r$ was trivial. He also found a similar inequality for the number of solutions of equation (38.43) when $b \ne 0$. We note, as did Weil, that Hua and Vandiver independently obtained (38.51).[49]

Weil next considered the solutions of the equation

$$a_0 x_0^n + a_1 x_1^n + \cdots + a_r x_r^n = 0, \tag{38.56}$$

defined in r-dimensional projective space: the set of points $(c_0, c_1, \ldots, c_r)$ where $c_i \in F$, $i = 0, 1, \ldots, r$ and not all c_i are zero. Two points $(c_0, c_1, \ldots, c_r)$ and $(d_0, d_1, \ldots, d_r)$ are equivalent if there is a nonzero $\alpha \in F$ such that $c_i = \alpha d_i$, $i = 0, 1, \ldots, r$; this is clearly an equivalence relation. An equivalence class describes a point in projective space. It is also clear that the number of points in the r-dimensional projective space may be given as

$$\frac{q^{r+1} - 1}{q - 1} = q^r + q^{r-1} + \cdots + q + 1.$$

[49] Hua and Vandiver (1949).

If $\overline{N}$ denotes the number of solutions of (38.56) and N the number of solutions in the sense of (38.51), then manifestly $N = 1 + (q - 1)\overline{N}$. Using (38.51), Weil had

$$\overline{N} = 1 + q + \cdots + q^{r-1} + \eta_0(-1) \sum_{\eta_1,\ldots,\eta_r} \overline{\eta}_1(a_1)\cdots\overline{\eta}_r(a_r) J(\eta_1,\ldots,\eta_r), \quad (38.57)$$

where η_i, $i = 0, 1, \ldots, r$ are all nontrivial and $\eta_0\eta_1\cdots\eta_r = \chi_0$, the trivial character. He denoted by $\overline{N}_\nu$ the number of solutions of (38.56) in the extension F_ν of F of degree ν; he then calculated the series

$$\sum_{\nu=1}^{\infty} \overline{N}_\nu U^{\nu-1}. \quad (38.58)$$

For this purpose, he needed a result now called the Davenport–Hasse relation. We give some definitions required to state the result: Let F be a finite field with q elements and let E be a finite extension of F with q^ν elements. For $\alpha \in E$, the trace of α from E to F is defined by

$$Tr_{E/F}(\alpha) = \alpha + \alpha^q + \cdots + \alpha^{q^{\nu-1}}.$$

The norm of α from E to F is defined as

$$N_{E/F}(\alpha) = \alpha \cdot \alpha^q \cdots \alpha^{q^{\nu-1}}.$$

Next, if χ is a multiplicative character on F, then define χ' on F_ν by $\chi' = \chi \circ N_{F_\nu/F}$. And if ψ is an additive character on F, the set $\psi'_a = \psi \circ Tr_{F_\nu/F}$. Now if the Gauss sum belonging to a character χ on F is given by

$$g_\chi(x) = \sum_{x\in F} \chi(x)\psi(x), \quad (38.59)$$

then we have the corresponding Gauss sum $g'_{\chi'}$ belonging to the character χ' on F_ν:

$$g'_{\chi'}(y) = \sum_{y\in F_\nu} \chi'(y)\psi'(y). \quad (38.60)$$

The Davenport–Hasse relation[50] states

$$-g'_{\chi'} = (-g_\chi)^\nu. \quad (38.61)$$

We present Weil's proof of (38.61), since it simpler than that of Davenport and Hasse. For each monic polynomial of degree $n \geq 1$ with coefficients in K,

$$f(x) = X^n + c_1 X^{n-1} + \cdots + c_n,$$

---

[50] Davenport and Hasse (1935); see formula (0.8).

set

$$\lambda(f) = \chi(c_n)\psi(c_1).$$

Let deg f denote the degree of the polynomial f and let

$$g(x) = X^m + d_1 X^{m-1} + \cdots + d_m,$$

so that it is clear that

$$
\begin{aligned}
\lambda(fg) &= \chi(c_n d_n)\,\psi(c_1 + d_1) \\
&= \chi(c_n)\,\psi(c_1)\,\chi(d_n)\,\psi(d_1) \\
&= \lambda(f)\lambda(g).
\end{aligned}
\tag{38.62}
$$

Set P as an irreducible monic polynomial with coefficients in K and set U as an indeterminate. Weil gave the formal identity

$$1 + \sum_f \lambda(f)\,U^{\deg f} = \prod_P \left(1 - \lambda(P)\,U^{\deg P}\right)^{-1},
\tag{38.63}$$

to prove which, first note that every monic polynomial f is uniquely a product of irreducible polynomials P. Then using (38.62), the right-hand side of (38.63) can be expressed as

$$\prod_P \left(1 + \lambda(P)\,U^{\deg P} + \lambda(P^2)\,U^{\deg P^2} + \cdots\right) = 1 + \sum_f \lambda(f)\,U^{\deg f}.$$

Weil observed that the sum on the left-hand side of (38.63) for polynomials $f(\chi) = X + c$, of degree 1, would be

$$\sum_{\deg f = 1} \lambda(f)U = \left(\sum_{c \in K} \chi(c)\psi(c)\right) U$$

$$= g_\chi U.$$

Moreover, the sum on the left-hand side of (38.63), for monic polynomials of degree $n > 1$, was

$$\left(q^{n-2}\sum_{c_n} \chi(c_n) \sum_{c_1} \chi(c_1)\right) U.$$

Now because $\sum_{c_n} \chi(c_n) = 0$, all the coefficients of U^n for $n > 1$, on the left-hand side of (38.63) vanished, so that (38.63) would reduce to

$$1 + g_\chi U = \prod_P \left(1 - \lambda(P)\,U^{\deg P}\right)^{-1}.
\tag{38.64}$$

Similarly, Weil noted that if

$$f'(X) = X^n + d_1 X^{n-1} + \cdots + d_n$$

was a polynomial over $K' = K_\nu$, and

$$\lambda'(f') = \chi'(d_n)\,\chi'(d_1),$$

and U' denoted an indeterminate, then

$$1 + g'_{\chi'} U' = \prod_{P'} \left(1 - \lambda(P')\,U'^{\deg P'}\right)^{-1}, \tag{38.65}$$

where the product was taken over all irreducible polynomials P' on $K' = K_\nu$. Weil took P as a monic irreducible polynomial over K and set P' as an irreducible factor of P in K'. He proved that

$$\lambda'(P') = \lambda(P)^{\frac{\nu}{d}}. \tag{38.66}$$

To prove (38.66), Weil let $-\xi$ be a root of P' and noted that $K(\xi)$ was an extension of K of degree $= \deg P = m$, while $K'(\xi)$ was an extension of K' of degree $= \deg P' = m'$. Since $K'(\xi)$ was the smallest field containing $K(\xi)$ and K', its degree over K was the l.c.m. of m and ν and was thus equal to $\frac{m\nu}{d}$, where $d =$ g.c.d. (m, ν). This implied that $m' = \frac{m}{d}$; thus, P had d irreducible factors, each of degree $\frac{m}{d}$. Taking a and b to be the norm and trace, respectively, of ξ from $K(\xi)$ to K, Weil had

$$P(X) = X^n + bX^{n-1} + \cdots + a$$

and

$$\lambda(P) = \chi(a)\psi(b).$$

Recall that Weil took $-\xi$ to be a root of $P(x)$. Similarly, with a' and b' denoting the norm and trace, respectively, of ξ from $K'(\xi)$ to K', and where $N = N_{K'/K}$, $T = \mathrm{Tr}_{K'/K}$, he arrived at

$$\lambda(P') = \chi'(a')\,\psi'(b') = \chi(Na')\,\psi(Tb').$$

Finally, Weil completed the proof of (38.66) by observing that

$$Na' = N_{K'/K}\left(N_{K'(\xi)/K'}(\xi)\right) = N_{K(\xi)/K}\left(N_{K'(\xi)/K(\xi)}(\xi)\right)$$
$$= N_{K(\xi)/K}(\xi^{\frac{\nu}{d}}) = a^{\frac{\nu}{d}}$$

and, similarly,

$$Tb' = \frac{\nu}{d}\,b.$$

After changing U' to U^ν, relation (38.66) produced the factor on the right-hand side of (38.65):

$$\left(1 - \lambda(P)\, U^{\frac{\nu m}{d}}\right)^{-1}.$$

Since each of the d irreducible factors of P produced the same factor, the contribution by P to the right-hand side of (38.65) was

$$\left(1 - \lambda(P)\, U^{\frac{\nu m}{d}}\right)^{-d}, \tag{38.67}$$

and Weil noted that this could be written as

$$\prod_{\rho=0}^{\nu-1}\left(1 - \lambda(P)(\zeta^\rho U)^m\right)^{-1},$$

with ζ being any primitive νth root of unity. This is not difficult to prove; although Weil omitted the argument, we offer one here: Let $\nu = dt$ so that we have to show that

$$\left(1 - \lambda(P)^t\, U^{tm}\right)^d = \prod_{\rho=0}^{\nu-1}\left(1 - \lambda(P)\,(\zeta^\rho U)^m\right).$$

Let $\zeta = e^{\frac{2\pi i}{\nu}}$. Then

$$\zeta^{\rho m} = e^{\frac{2\pi i \rho m}{\nu}} = e^{\frac{2\pi i \rho m'}{t}}$$

and

$$\prod_{\rho=0}^{\nu-1}(1 - e^{\frac{2\pi i \rho m'}{t}}\lambda(P)U^m) = \prod_{\rho=0}^{t-1}(1 - e^{\frac{2\pi i \rho m'}{t}}\lambda(P)U^m)\prod_{\rho=t}^{2t-1}(1 - e^{\frac{2\pi i \rho m'}{t}}\lambda(P)U^m)\cdots$$
$$= (1 - \lambda(P)^t U^{tm})(1 - \lambda(P)^t U^{tm})\cdots$$
$$= (1 - \lambda(P)^t U^{tm})^d.$$

Equation (38.65) has thus been transformed into

$$1 - (-g'_{\chi'})U^\nu = \prod_{\rho=0}^{\nu-1}\prod_{P}\left(1 - \lambda(P)(\zeta^\rho U)^{\deg P}\right)^{-1}$$
$$= \prod_{\rho=0}^{\nu-1}(1 + g_\chi \zeta^\rho U)$$
$$= \prod_{\rho=0}^{\nu-1}\left(1 - \zeta^\rho(-g_\chi U)\right)$$
$$= 1 - (-g_\chi)^\nu U^\nu, \tag{38.68}$$

proving the Davenport–Hasse relation. A slightly simpler version of this proof of Weil, due to Paul Monsky, is given by Ireland and Rosen.[51] The simplification is brought about by proving

$$g_{\chi'} = \sum (\deg P) \lambda(P)^{\frac{v}{\deg P}},$$

where the sum is taken over all monic irreducible P whose degree divides v. With this result in hand, take the logarithmic derivative of (38.64) and multiply by U to arrive at

$$\frac{g_\chi U}{1 + g_\chi U} = \sum_P \frac{\lambda(P)(\deg P)U^{\deg P}}{1 - \lambda(P)U^{\deg P}},$$

or, after expanding as an infinite series,

$$\sum (-1)^{v-1} g_\chi^v \, U^v = \sum_P \left(\sum_{r=1}^{\infty} (\deg P) \lambda(P)^r U^{r \deg P} \right).$$

Now equate the coefficients of U^v to obtain the Davenport–Hasse formula.

Weil took the D-H formula, together with (38.53) and (38.57) to conclude that $\overline{N}_v$, given in (38.58), could be expressed as

$$\overline{N}_v = 1 + q^v + q^{2v} + \cdots + q^{v(r-1)}$$

$$+ (-1)^{r-1} \sum_{\eta_0, \ldots, \eta_r} \left(\frac{(-1)^{r+1}}{q} \overline{\eta}_0(a_0) \cdots \overline{\eta}_r(a_r) \, g(\eta_0) \cdots g(\eta_r) \right)^v.$$

Weil then observed that

$$\sum_{v=1}^{\infty} \overline{N}_v U^{v-1} = - \sum_{h=0}^{r-1} \frac{d}{dU} \log(1 - q^h U) + (-1)^r$$

$$\times \sum_{\eta_0, \ldots, \eta_r} \log \left(1 - \frac{(-1)^{r+1}}{q} \overline{\eta}_0(a_0) \cdots \overline{\eta}_r(a_r) \, g(\chi_0) \cdots g(\chi_r) U \right)$$

$$= \frac{d}{dU} \log Z(U).$$

Note that here we have taken

$$Z(U) = \frac{P(U)^{(-1)^r}}{(1 - U)(1 - qU) \cdots (1 - q^{r-1}U)}$$

and

$$P(U) = \prod_{\eta_0, \eta_1, \ldots, \eta_r} \left(1 - \frac{(-1)^{r+1}}{q} \overline{\eta}_0(a_0) \cdots \overline{\eta}_r(a_r) \, g(\eta_0) \cdots g(\eta_r) \, U \right).$$

[51] Ireland and Rosen (1982) pp. 164–165.

Recall that all the η_i, $i = 0, 1, \ldots, r$, are nontrivial, while their product, $\eta_0\eta_1 \cdots \eta_r$, is trivial.

This result was among those that lead André Weil at the end of his 1949 paper to formulate the famous Weil's conjectures.

38.9 Exercises

(1) Let P be a monic irreducible polynomial in $R = \mathbb{Z}_p[x]$, and let $|P|$ denote the number of elements in $\frac{R}{P}$. Set

$$\zeta_R(s) = \prod (1 - |P|^{-s})^{-1},$$

where the product is taken over all monic irreducible polynomials in R. Determine $\zeta_R(s)$ and compare your result with equation (38.4).

(2) The last entry in Gauss's diary, dated July 9, 1814, reads (in translation):

> I have made by induction the most important observation that connects the theory of biquadratic residues most elegantly with the lemniscatic functions. Suppose $a + bi$ is a prime number, $a - 1 + bi$ divisible by $2 + 2i$, the number of all solutions to the congruence
>
> $$1 \equiv xx + yy + xxyy \pmod{a + bi},$$
>
> including $x = \infty$, $y = \pm i$, $x = \pm 1$, $y = \infty$ is $= (a - 1)^2 + bb$.

Prove Gauss's theorem. Note that the diary was discovered in 1897 and published in 1903. See Ireland and Rosen (1982) pp. 166–168, where a proof using Gauss and Jacobi sums is given. In 1921, Herglotz gave the first proof of Gauss's last entry by using complex multiplication of elliptic functions. Chapter 10 of Lemmermeyer (2000) gives an excellent discussion of this topic, including useful historical notes. See also Weil (1979) vol. 3, p. 298, for some perceptive historical remarks, pointing out the connection between Gauss's diary entry and lemniscatic function.

(3) Let p be an odd prime, and let Q, R be irreducible polynomials of degrees π and ρ in $\mathbb{Z}_p[x]$. With f any polynomial in this ring, let $\left(\frac{f}{Q}\right)$ denote the unique element of $\mathbb{Z}_p^\times$ such that

$$f^{(|Q|-1)/2} = \left(\frac{f}{Q}\right) \pmod{Q}.$$

Show that

$$\left(\frac{R}{Q}\right)\left(\frac{Q}{R}\right) = \left(\frac{-1}{p}\right)^{\pi\rho}.$$

See Dedekind (1930) vol. I, pp. 56–59, for a proof of this analog of the law of quadratic reciprocity.

(4) Generalize the Euler totient function ϕ to the ring $\mathbb{Z}_p[x]$; state and prove a formula analogous to $\phi(m) = m(1 - \frac{1}{p_1}) \cdots (1 - \frac{1}{p_k})$, where $p_1, \ldots, p_k$

comprise all the distinct prime factors of the positive integer m. See Dedekind (1930) vol. I, pp. 50–51.

(5) State a generalization of Dirichlet's theorem on primes in an arithmetic progression to the ring $F_q[x]$, where F_q is a finite field with $q = p^n$ elements, p a prime. Rosen (2002) offers a statement and a proof of this theorem and a reference to Kornblum's paper. Compare Rosen's proof with that of Kornblum.

(6) Let $q = p^n$ and $a_0, a_1, \ldots, a_r$ be nonzero elements of F_q. Let $n_0, n_1, \ldots, n_r$ be positive integers, and let d_i denote the greatest common divisor of $q - 1$ and n_i. Let N_1 represent the number of solutions in F_q of the equation

$$a_0 x_0^{n_0} + a_1 x_1^{n_1} + \cdots + a_r x_r^{n_r} + 1 = 0.$$

Prove that

$$|N_1 - q^r| \leq (d_0 - 1) \cdots (d_r - 1) q^{\frac{r}{2}}.$$

See Weil (1979) vol. I, pp. 399–410. On the basis of some of his earlier theorems and this result, Weil made four conjectures for zeta functions of smooth projective varieties over a finite base field; see pp. 409–410 for a statement of these conjectures, one of which was the Riemann hypothesis.

(7) Define the Ramanujan $\tau(n)$ function by the formula

$$q \prod_{n=1}^{\infty} (1 - q^n)^{24} = \sum_{n=1}^{\infty} \tau(n) q^n.$$

Assuming the convergence of the series and the product, show that

$$\sum_{n=1}^{\infty} \tau(n) n^{-s} = \prod_{p} \left(1 - \tau(p) p^{-s} + p^{11-2s}\right)^{-1},$$

where the product is over all the primes. This result was conjectured by Ramanujan and proved in 1917 by Louis J. Mordell (1888–1972). See Hardy (1978) pp. 161–165. Ramanujan also conjectured that $|\tau(p)| \leq 2p^{\frac{11}{2}}$. This was deduced by Pierre Deligne from his 1974 proof of the characteristic p Riemann hypothesis.

38.10 Notes on the Literature

Gauss (1965) is an English translation by Arthur A. Clarke of Gauss's *Disquisitiones Arithmeticae*; the quotations in the text are taken from this translation. Due to space considerations, Gauss was unable to include in this work an eighth section on polynomials with coefficients from a finite field. A German translation of this section is available in Gauss (1981) pp. 589–629. Günther Frei (2007) gives an excellent treatment in English of Gauss's researches on finite fields. Peter Roquette's 2018 *The Riemann Hypothesis in Characteristic p in Historical Perspective* presents the work on this topic from Artin in 1921 to Weil in 1948.

Bibliography

Abel, N. 2007. *Abel on Analysis: Papers of N.H. Abel on Abelian and Elliptic Functions and the Theory of Series*. Heber City, UT: Kendrick Press. Translated by Phillip Horowitz from the second edition of Abel's *Oeuvres*.

Abel, N.H. 1826. Untersuchungen über die Reihe $1 + (m/1)x + (m(m-1)/2)x^2 + \cdots$. *J. Reine Angew. Math.*, **1**, 311–339.

Abel, N.H. 1965. *Oeuvres complètes*. New York: Johnson Reprint. Edited by L. Sylow and S. Lie.

Abhyankar, S.S. 1976. Historical ramblings in algebraic geometry and related algebra. *Am. Math. Monthly*, **83**, 409–448.

Acosta, D.J. 2003. Newton's rule of signs for imaginary roots. *Am. Math. Monthly*, **110**, 694–706.

Ahlfors, L.V. 1982. *Collected Papers*. Boston: Birkhäuser. Asst. Editor: R.M. Shortt.

Ahlgren, S., and Ono, K. 2001. Addition and counting: The arithmetic of partitions. *Notices A.M.S.*, **48**, 978–984.

Alder, H.L. 1969. Partition identities — from Euler to the present. *Am. Math. Monthly*, **76**, 733–746.

Alexander, J.W. 1915. Functions which map the interior of the unit circle upon simple regions. *Ann. of Math.*, **17**, 12–22.

Allaire, P., and Bradley, R.E. 2004. Symbolical algebra as a foundation for calculus: D.F. Gregory's contribution. *Hist. Math.*, **29**, 395–426.

Almkvist, G., and Berndt, B. 1988. Gauss, Landen, Ramanujan, the arithmetic-geometric mean, ellipses, π, and the Ladies Diary. *Am. Math. Monthly*, **95**, 585–607.

Altmann, S., and Ortiz, E.L. 2005. *Olinde Rodrigues and His Times*. Providence: A.M.S.

Anderson, G.W. 1991. A short proof of Selberg's generalized beta formula. *Forum. Math.*, **3**, 415–417.

Anderson, M., Katz, V., and Wilson, R. (eds). 2004. *Sherlock Holmes in Babylon*. Washington, D.C.: M.A.A.

Anderson, M., Katz, V., and Wilson, R. (eds). 2009. *Who Gave You the Epsilon?* Washington, D.C.: M.A.A.

Andrews, G. 1981. Ramanujan's "lost" notebook. III. The Rogers-Ramanujan continued fraction. *Adv. Math.*, **41**, 186–208.

Andrews, G. 1986a. Eureka! num $= \triangle + \triangle + \triangle$. *J. Num. Theory*, **23**, 285–293.

Andrews, G. 1986b. *q-Series: Their Development and Application in Analysis, Number Theory, Combinatorics, Physics, and Computer Algebra*. Providence: A.M.S.

Andrews, G. 1998. *The Theory of Partitions*. Cambridge University Press.

Andrews, G., and Garvan, F. 1988. Dyson's crank of a partition. *Bull. A.M.S.*, **18**, 167–171.

Andrews, G., Askey, R., Berndt, B., Ramanathan, K., and Rankin, R. (eds). 1988. *Ramanujan Revisited*. Boston: Acad. Press

Andrews, G., Askey, R., and Roy, R. 1999. *Special Functions*. Cambridge: Cambridge University Press.

Aomoto, K. 1987. Jacobi polynomials associated with Selberg's integral. *SIAM J. Math. Phys.*, **18**, 545–549.

Arakawa, T., Ibukiyama, T., and Kaneko, M. 2014. *Bernoulli Numbers and Zeta Functions*. New York: Springer.

Arbogast, L. 1800. *Du calcul des dérivations*. Strasbourg: Levrault.

Archimedes, and Heath, T.L. 1953. *The Works of Archimedes*. New York: Dover. Translated with commentary by T.L. Heath; originally published in 1897.

Arnold, V.I. 1990. *Huygens and Barrow, Newton and Hooke*. Boston: Birkhäuser. Translated by E.J.F. Primrose.

Arnold, V.I. 2007. *Yesterday and Long Ago*. New York: Springer.

Artin, E. 1964. *The Gamma Function*. New York: Holt, Reinhart and Winston. Translated by Michael Butler.

Ash, J.M. 1976. *Studies in Harmonic Analysis*. Washington, D.C.: M.A.A.

Askey, R. 1975. *Orthogonal Polynomials and Special Functions*. Philadelphia: SIAM.

Askey, R., and Gasper, G. 1976. Positive Jacobi polynomial sums, II. *Amer. J. Math*, **98**, 709–737.

Askey, R., and Ismail, M. 1980. The Rogers q-ultraspherical polynomials. Pages 175–182 of: *Approximation Theory III*. New York: Acad. Press. Edited by E. W. Cheney.

Askey, R., and Ismail, M. 1983. A generalization of untraspherical polynomials. Pages 55–78 of: *Studies in Pure Mathematics, to the Memory of Paul Turán*. Basel: Birkhäuser. Edited by P. Erdös.

Askey, R., and J., Wilson. 1985. *Some Basic Hypergeometric Orthogonal Polynomials That Generalize Jacobi Polynomials*. Providence: Memoirs of the A.M.S.

Atkin, A.O.L. 1967. Proof of a Conjecture of Ramanujan. *Glasgow Math. J.*, **8**, 14–32.

Atkin, A.O.L. 1968. Multiplicative congruence properties. *Proc. London Math. Soc.*, **18**, 563–576.

Atkin, A.O.L., and Lehner, J. 1970. Hecke Operators on $\Gamma_0(m)$. *Math. Ann.*, **185**, 134–160.

Atkin, A.O.L., and Swinnerton-Dyer, P. 1954. Some properities of partititons. *Proc. London Math. Soc.*, **4**, 84–106.

Babbage, C., and Herschel, J. 1813. *Memoirs of the Analytical Society*. Cambridge: Cambridge University Press.

Bäcklund, R. 1918. Über die Nullstellen der Riemannschen Zetafunktion. *Acta Math.*, **41**, 345–375.

Baernstein, A. (ed). 1986. *The Bieberbach Conjecture*. Providence: A.M.S.

Bag, A.K. 1966. Trigonometrical series in the Karanapaddhati and the probable date of the text. *Indian J. Hist. of Sci.*, **1**, 98–106.

Baillaud, B., and Bourget, H. (eds). 1905. *Correspondance d'Hermite et de Stieltjes*. Paris: Gauthier-Villars.

Baker, A. 1988. *New Advances in Transcendence Theory*. New York: Cambridge University Press.

Baker, A., and Masser, D.W. 1977. *Transcendence Theory*. New York: Academic Press.

Barnes, E.W. 1908. A new development of the theory of the hypergeometric function. *Proc. London Math. Soc.*, **6**(2), 141–177.

Baron, M.E. 1987. *The Origins of the Infinitesimal Calculus*. New York: Dover.

Barrow, I. 1735. *Geometrical Lectures*. London: Austen. Translated by E. Stone.

Barrow, I. 1916. *Geometrical Lectures of Isaac Barrow*. Chicago: Open Court. Translated with comprehensive introduction by J.M. Child.

Bateman, H. 1907. The correspondence of Brook Taylor. *Bibliotheca Math.*, **7**, 367–371.

Bateman, P.T., and Diamond, H.G. 1996. A hundred years of prime numbers. *Am. Math. Monthly*, **103**(9), 729–741. Reprinted in Anderson, Katz, and Wilson (2009), pp. 328–336.

Baxter, R.J. 1980. Hard hexagons: exact solution. *J. Phys.*, **A 13**, L61–L70. Letter to the editor.

Becher, H.W. 1980. Woodhouse, Babbage, Peacock, and modern algebra. *Hist. Math.*, **7**, 389–400.

Bell, E.T. 1937. *Men of Mathematics*. New York: Simon and Schuster.

Berggren, L., Borwein, J., and Borwein, P. (eds). 1997. *Pi: A Source Book*. New York: Springer-Verlag.

Berndt, B. 1985–1998. *Ramanujan's Notebooks*. New York: Springer-Verlag.

Berndt, B. 1998. *Gauss and Jacobi sums*. New York: Wiley.

Berndt, B., and Ono, K. 2001. Ramanujan's Unpublished Manuscript on the Partition and Tau Functions with Proof and Commentary. Pages 39–110 of: Foata, D., and Han, G.-N. (eds), *The Andrews Festschrift*. New York: Springer.

Bernoulli, D. 1753a. Réflexions et éclairissements sur les nouvelles vibrations des cordes exposées dans les Mémoires de l'Académie de 1747 et 1748. *Hist. Acad. Sci. Berlin*, **9**, 147–172.

Bernoulli, D. 1753b. Sur le mélange de plusieurs especes de vibrations simples isochrones, qui peuvent coexister dans un même système de corps. *Hist. Acad. Sci. Berlin*, **9**, 173–195.

Bernoulli, D. 1982–1996. *Die Werke von Daniel Bernoulli*. Basel: Birkhäuser.

Bernoulli, Ja. 1744. *Jacobi Bernoulli Basileensis, Opera*. Geneva: Cramer and Philibert.

Bernoulli, Ja. 1993–1999. *Die Werke von Jakob Bernoulli*. Basel: Birkhäuser.

Bernoulli, Ja., and Sylla, E.D. 2006. *The Art of Conjecturing, Translation of Ars Conjectandi*. Baltimore: Johns Hopkins University Press. Translated with comprehensive introduction by E.D. Sylla.

Bernoulli, Joh. 1696. Curvatura radii in diaphanis non uniformiformibus. *Acta Erud.*, **16**, 206–211. Reprinted in Bernoulli (1968) vol. 1, pp. 187–193.

Bernoulli, Joh. 1742. *Opera Omnia*. Lausanne; Geneva: Bousquet.

Bernoulli, Joh. 1968. *Opera omnia*. Hildesheim, Germany: G. Olms Verlag.

Bernoulli, N. 1738. Inquisitio in summam seriei $1 + \frac{1}{4} + \frac{1}{9} + \frac{1}{16} + \frac{1}{25} + \frac{1}{36} +$ etc. *Comment. Petropolitanae*, **10**, 19–21.

Bers, L. 1998. *Selected Works of Lipman Bers*. Providence: A.M.S. Edited by I. Kra and B. Maskit.

Bertrand, J. 1845. Mémoire sur le nombre de valeurs que peut prendre une fonction quand on y permute les lettres qu'elle renferme. *J. École Poly.*, **18**, 123–140.

Beukers, F., Brownawell, W.D., and Heckman, G. 1988. Siegel normality. *Ann. of Math.*, **127**, 279–308.

Bézout, É. 2006. *General Theory of Algebraic Equations*. Princeton: Princeton University Press. Translated by E. Feron.

Bhaskaracharya. 2018. *Translation of the Surya Siddhanta and of the Siddhanta Siromani*. London: Forgotten Books. First published 1861. Translated with commentary by B.D. Sastri and L. Wilkinson.

Bieberbach, L. 1916. Über die Koeffizienten derjenigen Potenzreihen, ... *S.B. Preuss. Akad. Wiss.*, **138**, 940–955.

Binet, J. 1839. Mémoire sur les intégrales définies Eulériennes. *J. École Poly.*, **16**, 123–343.

Bissell, C. C. 1989. Cartesian geometry: The Dutch contribution. *Math. Intelligencer*, **9**(4), 38–44.

Boas, R.P. 1954. *Entire Functions*. New York: Academic Press.

Boehle, K. 1933. Über die Transzendenz von Potenzen mit algebraischen Exponenten. *Math. Ann.*, **108**, 56–74.

Bogolyubov, N.N., Mikhaĭlov, G.K., and Yushkevich, A.P. (eds). 2007. *Euler and Modern Science*. Washington, D.C.: M.A.A.

Bohr, H., and Mollerup, J. 1922. *Laerebog i Matematisk Analyse*. Copenhagen: Jul. Gjellerups Forlag.

Bolibruch, A.A., Osipov, Yu.S., and Sinai, Ya.G. (eds). 2006. *Mathematical Events of the Twentieth Century*. New York: Springer.

Bolzano, B. 1930. *Functionenlehre*. Prague: Roy. Bohemian Acad. Sci. Edited by K. Rychlik.

Bolzano, B. 1980. A translation of Bolzano's paper on the intermediate value theorem. *Hist. Math.*, **7**, 156–185. Translated by S.B. Russ.

Bombieri, E., and Gubler, W. 2006. *Heights in Diophantine Geometry*. New York: Cambridge University Press.

Boole, G. 1839. Researches in the theory of analytical transformations, with a special application to the reduction of the general equation of the second order. *Cambridge Math. J.*, **2**, 34–78.

Boole, G. 1841. Exposition of a general theory of linear transformations, Parts I and II. *Cambridge Math. J.*, **3**, 1–20, 106–111.

Boole, G. 1844a. Notes on linear transformations. *Cambridge Math. J.*, **4**, 167–71.

Boole, G. 1844b. On a general method in analysis. *Phil. Trans. Roy. Soc. London*, **124**, 225–282.

Boole, G. 1845. Notes on linear transformations. *Cambridge Math. J.*, **6**, 106–113.

Boole, G. 1847. *The Mathematical Analysis of Logic*. London: George Bell.

Boole, G. 1877. *A Treatise on Differential Equations*. London: Macmillan.

Borel, É. 1896. Démonstration élémenatire d'un théorème de M. Picard sur les fonctions entières. *Comptes Rendus*, **122**, 1045–1048.

Borel, É. 1900. *Leçons sur les fonctions entières*. Paris: Gauthier-Villars.

Bornstein, M. 1997. *Symbolic Integration*. New York: Springer-Verlag.

Boros, G., and Moll, V. 2004. *Irresistible Integrals*. New York: Cambridge University Press.

Borwein, J., Bailey, D., and Girgensohn, R. 2004. *Experimentation in Mathematics: Computational Paths to Discovery*. Natick, MA: Peters.

Bos, H.J.M. 1974. Differentials, higher-order differentials and the derivative in the Leibnizian calculus. *Archive Hist. Exact Sci.*, **14**, 1–90.

Bos, H.J.M. 1996. Johann Bernoulli on Exponential Curves. *Nieuw Archief Wisk.*, **14**, 1–19.

Bottazzini, U. 1986. *The Higher Calculus*. New York: Springer-Verlag. Translated by W. Van Egmond.

Bourbaki, N. 1994. *Elements of the History of Mathematics*. New York: Springer-Verlag. Translated by J. Meldrum.

Boyer, C.B. 1943. Pascal's formula for the sums of the powers of integers. *Scripta Math.*, **9**, 237–244.

Boyer, C.B., and Merzbach, U.C. 1991. *A History of Mathematics*. New York: Wiley.

Bradley, R.E., and Sandifer, C.E. (eds). 2007. *Leonhard Euler: Life, Work and Legacy*. Amsterdam: Elsevier.

Bradley, R.E., and Sandifer, C.E. (eds). 2009. *Cauchy's Cours d'analyse*. New York: Springer. Translated with commentary by Bradley and Sandifer.

Brahmagupta. 1817. *Algebra, with arithmetic and mensuration, from the Sanscrit of Brahmegupta and Bháscara*. London: Murray. Translated with notes by H.T. Colebrooke.

Bressoud, D. 2002. Was calculus invented in India? *College Math. J.*, **33**(1), 2–13. Reprinted in Anderson, Katz, and Wilson (2004), 131–137.

Bressoud, D. 2007. *A Radical Approach to Real Analysis, second edition*. Washington, D.C.: M.A.A.

Bressoud, D. 2008. *A Radical Approach to Lesbesgue's Theory of Integration*. Cambridge: Cambridge University Press.

Bressoud, D., and Zeilberger, D. 1982. A short Rogers–Ramanunan bijection. *Discrete Math.*, **38**, 313–315.

Brezinski, C. 1991. *History of Continued Fractions and Padé Approximations*. New York: Springer.

Briggs, H. 1624. *Arithmetica Logarithmica*. London: W. Jones.

Briggs, H. 1633. *Trigonometria Britannica*. Gouda: Pierre Rammasen.

Brinkley, J. 1807. An investigation of the general term of an important series in the inverse method of finite differences. *Phil. Trans.*, **97**, 114–132.

Briot, C., and Bouquet, J.C. 1856. Recherches sur les propriétés des fonctions définies par des équations différentielles. *J. École Poly.*, **t. 21, cahier 36**, 133–198.

Briot, C., and Bouquet, J.C. 1859. *Théorie des fonctions doublement périodiques et, en particulier, des fonctions elliptiques*. Paris: Mallet-Bachelier.

Bronstein, M. 1997. *Symbolic Integration*. Heidelberg: Springer-Verlag.

Bronwin, B. 1849. On the determination of the coefficients in any series of sines and cosines of multiples of a variable angle from particular values of that series. *Phil. Magazine*, **34**, 260–268.

Brouncker, W. 1668. The squaring of the hyperbola, by an infinite series of rational numbers, together with its demonstration. *Phil. Trans.*, **3**, 645–649.

Browder, F.E. (ed). 1976. *Mathematical Developments Arising from Hilbert Problems*. Providence: A.M.S.

Buchler, J. 1955. *The Philosophy of Peirce: Selected Writings*. New York: Dover.

Budan de Boislaurent, F. 1822. *Nouvelle méthode pour la résolution des équations numériques* ... Paris: Dondey-Dupré.

Bühler, W.K. 1981. *Gauss: A Biographical Study*. New York: Springer-Verlag.

Bunyakovski, V. 1859. Sur quelques inégalités concernant les intégrales ordinaires et les intégrales aux différences finies. *Mém. de Acad. Sci. St.-Pétersbourg*, **1**, 1–18.

Burn, R.P. 2001. Alphose Antonio de Sarasa and logarithms. *Hist. Math.*, **28**, 1–17.

Burnside, W.S., and Panton, A.W. 1960. *The Theory of Equations*. New York: Dover.

Butzer, P.L., and Sz.-Nagy, B. 1974. *Linear Operators and Approximation II*. Basel: Birkhäuser.

Cahen, E. 1894. Sur la fonction $\zeta(s)$ de Riemann et sur des fonctions analogues. *Ann. Sci. École Norm. Sup.*, **11**, 75–164.

Cajori, F. 1913. *A History of Mathematics*. New York: Macmillan.

Cajori, F. 1993. *A history of mathematical notations*. New York: Dover. Two volumes bound as one.

Campbell, G. 1728. A method of determining the number of impossible roots in affected aequations. *Phil. Trans. Roy. Soc.*, **35**, 515–531.

Campbell, P.J. 1978. The origin of "Zorn's Lemma" *Hist. Math.*, **5**, 77–89.

Cannon, J.T., and Dostrovsky, S. 1981. *The Evolution of Dynamics: Vibration Theory from 1687 to 1742*. New York: Springer-Verlag.

Cantor, G. 1870a. Beweis, dass eine fur̈ jeden reellen Werth von x durch eine trigonometrische Reihe gegebene Function $f(x)$ sich nur auf eine einzige Weise in dieser Form darstellen lässt. *J. Reine Angew. Math*, **72**, 139–142.

Cantor, G. 1870b. Über einen die trigonometrischen Reihen betreffenden Lehrsatz. *J. Reine Angew. Math*, **72**, 130–138. Reprinted in Cantor (1932) pp. 71–79.

Cantor, G. 1871. Notiz zu dem vorangehenden Aufsatze. *J. Reines Angew. Math*, **73**, 294–296.

Cantor, G. 1872. Über die Ausdehnung eines Satzes aus der Theorie der trigonometrischen Reihen. *Math. Ann.*, **5**, 123–132.

Cantor, G. 1932. *Gesammelte Abhandlungen*. Berlin: Springer. Edited by E. Zermelo.

Carathéodory, C. 1912. Untersuchungen über die konformen Abbildungen von festen und veränderlichen Gebieten. *Math. Ann.*, **72**, 107–144.

Cardano, G. 1993. *Ars Magna or the Rules of Algebra*. New York: Dover. Translated by T.R. Witmer.

Carleson, L. 1966. On convergence and growth of partial sums of Fourier series. *Acta. Math.*, **116**, 135–157.

Cartier, P. 2000. Mathemagics. Pages 6–67 of: Planat, M. (ed), *Lecture Notes in Physics, vol. 550*. Berlin: Springer.

Cauchy, A.-L. 1823. *Résumé des leçons données à l'École Royale Polytechnique sur le calcul infinitésimal*. Paris: De Bure.

Cauchy, A.-L. 1827. Mémoire sur les intégrales définies. *Mém. Acad. Roy. Sci.*, **1**, 601–799. First presented to the Academy in 1814. Reprinted in Cauchy's *Oeuvres complètes* (1), vol. 1, pp. 329–506.

Cauchy, A.-L. 1829. *Calcul différentiel*. Paris: De Bure.

Cauchy, A.-L. 1840–1841. *Exercices d'analyse et de physique mathématique*. Paris: Bachelier.

Cauchy, A.-L. 1843a. Mémoire sur les fonctions dont plusieurs valuers sont liées entre elles par une équation linéaire, et sur diverses ... *Comptes Rendus*, **17**, 523–531. Reprinted in Cauchy (1882–1974), sér. 1, vol. 8, pp. 42–50.

Cauchy, A.-L. 1843b. Sur l'emploi légitime des séries divergentes. *Comptes Rendus*, **17**, 370–376.

Cauchy, A.-L. 1853. Note sur les séries convergentes dont les divers termes sont des fonctions continues ... *Comptes Rendus*, **36**, 454–459.

Cauchy, A.-L. 1882–1974. *Oeuvres complètes*. Paris: Gauthier-Villars.

Cauchy, A.-L. 1989. *Analyse algébrique*. Paris: Gabay.

Cayley, A. 1843. On the theory of determinants. *Trans. Cambridge Phil. Soc.*, **8**, 1–16. Reprinted in Cayley (1889–1898) vol. 1, pp. 63–79.

Cayley, A. 1845a. Mémoire sur les fonctions doublement périodiques. *J. Math. Pures Appl.*, **10**, 385–410.

Cayley, A. 1845b. On the theory of linear transformations. *Cambridge Math. J.*, **1**, 193–209. Reprinted in Cayley (1889–1898) vol. 1, pp. 80–94.

Cayley, A. 1846. On linear transformations. *Cambridge and Dublin Math. J.*, **1**, 104–122. Reprinted in Cayley (1889–1898) vol. 1, pp. 95–112.

Cayley, A. 1854. An introductory memoir upon quantics. *Phil. Trans. Roy. Soc. London*, **144**, 244–258. Reprinted in Cayley (1889–1898) vol. 2, pp. 221–234.

Cayley, A. 1855. A second memoir upon quantics. *Phil. Trans. Roy. Soc. London*, **146**, 101–126. Reprinted in Cayley (1889–1898) vol. 2, pp. 250–275.

Cayley, A. 1856. A third memoir upon quantics. *Phil. Trans. Roy. Soc. London*, **146**, 627–647.

Cayley, A. 1889–1898. *Collected Mathematical Papers*. Cambridge: Cambridge University Press.

Cayley, A. 1895. *Elliptic Functions*. London: Bell.

Cesàro, E. 1890. Sur la multiplication des séries. *Bull. Sci. Math.*, **14**, 114–20.

Chabert, J.-L. 1999. *A History of Algorithms: From the Pebble to the Microchip*. New York: Springer. Translated by C. Weeks.

Chakravarti, G. 1932. Growth and development of permutations and combinations in India. *Bull. Calcutta Math. Soc.*, **24**, 79–88.

Chandrasekhar, S. 1995. *Newton's Principia for the Common Reader*. Oxford: Oxford University Press.

Charzynski, Z., and Schiffer, M. 1960. A new proof of the Bieberbach conjecture for the fourth coefficient. *Arch. Rational Mech. Anal.*, **5**, 187–193.

Chebyshev, P. 1848. Sur la fonction qui détermine la totalité des nombres premiers inférieurs à une limite donnée. *Mém. savants étrangers Acad. Sci. St. Petersbourg*, **6**, 1–19.

Chebyshev, P. 1850. Mémoire sur nombres premiers. *Mém. savants étrangers Acad. Sci. St. Petersbourg*, **1**, 17–33.

Chebyshev, P. 1858. Sur une nouvelle série. *Bull. Phys. Math. Acad. St. Petersburg*, **17**, 257–261.

Chebyshev, P. L. 1899–1907. *Oeuvres de P.L. Tchebychef*. St. Petersburg: Acad. Impériale Sci.

Cheney, E. (ed). 1980. *Approximation Theory III*. New York: Academic Press.

Cherry, W., and Ye, Z. 2001. *Nevanlinna Theory of Value Distribution*. New York: Springer.

Chowla, S. 1934. Congruence properties of partitions. *J. London Math. Soc.*, **9**, 247.

Christoffel, E.B. 1867. Sul problema delle temperature stazionarie e la rappresentazione di una superficie. *Annali Mat. Pura Appl.*, **1**, 89–103. Translated by C. Formenti.

Chudnovsky, D.V., and Chudnovsky, G.V. 1988. Approximations and complex multiplication according to Ramanujan. Pages 375–472 of: Andrews, G., Askey, R., Berndt, B., Ramanathan, K.G., and Rankin, R. (eds), *Ramanujan Revisited: Proceedings of the Ramanujan Centenary Conference held at the University of Illinois, Urbana-Champaign, Illinois, June 1–5, 1987*. Boston: Academic Press.

Clairaut, A.-C. 1734. Solution de plusieurs problèmes où il s'agit de trouver des courbes ... *Hist. Acad. Roy. Sci.*, 196–215.

Clairaut, A.-C. 1739. Recherches générales sur le calcul intégral. *Mém. Acad. Roy. Sci.*, **1**, 425–436.

Clairaut, A.-C. 1740. Sur l'intégration ou la construction des équations différentielles du premier ordre. *Mém. Acad. Roy. Sci.*, **2**, 293–323.

Clairaut, A.-C. 1754. Sur l'oribit apparente du Soleil autour de la terre, ... *Mém Hist. Acad. Sci. Paris*, **9**, 521–565.

Clarke, F. M. 1929. *Thomas Simpson and his Times*. Baltimore: Waverly Press.

Clausen, T. 1828. Ueber die Fälle, wenn die Reihe von der Form $y = 1 + \cdots$ etc. ein Quadrat von der Form $z = 1 + \cdots$ etc. hat. *J. Reine Angew. Math.*, **3**, 89–91.

Clausen, T. 1832. Über die Function $\sin\phi + \frac{1}{2^2}\sin 2\phi + \frac{1}{3^2}\sin 3\phi +$ etc. *J. Reine Angew. Math.*, **8**, 298–300.

Clausen, T. 1840. Theorem. *Astronom. Nach.*, **17**, 351–352.

Clausen, T. 1858. Beweiss des von Schlömilch ... *Archive Math. Phys.*, **30**, 166–169.

Cohen, H. 2007. *Number Theory, Volume II: Analytic and Modern Tools*. New York: Springer.

Cooke, R. 1984. *The Mathematics of Sonya Kovalevskaya*. New York: Springer-Verlag.

Cooke, R. 1993. Uniqueness of trigonometric series and descriptive set theory 1870–1985. *Archive Hist. Exact Sci.*, **45**, 281–334.

Cooper, S. 2006. The quintuple product identity. *Int. J. Number Theory*, **2**, 115–161.

Corry, L. 2004. *Modern Algebra and the Rise of Mathematical Structures*. Basel: Birkhäuser.

Cotes, R. 1714. Logometria. *Phil. Trans.*, **29**, 5–45.

Cotes, R. 1722. *Harmonia Mensurarum*. Cambridge: Cambridge University Press. Edited by Robert Smith.

Cox, D.A. 1984. The arithmetic-geometric mean of Gauss. *L'enseignement math.*, **30**, 275–330.

Cox, D.A. 2004. *Galois Theory*. Hoboken: Wiley.

Craig, J. 1685. *Methodus figurarum lineis rectis et curvis*. London: Pitt.

Craik, A.D.D. 2000. James Ivory, F.R.S., mathematician: "The most unlucky person that ever existed." *Notes and Records Roy. Soc. London*, **54**, 223–247.

Craik, A.D.D. 2005. Prehistory of Faà di Bruno's formula. *Am. Math. Monthly*, **112**, 119–130.

Crilly, T. 2006. *Arthur Cayley*. Baltimore: Johns Hopkins University Press.

d'Alembert, J. 1747. Recherches sur la courbe que forme une corde tenduë mise en vibration. *Hist. Acad. Roy. Sci. Belles-Lettres, Berlin*, **3**, 214–219, 220–249.

d'Alembert, J. 1761–1780. *Opuscules mathématiques*. Paris: David.

Darboux, G. 1875. Mémoire sur les fonctions discontinues. *Ann. École Norm. Sup.*, **4**, 57–112.

Datta, B., and Singh, A.N. 1962. *History of Hindu Mathematics*. Bombay: Asia Pub. House.

Datta, B.B. 1929. The Jaina school of mathematics. *Calcutta Math. Soc.*, **21**, 115–145.

Dauben, J. 1979. *Georg Cantor*. Princeton: Princeton University Press.

Davenport, H. 1980. *Multiplicative Number Theory*. New York: Springer-Verlag.

Davenport, H., and Hasse, H. 1935. Die Nullstellen der Kongruenz-zetafunktionen in gewissen zyklischen Fällen. *J. Reine Angew. Math.*, **172**, 151–182.

Davis, P.J. 1959. Leonhard Euler's Integral: A historical profile of the gamma function. *Am. Math. Monthly*, **66**, 849–869.

de Beaune, F., Girard, A., and Viète, F. 1986. *The Early Theory of Equations: On Their Nature and Constitution*. Annapolis, MD: Golden Hind Press. Translated by R. Schmidt.

de Branges, L. 1985. A proof of the Bieberbach conjecture. *Acta Math.*, **157**, 137–162.

de Moivre, A. 1697. A method of raising an infinite multinomial to any given power, or extracting any given root of the same. *Phil. Trans.*, **19**, 619–625.

de Moivre, A. 1698. A method of extracting the root of an infinite equation. *Phil. Trans.*, **20**, 190–193.

de Moivre, A. 1707. Aequationum quarundam potestatis tertiae, ... *Phil. Trans.*, **25**, 2368–2371.

de Moivre, A. 1730a. *Miscellanea analytica de seriebus et quadraturis*. London: Tonson and Watts.

de Moivre, A. 1730b. *Miscellaneis analyticis supplementum*. London: Tonson and Watts.

de Moivre, A. 1967. *The Doctrine of Chances*. New York: Chelsea.

Dedekind, R. 1857. Abriss einer Theorie der höheren Kongruenzen in Bezug auf einen reellen Primzahl-Modulus. *J. Reine Angew. Math.*, **54**, 1–26. Reprinted in Dedekind (1930) vol. 1, pp. 40–67.

Dedekind, R. 1872. *Stetigkeit and irrationale Zahlen*. Braunschweig: Vieweg.

Dedekind, R. 1877. Schreiben an Herr Borchardt über die Theorie der elliptischen Modulfunktionen. *J. Reine Angew. Math.*, **83**, 265–292. Reprinted in Dedekind's *Werke* (1930) vol. 1 pp. 174–201.

Dedekind, R. 1930. *Gesammelte Mathematische Werke*. Braunschweig: F. Vieweg. Edited by R. Fricke, E. Noether and Ø. Ore.

Dedekind, R. 1963. *Essays on the Theory of Numbers*. New York: Dover. Translated by W.W. Beman.

Dedekind, R., and Weber, H. 2012. *Theory of Algebraic Functions of One Variable*. Providence: A.M.S. Translated by J. Stillwell.

Delone, B.N. 2005. *The St. Petersburg School of Number Theory*. Providence: A.M.S.

Descartes, R. 1897–1913. *Oeuvres*. Paris: Léopold Cerf. edited by C. Adam and P. Tannery.

Descartes, R. 1954. *La Géométrie*. New York: Dover. Translated by D.E. Smith and M.L. Latham.

Dieudonné, J. 1931. Sur les fonctions univalentes. *Comptes Rendus*, **192**, 1148–1150.

Dieudonné, J. 1981. *History of Functional Analysis*. Amsterdam: Elsevier.

Dini, U. 1878. *Fondamenti per la teorica delle funzione de variabili reali*. Pisa: Nistri.

Dirichlet, P.G.L. 1829a. Note sur les intégrales définies. *J. Reine Angew. Math*, **4**, 94–98.

Dirichlet, P.G.L. 1829b. Sur la convergence des séries trigonométriques qui servent à représenter une fonction arbitraire entres des limites données. *J. Reine Angew. Math.*, **4**, 157–169. Reprinted in Dirichlet's *Werke* vol. 1, pp. 117–132.

Dirichlet, P.G.L. 1837. Über die Darstellung ganz willkürlicher Functionen durch Sinus- und Cosinusreihen. *Reper. der Physik*, **1**, 152–174. Reprinted in Dirichlet's *Werke*, vol. 1, pp. 133–160.

Dirichlet, P.G.L. 1839a. Sur une nouvelle méthode pour la détermination des intégrales multiples. *J. Math. Pures Appl.*, **4**, 164–168. Reprinted in Dirichlet's *Werke* vol. 1, pp. 377–380.

Dirichlet, P.G.L. 1839b. Ueber eine neue Methode zur Bestimmung vielfacher Integrale. *Akad. Wiss. Berlin von 1839*, 18–25. Reprinted in Dirichlet's *Werke* vol. 1, pp. 381–390; see pp. 393–410 for later expanded version from 1841.

Dirichlet, P.G.L. 1840. Recherches sur diverses applications de l'analyse infinitésimale à la théorie des nombres. *J. Reine Angew. Math.*, **19, 21**, (19): 324–369, (21): 1–12, 134–155. Reprinted in Dirichlet's *Werke* vol. 1, pp. 411–496.

Dirichlet, P.G.L. 1863. Démonstration d'un théorème d'Abel. *J. Math. Pures App.*, **7**(2), 253–255. Also in *Werke* vol. I, pp. 305–306.

Dirichlet, P.G.L. 1969. *Mathematische Werke*. New York: Chelsea.

Dirichlet, P.G.L., and Dedekind, R. 1999. *Lectures on Number Theory*. Providence: A.M.S. Translated by John Stillwell.

Dörrie, H. 1965. *100 Great Problems of Elementary Mathematics*. New York: Dover. Translated by D. Antin.

Douglas, R.G. 1972. *Banach Algebra Techniques in Operator Theory*. New York: Academic Press.

Doxiadis, A., and Mazur, B. (eds). 2012. *Circles Disturbed*. Princeton: Princeton University Press.

Dugac, P. 1973. Éléments d'analyse de Karl Weierstrass. *Archive Hist. Exact. Sci.*, **10**, 41–176.

Duke, W., and Tschinkel, Y. (eds). 2005. *Analytic Number Theory: A Tribute to Gauss and Dirichlet*. Providence: A.M.S.

Dunham, W. 1990. *Journey Through Genius*. New York: Wiley.

Dunnington, G. 2004. *Gauss: Titan of Science*. Washington, D.C.: M.A.A.

Dupré, A. 1859. *Examen d'une proposition de Legendre relative à la théorie des nombres*. Paris: MalletBachelier.

Duren, P.L. 1983. *Univalent Functions*. New York: Springer-Verlag.

Dutka, J. 1984. The early history of the hypergeometric series. *Archive Hist. Exact Sci.*, **31**, 15–34.

Dutka, J. 1991. The early history of the factorial function. *Archive Hist. Exact Sci.*, **43**, 225–249.

Dyson, F. 1944. Some guesses in the theory of partitions. *Eureka (Cambridge)*, **8**, 10–15.

Dyson, F. 1962. Statistical theory of the energy levels of complex systems. *J. Math. Phys.*, **3**, 140–156.

Edwards, A.W.F. 1986. A quick route to sums of powers. *Am. Math. Monthly*, **93**, 451–455.

Edwards, A.W.F. 2002. *Pascal's Arithmetical Triangle*. Baltimore: Johns Hopkins University Press.

Edwards, H.M. 1977. *Fermat's Last Theorem*. Berlin: Springer-Verlag.

Edwards, H.M. 1984. *Galois Theory*. New York: Springer-Verlag.

Edwards, H.M. 2001. *Riemann's Zeta Function*. New York: Dover.

Edwards, J. 1954a. *An Elementary Treatise on the Differential Calculus*. London: Macmillan.

Edwards, J. 1954b. *Treatise on Integral Calculus*. New York: Chelsea.

Eie, M. 2009. *Topics In Number Theory*. Singapore: World Scientific.

Eisenstein, G. 1844. Beiträge zur Kreistheilung. *J. Reine Angew. Math.*, **27**, 269–278.

Eisenstein, G. 1975. *Mathematische Werke*. New York: Chelsea.

Elliott, E.B. 1964. *An Introduction to the Algebra of Quantics*. New York: Chelsea.

Ellis, D.B., Ellis, R., and Nerurkar, M. 2001. The topological dynamics of semigroup actions. *Trans. A.M.S.*, **353**, 1279–1320.

Ellis, R.L. 1845. Memoir to D.F. Gregory. *Cambridge and Dublin Math. J.*, **4**, 145–152.

Elstrodt, J. 2005. The Life and Work of Gustav Lejeune Dirichlet (1805–1859). In: Duke, W., and Tschinkel, Y. (eds), *Analytic Number Theory: A Tribute to Gauss and Dirichlet*. Providence: A.M.S.

Emch, G.G., Sirdhara, R., and Srinivas, M.D. (eds). 2005. *Contributions to the History of Indian Mathematics*. New Delhi: Hindustan Book Agency and Springer.

Eneström, G. 1897. Sur la découverte de l'intégrale complète des équations différentielles linéaires à coefficients constants. *Bib. Math.*, **11**, 43–50.

Eneström, G. 1905. Der Briefwechsel zwischen Leonhard Euler und Johann I Bernoulli. *Bib. Math.*, **6**, 16–87.

Engelsman, S.B. 1984. *Families of Curves and the Origins of Partial Differentiation*. Amsterdam: North-Holland.

Enros, P. 1983. The analytical society (1812–1813). *Hist. Math.*, **10**, 24–47.

Erdős, P. 1932. Beweis eines Satzes von Tschebyschef. *Acta. Sci. (Szeged)*, **5**, 194–198.

Erdős, P. 1949. On a new method in elementary number theory which leads to an elementary proof of the prime number theorem. *Proc. Nat. Acad. Sci. USA*, **35**, 374–384.

Erman, A. (ed). 1852. *Briefwechsel zwischen Olbers und Bessel*. Leipzig: Avenarius and Mendelssohn.

Euler, L. 1911–2000. *Leonhardi Euleri Opera omnia*. Berlin: Teubner.

Euler, L. 1985. An essay on continued fractions. *Math. Syst. Theory*, **18**, 295–328. Translated by M. Wyman and B. Wyman.

Euler, L. 1988. *Introduction to Analysis of the Infinite*. New York: Springer-Verlag. Translated by J.D. Blanton.

Euler, L. 2000. *Foundations of Differential Calculus. English Translation of First Part of Euler's Institutiones calculi differentialis*. New York: Springer-Verlag. Translated by J.D. Blanton.

Faà di Bruno, F. 1857. Note sur une nouvelle formule de calcul différentiel. *Quarterly J. Pure Appl. Math.*, **1**, 359–360.

Fagnano, G.C. 1911. *Opere matematiche*. Rome: Albrighi.

Farkas, H.M., and Kra, I. 2001. *Theta Constants, Riemann Surfaces and the Modular Group*. Providence: A.M.S.

Fatou, P. 1906. Séries trigonométriques et séries de Taylor. *Acta Math.*, **30**, 335–400.

Favard, J. 1935. Sur les polynômes de Tchebicheff. *Comptes Rendus*, **200**, 2052–2053.

Feigenbaum, L. 1985. Taylor and the method of increments. *Archive Hist. Exact Sci.*, **34**, 1–140.

Feingold, M. 1990. *Before Newton*. Cambridge: Cambridge University Press.

Feingold, M. 1993. Newton, Leibniz and Barrow too. *Isis*, **84**, 310–338.

Fejér, L. 1900. Sur les fonctions bornées et intégrables. *Comptes Rendus*, **131**, 984–987.

Fejér, L. 1904. Untersuchungen über Fouriersche Reihen. *Math. Ann.*, **58**, 51–69.

Fejér, L. 1908. Sur le développement d'une fonction arbitraire suivant les fonctions de Laplace. *Comptes Rendus*, **146**, 224–227.

Fejér, L. 1970. *Gesammelte Arbeiten*. Basel: Birkhäuser. Edited by Paul Turán.

Fekete, M., and Szegő, G. 1933. Eine Bemerkung über ungerade schlichte Funktionen. *J. London Math. Soc.*, **8**, 85–89.

Feldheim, E. 1941. Sur les polynômes généralisés de Legendre. *Bull. Acad. Sci. URSS. Ser. Math. [Izvestia Acad. Nauk SSSR]*, **5**, 241–254.

Ferraro, G. 2004. Differentials and differential coefficients in the Eulerian foundations of the calculus. *Hist. Math.*, **31**, 34–61.

Ferreirós, J. 1993. On the relations between Georg Cantor and Richard Dedekind. *Hist. Math.*, **20**, 343–63.

FitzGerald, C.H. 1972. Quadratic inequalities and coefficient estimates for schlicht functions. *Arch. Rational Mech. Anal.*, **46**, 356–368.

FitzGerald, C.H. 1985. The Bieberbach conjecture: Retrospective. *Notices A.M.S.*, **32**, 2–6.

Foata, D., and Han, G.-N. 2001. The triple, quintuple and sextuple product identities revisited. Pages 323–334 of: Foata, D., and Han, G.-N. (eds), *The Andrews Festschrift: Seventeen Papers on Classical Number Theory and Combinatorics*. New York: Springer.

Fomenko, O.M., and Kuzmina, G.V. 1986. The last 100 days of the Bieberbach conjecture. *Math. Intelligencer*, **8**, 40–47.

Forrester, P.J., and Warnaar, S.O. 2008. The Importance of the Selberg Integral. *Bull. A.M.S.*, **45**, 498–534.

Fourier, J. 1955. *The Analytical Theory of Heat*. New York: Dover. Translated by A. Freeman.

Français, J.F. 1812–1813. Analise transcendante. Memoire tendant à démontrer la légitimité de la séparation des échelles de différentiation et d'intégration des fonctions qu'elles affectent. *Annales Gergonne*, **3**, 244–272.

Frei, G. 2007. The unpublished section eight: On the way to function fields over a finite field. Pages 159–198 of: Goldstein, C., Schappacher, N., and Schwermer, J. (eds), *The Shaping of Arithmetic after C.F. Gauss's Disquisitiones Arithmeticae*. New York: Springer.

Friedelmeyer, J.P. 1994. *Le calcul des dérivations d'Arbogast dans le projet d'algébrisation de l'analyse à fin du xviiie siècle*. Nantes, France: U. Nantes.

Frobenius, G. 1878. Über lineare Substitutionen und bilineare Formen. *J. Reine Angew. Math*, **84**, 1–63.

Frobenius, G. 1880. Ueber die Leibnizsche Reihe. *J. Reine Angew. Math.*, **89**, 262–264.

Fuss, P.H. (ed). 1968. *Correspondance mathématique et physique*. New York: Johnson Reprint.

Galois, É. 1830. Sur la théorie des nombres. *Bull. Sci. Math. Phys. Chem.*, **13**, 428–435.

Galois, É. 1897. *Oeuvres mathématiques*. Paris: Gauthier-Villars.

Garabedian, P., and Schiffer, M. 1955. A proof of the Bieberbach conjecture for the fourth coefficient. *J. Rational Mech. Anal.*, **4**, 427–465.

Gårding, L. 1994. *Mathematics in Sweden before 1950*. Providence: A.M.S.

Garsia, A.M., and Milne, S.C. 1981. A Rogers-Ramanujan bijection. *J. Combin. Theory*, **31**, 289–339.

Gauss, C.F. 1813. *Disquisitiones generales circa–seriem infinitam*. Gottingen: Comm. Soc. Reg. Gott: II. Reprinted in *Werke*, vol. 3, pp. 123–162.

Gauss, C.F. 1815. *Methodus nova integralium valores per approximationem inveniendi*. Göttingen: Dieterich. This monograph was presented to the Göttingen Society in 1814. It was reprinted in Gauss's *Werke* vol. 3, pp. 163–196.

Gauss, C.F. 1863–1927. *Werke*. Leipzig: Teubner.

Gauss, C.F. 1965. *Disquisitiones Arithmeticae (An English Translation)*. New Haven, Conn.: Yale University Press. Translated by A.A. Clarke.

Gauss, C.F. 1981. *Arithmetische Untersuchungen*. New York: Chelsea.

Gautschi, W. 1986. My involvement in de Branges's proof. Pages 205–211 of: Baernstein, A. (ed), *The Bieberbach Conjecture*. Providence: A.M.S.

Gegenbauer, L. 1884. Zur Theorie der Functionen $C_n^\nu(x)$. *Denkschriften Akad. Wiss. Wien, Math. Naturwiss. Klasse*, **48**, 293–316.

Gelfand, I.M. 1988. *Collected Papers*. New York: Springer-Verlag.

Gelfand, I.M., Kapranov, M.M., and Zelevinsky, A.V. 1994. *Discriminants, Resultants, and Multidimensional Determinants*. Boston: Birkhäuser.

Gelfond, A.O. 1934. Sur le septième problème de Hilbert. *Dok. Akad. Nauk SSSR*, **2**, 1–6.

Gelfond, A.O. 1939. In Russian: On the approximation by algebraic numbers of the ratio of the logarithms of two algebraic numbers. *Izvestia Akad. Nauk. SSSR*, **5–6**, 509–518.

Gelfond, A.O. 1960. *Transcendental and Algebraic Numbers*. New York: Dover. Translated by L. Boron.

Gelfond, A.O., and Linnik, Yu.V. 1966. *Elementary Methods in the Analytic Theory of Numbers*. Cambridge, MA: MIT Press. Translated by D.E. Brown.

Georgiadou, M. 2004. *Constantine Carathéodory*. New York: Springer-Verlag.

Gerhardt, K.I. (ed). 1899. *Der Briefwechsel von Gottfried Wilhelm Leibniz mit Mathematikern*. Berlin: Mahler and Müller.

Girard, A. 1884. *Invention nouvelle en l'algebre*. Leiden: Haan.

Glaisher, J.W.L. 1878. Series and products for π and powers of π. *Messenger Math.*, **7**, 75–80.

Glaisher, J.W.L. 1883. A theorem in partitions. *Messenger Math.*, **12**, 158–170.

Goethe, N., Beeley, P., and Rabouin, D. (eds). 2015. *G.W. Leibniz, Interrelation between Mathematics and Philosophy*. Dordrecht: Springer.

Goldstein, C., Schappacher, N., and Schwermer, J. (eds). 2007. *The Shaping of Arithmetic after C.F. Gauss's Disquisitiones Arithmeticae*. New York: Springer.

Goldstein, L. J. 1973. A history of the prime number theorem. *Am. Math. Monthly*, **80**, 599–615. Correction, p. 1115. Reprinted in Anderson, Katz, and Wilson (2009), pp. 318–327.

Goldstine, H.H. 1977. *A History of Numerical Analysis*. New York: Springer-Verlag.

Gong, S. 1999. *The Bieberbach Conjecture*. Providence: A.M.S.

Gonzalez-Velasco, E.A. 2011. *Journey through Mathematics*. New York: Springer.

Good, I.J. 1970. Short proof of a conjecture of Dyson. *J. Math. Phys.*, **11**, 1884.

Gordan, P. 1868. Beweis, dass jede Covariante und Invariante einer binären Form eine ganze Function mit numerischen Coefficienten einer endlichen Anzahl solcher Formen ist. *J. Reine Angew. Math.*, **69**, 323–354.

Gordon, B. 1961. A combinatorial generalization of the Rogers-Ramanujan identities. *Amer. J. Math.*, **83**, 393–399.

Gouvêa, F.Q. 1994. A marvelous proof. *Am. Math. Monthly*, **101**, 203–222.

Gowing, R. 1983. *Roger Cotes*. New York: Cambridge University Press.

Grabiner, J.V. 1981. *The Origins of Cauchy's Rigorous Calculus*. Cambridge, MA: MIT Press.

Grabiner, J.V. 1990. *The Calculus as Algebra*. New York: Garland Publishing.

Grace, J.H., and Young, A. 1965. *The Algebra of Invariants*. New York: Chelsea.

Graham, G., Rothschild, B.L., and Spencer, J.H. 1990. *Ramsey Theory*. New York: Wiley.

Grattan-Guinness, I. 1972. *Joseph Fourier 1768–1830*. Cambridge, MA: MIT Press.

Grattan-Guinness, I. (ed). 2005. *Landmark Writings in Western Mathematics*. Amsterdam: Elsevier.

Graves, R.P. 1885. *Life of Sir William Rowan Hamilton*. London: Longmans.

Gray, J. 1986. *Linear Differential Equations and Group Theory from Riemann to Poincaré*. Boston: Birkhäuser.

Gray, J. 1994. On the history of the Riemann mapping theorem. *Rendiconti Circolo Matematico Palermo*, **34**, 47–94.

Gray, J., and Parshall, K.H. (eds). 2007. *Episodes in the History of Modern Algebra (1800–1950)*. Providence: A.M.S.

Green, G. 1970. *Mathematical Papers*. New York: Chelsea. Edited by N.M. Ferrers.

Greenberg, J.L. 1995. *The Problem of the Earth's Shape from Newton to Clairaut*. Cambridge: Cambridge University Press.

Gregory, D.F. 1865. *The Mathematical Writings of Duncan Farquharson Gregory*. Cambridge: Deighton, Bell. Edited by W. Walton.

Gregory, J. 1668. *Exercitationes Geometricae*. London: Godbid.

Grinshpan, A.Z. 1972. Logarithmic coefficients of functions in the class S, English translation. *Siberian Math J.*, **13**, 793–801.

Gronwall, T.H. 1913. On the degree of convergence of Laplace's series. *Trans. A.M.S.*, **15**, 1–30.

Gronwall, T.H. 1914. Some remarks on conformal representation. *Ann. Math.*, **16**, 72–76.

Grootendorst, A.W., and van Maanen, J.A. 1982. Van Heuraet's letter (1659) on the rectification of curves. *Nieuw Archief Wiskunde*, **30**(3), 95–113.

Grunsky, H. 1939. Koeffizientenbedingungen für schlicht abbildende meromorphe Funktionen. *Math. Zeit.*, **45**, 29–61.

Gudermann, C. 1838. Theorie der Modular-Functionen und der Modular-Integrale. *J. Reine Angew. Math.*, **18**, 220–258.

Guicciardini, N. 1989. *The Development of Newtonian Calculus in Britain 1700–1800*. Cambridge: Cambridge University Press.

Gunson, J. 1962. Proof of a conjecture of Dyson in the statistical theory of energy levels. *J. Math. Physics*, **3**, 752–753.

Gupta, R.C. 1969. Second order interpolation in Indian mathematics up to the fifteenth century. *Ind. J. Hist. Sci.*, **4**, 86–98.

Gupta, R.C. 1977. Paramesvara's rule for the circumradius of a cyclic quadrilateral. *Hist. Math.*, **4**, 67–74.

Hadamard, J. 1893. Étude sur les propriétés des fonctions entières et en particulier d'une fonction considerée par Riemann. *J. Math. Pures Appl.*, **4**, 171–215.

Hadamard, J. 1896. Sur la distribution des zéros de la fonction $\gamma(s)$ et ses conséquences arithmétiques. *Bull. Soc. Math. France*, **24**, 199–220.

Hadamard, J. 1899. Théorème sur séries entières. *Acta Math.*, **22**, 55–64.

Haimo, D.T. 1968. *Orthogonal Expansions and Their Continuous Analogues*. Carbondale: Southern Illinois University Press.

Hald, A. 1990. *A History of Probability and Statistics and Their Applications Before 1750*. New York: Wiley.

Halley, E. 1695. A most compendious method for constructing the logarithms ... *Phil. Trans.*, **19**, 58–67.

Hamel, G. 1905. Eine Basis aller Zahlen und die unstetiggen Lösungen der Funktionalgleichung: $f(x + y) = f(x) + f(y)$. *Math. Ann.*, **60**, 459–462.

Hamilton, W.R. 1835. *Theory of Conjugate Functions or Algebraic Couples*. Dublin: Philip Dixon Hardy.

Hamilton, W.R. 1945. Quaternions. *Proc. Roy. Irish Acad.*, **50**, 89–92.

Hankel, H. 1864. Die Eulerschen Integrale bei unbeschränkter Variabilität des Arguments. *Zeit. Math. Phys.*, **9**, 1–21.

Hardy, G.H. 1905. *The Integration of Functions of a Single Variable*. Cambridge: Cambridge University Press.

Hardy, G.H. 1929. Prolegomena to a chapter on inequalities. *J. London Math. Soc.*, **1**, 61–78.

Hardy, G.H. 1937. *A Course in Pure Mathematics*. Cambridge: Cambridge University Press.

Hardy, G.H. 1949. *Divergent Series*. Oxford: Clarendon.

Hardy, G.H. 1966–79. *Collected Papers*. Oxford: Clarendon.

Hardy, G.H. 1978. *Ramanujan*. New York: Chelsea.

Hardy, G.H., and Heilbronn, H. 1914. Edmund Landau. *J. London Math. Soc.*, **13**, 302–310.

Hardy, G.H., and Littlewood, J.E. 1913. Tauberian theorems concerning power series of positive terms. *Messenger Math.*, **42**, 191–192.

Hardy, G.H., and Littlewood, J.E. 1914. Tauberian theorems concerning power series and Dirichlet series whose coefficients are positive. *Proc. London Math. Soc.*, **13**, 174–191.

Hardy, G.H., and Littlewood, J.E. 1918. Contributions to the theory of the Riemann zeta-function and the theory of the distribution of primes. *Acta Math.*, **41**, 119–196.

Hardy, G.H., and Littlewood, J.E. 1921. On a Tauberian theorem for Lambert's series, and some fundamental theorems in the analytic theory of numbers. *Proc. London Math. Soc.*, **19**, 21–29.

Hardy, G.H., and Littlewood, J.E. 1922. Some problems of 'partitio numerorum' IV. *Math. Zeit.*, **12**, 161–188.

Hardy, G.H., Littlewood, J., and Pólya, G. 1967. *Inequalities*. Cambridge: Cambridge University Press.

Harkness, J., and Morley, F. 1898. *Introduction to the Theory of Analytic Functions*. London: Macmillan.

Harriot, T., and Stedall, J. 2003. *The Greate Invention of Algebra: Thomas Harriot's Treatise on Equations*. Oxford: Oxford University Press. Introduction and commentary by J. Stedall.

Harriot, T., Beery, J., and Stedall, J. 2009. *Thomas Harriot's Doctrine of Triangular Numbers: 'The Magisteria Magna.'* Zürich: European Math. Soc. Extensive background and commentary by Beery and Stedall.

Hasse, H., and Davenport, H. 2014. *Manuskripte Hasse-Davenport*. Heidelberg: Heidelberg University Communication between Davenport and Hasse: heidelberg.de.

Hawking, S. 2005. *God Created the Integers*. Philadelphia: Running Press.

Hawkins, T. 1975. *Lebesgue Theory*. New York: Chelsea.

Hayman, W.K. 1964. *Meromorphic Functions*. Oxford: Clarendon.

Hayman, W.K. 1994. *Multivalent Functions*. Cambridge: Cambridge University Press.

Heegner, K. 1952. Diophantische Analysis und Modulfunktionen. *Math. Zeit.*, **56**, 227–253.

Heine, E. 1847. Untersuchungen über die Reihe ... *J. Reine Angew. Math.*, **34**, 285–328.

Heine, E. 1870. Über trigonometrische Reihen. *J. Reine Angew. Math.*, **71**, 353–365.

Heine, E. 1872. Die Elemente der Functionenlehre. *J. Reine Angew. Math.*, **74**, 172–188.

Heine, E. 1878. *Handbuch der Kugelfunctionen*. Berlin: Reimer.

Hellegourarch, Y. 2002. *Invitation to the Mathematics of Fermat-Wiles*. London: Acad. Press. Translated by Leila Schneps.

Hermann, J. 1716. De variationibus chordarum tensarum disquisitio, ... *Acta Erud.*, 370–377.

Hermann, J. 1719. Solution duorum problematum ... *Acta Erud.*, **August, 1719**, 351–361.

Hermite, C. 1848. Note sur la théorie des fonctions elliptiques. *Cam. and Dub. Math. J.*, **3**, 54–56.

Hermite, C. 1873. *Cours d'analyse*. Paris: Gauthier-Villars.

Hermite, C. 1891. *Cours de M. Hermite (rédigé en 1882 par M. Andoyer)*. Ithaca: Cornell University Library Reprint.

Hermite, C. 1905–1917. *Oeuvres*. Paris: Gauthier-Villars. Edited by É. Picard.

Herschel, J.F.W. 1820. *A Collection of Examples of the Applications of the Calculus of Finite Differences*. Cambridge: Cambridge University Press.

Hewitt, E., and Hewitt, R. E. 1980. The Gibbs-Wilbraham phenomenon. *Archive Hist. Exact Sci.*, **21**, 129–160.

Hickerson, D. 1988. A proof of the mock theta conjectures. *Inventiones Math.*, **94**, 639–660.

Hilbert, D. 1893. Über die Transzendenz der Zahlen e und π. *Gött. Nachr.*, 113–116.

Hilbert, D. 1897. Zum Gedächtnis an Karl Weierstrass. *Gött. Nach.*, 60–69.

Hilbert, D. 1906. Grundzüge einer allgemeinen Theorie der linearen Integralgleichungen, vierte Mitteilung. *Gött. Nach.*, 157–227.

Hilbert, D. 1970. *Gesammelte Abhandlungen*. Berlin: Springer.

Hilbert, D. 1978. *Hilbert's Invariant Theory Papers*. Brookline, MA: Math. Sci. Press. Translated by M. Ackerman with Commentary by R. Hermann.

Hilbert, D. 1993. *Theory of Algebraic Invariants*. New York: Cambridge University Press. Translated by R. C. Laubenbacher.

Hindenburg, C.F. 1778. *Methodus nova et facilis serierum infinitarum exhibendi dignitates exponentis indeterminati*. Leipzig: Langenhem.

Hobson, E.W. 1957a. *The Theory of Functions of a Real Variable*. New York: Dover.

Hobson, E.W. 1957b. *A Treatise on Plane and Advanced Trigonometry*. New York: Dover.

Hoe, J. 2007. *A Study of the Jade Mirror of the Four Unknowns*. Christchurch, N.Z.: Mingming Bookroom.

Hofmann, J.E. 1974. *Leibniz in Paris*. Cambridge: Cambridge University Press. Translated by A. Prag and D.T. Whiteside.

Hofmann, J.E. 1990. *Ausgewählte Schriften*. Zürich: Georg Olms Verlag. Edited by C. Scriba.

Hofmann, J.E. 2003. *Classical Mathematics*. New York: Barnes and Noble.

Hölder, O. 1889. Über einen Mittelwertsatz. *Gött. Nach.*, 38–47.

Horiuchi, A.T. 1994. The *Tetsujutsu sankei* (1722), an 18th century treatise on the methods of investigation in mathematics. Pages 149–164 of: Sasaki, C., Sugiura, M., and Dauben, J.W. (eds), *The Intersection of History and Mathematics*. Basel: Birkhäuser.

Horowitz, D. 1978. A further refinement for coefficient estimates of univalent functions. *Proc. A.M.S.*, **71**, 217–221.

Horowitz, E. 1969. *Algorithms for Symbolic Integration of Rational Functions*. PhD thesis, University of Wisconsin, Madison.

Hua, L.K. 1981. *Starting with the Unit Circle*. New York: Springer-Verlag. Translated by K. Weltin.

Hua, L.K., and Vandiver, H.S. 1948. On the existence of solutions of certain equations in a finite field. *Proc. Nat. Acad. Sci. USA*, **34**, 258–263.

Hua, L.K., and Vandiver, H.S. 1949. On the existence of solutions of certain equations in a finite field. *Proc. Nat. Acad. Sci. USA*, **35**, 94–99.

Hutton, C. 1812. *Tracts on mathematical and philosophical subjects*. London: Rivington.

Ireland, K., and Rosen, M. 1982. *A Classical Introduction to Modern Number Theory*. New York: Springer-Verlag.

Ivory, J. 1796. A new series for the rectification of the ellipses. *Trans. Roy. Soc. Edinburgh*, **4**, 177–190.

Ivory, J. 1812. On the attractions of an extensive class of spheroids. *Phil. Trans. Roy. Soc. London,* **102**, 46–82.

Ivory, J. 1824. On the figure requisite to maintain the equilibrium of a homogeneous fluid mass that revolves upon an axis. *Phil. Trans. Roy. Soc. London,* **114**, 85–150.

Ivory, J., and Jacobi, C.G.J. 1837. Sur le développement de $(1 - 2xz + z^2)^{-1/2}$. *J. Math. Pures App.,* **2**, 105–106.

Jackson, F.H. 1910. On q-definite integrals. *Quart. J. Pure App. Math.,* **41**, 193–203.

Jacobi, C.G.J. 1826. Ueber Gauss' neue Methode, die Werthe der Integrale näherungsweise zu finden. *J. Reine Angew. Math.,* **1**, 301–308. Reprinted in Jacobi (1969) vol. 6, pp. 3–11.

Jacobi, C.G.J. 1833. Demonstratio Formulae. *J. Reine Angew. Math.,* **11**, 307. Reprinted in Jacobi (1969) vol. 6, pp. 62–63.

Jacobi, C.G.J. 1834. De usu legitimo formulae summatoriae Maclaurinianae. *J. Reine Angew. Math.,* **12**, 263–272. Reprinted in Jacobi (1969) vol. 6, pp. 64–75.

Jacobi, C.G.J. 1837. Über die Kreistheilung und ihre Anwendung auf die Zahlentheorie. *Monats. Akad. Wiss. Berlin,* 127–136. Reprinted in 1846 in Crelle's Journal; in Jacobi's Werke vol. 6, pp. 254–274; French translation published in 1856 in *Nouv. Ann. Math.*

Jacobi, C.G.J. 1841. De determinantibus functionalibus. *J. Reine Angew. Math.,* **22**, 319–359. Reprinted in Jacobi (1969) vol. 3, pp. 393–438.

Jacobi, C.G.J. 1846. Über einige der Binomialreihe Analoge Reihen. *J. Reine Angew. Math,* **32**, 197–204. Reprinted in Jacobi (1969) vol. 6, pp. 163–173.

Jacobi, C.G.J. 1847. De seriebus ac differentiis observatiunculae. *J. Reine Angew. Math.,* **36**, 135–142. Republished in Jacobi (1969) vol. 4, pp. 174–182.

Jacobi, C.G.J. 1859. Untersuchungen über die Differentialgleichung der hypergeometrische Reihe. *J. Reine Angew. Math,* **56**, 149–165. Reprinted in Jacobi (1969) vol. 6, pp. 184–202.

Jacobi, C.G.J. 1969. *Mathematische Werke.* New York: Chelsea.

Jahnke, H.N. 1993. Algebraic analysis in Germany, 1780–1849: Some mathematical and philosophical issues. *Hist. Math.,* **20**, 265–284.

James, I.M. 1999. *History of Topology.* Amsterdam: Elsevier.

Jensen, J. 1899. Sur un nouvel et important théorèm de la théorie des fonctions. *Acta Math.,* **22**, 359–364.

Jensen, J. 1906. Sur les fonctions convexes et les inégalités entre les valeurs moyennes. *Acta Math.,* **30**, 175–193.

Johnson, W.P. 2002. The curious history of Faà di Bruno's formula. *Am. Math. Monthly,* **109**, 217–234.

Johnson, W.P. 2007. The Pfaff/Cauchy derivative and Hurwitz type extensions. *Ramanujan J. Math.,* **13**, 167–201.

Jordan, C. 1979. *Calculus of Finite Differences.* New York: Chelsea.

Jyesthadeva. 2009. *Ganita-Yukti-Bhasa of Jyesthadeva, 2 volumes.* New Delhi: Hindustan Book Agency and Springer. Jyesthadeva's text translated from Malayalam into English by K. V. Sarma, with notes by editors K. Ramasubramanian, M.D. Srinivas, and M.S. Sriram.

Kac, M. 1936–1937. Une remarque sur les équations fonctionnelles. *Comment. Math. Helv.,* **9**, 170–171.

Kac, M. 1965. A Remark on Wiener's Tauberian Theorem. *Proc. A.M.S.,* **16**, 1155–1157.

Kac, M. 1979. *Selected Papers.* Cambridge, MA: MIT Press.

Kac, M. 1987. *Enigmas of Chance: An Autobiography.* Berkeley: U. Calif. Press.

Kalman, D. 2009. *Polynomia and Related Realms.* Washington, D.C.: M.A.A.

Karamata, J. 1930. Über die Hardy-Littlewoodschen Umkehrungen des Abelschen Stetigkeitssatzes. *Math. Zeit.,* **32**, 319–320.

Katz, N.M. 1976. An overview of Deligne's proof of the Riemann hypothesis for varieties over finite fields. Pages 275–305 of: Browder, F.E. (ed), *Mathematical Developments Arising from Hilbert Problems.* Providence: A.M.S.

Katz, V.J. 1979. The history of Stokes' theorem. *Math. Mag.,* **52**, 146–156.

Katz, V.J. 1982. Change of variables in multiple integrals: Euler to Cartan. *Math. Mag.,* **55**, 3–11.

Katz, V.J. 1985. Differential forms—Cartan to de Rham. *Archive Hist. Exact Sci.*, **33**, 307–319.

Katz, V.J. 1987. The calculus of the trigonometric functions. *Hist. Math.*, **14**, 311–324.

Katz, V.J. 1995. Ideas of calculus in Islam and India. *Math. Mag.*, **3**(3), 163–174. Reprinted in Anderson, Katz, and Wilson (2004), pp. 122–130.

Katz, V.J. 1998. *A History of Mathematics: An Introduction*. Reading, MA: Addison-Wesley.

Khinchin, A.Y. 1998. *Three Pearls of Number Theory*. New York: Dover.

Khrushchev, S. 2008. *Orthogonal Polynomials and Continued Fractions*. Cambridge: Cambridge University Press.

Kichenassamy, S. 2010. Brahmagupta's derivation of the area of a cyclic quadrilateral. *Hist. Math.*, **37**(1), 28–61.

Kikuchi, D. (ed). 1891. *Memoirs on Infinite Series*. Tokyo: Tokio Math. Phys. Soc. Translation by D. Kikuchi.

Kinkelin, H. 1861–1862. Allgemeine Theorie der harmonischen Reihen, mit Anwendung auf die Zahlentheorie. *Programm der Gewerbeschule Basel*, 1–32.

Klein, F. 1911. *Lectures on Mathematics*. New York: Macmillan.

Klein, F. 1933. *Vorlesungen über die hypergeometrische Funktion*. Berlin: Springer.

Klein, F. 1979. *Development of Mathematics in the 19th Century*. Brookline, MA: Math. Sci. Press. Translated by M. Ackerman.

Knoebel, A., Laubenbacher, R., Lodder, J., and Pengelley, D. 2007. *Mathematical Masterpieces*. New York: Springer.

Knopp, K. 1990. *Theory and Application of Infinite Series*. New York: Dover.

Knuth, D. 1992. Two notes on notation. *Am. Math. Monthly*, **99**, 403–422. Reprinted as chapter 2 in Knuth (2003). Addendum 'Stirling Numbers' published in the *Monthly*, vol. 102, p. 562.

Knuth, D. 1993. Johann Faulhaber and Sums of Powers. *Math. of Computation*, **61**, 277–294.

Knuth, D. 2003. *Selected Papers*. Stanford, CA: Center for the Study of Language and Information (CSLI),.

Knuth, D. 2011. *The Art of Computer Programming, vol. 4A: Combinatorial Algorithms, part 1*. New York: Addison-Wesley.

Koebe, P. 1907–1908. Über die Uniformisierung beliebiger analytischer Kurven. *Gött. Nach.*, 1907: 191–210, 633–649; 1908: 337–358.

Kolmogorov, A.N. 1923. Une série de Fourier-Lebesgue divergente presque partout. *Fund. Math.*, **4**, 324–328.

Kolmogorov, A.N., and Yushkevich, A.P. (eds). 1998. *Mathematics of the 19th Century: Vol. III: Function Theory According to Chebyshev; Ordinary Differential Equations; Calculus of Variations; Theory of Finite Differences*. Basel: Birkhäuser.

Koppelman, E. 1971. The calculus of operations and the rise of abstract algebra. *Archive Hist. Exact Sci.*, **8**, 155–242.

Korevaar, J. 2004. *Tauberian Theory*. New York: Springer.

Kramp, C. 1796. Coefficient des allgemeinen Gliedes jeder willkürlichen Potenz eines Infinitinomiums ... Pages 91–122 of: Hindenburg, C. (ed), *Der polynomische Lehrsatz*. Leipzig: Fleischer.

Kronecker, L. 1968. *Mathematische Werke*. New York: Chelsea. Edited by K. Hensel.

Kubota, K.K. 1977. Linear functional equations and algebraic independence. Pages 227–229 of: Baker, A., and Masser, D.W. (eds), *Transcendence Theory*. New York: Academic Press.

Kummer, E.E. 1836. Über die hypergeometrische Reihe. *J. Reine Angew. Math.*, **15**, 39–83, 127–172.

Kummer, E.E. 1840. Über die Transcendenten, welche aus wiederholten Integrationen rationaler Formeln entstehen. *J. Reine Angew. Math.*, **21**, 74–90, 193–225, 328–371.

Kummer, E.E. 1847. Beitrag zur Theorie der Function $\Gamma(x)$. *J. Reine Angew. Math.*, **35**, 1–4.

Kummer, E.E. 1975. *Collected Papers*. Berlin: Springer-Verlag. Edited by A. Weil.

Kung, J. (ed). 1995. *Gian-Carlo Rota on Combinatorics*. Boston: Birkhäuser.

Kuzmin, R. 1930. In Russian: On a new class of transcendental numbers. *Izvestia Akad. Nauk. SSSR*, **3**, 583–597.

Kuzmin, R. 1938. On the transcendental numbers of Goldbach. *Trudy Leningrad Indust. Inst.*, **1–5**, 28–32.

Lacroix, S., Babbage, C., Herschel, J., and Peacock, G. 1816. *An elementary treatise on the Differential and Integral Calculus.* Cambridge: Cambridge University Press.

Lacroix, S.F. 1800. *Traité des différences et des séries.* Paris: Duprat.

Lacroix, S.F. 1819. *Traité du calcul différentiel et du calcul intégral.* Vol. 3. Paris: Courcier.

Lagrange, J.L. 1770–1771. Réflexions sur la résolution algébraic des équations. *Nouv. Mém. Acad. Roy. Sci. Berlin,* 1770 [1772]: 134–215; 1771 [1773]: 138–253. Reprinted in Lagrange's *Oeuvres,* vol. 3, pp. 205–421.

Lagrange, J.L. 1771. Démonstration d'un théorèm nouveau concernant les nombres premiers. *Nouv. Mém. Acad. Roy. Sci. Belles-Let., 1771,* 125–137. Reprinted in *Oeuvres,* vol. 3, pp. 425–438.

Lagrange, J.L. 1772. Sur une nouvelle espèce de calcul relatif à la différentiation et à l'intégration des quantités variables. *Nouv. Mém. Acad. Roy. Sci. Belle-lettres Berlin,* 441–476.

Lagrange, J.L. 1781. Sur la construction des cartes géographique. *Nouv. Mém. Acad. Roy. Sci. Berlin,* 161–210. Reprinted in Lagrange (1867–1892) vol. 4, pp. 637–692.

Lagrange, J.L. 1797. *Théorie des fonctions analytiques.* Paris: Imprim. de la Répub.

Lagrange, J.L. 1867–1892. *Oeuvres.* Paris: Gauthier-Villars. Edited by J. Serret.

Laguerre, E. 1882. Sur quelques équations transcendantes. *Comptes Rendus,* **94,** 160–163.

Laguerre, E. 1972. *Oeuvres.* New York: Chelsea. Edited by C. Hermite, H. Poincaré, and E. Rouché.

Lambert, J.H. 1768. Mémoire sur quelques propriétés remarquables des quantités transcendentes circulaires et logarithmiques. *Hist. Akad. Berlin, 1761,* **17,** 265–322. See Lambert's *Opera mathematica* (1946, 1948), Zurich: Füssli, vol. I, pp. 194–212 and vol. 2, pp. 112–159.

Landau, E. 1903. Neuer Beweis des Primzahlsatzes und Beweis des Primidealsatzes. *Math. Ann.,* **56,** 645–670.

Landau, E. 1904. Über eine Verallgemeinerung des Picardschen Satzes. *S.B. Preuss. Akad. Wiss.,* **38,** 1118–1133.

Landau, E. 1906. Euler und die Funktionalgleichung der Riemannschen Zetafunktion. *Bib. Math.,* **7,** 69–79.

Landau, E. 1907a. Sur quelques généralisations du théorème de M. Picard. *Ann. École Norm. Sup.,* **24,** 179–201.

Landau, E. 1907b. Über die Konvergenz einiger Klassen von unendlichen Reihen am Rànde des Konvergenzgebietes. *Monatsh. Math. Phys.,* **18,** 8–28.

Landau, E. 1907c. Über einen Konvergenzsatz. *Gött. Nach.,* **8,** 25–27.

Landau, E. 1910. Über die Bedeutung einiger neuen Grentzwertsätze der Herren Hardy und Axer. *Prace Mat.-Fiz.,* **21,** 97–177.

Landen, J. 1758. *A Discourse Concerning The Residual Analysis.* London: Nourse.

Landen, J. 1760. A new method of computing the sums of certain series. *Phil. Trans. Roy. Soc. London,* **51,** 553–565.

Landen, J. 1771. A Disquisition concerning Certain Fluents, which are Assignable by the Arcs of the Conic Sections ... *Phil. Trans.,* **61,** 298–309.

Landen, J. 1775. An Investigation of a General Theorem for Finding the Length of Any Arc of Any Conic Hyperbola ... *Phil. Trans.,* **65,** 283–289.

Landis, E.M. 1993. About mathematics at Moscow State University in the late 1940s and early 1950s. Pages 55–73 of: Zdravkovska, S., and Duren, P. (eds), *Golden Years of Moscow Mathematics.* Providence: A.M.S.

Lanzewizky, I.L. 1941. Über die orthogonalität der Fejér-Szegőschen polynome. *D. R. Dokl. Acad. Sci. URSS,* **31,** 199–200.

Laplace, P.S. 1782. Mémoire sur les suites. *Mém. Acad. Roy. Sci. Paris,* 207–309.

Laplace, P.S. 1812. *Théorie analytique des probabilités.* Paris: Courcier.

Laplace, P.S. 1814. *Théorie analytique des probabilités, seconde édition, revue et augmentée par l'auteur.* Paris: Courcier.

Lascoux, A. 2003. *Symmetric Functions and Combinatorial Operators on Polynomials.* Providence: A.M.S.

Laudal, O.A., and Piene, R. (eds). 2002. *The Legacy of Niels Henrik Abel.* Berlin: Springer.

Laugwitz, D. 1999. *Bernhard Riemann 1826–1866.* Boston: Birkhäuser. Translated by A. Shenitzer.

Lebedev, N.A., and Milin, I.M. 1965. In Russian: An inequality. *Vestnik Leningrad University*, **20**, 157–158.

Lebesgue, H. 1902. Intégrale, longueur, aire. *Annali Mat. Pura. App.*, **7**, 231–359.

Lebesgue, H. 1906. *Leçons sur les séries trigonométriques*. Paris: Gauthier-Villars.

Lebesgue, V.A. 1837. Researches sur les nombres. *J. Math Pures Appl.*, **2**, 253–292.

Lebesgue, V.A. 1838. Researches sur les nombres. *J. Math Pures Appl.*, **3**, 113–144.

Legendre, A.M. 1794. *Éléments de Géométrie avec des notes*. Paris: Didot.

Legendre, A.M. 1811–1817. *Exercices de calcul intégral*. Paris: Courcier.

Legendre, A.M. 1825–1828. *Traité des fonctions elliptiques*. Paris: Huzard-Courcier.

Legendre, A.M. 1830. *Théorie des nombres*. Paris: Didot.

Legendre, A.M. 1885. Recherches sur l'attraction des sphéroïdes homogènes. *Mém. Math. Phys. prés. à Acad. Roy. Sci.*, **10**, 411–435.

Leibniz, G.W. 1684. Nova methodus pro maximis et minimis, itemque tangentibus, ... *Acta Erud.*, **3**, 467–473. Reprinted in Leibniz (1971) vol. 5, pp. 220–225.

Leibniz, G.W. 1713. Epistola ad V. Cl. Christianium Wolfium. *Acta Erud.*, **Supp. 5, 1713**, 264–270.

Leibniz, G.W. 1920. *The Early Mathematical Manuscripts of Leibniz*. Chicago: Open Court. Translated with extensive notes by J.M. Child.

Leibniz, G.W. 1971. *Mathematische Schriften*. Hildesheim, Germany: Georg Olms Verlag. edited by K. Gerhardt.

Leibniz, G.W., and Bernoulli, Joh. 1745. *Commercium philosophicum et mathematicum*. Geneva: Bosquet.

Leibniz, G.W., and Knobloch, E. 1993. *De quadratura arithmetica circuli ellipseos et hyperbolae cujus corollarium est trigonometria sine tabulis*. Göttingen: Vandenhoeck and Ruprecht. Presented with commentary by E. Knobloch; translation into German by O. Hamborg.

Lemmermeyer, F. 2000. *Reciprocity Laws*. New York: Springer-Verlag.

Lewin, L. 1981. *Polylogarithms and Associated Functions*. Amsterdam: Elsevier.

Li, Y., and Du, S. 1987. *Chinese mathematics: A concise history*. Oxford: Clarendon. Translated by J. Crossley and A. Lun.

Lindelöf, E. 1902. Mémoire sur la théorie des fonctions entières de genre fini. *Acta Soc. Sci. Fennicae*, **31**, 1–79.

Lindemann, F. 1882. Über die Zahl π. *Math. Ann.*, **20**, 213–225.

Liouville, J. 1832. Sur quelques questions de géométrie et de mécanique, et sur un nouveau genre de calcul pour résoudre ces questions. *J. École Polytech.*, **13**, 1–69.

Liouville, J. 1837a. Note sur le développement de $(1 - 2xz + z^2)^{-1/2}$. *J. Math. Pures App.*, **2**, 135–139.

Liouville, J. 1837b. Sur la sommation d'une série. *J. Math. Pures Appl.*, **2**, 107–108.

Liouville, J. 1839. Note sur quelques intégrales définies. *J. Math. Pures Appl*, **4**, 225–235.

Liouville, J. 1841. Remarques nouvelles sur l'équation de Riccati. *J. Math. Pures Appl.*, **6**, 1–13.

Liouville, J. 1851. Sur des classes très étendues de quantités dont la valeur n'est ni algébrique, ni même réductible à des irrationnelles algébriques. *J. Math. Pures Appl.*, **16**, 133–142.

Liouville, J. 1857. Sur l'expression $\phi(n)$, qui marque combien la suite $1, 2, 3, \ldots, n$ contient de nombres premiers à n. *J. Math. Pures Appl.*, **2**, 110–112.

Liouville, J. 1880. Leçons sur les fonctions doublement périodiques. *J. Reine Angew. Math.*, **88**, 277–310.

Lipschitz, R. 1889. Untersuchungen der Eigenschaften einer Gattung von unendlichen Riehen. *J. Reine Angew. Math.*, **105**, 127–156.

Littlewood, J.E. 1911. The converse of Abel's theorem on power series. *Proc. London Math. Soc.*, **9**, 434–448.

Littlewood, J.E. 1925. On inequalities in the theory of functions. *Proc. London Math. Soc.*, **23**, 481–519. Reprinted in Littlewood (1982) vol. 2, pp. 963–1004.

Littlewood, J.E. 1982. *Collected Papers*. Oxford: Clarendon.

Littlewood, J.E. 1986. *Littlewood's Miscellany*. Cambridge: Cambridge University Press.

Littlewood, J.E., and Paley, R. 1932. A proof that an odd schlicht function has bounded coefficients. *J. London Math. Soc.*, **7**, 167–169. Reprinted in Littlewood (1982) vol. 2, pp. 1046–1048.

Loewner, C. 1988. *Collected Papers*. Boston: Birkhäuser. Edited by L. Bers.

Lototskii, A.V. 1943. Sur l'irrationalité d'un produit infini. *Mat. Sb.*, **12**, 262–272.

Lusin, N. 1913. Sur la convergence des séries trigonométriques de Fourier. *Comptes Rendus.*, **156**, 1655–1658.

Lützen, J. 1990. *Joseph Liouville*. New York: Springer-Verlag.

Maclaurin, C. 1729. A second letter to Martin Folkes, Esq.: Concerning the roots of equations, with the demonstration of other rules in algebra. *Phil. Trans. Roy. Soc.*, **36**, 59–96.

Maclaurin, C. 1742. *A Treatise of Fluxions*. Edinburgh: Ruddimans.

Maclaurin, C. 1748. *A Treatise of Algebra*. London: Millar and Nourse.

Macmahon, P. 1915–1916. *Combinatory Analysis*. Cambridge: Cambridge University Press.

MacMahon, P. A. 1978. *Collected Papers*. Cambridge, MA: MIT Press. Edited by G. Andrews.

Mahlburg, K. 2005. Partition congruences and the Andrews-Garvan-Dyson crank. *Proc. Nat. Acad. Sci. USA*, **102 (43)**, 15373–15376.

Mahler, K. 1982. Fifty years as a mathematician. *J. Number Theory*, **14**, 121–155. Corrected version on the online Kurt Mahler Archive.

Mahnke, D. 1912–1913. Leibniz auf der Suche nach einer allgemeinen Primzahlgleichung. *Bib. Math.*, **13**, 29–61.

Mahoney, M.S. 1973. *The Mathematical Career of Pierre de Fermat (1601–1665)*. Princeton: Princeton University Press.

Malet, A. 1993. James Gregorie on tangents and the Taylor rule. *Archive Hist. Exact Sci*, **46**, 97–138.

Malmsten, C. J. 1849. De integralibus quibusdam definitis. *J. Reine Angew. Math.*, **38**, 1–39.

Manders, K. 2006. Algebra in Roth, Faulhaber and Descartes. *Hist. Math.*, **33**, 184–209.

Manning, K. R. 1975. The emergence of the Weierstrassian approach to complex analysis. *Archive Hist. Exact Sci.*, **14**, 297–383.

Maor, E. 1998. *Tirgonometric Delights*. Princeton: Princeton University Press.

Martzloff, J.C. 1997. *A History of Chinese Mathematics*. New York: Springer.

Masani, P.R. 1990. *Norbert Wiener*. Basel: Birkhäuser.

Mathews, G.B. 1961. *Theory of Numbers*. New York: Chelsea.

Maxwell, J.C. 1873. *A Treatise on Electricity and Magnetism*. Oxford: Clarendon.

Maz'ya, V., and Shaposhnikova, T. 1998. *Jacques Hadamard*. Providence: A.M.S., London Math. Soc. Translated by P. Basarab-Horwath.

McClintock, E. 1881. On the remainder of Laplace's series. *Amer. J. Math.*, **4**, 96–97.

Mclarty, C. 2012. Hilbert on Theology and its Discontents: The Origin of Myth in Modern Mathematics. Pages 105–129 of: Doxiadis, A., and Mazur, B. (eds), *Circles Disturbed*. Princeton: Princeton University Press.

Mehta, M.L., and Dyson, F.J. 1963. Statistical theory of energy levels of complex systems: V. *J. Math. Phys.*, **4**, 713–719.

Meijering, E. 2002. A chronology of interpolation: from ancient astronomy to modern signal and image processing. *Proc. I.E.E.E.*, **90**, 319–342.

Meixner, J. 1934. Orthogonale Polynomsysteme mit einer besonderen Gestalt der erzeugenden Funktion. *J. London Math. Soc.*, **9**, 6–13.

Mengoli, P. 1650. *Novae quadraturae arithmeticae seu de additione fractionum*. Bologna: Montij.

Méray, C. 1869. Remarques sur la nature des quantités définies par la condition de servir de limites à des variables données. *Revue des Soc. Savantes*, **4**, 280–289.

Méray, C. 1888. Valeur de l'intégrale définie $\int_0^\infty e^{-x^2} dx$ déduite de la formule de Wallis. *Bull. Sci. Math.*, **12**, 174–176.

Mercator, N. 1668. *Logarithmotechnia*. Godbid.

Mertens, F. 1874a. Ein Beitrag zur analytischen Zahlentheorie. *J. Reine Angew. Math.*, **78**, 46–62.

Mertens, F. 1874b. Ueber einige asymptotische Gesetze der Zahlentheorie. *J. Reine Angew. Math.*, **77**, 289–338.

Mertens, F. 1875. Über die Multiplicationsregel für zwei unendliche Reihen. *J. Reine Angew. Math.*, **79**, 182–184.

Mertens, F. 1895. Über das nichtverschwinden Dirichletscher Reihen mit reellen Gliedern. *S.B. Kais. Akad. Wiss. Wien*, **104**(Abt. 2a), 1158–1166.

Mertens, F. 1897. Über Dirichlet's Beweis des Satzes, dass jede unbegrenzte ganzzahlige arithmetische Progression, deren Differentz zu ihren Gliedern teilerfreund ist, unendliche viele Primzahlen darstellt. *S.B. Kais. Akad. Wiss. Wien*, **106**, 254–286.

Merzbach, U. 2018. *Dirichlet: A Mathematical Biography*. Cham: Birkhäuser.

Meschkowski, H. 1964. *Ways of Thought of Great Mathematicians*. San Francisco: Holden-Day. Translated by J. Dyer-Bennett.

Mikami, Y. 1914. On the Japanese theory of determinants. *Isis*, **2**, 9–36.

Mikami, Y. 1974. *The Development of Mathematics in China and Japan*. New York: Chelsea.

Milin, I.M. 1964. The area method in the theory of univalent functions, English translation. *Soviet Math. Dokl.*, **5**, 78–81.

Milin, I.M. 1986. Comments on the Proof of the Conjecture on Logarithmic Coefficients. Pages 109–112 of: Baernstein, A. (ed), *The Bieberbach Conjecture*. Providence: A.M.S.

Milne-Thomson, L.M. 1981. *The calculus of finite differences*. New York: Chelsea.

Mittag-Leffler, G. 1923. An introduction to the theory of elliptic functions. *Ann. of Math.*, **24**, 271–351.

Miyake, K. 1994. The establishment of the Takagi-Artin class field theory. Pages 109–128 of: Sasaki, C., Sugiura, M., and Dauben, J.W. (eds), *The Intersection of History and Mathematics*. Basel: Birkhäuser.

Möbius, A.F. 1832. Über eine besondere Art von Umkehrung der Reihen. *J. Reine Angew. Math.*, **9**, 105–123.

Moll, V. 2002. The evaluation of integrals: A personal story. *Notices A.M.S.*, 311–317.

Monsky, P. 1994. Simplifying the proof of Dirichlet's theorem. *Am. Math. Monthly*, **100**, 861–862.

Montmort, P.R. de. 1717. De seriebus infinitis tractatus. *Phil. Trans. Roy. Soc.*, **30**, 633–675.

Moore, G.H. 1982. *Zermelo's Axiom of Choice*. New York: Springer-Verlag.

Morrison, P., and Morrison, E. (eds). 1961. *Charles Babbage and His Calculating Engines: Selected Writings by Charles Babbage*. New York: Dover.

Muir, T. 1960. *The Theory of Determinants in the Historical Order of Development*. New York: Dover.

Mukhopadhyay, A. 1889a. The geometric interpretation of Monge's differential equation to all conics. *J. Asiatic Soc. of Bengal*, **58**, 181–185.

Mukhopadhyay, A. 1889b. On a curve of aberrancy. *J. Asiatic Soc. of Bengal*, **59**, 61–63.

Mukhopadhyay, S. 1909. New methods in the geometry of the plane arc. *Bull. Calcutta Math. Soc.*, **1**, 31–37.

Mukhopadhyaya (Mookerjee), A. 1998. *A Diary of Asutosh Mookerjee*. Calcutta: Mitra and Ghosh.

Mullin, R., and Rota, G.-C. 1970. On the Foundations of Combinatorial Theory: III. Theory of Binomial Enumeration. Pages 167–213 of: Harris, B. (ed), *Graph Theory and its Applications*. New York: Acad. Press. Reprinted in 1995 in *Gian-Carlo Rota on Combinatorics*, Ed. J. Kung, Boston: Birkhäuser, pp. 118–147.

Murphy, R. 1833a. *Elementary Principles of the Theories of Electricity, Heat, and Molecular Actions, Part I*. Cambridge: Cambridge University Press.

Murphy, R. 1833b. On the inverse method of definite integrals, with physical applications. *Trans. Cambridge Phil. Soc.*, **4**, 353–408.

Murphy, R. 1833c. Resolution of Algebraical Equations. *Trans. Cambridge Phil. Soc.*, **4**, 125–153.

Murphy, R. 1835. Second memoir on the inverse method of definite integrals. *Trans. Cambridge Phil. Soc.*, **5**, 113–148.

Murphy, R. 1837. First memoir on the theory of analytic operations. *Phil. Trans. Roy. Soc. London*, **127**, 179–210.

Murphy, R. 1839. *A Treatise on the Theory of Algebraical Equations*. London: Society for Diffusion of Useful Knowledge.

Mustafy, A. K. 1966. A new representation of Riemann's zeta function and some of its consequences. *Norske Vid. Selsk. Forh.*, **39**, 96–100.

Mustafy, A.K. 1972. On a criterion for a point to be a zero of the Riemann zeta function. *J. London Math. Soc. (2)*, **5**, 285–288.

Narasimhan, R. 1991. The coming of age of mathematics in India. Pages 235–258 of: Hilton, P., Hirzebruch, F., and Remmert, R. (eds), *Miscellanea Mathematica*. New York: Springer.

Narayana. 2001. The *Ganita Kamudi* of Narayana Pandit, Chapter 13. *Ganita Bharati*, **23**, 18–82. English translation with notes by Parmanand Singh.

Narkiewicz, W. 2000. *The Development of Prime Number Theory*. New York: Springer.

Needham, J. 1959. *Science and Civilization in China, vol. 3: Mathematics and the Sciences of the Heavens and the Earth*. New York: Cambridge University Press.

Nesterenko, Yu. V. 2006. Hilbert's seventh problem. Pages 269–282 of: Bolibruch, A. A., Osipov, Yu. S., and Sinai, Ya. G. (eds), *Mathematical Events of the Twentieth Century*. Berlin: Springer. Translated by L.P. Kotova.

Neuenschwander, E. 1978. The Casorati–Weierstrass theorem. *Hist. Math.*, **5**, 139–166.

Neumann, O. 2007a. Cyclotomy: From Euler through Vandermonde to Gauss. Pages 323–362 of: Bradley, R.E., and Sandifer, C.E. (eds), *Leonhard Euler: Life, Work and Legacy*. Amsterdam: Elsevier.

Neumann, O. 2007b. The *Disquisitiones Arithmeticae* and the theory of equations. Pages 107–128 of: Goldstein, C., Schappacher, N., and Schwermer, J. (eds), *The Shaping of Arithmetic after C. F. Gauss's Disquisitiones Arithmeticae*. New York: Springer.

Neumann, P. (ed). 2011. *The mathematical writings of Évariste Galois*. Zurich: European Math, Soc. Translated with commentary by P. Neumann.

Nevai, P. 1990. *Orthogonal Polynomials: Theory and Practice*. Dordrecht: Kluwer.

Nevanlinna, R. 1925. Zur Theorie der meromorphen Funktionen. *Acta Math.*, **46**, 1–99.

Nevanlinna, R. 1974. *Le théorème de Picard–Borel et la théorie des fonctions méromorphes*. New York: Chelsea.

Nevanlinna, R., and Paatero, V. 1982. *Introduction to Complex Analysis, 2nd edition*. Providence: A.M.S. Chelsea. Translated by T. Kövari and G.S. Goodman.

Newman, F.W. 1848. On $\Gamma(a)$, especially when a is negative. *Cambridge and Dublin Math. J.*, **3**, 57–63.

Newton, I. 1959–1960. *The Correspondence of Isaac Newton*. Cambridge: Cambridge University Press. Edited by H.W. Turnbull.

Newton, I. 1964–1967. *The Mathematical Works of Isaac Newton*. New York: Johnson Reprint. Edited with introduction by D.T. Whiteside.

Newton, I. 1967–1981. *The Mathematical Papers of Isaac Newton*. Cambridge: Cambridge University Press. Edited by D.T. Whiteside.

Nicole, F. 1717. Traité du calcul des différences finies. *Hist. Acad. Roy. Sci.*, 7–21.

Nicole, F. 1727. Méthode pour sommer une infinité de suites nouvelles, dont on ne peut trouver les sommes par les Méthodes connues. *Mém. Acad. Roy. Sci. Paris*, 257–268.

Nikolić, A. 2009. The story of majorizability as Karamata's condition of convergence for Abel summable series. *Hist. Math.*, **36**, 405–419.

Nilakantha. 1977. *Tantrasangraha of Nilakantha Somayaji, with Yuktidipika and Laghu-vivrti*. Hoshiarpur, India: Punjab University Edited by K.V. Sarma.

Nörlund, N.E. 1923. Mémoire sur le calcul aux différences finies. *Acta Math.*, **44**, 71–212.

Nörlund, N.E. 1924. *Vorlesungen über Differenzenrechnung*. Berlin: Springer.

Ogawa, T., and Morimoto, M. 2018. *Mathematics of Takebe Katahiro and History of Mathematics in East Asia*. Tokyo: Math. Soc. Japan.

Olver, P.J. 1999. *Classical Invariant Theory*. Cambridge: Cambridge University Press.

Ore, Ø. 1974. *Niels Henrik Abel: Mathematician Extraordinary*. Providence: A.M.S.

Ostrogradsky, M.V. 1845. De l'integration des fractions rationelles. *Bull. Physico-Math. Acad. Sci St. Pétersbourg*, **4**, 145–167 and 268–300.

Ozhigova, E.P. 2007. The part played by the Petersburg Academy of Sciences (the Academy of Sciences of the USSR) in the publication of Euler's collected works. Pages 53–74 of: Bogolyubov, N.N., Mikhaïlov, G.K., and Yushkevich, A.P. (eds), *Euler and Modern Science*. Washington, D.C.: M.A.A.

Papperitz, E. 1889. Ueber die Darstellung der hypergeometrischen Transcendenten durch eindeutige Functionen. *Math. Ann.*, **34**, 247–296.

Parameswaran, S. 1983. Madhava of Sangamagramma. *J. Kerala Studies*, **10**, 185–217.

Patterson, S.J. 2007. Gauss sums. Pages 505–528 of: Goldstein, C., Schappacher, N., and Schwermer, J. (eds), *The Shaping of Arithmetic after C. F. Gauss's Disquisitiones Arithmeticae*. New York: Springer.

Peacock, G. 1843–1845. *Treatise on Algebra*. Cambridge: Cambridge University Press.

Peano, G. 1973. *Selected Works*. London: Allen and Unwin. Edited by H.C. Kennedy.

Peirce, B. 1881. Linear Associative Algebra, with Notes and Addenda by C.S. Peirce. *Amer. J. Math.*, **4**, 97–229.

Pepper, J.V. 1968. Harriot's calculation of the meridional parts as logarithmic tangents. *Archive Hist. Exact Sci.*, **4**, 359–413.

Perron, O. 1929. *Die Lehre von den Kettenbrüchen, second edition*. Leipzig: Teubner.

Peters, C.A. (ed). 1860–1865. *Briefwechsel zwischen C.F. Gauss und H.C. Schumacher, vols. 1–5*. Altona: Esch.

Petrovski, I.G. 1966. *Ordinary Differential Equations*. Englewood Cliffs, N.J.: Prentice-Hall. Translated by R.A. Silverman.

Pfaff, J. 1797a. *Disquisitiones analyticae*. Helmstadii: Fleckheisen.

Pfaff, J. 1797b. Observationes analyticae ad L. Euleri Institutiones Calculi Integralis. *Supplement IV, Historie de 1793, Nova Acta Acad. Sci. Petropolitanae*, **XI**, 38–57.

Picard, É. 1879. Sur une propriété des fonctions entières. *Comptes Rendus*, **88**, 1024–1027.

Pick, G. 1915. Über eine Eigenschaft der konformen Abbildung kreisförmiger Bereiche. *Math. Ann.*, **77**, 1–6.

Pieper, H. (ed). 1998. *Korrespondenz zwischen Legendre und Jacobi*. Leipzig: Teubner.

Pieper, H. 2007. A network of scientific philanthropy: Humboldt's relations with number theorists. Pages 201–234 of: Goldstein, C., Schappacher, N., and Schwermer, J. (eds), *The Shaping of Arithmetic after C. F. Gauss's Disquisitiones Arithmeticae*. New York: Springer.

Pierpoint, W. S. 1997. Edward Stone (1702–1768) and Edmund Stone (1700–1768): Confused Identities Resolved. *Notes and Records Roy. Soc. London*, **51**, 211–217.

Pierpont, J. 1904. The history of mathematics in the nineteenth century. *Bull. A.M.S.*, **11**, 136–159. Reprinted in 2000 in the *Bulletin*, vol. 37, pp. 9–24.

Pietsch, A. 2007. *History of Banach Spaces and Linear Operators*. Boston: Birkhäuser.

Pingree, D. 1970–1994. Census of the exact sciences in Sanskrit. *Amer. Phil. Soc.*, **81, 86, 111, 146, 213**.

Pitt, H.R. 1938. General Tauberian theorems. *Proc. London Math. Soc.*, **44**, 243–288.

Plofker, K. 2009. *Mathematics in India*. Princeton: Princeton University Press.

Poincaré, H. 1883. Sur les fonctions entières. *Bull. Soc. Math. France*, **11**, 136–144.

Poincaré, H. 1886. Sur les intégrales irrégulières des équations linéaires. *Acta Math.*, **8**, 295–344.

Poincaré, H. 1907. Sur l'uniformisation des fonctions analytiques. *Acta Math.*, **31**, 1–64.

Poincaré, H. 1985. *Papers on Fuchsian Functions*. New York: Springer-Verlag. Translated by J. Stillwell.

Poisson, S.D. 1808. Mémoire sur la propagation de la chaleur dans les corps solides (extrait). *Nouv. Bull. Sci. Soc. Philom. Paris*, **1**, 112–116.

Poisson, S.D. 1823. Suite du mémoire sur les intégrales définies et sur la sommation des séries. *J. École Poly.*, **12**, 404–509.

Poisson, S.D. 1826. Sur le calcul numérique des intégrales définies. *Mém. Acad. Sci. France*, **6**, 571–602.

Polignac, A. de. 1857. Recherches sur les nombres premiers. *Comptes Rendus.*, **45**, 575–580.

Pólya, G. 1974. *Collected Papers*. Cambridge, MA: MIT Press. Edited by Ralph Boas.

Pommerenke, C. 1985. The Bieberbach conjecture. *Math. Intelligencer*, **7**(2), 23–25; 32.

Popoff, A. 1861. Sur le reste de la série de Lagrange. *Comptes Rendus*, **53**, 795–798.

Prasad, G. 1931. *Six Lectures on the Mean-Value Theorem of the Differential Calculus*. Calcutta: Calcutta University Press.

Prasad, G. 1933. *Some Great Mathematicians of the Nineteenth Century*. Benares, India: Benares Mathematical Society.

Pringsheim, A. 1900. Zur Geschichte des Taylorschen Lehrsatzes. *Bibliotheca Math.*, **3**, 433–479.

Probst, S. 2015. Leibniz as reader and second inventor: the cases of Barrow and Mengoli. Pages 111–134 of: Goethe, N., Beeley, P., and Rabouin, D. (eds), *G.W. Leibniz, Interrelation between Mathematics and Philosophy*. Dordrecht: Springer.

Purkert, W., and Ilgauds, H.J. 1985. *Georg Cantor*. Leipzig: Vieweg and Teubner.

Rabuel, C. 1730. *Commentaires sur la géométrie de M. Descartes*. Lyon: M. Duplain.

Rajagopal, C.T. 1949. A neglected chapter of Hindu mathematics. *Scripta Math.*, **15**, 201–209.

Rajagopal, C.T., and Rangachari, M.S. 1977. On the untapped source of medieval Keralese mathematics. *Archive Hist. Exact Sci.*, **18**, 89–102.

Rajagopal, C.T., and Rangachari, M.S. 1986. On medieval Keralese mathematics. *Archive Hist. Exact Sci.*, **35**, 91–99.

Rajagopal, C.T., and Vedamurtha Aiyar, T.V. 1951. On the Hindu proof of Gregory's series. *Scripta Math.*, **17**, 65–74.

Rajagopal, C.T., and Venkataraman, A. 1949. The sine and cosine power series in Hindu mathematics. *J. Roy. Asiatic Soc. Bengal, Sci.*, **15**, 1–13.

Ramanujan, S. 1911. Some properties of Bernoulli numbers. *J. Indian Math. Soc.*, **3**, 219–234.

Ramanujan, S. 1917. A series for Euler's constant γ. *Messenger Math.*, **46**, 73–80.

Ramanujan, S. 1919a. Proof of certain identities in combinatory analysis. *Proc. Camb. Phil. Soc.*, **19**, 214–216. Reprinted in Ramanujan (2000), pp. 214–215.

Ramanujan, S. 1919b. Some properties of $p(n)$, the number of partitions of n. *Proc. Camb. Phil. Soc.*, **19**, 207–210. Reprinted in Ramanujan (2000), pp. 210–213.

Ramanujan, S. 1988. *The Lost Notebook and Other Unpublished Papers*. Delhi: Narosa Publishing House. Introduction by G. Andrews.

Ramanujan, S. 2000. *Collected Papers*. Providence: A.M.S. Chelsea. Edited by G.H. Hardy, P.V. Seshu Aiyar, B. M. Wilson, with extensive commentary by B. Berndt.

Rashed, R. 1970. al-Karaji. Pages 240–246 of: *Dictionary of scientific biography, volume 7*. New York: Scribners.

Rashed, R. 1980. Ibn al-Haytham et le théorème de Wilson. *Archive Hist. Exact Sci.*, **22**, 305–321.

Rassias, T.M., Srivastava, H.M., and Yanushauskas, A. 1993. *Topics in Polynomials of One and Several Variables and Their Applications*. Singapore: World Scientific.

Raussen, M., and Skau, C. 2010. Interview with Mikhail Gromov. *Notices A.M.S.*, **57**, 391–403.

Remmert, R. 1991. *Theory of Complex Functions*. New York: Springer-Verlag. Translated by R. Burckel.

Remmert, R. 1996. Wielandt's theorem about the Γ-function. *Am. Math. Monthly*, **103**, 214–220.

Remmert, R. 1998. *Classical Topics in Complex Function Theory*. New York: Springer-Verlag. Translated by Leslie Kay.

Riccati, J. 1724. Animadversationes in aequationes differentiales. *Acta Erud.*, **Supp. 8, 1724**, 66–73.

Riemann, B. 1859. Ueber die Anzahl der Primzahlen unter einer gegebenen Grösse. *Monatsber. Berlin. Akad.*, 671–680.

Riemann, B. 1899. *Elliptische Functionen. Vorlesungen von Bernhard Riemann. Mit Zusätzen herausgegeben von Hermann Stahl*. Leipzig: Teubner. Edited by H. Stahl.

Riemann, B. 1990. *Gesammelte Mathematische Werke*. New York: Springer-Verlag. Edited by R. Dedekind, H. Weber, R. Narasimham, and E. Neuenschwander.

Riesz, F. 1909. Sur les opérations fonctionnelles linéaires. *Comptes Rendus*, **149**, 974–977.

Riesz, F. 1910. Untersuchungen über Systeme integrierbarer Funktionen. *Math. Ann.*, **69**, 449–497. Reprinted in Riesz (1960) vol. 1, pp. 441–489.

Riesz, F. 1913. *Les systèmes d'équations linéaires a une infinité d'inconnues*. Paris: Gauthier-Villars.

Riesz, F. 1960. *Oeuvres complètes*. Budapest: Académie des Sciences de Hongrie. Edited by Á. Császár.

Riesz, M. 1928. Sur les fonctions conjugées. *Math. Z.*, **27**, 218–44.

Rigaud, S.P. (ed). 1841. *Correspondence of Scientific Men of the Seventeenth Century*. Oxford: Oxford University Press.

Robertson, M.S. 1936. A remark on the odd schlicht functions. *Bull. A.M.S.*, **42**, 366–370.

Robins, B. 1727. A demonstration of the 11th proposition of Sir Isaac Newton's treatise on Quadrature. *Phil. Trans. Roy. Soc.*, **34**, 230–236.

Rodrigues, O. 1816. Mémoire sur l'attraction des sphéroids. *Correspondance École Poly.*, **3**, 361–385.

Rodrigues, O. 1839. Note sur les inversions, ou dérangements produits dans les permutations. *J. Math. Pures Appl.*, **4**, 236–240.

Rogers, L.J. 1888. An extension of a certain theorem in inequalities. *Messenger Math.*, **17**, 145–150.

Rogers, L.J. 1893a. Note on the transformation of an Heinean series. *Messenger Math.*, **23**, 28–31.

Rogers, L.J. 1893b. On a three-fold symmetry in the elements of Heine's series. *Proc. London Math. Soc.*, **24**, 171–179.

Rogers, L.J. 1893c. On the expansion of some infinite products. *Proc. London Math. Soc.*, **24**, 337–352.

Rogers, L.J. 1894. Second memoir on the expansion of certain infinite products. *Proc. London Math. Soc.*, **25**, 318–343.

Rogers, L.J. 1895. Third memoir on the expansion of certain infinite products. *Proc. London Math. Soc.*, **26**, 15–32.

Rogers, L.J. 1907. On function sum theorems connected with the series $\sum_{n=1}^{\infty} \frac{x^n}{n^2}$. *Proc. London Math. Soc.*, **4**, 169–189.

Rogers, L.J. 1917. On two theorems of combinatory analysis and some allied identities. *Proc. London Math. Soc.*, **16**, 321–327.

Rolle, M. 1690. *Traité d'algebre*. Paris: Michallet.

Rolle, M. 1691. *Démonstration d'une Méthode pour résoudre les égalitez de tous les degréz*. Paris: Cusson.

Roquette, P. 2002. The Riemann hypothesis in characteristic p, its origin and development. Part I. The formation of the zeta-functions of Artin and of F. K. Schmidt. *Mitt. Math. Ges. Hamburg*, **21**, 79–157.

Roquette, P. 2004. The Riemann hypothesis in characteristic p, its origin and development. Part II. The first steps by Davenport and Hasse. *Mitt. Math. Ges. Hamburg*, **23**, 5–74.

Roquette, P. 2018. *The Riemann Hypothesis in Characteristic p in Historical Perspective*. Switzerland: Springer Nature.

Rosen, M. 2002. *Number Theory in Function Fields*. New York: Springer.

Rothe, H.A. 1811. *Systematisches Lehrbuch der Arithmetik*. Erlangen: Barth.

Rowe, D.E., and McCleary, J. 1989. *The History of Modern Mathematics*. Boston: Academic Press.

Roy, R. 1990. The discovery of the series formula for π by Leibniz, Gregory and Nilakantha. *Math. Mag.*, **63**(5), 291–306. Reprinted in Anderson, Katz, and Wilson (2004), pp. 111–121.

Roy, R. 1993. The Work of Chebyshev on Orthogonal Polynomials. Pages 495–512 of: Rassias, T., Srivastava, H., and Yanushauskas, A. (eds), *Topics in Polynomials of one and Several Variables and their Applications*. Singapore: World Scientific.

Roy, R. 2017. *Elliptic and Modular Functions*. Cambridge: Cambridge University Press.

Ru, M. 2001. *Nevanlinna Theory and its Relation to Diophantine Approximation*. Singapore: World Scientific.

Rudin, W. 1966. *Real and Complex Analysis*. New York: McGraw-Hill.

Saalschütz, L. 1890. Eine Summationsformel. *Zeit. Math. Phys.*, **35**, 186–188.

Salmon, G. 1879. *A Treatise on the Higher Plane Curves, third edition*. Dublin: Hodges, Foster, and Figgis.

Sandifer, C.E. 2007. *The Early Mathematics of Leonhard Euler*. Washington, D.C.: M.A.A.

Sarma, K.V. 1972. *A History Of The Kerala School Of Hindu Astronomy.* Hoshiarpur, India: Punjab University

Sarma, K.V., and Hariharan, S. 1991. Yuktibhasa of Jyesthadeva. *Indian J. Hist. Sci.*, **26**(2), 185–207.

Sasaki, C. 1994. The adoption of Western mathematics in Meiji Japan, 1853–1903. Pages 165–186 of: Sasaki, C., Sugiura, M., and Dauben, J.W. (eds), *The Intersection of History and Mathematics.* Basel: Birkhäuser.

Sasaki, C., Sugiura, M., and Dauben, J. W. (eds). 1994. *The Intersection of History and Mathematics.* Basel: Birkhäuser.

Schaar, M. 1850. Recherches sur la théorie des résidues quadratiques. *Mém. couronnés et mém. savants étrangers Acad. Roy. Sci. Belgique*, **25**, 1–20.

Scharlau, W., and Opolka, H. 1984. *From Fermat to Minkowski: Lectures on the Theory of Numbers and its Historical Development.* New York: Springer.

Schellbach, K. 1854. Die einfachsten periodischen Functionen. *J. Reine Angew. Math.*, **48**, 207–236.

Schlömilch, O. 1843. Einiges über die Eulerischen Integrale der zweiten Art. *Archiv Math. Phys.*, **4**, 167–174.

Schlömilch, O. 1847. *Handbuch der Differenzial- und Integralrechnung.* Greifswald, Germany: Otte.

Schlömilch, O. 1849. Uebungsaufgaben für Schüler, Lehrsatz von dem Herrn. Prof. Dr. Schlömilch. *Archiv Math. Phys.*, **12**, 415.

Schlömilch, O. 1858. Ueber eine Eigenschaft gewisser Reihen. *Zeit. Math. Phys.*, **3**, 130–132.

Schneider, I. 1968. Der Mathematiker Abraham de Moivre (1667–1754). *Archive Hist. Exact Sci*, **5**, 177–317.

Schneider, I. 1983. Potenzsummenformeln im 17. Jahrhundert. *Hist. Math.*, **10**, 286–296.

Schneider, T. 1934. Transzendenzuntersuchungen periodischer Funktionen: I. Transzendenz von Potenzen; II. Transzendenzeigenschaften elliptischer Funktionen. *J. Reine Angew. Math.*, **172**, 65–74.

Schoenberg, I.J. 1988. *Selected Papers.* Boston: Birkhäuser.

Schönemann, T. 1845. Grundzüge einer allgemeinen Theorie der höheren Congruenzen, deren Modul eine reelle Primzahl ist. *J. Reine Angew. Math.*, **31**, 269–325.

Schur, I. 1917. Ein Beitrag zur additiven Zahlentheorie der Kettenbrüche. *S.B. Preuss. Akad. Wiss. Phys.-Math.*, 302–321.

Schur, I. 1929. Einige Sätze über Primzahlen mit Anwendung auf Irreduzibilitätsfragen. *S-B Akad. Wiss. Berlin Phys. Math. Klasse*, 125–136.

Schwarz, H.A. 1885. Über ein die Flächen kleinsten Flächeninhalts betreffends Problem der Variationsrechnung. *Acta Soc. Scient. Fenn.*, **15**, 315–362. Reprinted in Schwarz (1972) vol. 1, pp. 223–269.

Schwarz, H.A. 1893. *Formeln und Lehrsätze zum Gebrauche der elliptischen Funktionen.* Berlin: Springer.

Schwarz, H.A. 1972. *Abhandlungen.* New York: Chelsea.

Schweins, F. 1820. *Analysis.* Heidelberg: Mohr und Winter.

Schwering, K. 1899. Zur Theorie der Bernoulli'schen Zahlen. *Math. Ann.*, **52**, 171–173.

Scriba, C.J. 1961. Zur Lösung des 2. Debeauneschen Problems durch Descartes. *Archive Hist. Exact Sci.*, **1**, 406–419.

Scriba, C.J. 1964. The inverse method of tangents. *Archive Hist. Exact Sci.*, **2**, 113–137.

Seal, H.L. 1949. The historical development of the use of generating functions in probability theory. *Bull. Assoc. Actuaires Suisses*, **49**, 209–228.

Segal, S.L. 1978. Riemann's example of a continuous "nondifferentiable" function. *Math. Intelligencer*, **1**, 81–82.

Seidel, P.L. 1847. Note über eine Eigenschaft der Reihen, welche discontinuirliche Functionen darstellen. *Abhand. Math. Phys. Klasse der Kgl. Bayrischen Akad. Wiss.*, **5**, 381–394.

Seki, T. 1974. *Takakazu Seki's Collected Works Edited with Explanations.* Osaka: Kyoiku Tosho. Edited by A. Hirayama, K. Shimodaira and H. Hirose; translated by Jun Sudo.

Selberg, A. 1941. Über einen Satz von A. Gelfond. *Arch. Math. Naturvidenskab*, **44**, 159–170. Reprinted in Selberg's *Collected Papers* vol. 1, pp. 62–73.

Selberg, A. 1944. Bemerkninger om et multipelt integral. *Norsk. Mat. Tidskr.*, **26**, 71–78. Republished in Selberg (1989) vol. 1, pp. 204–211.

Selberg, A. 1949. An elementary proof of the prime number theorem. *Ann. of Math.*, **50**, 305–313.

Selberg, A. 1989. *Collected Papers*. New York: Springer-Verlag.

Sen Gupta, D.P. 2000. Sir Asutosh Mookerjee—educationist, leader and institution-builder. *Current Sci.*, **78**, 1566–1573.

Serret, J.A. 1854. *Cours d'algèbre supérieure*. Paris: Mallet-Bachelier.

Serret, J.A. 1868. *Cours de calcul différentiel et intégral*. Paris: Gauthier-Villars.

Serret, J.A. 1877. *Cours d'algèbre supérieure, 4 é*. Paris: Gauthier-Villars.

Shah, S.M. 1948. A note on uniqueness sets for entire functions. *Proc. Indian Acad. Sci., Sect. A*, **28**, 519–526.

Shidlovskii, A.B. 1989. *Transcendental Numbers*. Berlin: de Gruyter. Translated by N. Koblitz.

Shimura, G. 2007. *Elementary Dirichlet Series and Modular Forms*. New York: Springer.

Shimura, G. 2008. *The Map of My Life*. New York: Springer.

Siegel, C.L. 1932. Über die Perioden elliptischer Funktionen. *J. Reine Angew. Math.*, **167**, 62–69.

Siegel, C.L. 1949. *Transcendental Numbers*. Princeton: Princeton University Press.

Siegel, C.L. 1969. *Topics In Complex Function Theory*. New York: Wiley.

Simmons, G.F. 1992. *Calculus Gems*. New York: McGraw-Hill.

Simon, B. 2005. OPUC on One Foot. *Bull. A.M.S.*, **42**, 431–460.

Simpson, T. 1743. *Mathematical Dissertations*. London: Woodward.

Simpson, T. 1750. *The Doctrine and Application of Fluxions*. London: Nourse.

Simpson, T. 1759. The invention of a general method for determining the sum of every second, third, fourth, or fifth, etc. term of a series, taken in order; the sum of the whole being known. *Phil. Trans. Roy. Soc.*, **50**, 757–769.

Simpson, T. 1800. *A Treatise of Algebra*. London: Wingrave.

Sluse, R.F. 1672. An extract of a letter ... concerning his short and easie method of drawing tangents to all geometical curves. *Phil. Trans. Roy. Soc.*, **7**, 5143–5147.

Smith, D.E. (ed). 1959. *A Source Book in Mathematics*. New York: Dover.

Smith, D.E., and Mikami, Y. 1914. *A History of Japanese Mathematics*. Chicago: Open Court.

Smith, H.J.S. 1875. On the Integration of Discontinuous Functions. *Proc. London Math. Soc.*, **6**, 140–153. Reprinted in Smith (1965) vol. 2, pp. 86–100.

Smith, H.J.S. 1965a. *Collected Mathematical Papers*. New York: Chelsea. Edited by J.W.L. Glaisher. Volume 1 includes Smith's *Report on the Theory of Numbers*.

Smith, H.J.S. 1965b. *Report on the Theory of Numbers*. New York: Chelsea.

Snow, J.E. 2003. Views on the real numbers and the continuum. *Rev. Mod. Logic*, **9**, 95–113.

Somayaji, Putumana. 2018. *Karanapaddhati of Putumana Somayaji*. Singapore: Springer Nature. Edited and translated with commentary by V. Pai, K. Ramasubramanian, M.S. Sriram and M.D. Srinivas.

Spence, W. 1809. *An Essay on the Theory of the Various Orders of Logarithmic Transcendents*. London: Murray.

Spence, W. 1819. *Mathematical Essays*. London: Whittaker. Edited by J.F.W. Herschel.

Spiess, O. (ed). 1955. *Der Briefwechsel von Johann Bernoulli*. Basel: Birkhäuser.

Spiridonov, V. 2013. Elliptic hypergeometric functions. Pages 577–606 of: *Special Functions, Russian edition*. Moscow: Cambridge University Press and MCCME. This article was written in Russian as an additional complementary chapter to the Russian edition of *Special Functions*, by Andrews, Askey, Roy.

Sridharan, R. 2005. Sanskrit prosody, Pingala sutras and binary arithmetic. Pages 33–62 of: Emch, G.G. et al. (ed), *Contributions to the history of Indian mathematics*. New Delhi: Hindustan Book Agency and Springer.

Srinivasiengar, C.N. 1967. *The History of Ancient Indian Mathematics*. Calcutta: World Press.

Stäckel, P. 1908. Eine vergessen Abhandlung Leonhard Eulers über die Summe der reziproken Quadrate der natürlichen Zalen. *Biblio. Math.*, **8**, 37–60.

Stäckel, P., and Ahrens, W. (eds). 1908. *Briefwechsel zwischen C.G.J. Jacobi und P.H. Fuss*. Leipzig: Teubner.

Stark, H.M. 1967. A complete determination of the complex quadratic fields with class-number one. *Michigan Math. J.*, **14**, 1–27.

Stark, H.M. 1969. On the 'gap' in a theorem of Heegner. *J. Number Theory*, **1**, 16–27.

Stedall, J. 2000. Catching proteus: The collaborations of Wallis and Brouncker, I and II. *Notes and Records Roy. Soc. London*, **54**, 293–331.

Steele, J.M. 2004. *The Cauchy-Schwarz Master Class*. Cambridge: Cambridge University Press.

Steffens, K.-G. 2006. *The History of Approximation Theory*. Boston: Birkhäuser.

Stephens, L., and Lee, S. (eds). 1908. *Dictionary of National Biography, 22 volumes*. New York: Macmillan.

Stickelberger, L. 1890. Über eine Verallgemeinerung von der Kreistheilung. *Math. Ann.*, **37**, 321–367.

Stieltjes, T.J. 1885a. Sur certains polynômes qui vérifient une équation différentielle linéaire du second ordre et sur la théorie des fonctions de Lamé. *Acta Math.*, **6**, 321–326. Reprinted in Stieltjes's 1993 *Collected Papers* vol. 1, pp. 522–527.

Stieltjes, T.J. 1885b. Sur les polynômes de Jacobi. *Comptes Rendus*, **100**, 620–622. Reprinted in Stieltjes' *Collected Papers* vol. 1, pp. 530–532.

Stieltjes, T.J. 1886a. Recherches sur quelques séries semi-convergentes. *Ann. Sci. École Norm.*, **3**, 201–258. Reprinted in Stieltjes's *Collected Papers* vol. 2, pp. 2–58.

Stieltjes, T.J. 1886b. Sur les racines de l'équation $X_n = 0$. *Acta Math.*, **9**, 385–400. Reprinted in Stieltjes's 1993 *Collected Papers* vol. 2, pp. 77–92.

Stieltjes, T.J. 1889. Sur le développement de $\log \Gamma(a)$. *J. Math. Pures Appl.*, **[4] 5**, 425–444. Reprinted in Stieltjes's *Collected Papers* vol. 2, pp. 215–234.

Stieltjes, T.J. 1890. Note sur l'intégrale $\int_0^\infty e^{-u^2} du$. *Nouv. Ann. Math. Paris*, **9**, 479–480. Reprinted in Stieltjes's Oeuvres Complètes vol. 2, pp. 263–264.

Stieltjes, T.J. 1993. *Collected Papers*. New York: Springer-Verlag. Edited by G. van Dijk.

Stirling, J. 1717. *Lineae Tertii Ordinis Neutonianae*. Oxford: Whistler.

Stirling, J. 1719. Methodus differentialis Newtoniana illustrata. *Phil. Trans.*, **30**, 1050–1070.

Stirling, J. 1730. *Methodus differentialis*. London: Strahan.

Stirling, J., and Tweddle, I. 2003. *James Stirling's Methodus differentialis, An Annotated Translation of Stirling's Text*. London: Springer.

Stokes, G.G. 1849. On the critical values of the sums of periodic series. *Trans. Cambridge Phil. Soc.*, **8**, 533–583.

Stone, E. 1730. *The Method of Fluxions both Direct and Inverse*. London: W. Innys.

Stone, M. 1932. *Linear transformations in Hilbert Spaces and their applications to analysis*. Providence: A.M.S.

Strichartz, R.S. 1995. *The Way of Analysis*. London: Jones and Bartlett.

Struik, D.J. 1969. *A Source Book in Mathematics*. Cambridge, MA: Harvard University Press.

Struik, D.J. 1987. *A Concise History of Mathematics*. New York: Dover.

Stubhaug, A. 2000. *Niels Henrik Abel and His Times*. New York: Springer.

Sturm, C. 1829. Analyse d'un mémoire sur la résolution des équations numériques. *Bull. Sci. Férussac*, **11**, 419.

Sturmfels, B. 2008. *Algorithms on Invariant Theory*. Wien: Springer.

Sylvester, J.J. 1852. On the principles of the calculus of forms. *Cambridge and Dublin Math. J.*, **7**, 52–97, 179–217. Reprinted in Sylvester (1973) vol. 1, pp. 284–327, 328–363.

Sylvester, J.J. 1853a. On a theory of the syzygetic relations of two rational integral functions ... *Phil. Trans.*, **143**, 407–548. Reprinted in Sylvester (1973) vol. 1, pp. 429–586.

Sylvester, J.J. 1853b. On Mr. Cayley's impromptu demonstration of the rule for determining at sight the degree ... *Phil. Mag.*, **5**, 199–202. Reprinted in Sylvester (1973) vol. 1, pp. 595–598.

Sylvester, J.J. 1865. On an elementary proof and generalization of Sir Isaac Newton's hitherto undemonstrated rule for the discovery of imaginary roots. *Proc. London Math. Soc.*, **1**, 1–16. Reprinted in Sylvester's *Collected Mathematical Papers* (1773) vol. 2, pp. 498–513.

Sylvester, J.J. 1869. On a new continued fraction applicable to the quadrature of the circle. *Phil. Magazine*, **37**, 373–375. Republished in Sylvester's *Collected Mathematical Papers* vol. 2, pp. 691–693.

Sylvester, J.J. 1878. Proof of the hitherto undemonstrated theorem of invariants. *Phil. Mag.*, **5**, 178–188. Reprinted in Sylvester (1973) vol. 3, pp. 117–126.

Sylvester, J.J. 1882. A constructive theory of partitions, arranged in three acts, an interact, and an exodion. *Amer. J. Math*, **5**, 251–330. Reprinted in Sylvester (1973) vol. 4, pp. 1–83.

Sylvester, J.J. 1973. *Mathematical Papers*. New York: Chelsea. Edited by H. Baker.

Szegő, G. 1926. Ein Beitrag zur Theorie der Thetafunktionen. *S.B. Preuss. Akad. Wiss. Phys-Math.*, 242–252. Reprinted in Szegő (1982) vol. 1, pp. 795–805.

Szegő, G. 1975. *Orthogonal Polynomials*. Providence: A.M.S.

Szegő, G. 1982. *The Collected Papers of Gabor Szegő*. Boston: Birkhäuser. Edited by R. Askey.

Szekeres, G. 1968. A combinatorial interpretation of Ramanujan's continued fraction. *Canadian Math. Bull.*, **11**, 405–408.

Takagi, T. 1990. *Collected Papers*. Tokyo: Springer-Verlag. Edited by S. Iyanaga, K. Iwasawa, K. Kodaira, and K. Yosida.

Takase, M. 1994. Three aspects of the theory of complex multiplication. Pages 91–108 of: Sasaki, C., Sugiura, M., and Dauben, J.W. (eds), *The Intersection of History and Mathematics*. Basel: Birkhäuser.

Tannery, J., and Molk, J. 1972. *Éléments de la théorie des fonctions elliptiques, 4 vols*. New York: Chelsea.

Tauber, A. 1897. Ein Satz aus der Theorie der unendlichen Reihen. *Monats. Math. und Phys.*, **8**, 273–277.

Taylor, B. 1715. *Methodus incrementorum directa et inversa*. London: W. Innys.

Taylor, B., and Feigenbaum, L. 1981. *Brook Taylor's "Methodus incrementorum": A Translation with Mathematical and Historical Commentary*. Ph.D. thesis, Yale University.

Thomae, J. 1869. Beiträge zur Theorie der durch die Heinsche Reihe ... *J. Reine Angew. Math.*, **70**, 258–281.

Thomson, W., and Tait, P.G. 1890. *Treatise on Natural Philosophy*. Cambridge: Cambridge University Press.

Tignol, J.-P. 1988. *Galois' Theory of Algebraic Equations*. New York: Wiley.

Titchmarsh, E.C., and Heath-Brown, D.R. 1986. *The Theory of the Riemann-Zeta Function*. Oxford: Oxford University Press.

Truesdell, C. 1960. *The Rational Mechanics of Flexible or Elastic Bodies, 1638–1788. Introduction to Vols. 10 and 11, Second Series of Euler's Opera omnia*. Zurich: Orell Füssli Turici.

Truesdell, C. 1984. *An Idiot's Fugitive Essays on Science*. New York: Springer-Verlag.

Tucciarone, J. 1973. The development of the theory of summable divergent series from 1880 to 1925. *Archive Hist. Exact Sci.*, **10**, 1–40.

Turán, P. 1990. *Collected Papers*. Budapest: Akadémiai Kiadó. Edited by P. Erdős.

Turnbull, H.W. 1933. James Gregory: A study in the early history of interpolation. *Proc. Edinburgh Math. Soc.*, **3**, 151–172.

Turnbull, H.W. (ed). 1939. *James Gregory Tercentenary Memorial Volume*. London: Bell.

Tweddle, I. 1984. Approximating $n!$ Historical origins and error analysis. *Amer. J. Phys.*, **52**, 487–488.

Tweddle, I. 1988. *James Stirling: "This about series and such things."* Edinburgh: Scottish Acad. Press.

Tweedie, C. 1917–1918. Nicole's contributions to the foundations of the calculus of finite differences. *Proc. Edinburgh Math. Soc.*, **36**, 22–39.

Tweedie, C. 1922. *James Stirling: A Sketch of His Life and Works along with His Scientific Correspondence*. Oxford: Oxford University Press.

Valiron, G. 1913. Sur les fonctions entières d'ordre fini et d'ordre nul, et en particulier les fonctions à correspondence régulière. *Ann. Fac. Sci. Toulouse*, **5**, 117–257.

Valiron, G. 1949. *Lectures on the General Theory of Integral Functions*. New York: Chelsea.

Vallée-Poussin, C.J. de la. 1896a. Demonstration simplifée du théorème de Dirichlet sur la progression arithmétique. *Mém. Acad. Roy. Soc. Bruxelles*, **53**, 6–8.

Vallée-Poussin, C.J. de la. 1896b. Recherches analytiques sur la théorie des nombres premiers, I–III. *Ann. Soc. Sci. Bruxelles*, **20**, 183–256, 281–362, 363–397.

Van Brummelen, G. 2009. *The Mathematics of the Heavens and the Earth*. Princeton: Princeton University Press.

Van Brummelen, G., and Kinyon, M. (eds). 2005. *Mathematics and the Historian's Craft*. New York: Springer.

Van Maanen, J.A. 1984. Hendrick van Heuraet (1634–1660?): His life and work. *Centaurus*, **27**, 218–279.

van Rootselaar, B. 1964. Bolzano's Theory of Real Numbers. *Archive Hist. Exact Sci.*, **2**, 168–180.

Vandermonde, T.A. 1772. Mémoire sur des irrationnelles de différents ordres avec une application au cercle. *Hist. Acad. Roy. Sci. Paris pour 1772*, 489–498.

Varadarajan, V.S. 2006. *Euler Through Time*. Providence: A.M.S.

Venkatachaliengar, K., and Cooper, S. 2011. *Development of Elliptic Functions According to Ramanujan*. Singapore: World Scientific.

Viète, F. 1593. *Variorum de rebus mathematicis responsorum, Liber. VIII*. Turonis: Iamettium Mettayer.

Viète, F. 1983. *The Analytic Art*. Kent, OH: Kent State University Press. Translated by T.R. Witmer.

Vlăduţ, S.G. 1991. *Kronecker's Jugendtraum and Modular functions*. New York: Gordon and Breach. Translated by M. Tsfasman.

von Kowalevsky, S. (Kovalevskaya). 1875. Zur Theorie der partiellen Differentialgleichungen. *J. Reine Angew. Math.*, **80**, 1–32.

von Staudt, K.G.C. 1840. Beweis eines Lehrsatzes, die Bernoullischen Zahlen betreffend. *J. Reine Angew. Math.*, **21**, 372–374.

Vorselman de Heer, P. 1833. *Specimen inaugurale de fractionibus continuis*. PhD thesis, Utrecht.

Wagstaff, S. S. 1981. Ramanujan's paper on Bernoulli numbers. *J. Indian Math. Soc.*, **45**, 49–65.

Wali, K.C. 1991. *Chandra: A Biography of S. Chandrasekhar*. Chicago: University Chicago Press.

Walker, J.J. 1891. On the influence of applied on the progress of pure mathematics. *Proc. London Math. Soc.*, **22**, 1–18.

Wallis, J. 1656. *Arithmetica Infinitorum*. Oxford: Lichfield.

Wallis, J. 1659. *Tractatus duo, prior, de cycloide ...* Oxford: Lichfield.

Wallis, J. 1668. Logarithmotechnia Nicolai Mercatoris. *Phil. Trans.*, **3**, 753–759.

Wallis, J. 1685. *A Treatise of Algebra both Historical and Practical*. London: Oxford University

Wallis, J. 1693–1699. *Opera Mathematica, 3 vols*. Oxford: T. Sheldoniano.

Wallis, J., and Stedall, J. 2004. *The Arithmetic of Infinitesimals*. New York: Springer. Translation with notes of Wallis's *Arithmetica Infinitorum* by J.A. Stedall.

Waring, E. 1779. Problems concerning interpolations. *Phil. Trans. Roy. Soc. London*, **69**, 59–67.

Waring, E. 1991. *Meditationes Algebraicae*. Providence: A.M.S. Translated by D. Weeks.

Watson, G.N. 1933. The marquis and the land-agent. *Math. Gazette*, **17**, 5–17.

Watson, G.N. 1938. Ramanujans Vermutung über Zerfällungsanzahlen. *J. Reine Angew. Math.*, **179**, 97–128.

Weber, H. 1895. *Lehrbuch der Algebra*. Braunschweig: Vieweg.

Weierstrass, K. 1856. Über die Theorie der analytischen Facultäten. *J. Reine Angew. Math*, **51**, 1–60. Reprinted in Weierstrass (1894–1927) vol. 1, pp. 153–221.

Weierstrass, K. 1885. Zu Lindemann's Abhandlung "Über die Ludolphsche Zahl." *S.B. Preuss. Akad. Wiss.*, 1067–1085. Reprinted in Weierstrass's *Werke*, vol. 2, pp. 341–361.

Weierstrass, K. 1894–1927. *Mathematische Werke*. Berlin: Mayer and Müller.

Weil, A. 1946. *Foundations of Algebraic Geometry*. Providence: A.M.S.

Weil, A. 1949. Numbers of solutions of equations in finite fields. *Bull. A.M.S.*, **55**, 497–508.

Weil, A. 1974. Two lectures on number theory, past and present. *Enseign. Math.*, **20**, 87–110.

Weil, A. 1976. *Elliptic Functions According to Eisenstein and Kronecker*. New York: Springer.

Weil, A. 1979. *Collected Papers*. New York: Springer-Verlag.

Weil, A. 1984. *Number Theory: An Approach through History from Hammurapi to Legendre*. Boston: Birkhäuser.

Weil, A. 1989a. On Eisenstein's copy of the *Disquisitiones*. *Adv. Studies Pure Math.*, **17**, 463–469.

Weil, A. 1989b. Prehistory of the zeta function. Pages 1–9 of: Aubert, K.E., Bombieri, E., and Goldfeld, D. (eds), *Number Theory, Trace Formulas, and Discrete Groups*. Boston: Academic Press.

Weil, A. 1992. *The Apprenticeship of a Mathematician*. Boston: Birkhäuser.

Westfall, R. 1980. *Never at Rest*. Cambridge: Cambridge University Press.

Whiteside, D.T. 1961a. Henry Briggs: The binomial theorem anticipated. *Math. Gazette*, **45**, 9–12.

Whiteside, D.T. 1961b. Patterns of mathematical thought in the later seventeenth century. *Archive Hist. Exact Sci.*, **1**, 179–388.

Whittaker, E.T., and Robinson, G. 1949. *The Calculus of Observations*. London: Blackie and Son.

Whittaker, E.T., and Watson, G.N. 1927. *A Course of Modern Analysis*. Cambridge: Cambridge University Press.

Wiener, N. 1958. *The Fourier Integral and Certain of Its Applications*. New York: Dover.

Wiener, N. 1979. *Collected Works*. Cambridge, MA: MIT Press. Edited by P. Masani.

Wilbraham, H. 1848. On a certain periodic function. *Cambridge and Dublin Math. J.*, **3**, 198–201.

Wilf, H.S. 2001. The number-theoretic content of the Jacobi triple product identity. Pages 227–230 of: Foata, D., and Han, G.-N. (eds), *The Andrews Festschrift: Seventeen Papers on Classical Number Theory and Combinatorics*. New York: Springer.

Wilson, K. 1962. Proof of a conjecture of Dyson. *J. Math. Physics*, **3**, 1040–1043.

Wintner, A. 1929. *Spektral Theorie der unendlichen Matrizen*. Leipzig: Hirzel.

Woodhouse, R. 1803. *The Principles of Analytical Calculation*. Cambridge: Cambridge University Press.

Yadegari, M. 1980. The binomial theorem: A widespread concept in medieval Islamic mathematics. *Hist. Math.*, **7**, 401–406.

Yandell, B.H. 2002. *The Honors Class*. Natick, MA: Peters.

Young, G.C., and Young, W.H. 1909. On derivatives and the theorem of the mean. *Quart. J. Pure Appl. Math.*, **40**, 1–26.

Young, G.C., and Young, W.H. 2000. *Selected Papers*. Lausanne, Switzerland: Presses Polytechniques. Edited by S.D. Chatterji and H. Wefelscheid.

Young, W.H. 1909. A note on a trigonometrical series. *Messenger Math.*, **38**, 44–48.

Yushkevich, A.P. 1964. *Geschichte der Mathematik in Mittelalter*. Leipzig: Teubner.

Yushkevich, A.P. 1971. The concept of function up to the middle of the 19th century. *Archive Hist. Exact Sci.*, **16**, 37–85.

Zdravkovska, S., and Duren, P. 1993. *Golden Years of Moscow Mathematics*. Providence: A.M.S.

Zolotarev, E. 1876. Sur la série de Lagrange. *Nouvelles Ann. Math.*, **15**, 422–423.

Index

Abel, Niels Henrik, 4, 119, 120, 131, 132, 182–184, 186, 188, 216, 240, 242, 244, 255–258, 260–267, 269, 271, 272, 279, 288, 297, 301, 309, 315, 341, 345
Abel mean, 182
Abhyankar, Shreeram, 160
absolutely convergent Fourier series, 192, 208, 209
Acta Eruditorum, 8
Adams, John Couch, 36
addition formula for elliptic functions, 242, 243, 257, 282, 317
addition formula for elliptic integrals, 217
Ahlfors, Lars, 347, 368, 386
Ahlgren, Scott, 42
Alder, H. L., 38, 39
Alexander, J. W., 367, 385
algebraically independent numbers, 329, 339
Almkvist, Gert, 254
Altmann, Simon, 32
Analyse algébrique, 210
analytic continuation, 186, 239
analytical engine, 160
Andrews, George, 12, 32, 41, 45, 51, 59, 60, 90
Appell, Paul, 344
Arbogast, Louis, 72
area inequalities, 367, 372
arithmetic-geometric mean, 231, 236, 237, 239, 240, 249, 250, 254, 255
Arnold, Vladimir, 187
Aronhold, Siegfried, 155, 157
Ars Conjectandi, 8
Artin, Emil, 390, 421
Artin, M., 409
Artmann, B., 98
Aryabhatyabhasya, 318
Ashworth, Margaret, 42
Askey, Richard, 25, 68, 87, 88, 384, 385
Askey–Gasper inequality, 385

Astronomische Nachrichten, 276
Atkin, A. O. L., 41, 42, 59

Babbage, Charles, 160
Bäcklund, R. J., 350, 351, 353
Baernstein, Albert, 385, 386
Bailey, W. N., 90
Baker, Alan, 330, 339
Baker, H. F., 58
Banach, Stefan, 208
Banach algebras, 192, 208, 209
Barnes, E. W., 58
Bartels, Johann, 240, 241
Baxter, R. J., 66
Bazilevich, I. E., 369
Bell, Eric Temple, 154
Berggren, Lennart, 254
Berndt, Bruce, 31, 254
Bernoulli, Daniel, 217, 319, 320
Bernoulli, Jakob, 4, 7, 8, 30, 213, 214, 219, 220, 223
Bernoulli, Johann, 34, 213, 214, 216, 217
Bernoulli numbers, 210
Bers, Lipman, 386
Bertrand, Joseph, 124, 133, 137, 148
Bertrand's conjecture, 124, 125, 133, 136, 147, 148
Bessel, F. Wilhelm, 124, 239, 326, 327, 329
beta integral, 2, 63, 220, 399
Betti, Enrico, 279
Beukers, Frits, 329
Bézout, Étienne, 158, 159
Bhaskara, 108
Bieberbach, Ludwig, 363, 366, 367, 373, 374, 376, 379–381
Bieberbach conjecture, 363, 366–368, 370, 373–375, 380, 381, 384
binomial coefficients, 16
binomial theorem, 7, 24, 70, 214, 237, 240, 390, 398
biquadratic reciprocity, 300
Bloch, André, 346

451